D0948764

Biology Teachers' Handbook

Biology Teachers' Handbook

BIOLOGICAL SCIENCES CURRICULUM STUDY PUBLICATIONS

Biological Science: Patterns and Processes
    Holt, Rinehart and Winston, New York
    BSCS Unit Tests and *Final Examination* for *Biological Science:*
    Patterns and Processes, The Psychological Corporation, New York
Biological Science: Molecules to Man (BSCS Blue Version)
    Houghton Mifflin Company, Boston
Biological Science: an Inquiry into Life (BSCS Yellow Version)
    Harcourt, Brace and World, New York
High School Biology: BSCS Green Version
    Rand McNally and Company, Chicago
    BSCS Quarterly Achievement Tests and *Comprehensive Final*
    (1968 Editions), available from the Version publishers
    BSCS Quarterly Achievement Tests (1963 Editions), available
    from the Version publishers
    BSCS Comprehensive Finals (1963 Editions), The Psychological
    Corporation, New York
    Processes of Science Test (for all Versions), The Psychological
    Corporation, New York
Biological Science: Interaction of Experiments and Ideas
    A BSCS Second Course, *Prentice-Hall, Englewood Cliffs, New Jersey*
    BSCS Quarterly Tests and *Final Examination* for *Biological Science:*
    Interaction of Experiments and Ideas, Prentice-Hall, Englewood
    Cliffs, New Jersey
BSCS Laboratory Blocks (13 titles)
    D. C. Heath and Company, Boston
    BSCS Test and Resource Booklet, D. C. Heath and Company, Boston
Research Problems in Biology: Investigations for Students (Series 1, 2, 3, 4)
    Doubleday and Company, Garden City, New York
BSCS Self-Instructional Programs (4 titles)
    Silver Burdett Company, Morristown, New Jersey
Innovations in Equipment and Techniques for the Biology Teaching Laboratory
    D. C. Heath and Company, Boston
Biology Teachers' Handbook
    John Wiley and Sons, New York
BSCS Pamphlet Series (24 titles)
    BSCS, Boulder, Colorado
BSCS Single Topic Inquiry Films (40 titles)
    Harcourt, Brace and World, New York
    Houghton Mifflin Company, Boston
    Rand McNally and Company, Chicago
BSCS Inquiry Slide Series (21 titles)
    Harcourt, Brace and World, New York
BSCS Bulletin Series (Nos. 1, 2, 3, *BSCS, Boulder, Colorado;*
    No. 4, Doubleday and Company, Garden City, New York)
BSCS Special Publications (7 titles) BSCS, Boulder, Colorado
BSCS Newsletter BSCS, Boulder, Colorado
BSCS International News Notes BSCS, Boulder, Colorado
BSCS Patterns of Life Series (8 titles, in press)
    Rand McNally and Company, Chicago
The Story of the BSCS (Information Film) BSCS, Boulder, Colorado
A BSCS Single Topic Inquiry Film Presentation (Teacher Preparation Film)
    BSCS, Boulder, Colorado

BIOLOGICAL SCIENCES CURRICULUM STUDY

Biology Teachers' Handbook

SECOND EDITION

EVELYN KLINCKMANN, *Supervisor*

John Wiley and Sons, Inc.

NEW YORK · LONDON · SYDNEY · TORONTO

CONTRIBUTORS TO THE SECOND EDITION

THOMAS J. CLEAVER
Biological Sciences Curriculum Study, Boulder, Colorado

MALCOLM CORRELL
University of Colorado, Boulder, Colorado

ARNOLD B. GROBMAN
Rutgers University, New Brunswick, New Jersey

BERT KEMPERS
Biological Sciences Curriculum Study, Boulder, Colorado

MANERT KENNEDY
Biological Sciences Curriculum Study, Boulder, Colorado

EVELYN KLINCKMANN
San Francisco College for Women, San Francisco, California

HAVEN KOLB
Hereford Junior-Senior High School, Parkton, Maryland

ADDISON E. LEE
The University of Texas at Austin, Austin, Texas

DAVID L. LEHMAN
Science Education Center, The University of Texas at Austin

HAROLD G. LIEBHERR
Nicolet High School, Milwaukee, Wisconsin

JOHN A. MOORE
University of California, Riverside, California

INGRITH D. OLSEN
University of Washington, Seattle, Washington

H. EDWIN STEINER, JR.
Science Education Center, The University of Texas at Austin

CLAUDE A. WELCH
Macalester College, St. Paul, Minnesota

CONTRIBUTORS TO THE FIRST EDITION

RICHARD BOYAJIAN
  The Laboratory School, The University of Chicago,
  Chicago, Illinois

PETER BURI
  San Francisco State College, San Francisco, California

MALCOLM CORRELL
  University of Colorado, Boulder, Colorado

G. J. FERGUSSON
  University of California, Los Angeles, California

FREDERICK L. FERRIS
  Educational Testing Service, Princeton, New Jersey

PAULA FOZZY
  The University of Chicago, Chicago, Illinois

PHILIP HANDLER
  Duke University, Durham, North Carolina

EVELYN KLINCKMANN
  San Francisco College for Women, San Francisco, California

L. M. ROHRBAUGH
  University of Oklahoma, Norman, Oklahoma

JOSEPH J. SCHWAB
  The University of Chicago, Chicago, Illinois

HENRY M. WALLBRUNN
  University of Florida, Gainesville, Florida

DALE S. WEIS
  The University of Chicago, Chicago, Illinois

ARTHUR E. WOODRUFF
  The University of Chicago, Chicago, Illinois

The Biological Sciences Curriculum Study (BSCS) was established in 1958 for the improvement of biological education at all levels. As a cooperative enterprise between the biological and educational communities, the BSCS has produced a variety of materials used at both the secondary and collegiate levels. It has been a major factor in reshaping the structure of the discipline of biology as evidenced by examining classroom materials available prior to 1960 and those that have appeared a decade later. The BSCS is dedicated to the production of creative and imaginative programs at the cutting edge of the discipline. Interdisciplinary studies and those concerning the social implications of biological knowledge now occupy the BSCS.

However, the single most important factor in reaching the student is the teacher. No matter how innovative or modern the materials or how creative the methodology, unless the classroom teacher has a comprehension of the goals of the program and a sympathy with them, the new programs can be vitiated by pedestrian handling to the point where only the materials are present in the classroom but not their philosophy nor their implementation.

Therefore, with full cognizance of the importance of the teacher, it was early decided that a handbook delineating modern aspects of content germane to biology and expatiating upon the philosophy of inquiry would be a valuable resource to the classroom teacher. With model methodologies and resource materials added, such a book might also serve in the training of teachers.

Initially, then, in 1960, an experimental volume entitled *Teacher's Commentary* was produced under the supervision of Dr. Joseph J. Schwab of the University of Chicago. Successive refinements of this initial paperback volume resulted in the commercial production in 1963 of the first edition of the *Biology Teachers' Handbook,* published by John Wiley and Sons. Feedback from individual classroom teachers and from teacher training institutions indicated that it was a valuable resource and, indeed, at that time, the only one to deal so extensively with the problems of teaching biology. Since 1963 it has been possible to go beyond what had been initially presented, and advances in content and methodology have dictated a second edition of the successful *Biology Teachers' Handbook.* This task has been under the supervision of Evelyn Klinckmann of San Francisco College for Women, who has been involved in the BSCS programs since their inception. Her understanding of classroom situations and her empathy for teachers have resulted in an updating and a restructuring of this volume to make it a unique contribution to pre-service preparation of biology teachers and a value for in-service programs.

As with all BSCS materials, we solicit the comments of those who use the volume as bases for future revisions. The comments should be sent to the Director, Biological Sciences Curriculum Study, Post Office Box 930, Boulder, Colorado 80302.

Arnold B. Grobman
Chairman of the Steering Committee
Biological Sciences Curriculum Study
Rutgers University
New Brunswick, New Jersey 08903

William V. Mayer, Director
Biological Sciences Curriculum Study
Post Office Box 930
Boulder, Colorado 80302

PREFACE

The revisions are designed to make this book more useful for teacher preparation. Section 1 not only provides a more complete description of the character of the BSCS approach but also includes descriptions of the many types of materials that have been produced and revised by BSCS since publication of the first edition. In Section 2 a complete Index to the Invitations has been added, which will facilitate their use. Section 3 expands the treatment of suggestions for teaching biology and emphasizes the activities designed to assist teachers in developing an enquiry approach to teaching. Section 4 (on background material) has been reviewed, and to some extent revised, to ensure that the material is consistent with current knowledge in the various fields. Section 5 (Appendices) has been updated with several revisions; the most extensive revision resulted in Appendix 4, "Techniques and Materials for the Biology Laboratory."

Comments and suggestions for future revision of this book are welcomed. Please send them to the BSCS headquarters in Boulder, Colorado.

Evelyn Klinckmann, SUPERVISOR

March 1969

PREFACE TO THE FIRST EDITION

In writing this book we have undertaken four tasks. In Section 1 we have tried to show the character of the BSCS approach to biology. In Section 2 we provide a special kind of teaching material, designed to serve one of the major objectives of BSCS—the teaching of biology as enquiry. In Section 3 we have tried to provide the teacher with that knowledge of the physical sciences, of statistics, and of recent biochemical discoveries necessary to the teaching and learning of modern biology. In Sections 4 and 5 we have provided information and materials that may be of assistance in the day-to-day teaching of modern biology and especially in teaching that uses BSCS materials.

Joseph J. Schwab, SUPERVISOR

Chicago, Illinois
June 1963

CONTENTS

Biology Teachers' Handbook

SECTION ONE

Background, Emphases, and Aims of BSCS Biology

CHAPTER 1

Origins of the BSCS Texts

Important characteristics of the Biological Sciences Curriculum Study (BSCS) biology texts have their origin in the history of the American science textbook, a relatively brief history extending from about 1890 to the present. Let us begin with a summary of the three developmental phases which mark that history.

In the first phase of the history of the American science textbook, extending from about 1890 to about 1929, the basic model for the conventional textbook was laid down. This basic model was determined by two factors: the state of the science at the time and the supposed goals of the high school student. We shall look at the consequences of these determiners in a moment.

In the second phase of its history, from about 1929 to 1957, the earlier textbook was extensively but not fundamentally modified. The modifications were brought about by a new determining factor—concern for the increasingly diverse abilities, interests, backgrounds, and intentions of high school students. Unfortunately, these modifications were often achieved at the expense of the best feature of the earlier, basic model—its correspondence with the state of the originating science. At the same time, the modifications were usually only a series of *ad hoc* emendations to the basic model, although it was this model itself that was being rendered obsolete by the changing makeup of the high school population.

In the third and current phase, of which BSCS is a part, two new developments are taking place. First, the basic model, left substantially unchanged in the second phase, is being radically reordered. Second, the factors that determined the earliest model on the one hand and those that led to the modifications of the second phase on the other are being brought to bear conjointly and in defensible relation with one another.

PHASE 1, THE BASIC MODEL FOR THE CONVENTIONAL TEXT

The basic model for the conventional science textbook was laid down for the high school of 1890–1929. This was a small high school by current standards, and by the same standards it was in many respects a homogeneous school. The distribution of the scholastic aptitudes of its students was more compact than the distribution of the aptitudes of today's high school population. There were fewer students at the extremes of the range, many more in the middle of the range. The same relative homogeneity held true for the social and economic backgrounds of the students.

In one vastly important respect, however, these early high schools were not homogeneous. They may have begun as primarily college-preparatory institutions, but they remained so only briefly. Many of their graduates would seek jobs in shops, banks, offices, and factories soon after graduation. Only some of them would go on to college. And only a few of these would continue in an academic life—to become teachers or scholars.

Nevertheless, the homogeneities that did exist seem to have veiled this heterogeneity of career from the sight of the curriculum makers. The curriculum was drawn largely from the traditional academic disciplines: mathematics, the sciences, Latin, literature, history, and English. This was a defensible curriculum for college entrance, but it was given as the curriculum for nearly all students—those who would not go on to college as well as those who would.

At first glance, this may seem an entirely sound policy, but it was not. It would seem to have the advantages of providing a common general education for all citizens, a common culture, and a common tradition. In fact, it provided a general education and a common tradition only for the students who went on to college.

It failed to provide a liberal, general education for all precisely because it was designed primarily for students who would go on to college. Its design seemed to *assume* that this further education would take place, for it provided, not a general and well-rounded education, but prerequisite courses, "propaedeutics," preliminaries. It required the colleges to complete its work.

This was true in different degrees for different areas of the curriculum. The Latin of Caesar and Virgil may not have been complete but it was liberal. The teaching of history served important liberal purposes. But the teaching of literature was stilted; mathematics was stunted;

and biology was the worst of the lot. The biology curriculum and its texts were almost entirely descriptive. They consisted almost wholly of a mass of disconnected facts and elementary generalizations.

Let us note, however, the outstanding virtue of these basic models in biology. They may have been a morass of isolated facts and primitive generalizations, but they came closer to being the "right" facts and generalizations for the biology of the time than we were to see in the next phase of American science education. The authors who selected them were working scientists or colleagues of working scientists, and they knew the state of the field.

THE SECOND PHASE

The second phase of development of the American high school was a phase of rank and reckless growth. Its student population vastly expanded in size, in range of scholastic aptitude, and in the range and variety of career intentions. The high school could no longer maintain a single broad objective for its work, not even a spurious common objective, such as the college preparation which dominated phase one. Instead, it had a mixed clientele, presenting many and mixed demands.

In the midst of this growing diversity the social sciences developed. These new sciences, especially sociology, educational psychology, and theories of personality, were grabbed, like straws, by educators. They hoped that these sciences would save them, solve their problems, show them how to handle the confusing diversity of purposes and clients which they faced. Investigators in these social science areas turned their attention to the different ways in which different people learned; to the differing attitudes and interests which grew out of different social and economic backgrounds; to the different career opportunities and career interests which characterized different groups in the community.

These developments soon led to a major shift in the content and emphasis of high school science texts. Their content was no longer mainly determined by the state of knowledge in the scientific field and the portion of the knowledge thought necessary for college work. Instead, many materials were omitted or emphasized on the basis of views as to what could be most readily taught. Still other material was modified to conform to theories of teaching and learning, regardless of the extent to which these modifications presented a distorted view of the subject as known by the scientists. Still other emphases departed from the structure of the science because of concern about special needs and special interests of different social classes in the community. Another large group of changes was nothing more than a bundle of concessions to the fact that the need for more and more teachers was being met by reducing both the quality and the quantity of their training.

THE THIRD PHASE

Here, then, is a rough picture of the pendulum swing which has characterized American education until recently. In the years before the First World War we had courses and textbooks in biology that were close to the structure of the science then current but concerned mainly with elementary facts and simple generalities about them. Further, these textbooks did little or nothing about the needs and conditions of teaching and learning or the special wants and needs of the high school population. In the second phase, the texts focused more and more on the learner and less and less on what there was in the sciences which might be learned. The elementary facts and generalizations of the earlier texts were still present, little was added to represent current development of the science, and what had been inherited from the basic models was modified and selected to fit local and passing needs of the nation and community.

This pendulum swing is clearly evidenced in a study by William Brownson at the University of Chicago.[1] He has found that better than 50% of the authors of high school science textbooks available in 1915 were in the roster of *American Men of Science*. In 1955, by contrast, the figure had dropped to less than 10%.

Clearly, neither extreme of this pendulum swing gives us good textbooks. The early textbooks, despite their narrow emphasis on the elementary facts and generalizations of the sciences, had the advantage of being in close contact with the sources of their information. New knowledge in the field and modifications of previous knowledge found their way relatively quickly into the high school textbook. By contrast, the later textbooks became more and more distantly related to their originating sources. Their authors lived and worked primarily in the environment of the school itself, and they were therefore more and more influenced by problems of teaching and learning. The line of communication between scientific enquiry on the one hand and the textbook on the other grew thinner and thinner. The gulf between scientific knowledge and the content of our textbooks grew wider and wider.

[1] Brownson, Wm. E., and Joseph J. Schwab. The textbook in science education. *The School Review*, 70:4, 71:2, Summer 1963.

On other matters the advantage was the other way around. The older textbooks were insensitive to changing wants and needs of the population and of our nation, whereas the new textbooks were extremely sensitive to these matters. The old textbooks used remote and often mistaken judgments about the ease and difficulty of materials, whereas the new textbooks were written out of immediate contact with the students who were to use them.

In brief, neither the old textbook nor the more recent one was satisfactory. Each had advantages, but each had weaknesses that might have been corrected by the influences which shaped the other. What was needed was a collaboration among the different competences responsible for the different texts—between the scientists on the one hand and the teachers on the other, between close contact with the field of knowledge and close contact with experience of and knowledge about teaching and education. Above all, what was needed was completion of the early basic model, its conversion from an elementary introduction to a three-dimensional model of our knowledge of biology.

The appearance of the first Russian satellite forced us into recognition of the need for this collaboration and rounding out. It wakened laymen, statesmen, and teachers to the need for bringing our textbooks into closer connection with the fields of knowledge. It wakened scientists to the necessity of learning something about educational needs and principles and to contribute to them.

As a result of this awakening, America has witnessed the development of still another of those collaborations, characterized by give-and-take and mutual respect, which are its unique contribution to ways of getting things done. Teachers have come out of the schools; educators have come out of colleges and universities; scientists have come out of their laboratories; the three groups together have begun to learn how to communicate and collaborate in producing better materials for our schools.

The BSCS texts are one result of this collaboration, because the BSCS accepted the obligation to bridge the gap between these indispensable sources of good educational endeavor.

Its aim was not merely to transcribe materials from the most recent scientific journals into textbook form but to select the materials most appropriate to the training of our youth; to develop and present these materials so as to contribute to the development of attitudes and skills as well as of knowledge; and to recognize the fact that, for many students, the high school is terminal. The materials were not to be confined to elementary facts and generalizations; they were to constitute something broader and larger—a reflection of the principles and emphases of the science as a whole.

The BSCS decided that the necessary collaboration among scientists, teachers, and educators was best done on a face-to-face basis. Jointly, scientists, teachers, and educators undertook the task of planning and writing text and laboratory materials that would best discharge the obligations they had accepted.

In summary, then, the BSCS texts represent a re-established communication among the major groups involved in education. They have tried to select the best and the most significant materials in biology for the purposes of public education. They have tried to develop these materials in forms that would best contribute to the attitudes and skills which are the objectives of American public education. They have tried to reflect the structure of biology as that science now exists. And they have tried to produce these materials in a way that will make them useful to the dedicated teachers of the American schools.

In the next chapter we shall examine the major changes in content which this collaboration has brought about.

CHAPTER

2

Content, Themes, and Objectives of the BSCS Materials

The major aim of BSCS biology is to present a valid image of current biological science. A field of knowledge can be viewed in terms of major dimensions that aid in describing what constitutes the field. Three of these dimensions are the subjects to be investigated, the major generalizations and conceptual schemes which result from and give focus to investigation, and the modes of investigation. For biology, the subjects of investigation can be organized in terms of *levels of biological organization*. Major generalizations have been identified in the form of *themes*, which permeate the BSCS materials. And modes of investigation are presented in terms of details of biological *enquiry*. Each of these is dealt with in this chapter.

THE LEVELS OF BIOLOGICAL ORGANIZATION

It is hardly necessary any more to point out that science not only grows but develops. There is accretion of new facts and discovery of new phenomena, but there is also something more. As new data are uncovered and new knowledge is formed, the older body of knowledge is re-formed: reorganized to allow for the new and to put old and new into connection with one another.

One of the most conspicuous ways in which biology reorganizes its knowledge is by changing the emphases that are put on different levels of biological organization. Ancient biology, handicapped by its lack of tools and techniques, could barely begin a study of gross anatomy. Therefore, its emphasis was on the whole organism—its appearance and its basic behavior—its functions and "faculties."

With the development of tools, techniques, and experience, biology entered an era of gross anatomizing. It began to know the larger parts

9

of the body, and emphasis shifted to the organization of tissues and organs. With this knowledge in hand and with further development of skills and techniques, physiology made its bow. We began to understand the functions and actions of gross parts of the body. But this constituted no change in the level of organization; the emphasis continued to be placed on organs and tissues.

In recent years the development of biology has taken it outward in both directions from this rough centering on organs and tissues. In the downward direction it has penetrated first to the cellular level and more recently to the molecular level.

Explorations at these levels have led to correction, enlargement, and reorganization of our knowledge of organs and tissues, because study of cellular architecture and molecular behavior has thrown a new light on the actions and functions of organs.

This reorganized knowledge of organs and tissues has, in turn, helped open to us certain higher levels of biological organization. We have returned to the ancient center of interest—the whole organism—and are gaining new knowledge of the behavior of that whole: how it learns; how it develops patterns of behavior without learning; how it carries on such fundamental processes as courting, mating, and nesting; migration; self-defense; and aggression.

We are also moving beyond the organism as an individual to a study of populations, treating the population as a subject in its own right and not merely as numbers of individuals. We are studying the growth and decline of populations, their sickness and health. Through population genetics we are even, in a sense, studying the mating and reproduction of populations.

Beyond the population of similar organisms, we are invading the organized community of diverse organisms, too. We are discovering some of its component parts, its diverse "organs" and how these "organs"—various kinds and types of organisms—act to fulfill the roles, occupy the niches, that constitute the community of which they are parts.

We have even begun to study the world biome as a whole, the totality of "life stuff," all kinds and varieties of organisms taken together and studied for their universal characteristics—how life, as such, originates and evolves; the directions in which evolution takes it; the necessary conditions for existence of life; and the repertory of ways in which living units can respond to varying conditions.

With biology advancing on these many fronts, our organized knowledge of living things has undergone a reformation. This reshaping includes, first, a shift in the emphases that are put on these various levels and, second, a change in how we understand the relations of one level to another.

One way, therefore, to view the character of the BSCS biology materials (particularly, the versions), is in terms of one aspect of this change—the changed emphasis on different levels. All of the versions include the seven levels; however, they differ with respect to the relative emphasis given to each. These differences, as well as the second aspect of the reshaping—our changing understanding of the relations of these levels to one another—are apparent in the synopses of materials presented in Chapter 3.

The Seven Levels.

Let us begin with a listing of the levels of biological organization, together with a definition of each by examples. We shall work with seven of them, as follows:

A. THE MOLECULAR LEVEL

Examples

1. Often the simple sugars are bonded together to form more complex sugars. Simple sugars can be oxidized to release energy. When they are, they form water and carbon dioxide.

2. The enzyme maltase acts as follows. The maltose molecule must collide with the enzyme molecule and adhere to it. . . .

3. Biological work uses *chemical-bond* energy, specifically, the energy present in molecules of one specific compound, adenosine triphosphate (ATP). The special character of this compound depends on the presence in it of high-energy phosphate bonds.

B. THE CELLULAR LEVEL

Examples

1. The site of all the enzymes that break the glucose molecule down from the pyruvic acid stage is the mitochondrion—a small body barely visible under the best light microscopes.

2. The most striking events in mitosis concern the chromosomes. They are nearly always invisible in nondividing cells. As mitosis gets under way, however, they become increasingly visible, with definite boundaries.

3. Materials pass from and into the interior of cells by way of the cell membrane. Three different processes may be involved in this transport across a membrane.

C. ORGAN AND TISSUE

Examples

1. Much of the blood flows from the heart to the body and back again in less than a minute.

2. We do not yet know just how the products of the thyroid gland do their work, but in some way they control the rate at which our tissues release energy.

3. Root systems perform two additional functions. They absorb water and nutrient salts from the soil and provide conducting tissue for the transport of these materials.

D. THE ORGANISM AS AN INDIVIDUAL

Examples

1. Studies have shown that innate behavior patterns are rarely a reaction to the environmental situation as a whole but only to a few parts of it. Other parts are entirely ignored.

2. The male stickleback isolates himself in the spring, develops a nuptial dress of whitish blue on the back and brilliant red on throat and belly, and selects a territory which it defends against other males.

3. Accumulated evidence makes it probable that the "imprinting" process is quite different from learning as we know it so far. For example, imprinting seems to occur—when it occurs—almost instantaneously. Neither periods of trial nor numerous repetitions are required.

E. THE POPULATION

Examples

1. Population size, then, may decrease by mortality or emigration and increase by natality and immigration.

2. At the University of Wisconsin, the mouse population increased rapidly at first until the living space became crowded. At this point, chasing, fighting, and cannibalism increased drastically. . . . Eventually, the mortality of the young reached 100%.

3. Sometimes unusual climatic conditions upset the equilibrium. A shift in weather conditions may result in greatly increased numbers, creating the conditions for an epidemic which will return the population to an earlier stage of growth.

F. THE COMMUNITY

Examples

1. A soil is a biological system that consists of a group of interact-

ing organisms living in a special sort of inorganic environment. It is an ecosystem. . . . In the total economy of the biological community bacteria are the primary decomposers. . . .

2. In the case of mosquitoes, men, and malaria, the parasite has a complex life history which involves both hosts. . . . Insects may serve as vectors of pathogens in a purely mechanical way.

3. Since the bulk of pond producers are microscopic, we would expect the first-order consumers to be small. And so they are. Many of the consumers are themselves a part of the plankton. They are protists—ciliates and flagellates.

G. WORLD BIOME

Examples

1. Already, the presence of so many human beings has brought stress to the lives of most kinds of plants and animals, from the arctic tundras to the equator and as far south as inhabitable land exists. Everywhere, the number of species is shrinking, the food webs are tearing. . . .

2. The whole natural system, in an analogous way, tends toward balance, equilibrium, at any given time; yet it is subject to changes, both short-term and long-term.

3. If, as we assumed, there was little or no oxygen in the atmosphere of the early earth, the evolution of early forms of life could not have occurred until oxygen was made available. We think that photosynthesis was the means by which free oxygen, over millions of years, was made available in the atmosphere.

THE UNIFYING THEMES

Men of different interests and points of view have contributed to the production of the BSCS materials. This diversity is reflected in the versions themselves: in their emphasis on different levels of biological organization, in differences in style, and in differing treatments of subjects within the field.

A unity, a common thread, however, binds together not only the various parts of each version but also the several versions themselves. This unity arises from an agreement among everyone concerned that nine themes, nine basic emphases, should be woven in and through every text.

These nine themes have been selected on the basis of two major determining factors. First, we have examined the content and the structure of modern biology. From the content we have tried to identify the characteristics and conceptions that provide the most comprehensive and reliable knowledge of living things as they are known to modern biology. By examining the structure of enquiry in biology we have tried to identify the procedures and conceptions that best characterize modern biological science.

Second, we have taken account of the needs and problems of students. We have asked ourselves what knowledge of living things, and what attitudes and skills relevant to biology, would contribute the most to students' personal lives and to the execution of their responsibilities as men and citizens. Thus the second factor determining our themes has been the needs of our nation and our fellow citizens. By using both factors jointly, we have tried to fulfill our dual obligation: to reflect the current state of biology and to construct a defensible general education in biology at the high school level.

The nine unifying themes on which we are agreed are listed next in an order and arrangement that suggest their interconnections. The first five concern the *content* of our texts. The last two concern the logical *structure* of our texts, the context of data and inference through which their content is conveyed. The remaining two (themes 6 and 7) are intermediate; they concern both structure and content. In the remainder of this chapter we describe each theme in detail.

1. Change of living things through time: evolution
    2. Diversity of type and unity of pattern in living things
    3. The genetic continuity of life
4. The complementarity of organism and environment
    5. The biological roots of behavior
6. The complementarity of structure and function
7. Regulation and homeostasis: preservation of life in the face of change
8. Science as enquiry
9. The history of biological conceptions

CHANGE OF LIVING THINGS THROUGH TIME: EVOLUTION

It is no longer possible to give a complete or even a coherent account of living things without the story of evolution.

On the one hand, many of the most striking characteristics of living things are *products* of the evolutionary process. We can make good sense and order of the similarities and differences among living things

efore we are studying the events of biological evolution as
eries.

evolution appears in organisms as a *present* phenomenon.
ot only inferred the course of evolution in the past from such
s all historians use, but we have also seen it occur in the
ent. Since Darwin's day our knowledge of the mechanisms
n has grown vastly in both scope and detail. We not only
fied some of the factors of this mechanism, but know how
what consequences they may have, and what auxiliary factors
ate or impede their action.

l word is necessary concerning our habit of referring to
of evolution." This usage is often taken to mean that
s but an envisaged possibility, something uncertain and
This interpretation, in turn, is due to a mistaken idea about
g of "theory" and its place in science. This mistaken idea
e as merely a process of verification. In that process of
it is mistakenly supposed that materials go through three
grees of certainty: a first stage, of complete doubt, called
second stage, of uncertainty, called theory; a third stage,
called fact or principle.

of "theory" no longer holds in science, if it ever did.
ce is not merely a process of verification of isolated items
of organization as well. In this twofold process, "theory"
the uncertain, the unverified, but rather to the coherent
d. When Mendel sought and found certain recurrent
the offspring of his garden peas and then saw that these
be related to the expansion of a binomial to the second
ognized and stated the components of a theory, a coherent
ter and form, data and conception. Evolution is a theory
yes—a body of *interrelated* facts. As new facts about
discovered, the organization may be changed in order
m, but this would *not* mean that the present organization
nown is unsound.

its pervasive and comprehensive character, evolution is
ee different ways in the BSCS materials. There are spe-
on evolution as the history of living things. There are
rs on evolution as a process. And third, evolution either
s process is interwoven in all other chapters where it has
treatment of cell chemistry, ecology, taxonomy, and so on.

only by reference to their evolutio
the particular environments in whi
the surface of the earth, the comin
development, even the chemistry
exchange it among their parts—all
explanation, in whole or in part, fr

On the other hand, another gr
things can be fully understood or
which evolution takes place. Th
events of meiosis and fertilizatio
It is only in terms of the contri
hancement and sorting out of a
we make sense of them. The san
esses that go under the name o
where the action and consequenc
isolation of populations, of the eff
groups.

Evolution, then, forms the wa
different ways. First, evolution
the sequence of unique events i
present has had its origin. Thi:
is a selection from many other
all the potentialities of life ha
all species that might have evolv
of actualities, selected in son
stitutes strong evidence about
the raw materials on which th
that the existence of such a hist

This history may well be
future. Through a study of si
that have not yet been actual
tentialities, together with kno
we have already made great
the animals, plants, and proti

Conversely, the study of si
what we cannot achieve and
in which one avenue of mov
may well "foreclose" the fu
bilities. Such series of deli
series—are now under serio

be long b
stochastic s

Second,
We have n
evidences a
living prese
of evolutio
have identi
they work,
may acceler

A specia
the "theory
evolution i
unproved.
the meaning
treats scienc
verification,
stages or de
hypothesis; a
or certainty,

This sense
Modern scien
but a process
refers, not to
and organize
ratios among
ratios could
power, he rec
union of mat
in this sense
evolution are
to include the
of facts now k

Because of
treated in thr
cific chapters
specific chapte
as history or a
a place: in the

DIVERSITY OF TYPE AND UNITY OF PATTERN IN LIVING THINGS

As we have indicated before, this theme is, in part, a special aspect of the theme of evolution. It is also, however, a theme in its own right. The extraordinary diversity of living forms and their adaptation to widely diverse environmental conditions are striking characteristics of life, entirely apart from their connection with evolution.

Another striking characteristic of life is the unity of pattern that cuts across these diversities. There is, first, the well-known similarity of pattern among such diverse parts as the limbs of different vertebrates. This parallelism is not, however, restricted to gross anatomy. It is evident at all levels of organization. There is the virtually universal role of ATP as the vehicle for energy transfer among all living things. There are, similarly, the universality of DNA and RNA as the materials of hereditary control, the common role of sugar as the fuel of life processes, the role of vitamins and coenzymes. At the cellular level, the unity of plan is seen in the remarkable unity of mitosis in cell division.

Because unity-diversity is both a theme in its own right and an aspect of evolution, it is treated in two ways in the BSCS texts. There are chapters specifically concerned with the variety and unity of living things. In addition, wherever it is feasible, particular diversities and unities are tied firmly to the mechanisms of adaptation through natural selection and of mutation as their originating sources.

THE GENETIC CONTINUITY OF LIFE

This theme, too, is a part of the theme of evolution. As E. B. Wilson expressed it, there is

". . . an uninterrupted series of cell-divisions extending backward from existing plants and animals to that remote and unknown period when vital organization assumed its present form. Life is a continuous stream. The death of the individual involves no breach of continuity in the series of cell-divisions by which the life of the race flows onwards. The individual body dies, it is true, but the germ cells live on, carrying with them, as it were, the traditions of the race from which they have sprung, and handing them on to their descendants." [1]

[1] E. B. Wilson, 1897. *The cell in development and inheritance.* Macmillan, London. Page 9.

As in the case of the theme of unity-diversity, the theme of continuity is also the theme of discontinuity. The replicating power of the genes is accompanied by occasional "errors" in replication. These "errors" constitute discontinuities in the stream of life, discontinuities to which we owe the variability, the diversity, basic to the process by which life has occupied so much of the world.

THE COMPLEMENTARITY OF ORGANISM AND ENVIRONMENT

This theme, too, is part of the theme of evolution, especially where it concerns the environment of the whole organism. In the BSCS texts, however, two further emphases find a place. First, there is emphasis on the interplay of "organism" and "environment" at all levels of biological organization. The chromosome and the chloroplast have their environment within the cell. The cell acts on and is acted on by neighboring cells, by the fluids which bathe it, and through the latter by cells remote from the neighborhood. In turn, the tissues and organs made of these cells have their environments. Finally, toward the other extreme, the individual organism exists in an environment of other organisms as well as in an environment composed of nonliving factors. It is a part of a population of its own kind. That population wanes and waxes in response to its environment as well as in response to internal factors, and in turn plays a part in a community of organisms that is part of the environment of the population.

Second, there is emphasis on the *reciprocal* relations of living unit and environment. The organism is not merely a passive recipient of stimuli from the environment, nor is it limited to stereotyped responses to environmental stimuli. It affects as much as it is affected. It modifies the environment and exploits it. This reciprocity is seen most vividly in what man has done to and with his environment in creating both the need for and the possibility of conservation. It is visible, too, in the phenomena of ecological succession and the relations of aquatic organisms to their medium, and it is dramatic in the behavior of such organisms as birds, larger predators, and many insects.

THE BIOLOGICAL ROOTS OF BEHAVIOR

It is important that the student understand that there may be limits to what he can do, limits imposed by his biology. As one of a certain kind of organism he is limited to the environmental conditions that his kind of organism requires. His exploitation of other environments (such as the environments of rapid acceleration, of weightlessness, of isolation, of other planets) may depend largely on his ingenuity in taking his own environment with him.

As an individual, he may be limited by conditions laid down by his particular heredity. Not all men do all things with equal facility. Each of our talents may be of great importance, but all men do not have the same talents in the same degree. The better we understand this the more effectively are we able to exploit what we are good at, overcome our limitations by cooperation, and profit from what we are.

The behavior of other organisms has a similar biological basis. In these other organisms, too, we see a vast range of behavior, with different degrees of flexibility and rigidity, from learned behavior to that which is largely unlearned. We have long known about simple unlearned behaviors, due almost wholly to interaction of the environment and of the organism as constituted by its heredity—auxin synthesis and movement in plants; the simple unlearned reflex and taxis in animals. We have also recently learned a great deal more about elaborate and lengthy patterns of behavior that owe little or nothing to "learning" as we commonly understand it: web and nest building, courting behavior in birds, aggression in fish, and the hunting and working patterns of dogs. Especially important is the behavior of groups, the biology of societies.

In brief, the BSCS texts emphasize behavior as arising not only from the experience of the individual but also from the "experience" of its forebears, the stored experience arising from variation and selection in evolution.

THE COMPLEMENTARITY OF STRUCTURE AND FUNCTION

This theme represents one of the oldest and most fruitful conceptions in the annals of biological enquiry. It originated with Aristotle, and in the hands of such men as William Harvey and later physiologists it provided the basic content of biology for many years.

The conception, *function,* grows from the idea of the organism as a well-organized system, an economy in which each part has a role in the continued operation of the whole. This role is the function of the part.

Long before the mechanism of evolution was understood, the well-organized character of life units was recognized and the functions of their parts investigated. With development of the theory of evolution, the conception of function underwent important changes. We no longer thought of the organism as a *perfect* organization but instead recognized the possibility of the vestigial, the novel, and the incompletely relevant organ and part. This did not mean, however, that the conception of function became obsolete. On the contrary, within the limitations required by our knowledge of evolutionary processes, we still sought evidence through which to understand each part in terms of its contribution to the whole.

In this conception, the *complementarity* of structure and function has two senses. First, it refers to the fact that what an organ *does* is dependent on its structure, that is, the character of its subparts and the pattern in which these subparts are arranged. The second meaning of the complementarity of structure and function develops from the first meaning. If what an organ does is dependent on its architecture, its structure, then that architecture can be used as evidence from which function can be inferred. Given a general idea of the economy of the whole, and the assumption that the structure of the part determines the part's capacity to perform its function; then, from data about structure, some idea of function can be obtained. It was in this way that Harvey inferred the pumping action of the heart. He examined the arrangement of the muscle fibers of the heart, the channels and valves connecting its hollow spaces, and the relations of these spaces to connecting vessels. It was on the basis of these data that he identified systole as the functional motion of the heart and was able to show that this systole ejected blood from the heart.

It goes without saying, of course, that this complementarity of structure and function applies to all levels of biological organization; it is not restricted to gross organs and parts. Just as Harvey worked with cardiac muscle fibers in relation to the heart, so can we work with segments of muscle molecules in relation to the function of muscle fiber. In the same way, the structure of a mitochondrion can be related to respiration and the structure of a chloroplast to photosynthesis.

This theme is treated in the BSCS materials in three places. First, we have taken pains to be sure that wherever structure or the fine-grained details of a process are discussed, these structures and details are not left hanging in isolation from the biology of the organism. Instead, such matters as the behavior of phosphate bonds, the structure of an organic molecule, the steps in a chemical change are, wherever feasible, related to the role they play in the economy of which they are parts.

Second, an entire group of "Invitations to Enquiry" is devoted explicitly to the theme of function and to the evidence by which function is inferred. "Invitations to Enquiry" are a relatively new kind of teaching material. They are printed here as Section 2. What they are like and how they can be used will be discussed later.

Third, there is a special laboratory "block" on the interdependence of structure and function. These blocks represent another relatively new departure in BSCS materials. (See Chapter 3.)

REGULATION AND HOMEOSTASIS

To a considerable degree, the conception of homeostasis and regulation is the offspring and heir of the conception of function. The latter gave us an understanding of the relation of parts to the whole, but it did so in a way which left both structure and function too rigid, too inflexibly fixed, to account for the capacity of the organism to adjust to change. It is desirable to know that the heart functions to pump blood in a circuit through the body. It is more desirable to know that the rate and amplitude of the heart-pump vary with the demands made by the body in response to changing conditions of the environment. It is still more desirable to understand something of the way in which the heart is made sensitive and responsive to such demands.

It is this kind of knowledge which is given us by the concepts of homeostasis and regulation. The former conception (for which credit goes mainly to Claude Bernard) includes the idea of a *stable internal environment* consisting of many different dynamic equilibriums. The aim of this conception was to guide research that would accomplish two things. It would identify the important dynamic equilibriums, such as the concentration of metal ions in the blood, the pH, and the glucose level. Then it would find the mechanisms by which these equilibriums were maintained in the face of change, such as the alternating storage and release of glucose by the liver as the concentration of blood sugar increases or decreases with intake or oxidation.

The concept of regulation goes a step further. Bernard's idea of homeostasis, useful as it was, still contained a large element of inflexibility. We were given a view of the continuing changes by which homeostases were maintained, but the homeostases remained *stases,* static. The mechanism was like one regulated by a thermostat whose setting could never change. It was clear, however, that the setting of our bodies' thermostats can change. As embryonic development proceeds, the regulation of many processes is "reset." Similarly, the basal metabolism varies with changes in climate. Muscle tone is altered by long-term changes in muscular activity. More extreme changes lead to still more dramatic examples of the flexibility of the organism. If one kidney or testis is lost, there may be compensatory overgrowth of the other. If a part of the brain is destroyed experimentally, another part can become involved in the relearning of competences lost in the loss of the damaged part.

Regulation, then, is a long-term and large-scale kind of homeostasis. We may think of it as changes in structure as well as function by which the overall organization of the body is maintained in the face of change.

As in the case of the theme of structure and function, BSCS materials handle this theme by two means. First, wherever feasible, the texts discuss structures in terms of their flexible characteristics, as well as in terms of their more constant states. Second, Invitations to Enquiry are used. Homeostasis is introduced in Invitations 24 and 25, where the feedback control of thyroid and other hormones is discussed. In addition to these Invitations, an entire group (Group V) is devoted exclusively to regulation and homeostasis.

SCIENCE AS ENQUIRY

This theme concerns the materials and means by which, ideally, all subjects would be treated in the BSCS texts. It also represents one of the most radical departures of the BSCS texts from conventional patterns.

If we examine a conventional high school text, we find that it consists mainly or wholly of a series of unqualified, positive statements. "There are so many kinds of mammals." "Organ A is composed of three tissues." "Respiration takes place in the following steps." "The genes are the units of heredity." "The function of A is X."

This kind of exposition (called rhetoric of conclusions) has long been the standard rhetoric of textbooks even at the college level. It has many advantages, not the least of which are simplicity and economy of space. Nevertheless, there are serious objections to it. Both by omission and by commission it gives a false and misleading picture of the nature of science.

By commission, a rhetoric of conclusions has two unfortunate effects on the student. First, it gives the impression that science consists of unalterable, fixed truths. Yet this is not the case. The accelerated pace of research in recent years has made it abundantly clear that scientific knowledge is revisionary. It is a temporary codex, continuously restructured as new data are related to old.

A rhetoric of conclusions also tends to convey the impression that science is complete. Hence, the fact that scientific investigation still goes on and at an ever-accelerated pace is left unaccounted for to the student.

The sin of omission by a rhetoric of conclusions can be stated thus: It fails to show that scientific knowledge is more than a simple report of things observed, that it is a body of knowledge forged slowly and tentatively from raw materials. It does not show that these raw materials, data, spring from planned observations and experiments. It does not show that the plans for experiment and observation arise from problems posed, and that these problems, in turn, arise from concepts which summarize our earlier knowledge. Finally, of great importance is the fact that a rhetoric of conclusions fails to show that scientists, like other men, are capable of error and that much of enquiry has been concerned with the correction of error.

Above all, a rhetoric of conclusions fails to show that our summarizing concepts are tested by the fruitfulness of the questions they suggest and through this testing are continually revised and replaced.

The essence, then, of a teaching of science as enquiry would be to show some of the conclusions of science in the framework of the way they arise and are tested. This would mean to tell the student about the problems posed and the experiments performed, to indicate the data thus found, and to follow the interpretation by which these data were converted into scientific knowledge.

The teaching of science as enquiry would also include a fair treatment of the doubts and the incompleteness of science and indicate the possibility that through the advance of enquiry scientific knowledge can change.

The BSCS materials use six ways of achieving this difficult aim. First, it has made liberal use of the most forthright expressions of uncertainty and incompleteness. Over and over again you will find phrases of the following kind: "We do not know." "We have been unable to discover how this happens." "The evidence about this is contradictory." "It is not certain how this happens." "The favored theory at the moment is as follows. . . ."

Second, where possible, BSCS has begun to replace rhetoric of conclusions by a narrative of enquiry. That is, our current views on a subject such as genetics are developed step by step through a description of the experiments performed, the data obtained, and the interpretations made of them. Whole chapters and major sections of chapters are developed in this way. Notable examples of such narrative of enquiry will be found in sections on genetics and development.

Third, the laboratory work has been organized to convey a sense of science as enquiry. Some of the exercises are of the traditional kind, serving the necessary traditional purpose of making clear and vivid materials expounded by the text. But many are of another kind. They are not illustrative but *investigative*. They treat problems for which the text does not provide the answers. They create situations in which the student may participate in the enquiry.

Fourth, the laboratory block programs have been designed to serve this end. The block programs engage the student in an investigation of a variety of biological problems. Each of these is real, a true introduction to scientific investigation. A given block may begin with materials familiar to scientists and with problems whose solutions have already been disclosed. As the series of problems progresses, however, they come nearer and nearer to the frontier of knowledge. Such blocks are pursued in depth over periods as long as six weeks.

Fifth, the Invitations to Enquiry are enlisted in this enterprise. Their function is similar to the function of the laboratory devoted to investigation. One Invitation may describe a problem and the experiment done to solve it and give the data which were found. It then "invites" the student to interpret the data to form a conclusion. Another Invitation may set the stage for the student to consider ways and means of controlling an uncontrolled factor. A third will provide a problem situation and data and invite the student to develop hypotheses to account for the data. Still another Invitation will invite the student to devise ways for testing one or another hypothesis.

Specifically, you will find that the entire first group of Invitations (Invitations 1–16) has been developed to introduce the student to enquiry.

Finally, the Single Topic Inquiry Films have been developed as a new teaching strategy for generating an understanding of biological enquiry. The films are similar to the Invitations in that they invite questions, raise problems, and present experimental data to promote active thinking on the part of the student. They are different in that the photographs themselves present much of the material to be dealt with by the student.

THE HISTORY OF BIOLOGICAL CONCEPTS

This theme is actually two themes, only one of which is, strictly speaking, a history of scientific ideas or conceptions. This strictly conceptual history is an aspect of the teaching of science as enquiry and has been dealt with before. We said there that the concepts through which scientists frame problems for research are in turn tested by the fruitfulness of these problems. Where the problems lead to experiments that yield coherent data, we are inclined to go on to further problems suggested by the conception. On the other hand, when a conception yields problems that lead to no data or incoherent data, we consider the conception inadequate or outworn and try to find others by which to replace, enlarge, or revise it. The progression from the conception of function through the concept of homeostasis to the concept of regulation is a good example of such a progressive enlargement as concepts are used, tested, and exploited.

The history of scientific concepts underlies much of our changing technology, agriculture, medicine, and management of natural resources, as well as our changing bodies of scientific knowledge.

This aspect of the history of conceptions is treated in the narratives of enquiry of the text and in the Invitations to Enquiry.

The second subtheme of the history of biology concerns man and events rather than conceptions in themselves. There is a human side to enquiry. Discoveries are made by *persons*. Furthermore, scientific research is not invariably the result of formal plan and clear purpose. Chance and intuition play their parts.

Such matters as these can play useful roles in conveying to students a realistic and understandable view of science and scientists. They can help to mitigate the extraordinarily unreal, antagonistic, and fantastic views of science and scientists which repeated studies show to have grown up in the minds of many persons. Hence, wherever possible, narrative of enquiry, Invitations to Enquiry, and other materials describe research in terms of the persons, the places, and the incidents involved.

OBJECTIVES FOR TEACHING ENQUIRY PROCESSES

Biological enquiry has been discussed in general terms as a major theme of BSCS biology. However, it also constitutes a major dimension of the field of biological science. Because an understanding of enquiry is so crucial to an adequate understanding of science, behavioral objectives for the teaching of enquiry processes have been formulated.[2] Whereas the Invitations to Enquiry (Section 2) provide discussion patterns for developing an understanding of biological enquiry, the ob-

[2] Developed by a joint committee consisting of personnel from the Mid-Continent Regional Educational Laboratory and BSCS. The material included here is one component of material published by the joint committee in the summer of 1969.

jectives present behaviors that are expected outcomes of enquiry activities. Thus the following section and the Invitations are two different but complementary tools for teaching biological enquiry.

The Substantive Knowledge of Biological Science [3]

The current substantive body of biological knowledge includes the knowledge of enquiry processes and the findings of these enquiries. Although it is true that they are inseparable and probably equally important, for many years in educational practice the findings of enquiry (content) have been heavily emphasized to the exclusion of enquiry processes. Recently, emphasis has been placed on the enquiry processes; however, many biology teachers are not familiar with them. For these reasons, it is appropriate that they should be given a great deal of attention. This attention should not exclude or foreshadow the importance of teaching the findings of scientific enquiries. Actually, teaching enquiry processes demands the teaching of content inseparable from process.

Behavioral Objectives

The following behavioral objectives (Table 1) are believed to represent a wide array of behaviors that are applicable to enquiry activities.

Success in any particular enquiry activity may require some but not all of the behaviors. There is no attempt to prescribe a definite sequence of behaviors for any enquiry activity or for any time during the year. However, for purposes of this document it was necessary to impose a logical order on the enquiry factors. In conducting biological research or carrying out enquiry processes in the classroom, the behaviors may not be followed in the order listed. Also, in actual practice, the enquiry factors are closely interrelated.

It is understood that at the beginning of the course the student will possess certain information and enquiry skills. These will be applied and built upon throughout the enquiry activities. But it should be recognized that some skills are more difficult to achieve than others. Students will vary in the extent to which they achieve all of the skills included in Table 1.

[3] The remainder of the material in this chapter is based on the report *Inquiry Objectives in the Teaching of Biology* published by the McREL-BSCS group in the summer of 1969. See Chapter Four of that report for objectives stated with greater specificity.

TABLE 1. *Objectives for Teaching Enquiry Processes in Biology*

I. ENQUIRY

Enquiry is defined as a set of activities directed to solving an open number of related problems in which the pupil has as his principal focus a productive enterprise leading to increased understanding and application.

There are at least four perspectives for viewing enquiry. The first perspective consists of the major factors in enquiry. The second perspective is made up of the major guiding principles of enquiry. The third perspective focuses on previously conducted enquiries in which the factors, as well as the guiding principles, are examined. The fourth perspective emphasizes attitudes and appreciations that are important to carrying out enquiry.

A. Factors in Enquiry

1. *Formulating a problem.* (Biologists formulate problems within identifiable contexts of the discipline.)

Related Behaviors

The student will:

a. Identify one or more discrepant events:

    (1) In a structured situation.

    (2) In an open or less structured situation.

b. Select (carve out) a problem for study according to the following criteria:

    (1) Utilization of findings from other sources (teacher, text, research reports, etc.) to select a problem. He will: (For example, in the case of research reports.)

        (a) Identify sources of information.

        (b) Analyze parts and relationships.

        (c) Describe it generally.

        (d) Evaluate the report on criteria.

Enquiry Factors

Related Behaviors

(2) Judgment of feasibility of the problem. He will:

    (a) Recognize there are problems that cannot be solved within man's knowledge, skill, and technology.

    (b) Select a problem that can be solved with his present knowledge and skills.

    (c) Select a problem that can be solved with considerations of limitations and availability of time, tools, observations.

(3) Selection of a problem which is important to him using criteria such as:

    (a) His interest in the problem.

    (b) His recognition that solution of the problem provides a means for solving other problems of greater interest.

    (c) Identification of the problem(s) as pertinent to his needs.

c. State the problem in researchable terms.

Related Behaviors

The student will:

a. Identify the elements of a problem on which hypotheses could be based.

b. Generate hypotheses about the critical elements in the situation.

c. Clarify the statement of hypotheses:

 (1) Eliminate duplication.

 (2) Determine if the hypothesis is testable.

 (3) Determine relevance of the hypothesis to the problem.

When appropriate, the student will:

a. Plan to test hypotheses.

 (1) Identify all of the variables possible.

 (2) Select a variable to be studied.

 (3) Establish proper controls.

b. Plan for:

 (1) Replication.

 (2) Systematic observation of descriptive data.

c. Plan procedures to yield wanted data.

 (1) Select the simplest technique to get data.

Enquiry Factors

2. *Formulating hypotheses.* (Hypotheses are herein defined as tentative answers to problems or questions that can be investigated. Not all investigations necessarily have stated, or even implied, hypotheses; this is most likely to be the case in studies conducted in the taxonomic mode.)

3. *Designing a study.* (a.(1)-a.(3) are stated in terms of the antecedent-consequent mode.)

Related Behaviors

(2) Select, as needed, the appropriate:

    (a) Sample.

    (b) Instruments and chemical procedures such as gauges, microscopes, pH indicators, etc.

    (c) Identification keys.

    (d) Mathematical procedures such as determination of area, volume, density, weight.

    (e) Mathematical operations such as addition, subtraction, multiplication, division.

(3) Identify sources of errors such as measurement, computations, tools and instruments, operator.

(4) Distinguish between random and systematic error.

(5) Plan procedures to minimize random and systematic error such as:

    (a) Measuring consistently.

    (b) Practicing a laboratory technique, e.g., counting yeast cells.

d. Plan a system for processing data to make it ready for interpretation.

    (1) Select appropriate techniques such as graphs, charts, figures for organizing data.

(2) Select appropriate statistical procedures such as determination of mean, mode, range, standard deviation, chi-square.

(3) Record findings or significant relationships in written and oral reports.

The student will:

a. Follow the plan for collecting, organizing, analyzing the data, and presenting the findings.

b. Use the tools properly.

c. Record data accurately. (Degree of accuracy determined by the nature of the problem.)

d. Review the tools and procedures used.

e. Revise procedures when indicated by results.

The student will:

a. Identify assumptions he has used in the study.

b. Use results of other studies to interpret the data or findings.

c. Employ reasoning skills, e.g.:

(1) Deductive—arriving at a particular implication from a generalization (general to specific).

Enquiry Factors

4. *Executing the plan of investigation.*

5. *Interpreting the data or findings.*

Related Behaviors

    (2) Inductive—arriving at a generalization from the evidence or arriving at an interpretation by reasoning from evidence (specific to general).

d. Use various means of presenting the data to bring out different features.

e. Examine collected data to determine its relevance to both the problem and hypotheses at hand and to other problems.

f. Identify conflicts or discrepancies in the data.

g. Draw tentative conclusions.

h. Avoid overgeneralizing the results:

    (1) Withhold judgment until sufficient data is available.

    (2) Restrict interpretations to limitations of the hypotheses being tested.

    (3) Restrict interpretations to limitations of assumptions.

    (4) Restrict interpretations to limitations of available evidence.

The student will:

a. Relate findings to varied personal interests and to the world at large.

b. Collate findings and make interpretations from several experiments.

6. *Synthesizing knowledge gained from the investigation.*

Related Behaviors

c. Apply knowledge gained to new situations: (Applications are made in light of limiting assumptions and conditions.)

    (1) Predict additional applications.

    (2) Speculate about potential applications.

    (3) Make predictions that suggest new research problems.

    (4) Determine the validity of the relationship between tentative conclusions and current theories.

d. Recognize new problems.

    (1) Identify additional research problems.

    (2) Design new investigations to test the validity of scientific models.

    (3) Design modified procedures for investigating the same problem.

e. Use theories, theoretical constructs and models as a means of relating and organizing new knowledge.

f. Recognize that evidence for or against a given theory may be inconclusive.

g. Recognize that a theory may or may not be testable at the time of formulation.

h. Recognize that several theories may be useful, each one making a unique contribution.

7. *Differentiating the Various Guiding Principles of Enquiry that Guide Scientific Studies Including His Own.* The student will:

a. Recognize the various guiding principles of enquiry.

(1) Taxonomic—This principle is based on collecting, organizing, and classifying data to develop a basis for formulating researchable questions. The typical mode of enquiry related to this guiding principle is the scheme for classification. Implicit in this guiding principle of enquiry are the following assumptions about the nature of biological phenomena:

    (a) There are different kinds of "things."

    (b) The kinds of "things" can be differentiated in terms of differences in observable characteristics.

(2) Antecedent—Consequent.[a] This principle is based on a conception of independent chains of cause and effect relationships. The typical mode of enquiry related to

Examples:

The student measures and compares the length of lima bean seeds and later develops a problem regarding the length of second generation seeds.

The student, by collecting and classifying bird calls, finds some that do not fit patterns. From this discrepant event, the student formulates a problem.

In a series of experiments with proper controls, the thyroid gland is removed and the following characteristics are noted: lower temperature, lower respiratory rate, placidity and obesity. The student would interpret from these find-

(continued)

[a] Invitations to Enquiry, Groups I and II.

34

Examples:

this guiding principle is the controlled experiment. Implicit in this guiding principle of enquiry are the following assumptions about the nature of biological phenomena:

(a) The whole organism consists of many separate parts.

(b) Each of the separate parts operates as an independent entity.

(3) Structure-Function.[b] This principle is based on a conception of the interrelationships of chains of causes and effects that contribute to a wholeness of an organism having a certain stable character or nature. Harvey's study of the circulation of the blood illustrates the mode of enquiry related to this principle. Implicit in this principle of enquiry are the following assumptions:

(a) That there are interrelationships of chains of causes and effects that contribute to a wholeness.

ings that the characteristics are due solely to the removal of the gland.

Gene *A* determines tallness and gene *a* determines shortness. Garden peas having *AA* or *Aa* genes will be tall as there is no interaction between gene *A* and gene *a*.

The regulatory process is essential to life. The student might determine the role of the thyroid gland.

The circulation of blood is vital to the body activities. The student determines how the heart and blood vessels are organized to accomplish this activity.

[b] Invitations to Enquiry, Group IV.

35

(b) That the "wholeness" of an organism has a certain stable character or nature.

(4) Regulation and Homeostasis.[b] This principle is based on the conception of the flexibility of organs to maintain an equilibrium among changes of parts (organs, tissues, etc.) in response to changes in external conditions coincident with a stability of the whole organism. Dynamic equilibrium is maintained by structural and/or functional changes *in degree* rather than kind. The thermostat model fits this guiding principle. Implicit in this guiding principle are the following assumptions:

(a) That parts are flexible and can change in response to external changes.

The sugar level in the blood remains in equilibrium despite the fact that no sugars are digested. The sugar content in the blood remains in equilibrium or balance despite the fact that excessive sugar is digested.

(b) That a dynamic equilibrium among changes of organs can be maintained.

(c) That a dynamic equilibrium among changes will not result in structural changes or major functional changes,

There are several concentrations which are right for the body depending on the condition and activity of many or all of the other chemicals, physical factors and structures.

[b] Invitations to Enquiry, Group V.

e.g., a transmitting membrane becoming secretory.

(5) Self-Regulatory System.[e] This principle is based on the conception of the flexibility of the organism as a whole, as well as the flexibility of the parts, to maintain equilibrium of changes to external or environmental changes. However, the conception of this principle emphasizes functional or structural changes *in kind* in the parts to make a new integrated whole. Implicit in this guiding principle are the following assumptions:

(a) The organism, as well as its parts, is flexible and may change in response to external changes.

When part of the brain is destroyed the adjacent parts may incorporate some or all of the functions of the incapacitated part.

(b) That a new dynamic equilibrium among changes can be maintained.

(c) That a dynamic equilibrium among changes may result in structural changes or major functional changes.

b. Determine which principle of enquiry is appropriate for a study.

Using this principle of enquiry can I collect the proper data?

[e] Interim Summary 5.

37

c. Describe how the study would have been different if I had used another principle of enquiry were used.

How would the data have been different if I had used another principle of enquiry?

B. Skills in Carrying Out Principles of Enquiry.

In biological science five guiding principles for formulating problems as illustrated on pages 00-00 have been identified. They are:

1. Taxonomic
2. Antecedent-Consequent
3. Structure-Function
4. Regulation and Homeostasis
5. Self-Regulatory System

New principles may well emerge in the future, but at the present time these principles represent a range of perspectives from which a biologist or student might view a given phenomenon. Depending upon which perspective he uses, there will be differences in certain basic activities which are necessary for him to carry out the enquiry. These activities are:

1. Asking of initial questions—What do I want to find out about the subject or phenomenon of interest?
2. Observation—What data should I look for to help answer that question?
3. Organization of observations—What should I do with the data to get the clearest answer to my question?

Thus the *principles* of enquiry become *modes* of enquiry when they are activated as methods in carrying out an enquiry. It is desirable that biology students begin to gain some understanding of these different modes, hence the following objectives:

(continued)

38

1. The student will carry out activities characteristic of the *Taxonomic mode.*

a. He will *ask questions* such as the following:

  (1) What are the similarities and differences among the organisms found in this pond water?

  (2) What are the kinds of plants found in Prairie State Park?

b. He will *make observations* necessary to answer such questions:

  (1a) See hairlike structures on the organisms found in the pond water.

  (1b) See internal structures of the organisms.

  (2a) See differences in leaf shapes.

  (2b) See differences in structure of flower parts.

c. He will *handle data* in (any) appropriate way(s):

  (1a) Make *comparisons* between observed characteristics.

  (1b) *Group* data according to observed similarities and differences.

  (2a) Use existing classification schemes for organisms under study.

2. The student will carry out activities characteristic of the Antecedent-Consequent mode:[d]

a. He will *ask questions* such as the following:

  (1) If the thyroid gland is removed from mice, what happens to their physical characteristics?

b. He will *make observations* necessary to answer such questions: e.g.,

  (1a) Measure temperature of mice before and after treatment.

  (1b) Measure respiratory rate of mice.

  (1c) Measure weight of mice.

c. He will *handle data* in appropriate ways:

  (1) Organize data into tables and/or graphs which show differences in physical characteristics of the mice before and after removal of the thyroid.

[d] *Invitations to Enquiry,* Groups I and II.

(2) What will occur if we remove *both* the islet tissue of the pancreas *and* the pituitary gland from mice?

(3) If I cross fruitflies with known genotypes for eye-color, can I expect results to show a simple dominant-recessive pattern?

(1d) Provide for appropriate controls.

(2) Observations similar to (1a) through (1d).

(3) Distinguish differences in eye color in F_1 and F_2 generations.

(2) Organize data in ways similar to (1).

(3) Organize data to show proportion of eye colors in F_1 and F_2 generations.

3. The student will carry out activities characteristic of the Structure-Function mode:[e]

a. He will *ask questions* such as the following:

(1) When bacteria invade a local area, such as by a scratch or puncture wound, what processes are set in motion which serve to protect the invaded organism?

(2) Is the migrating behavior of x species of bird ini-

b. He will *identify what observations* to make in order to answer such questions: e.g.,

(1a) Observations on changes in skin cells near wound.

(1b) Observations on changes in blood vessels near wound.

(1c) Observations on changes in blood cells.

(2a) Observations on changes in organs.

c. He will *identify ways* in which *data must be handled:*

(1) Find ways of showing relationships, or lack of them, between observations on (1a), (1b), (1c).

(2) Find ways of showing relationships, or lack of them,

[e]Invitations to Enquiry, Group IV.

40

tiated by changes in various organs? Are these changes in organs triggered by changes in environment?

(2b) Observations on factors in the environment, such as daylight, temperature, which may affect those organs; measurement of these factors.

between observations (2a) and (2b).

(3) What is the function of the heart in a mammal to the total organism?

(3a) Observations on changes in the structure, location and action of the various parts of the heart.

(3b) Also, observations of changes in neighboring organs following heart action.

(3) Find ways of showing relationships or lack of them regarding structure, location and action between observed parts (3a), (3b).

4. The student will recognize activities characteristic of the Regulation-Homeostasis mode:[f]

a. He will recognize *questions* such as the following:

b. He will recognize *types of observations* that must be made in order to answer such questions: e.g.,

c. He will *recognize ways in which data must be handled* to answer the question:

(1) How does daily rapid walking, then running, of a mile result in progressively less feeling of stress in breathing and muscular action?

(1a) Observations on breathing and circulation.

(1b) Observations on chemical and other changes during muscular activity.

(1c) Observations on feelings of stress related to each factor in (1a) and (1b).

(1) Find ways of relating the changes found in breathing, muscle activity, etc. to each other *and* to reduction in feelings of stress, by means of tables, graphs, mathematical analyses and inferences.

[f]Invitations to Enquiry, Group V.

41

(2) Why does the reduction of food intake in the human eventually result in no feelings of hunger or fatigue?

(3) Years after a forest fire which destroys most of the trees, why is there a similar forest in that place?

(2) Observations on presence of food in stomach, blood sugar level, etc., in relation to hunger pangs and fatigue.

(3a) Observations on changes in plant and animal species in that area over many years.

(3b) Observations on changes in environmental conditions related to changes in plant and animal species.

(2) Find ways of relating factors to each other and to the condition of the total organism.

(3) Find ways of relating changes in species and environmental conditions to each other and to the re-establishment of balances between them.

5. The student will recognize activities characteristic of the Self-Regulatory mode:g

a. He will recognize the types of questions which stem from this view of the problem situation:

(1) Will destroying a portion of the brain associated with a particular learned behavior make it impossible for the animal to relearn the behavior after recovery from surgery?

b. He will recognize types of data needed to answer such questions:

(1a) Data on the repertory of behaviors that might be learned with different parts of the brain destroyed.

(1b) Data on changes in function and possible changes in structure of undamaged parts of brain.

(1c) Data on amount and regions of undamaged brain necessary for relearning.

c. He will recognize ways of handling the data in order to find answers to the question:

(1) Find ways of relating the data to show relationships between factors of the whole, changes in the whole and evidences for a new integrated whole.

g Interim Summary 5.

6. The student when given a biological situation such as a pond will:

a. State a problem in terms of each of the principles of enquiry.

Examples

(1) What organisms (populations) are present in the pond? What is the size of each population? What is its distribution?

(2) What effect does light intensity (or other abiotic factors such as O_2, heat, etc.) have on the size and distribution of a particular population in the pond?

(3) What are the roles of the various populations in the food web of this pond? What changes in the web and in abiotic factors occur when there is an increase, or decrease, in the various populations?

(4) When sewage is introduced into the pond, what changes occur in the size and distribution of the populations and in the abiotic factors? Is a new balance in the food web established?

(5) How much sewage may be introduced before there is a change in the original food web and abiotic factors, i.e., before a "new pond" is established?

b. Describe the experimental design appropriate to each type of problem.

Examples

(1) Identify the populations present. Use sampling techniques to determine size and distribution of each population.

(2) Measure light intensity (or other abiotic factors) at various depths and location; relate these findings to size and distribution of the specified population.

(3) Observations to determine food web. Introduce additional members of a specific population, or decrease their number, and determine changes in the other populations and in abiotic factors.

(4) Introduce measured amounts of sewage and observe changes in abiotic factors, the populations, and their interrelationships.

(5) Extrapolate from results of investigation (4) to predict change (s) in original food web and abiotic factors.

43

II. ENQUIRY INTO ENQUIRY

There are means other than direct laboratory experiences to involve students in enquiry. One of these consists of allowing students to critique scientific papers, abstracts, or other reports to discover the variety of logical patterns in scientific investigations carried out by scientists and science students. This type of activity can be challenging and can help students develop skills and appreciations of critical reading and thinking that are generalizable to other types of communication. These skills and appreciations are very important when it is recognized that the majority of students enrolled in biology will become consumers of science rather than scientists.

Given one or more scientific papers, reports, or abstracts, the student will:

A. Analyze the research for its basic parts:

The student might ask:

1. The problem.

What question was the investigator concerned with? Was the problem stated explicitly?
How did he delimit it, e.g., organisms studied, factors looked for, time over which data was collected?

2. The hypotheses.

What initial guesses did he make?
How did he select among these?
What consequences, if any, did he predict from his best guess?

3. The experimental design.

What plan was used to answer the question?
What were the experimental factors?
What other variables were identified?
What controls did he use?

4. The procedures for collecting the data.

What procedures were used?
What instruments, tools, etc., were used?
What procedures were made for replication?

5. The kind(s) of data collected.

Were verbal descriptions, diagrams, etc., constructed?
Were qualitative and quantitative measures taken?
Were sampling techniques used?

6. The ways of organizing the data.

How were the data organized?
Were graphs, charts, and/or statistical procedures used?

7. The interpretations and/or conclusions.

What assumptions were made in making the interpretation?
Were discrepancies or conflicts in the data identified?
Were interpretations made within the limits of the original problem and hypothesis?

8. The assumptions, both explicit and implicit, that guided the researcher in formulating the problem and interpreting the data.

What ideas about the nature of the problem situation does he take for granted? E.g., single factor, multiple factors, discrete entities, dynamic system, stable system, linear causal chains, cyclical causal chains.

B. Analyze the relationships among the basic parts such as:

1. Between the data sought and the problem.

Are the data sought consistent with the way the problem is stated? Are the means of collecting the data consistent with the way the problem is delimited?

2. Between the interpretations and the kinds of data obtained.

Do the interpretations "fit" the data? Do they go beyond the data? Are they more limited than the data warrant?

3. Between the interpretations and the way the problem was formulated.

Do the interpretations provide an answer to the problem as stated? Do they show the need for additional data in order to adequately investigate the stated problem?

45

The student might ask:

D. Compare several research reports that deal with closely related problems yet give similar results, different results, or conflicting results. Given two or more of such reports, he will:

If the reports deal with the same or similar problems, how do they differ with respect to general approach, formulation of hypotheses, and experimental design?

1. State how they differ with respect to (b) through (f) above and identify variations and possible causes for variance.

Can defensible choices between the approaches be made? Can their basic conceptions be reconciled?

2. Determine if and how *different solutions* (if present) to the problem can be combined into a synthesis of understanding.

If there are numerous solutions to the problems presented, can the conclusions be combined in a synthesis of understanding?

3. Identify some implications of the interpretations to other problems as well as the one at hand.

If the papers present apparently contradictory or inconsistent conclusions, can these be reconciled in some way?

E. Apply skills gained from analyzing research reports to other kinds of reports.

What is the report about? What does it say? What are the meanings of key terms?

AFFECTIVE OR ATTITUDINAL QUALITIES OF ENQUIRY BEHAVIORS

Among behaviors important to success at enquiry are those sometimes termed *affective* or *attitudinal*. Although a variety of these behaviors seem to pervade the entire enquiry process, no consistent method of identifying, describing or measuring their extent is now available. Their importance, however, is not to be negated or minimized. As more is known about them, they will be accorded more specific and detailed attention and treatment.

Many of these behaviors are difficult to identify because no verifiable psychomotor response can be associated with them. Yet, certain observable behaviors may reasonably be taken as indicative of affect or as expressions of attitude. To that extent, a sampling of those behaviors have been included here as desirable objectives of the enquiry experience.

No attempt has been made to be all inclusive, to prescribe sequence or to associate these behaviors with particular activities. It is assumed that students will demonstrate these and other desirable behaviors to varying degrees at the beginning of the course and will apply and build on them throughout their enquiry experience. It is further assumed that the student perceives his role in the enquiry process as one of a voluntary active participant rather than a passive receptor.

Attitude or Quality

I. Curiosity

Related Observable Behaviors

The Student: A. Expresses a desire to investigate new things or ideas.

B. Expresses a desire for additional information.

C. Asks for evidence to support conclusions made from scientific materials.

D. Expresses interest in scientific issues in the public domain.

E. Expresses a desire for explanations.

II. Openness

The Student: A. Demonstrates willingness to subject data and/or opinions to criticism and evaluation by others.

B. Seeks and considers new evidence.

C. Expresses the realization that knowledge is incomplete.

D. Expresses knowledge of the tentative nature of conclusions as products of science.

49

50

Attitude or Quality

III. Reality Orientation

Related Observable Behaviors

The Student: A. Demonstrates knowledge and acceptance of his limitations.

B. Expresses awareness that change is the rule rather than the exception.

C. Expresses awareness of several sources of knowledge.

D. Expresses awareness of the fallibility of human effort.

E. Expresses belief in science as a means of influencing the environment.

F. Does not alter his data.

G. Demonstrates the realization that research in science requires hard work.

H. Demonstrates awareness of the limitations of present knowledge.

I. Expresses awareness of the historic development of patterns of enquiry, and of the processes and characteristics of science.

J. Demonstrates belief that the search for desirable novelty should be tempered by awareness and understanding of traditional concepts.

Related Observable Behaviors

Attitude or Quality

IV. Risk-taking

The Student:
A. Willingly subjects himself to possible criticism and/or failure.

B. Expresses his opinions, feelings or criticisms regardless of the presence of authority.

C. Participates freely in class discussions.

D. Indicates a willingness to try new approaches.

V. Objectivity

The Student:
A. Indicates a preference for statements supported by evidence over unsupported opinion.

B. Indicates a preference for scientific generalizations that have withstood the test of critical review.

VI. Precision

The Student:
A. Indicates a preference for coherent statements.

B. Seeks definitions of important words.

C. Demonstrates sensitivity to the appropriateness of general and/or specific statements in a given context.

D. Expresses the need to examine a problem from more than one point of view.

Attitude or Quality	Related Observable Behaviors
VII. Confidence	The Student: A. Expresses confidence that he can achieve success at enquiry.
	B. Demonstrates willingness to take "intuitive leaps."
VIII. Perseverance	The Student: A. Pursues a problem to its solution or to a practical point of termination.
IX. Satisfaction	The Student: A. Expresses satisfaction with the process of enquiry.
	B. Expresses confidence that his enquiry experiences will enable him to attain future goals.
X. Respect for Theoretical Structures	The Student: A. Demonstrates awareness of the importance of models, theories and concepts as means of relating and organizing new knowledge.
	B. Demonstrates awareness of the importance of currently accepted theories and concepts as a framework or basis for the emergence of new knowledge.
	C. Demonstrates awareness of the importance of scientific procedures to the generation of new knowledge, theories and concepts.

Attitude or Quality

Related Observable Behaviors

XI. Responsibility

The Student:
A. Is active in helping to identify and establish learning goals.

B. Demonstrates willingness to work beyond the assignment.

C. Insists upon adequate evidence on which to base conclusions.

D. Suggests changes to improve procedure.

E. Shows respect for the contributions of others.

F. Demonstrates willingness to share knowledge with others.

G. Offers a rationale for criticism.

H. Initiates action for the benefit of the group.

XII. Consensus and collaboration

The Student:
A. Demonstrates willingness to change from one idiom, style or frame of reference when working with others.

B. Calls upon other talent from within the group when opinions and help are needed.

C. Seeks clarification of another person's point of view or frame of reference.

Emphases of the BSCS Publications

The materials produced by BSCS have taken several forms. There are the four versions—*Biological science: molecules to man; High school biology, BSCS green version; Biological science; an inquiry into life;* and *Biological science: patterns and processes*—each of which constitutes a beginning biology course for the high school student. There are materials for the advanced science student—*Biological science: interaction of experiments and ideas* and *Research problems in biology: investigations for students, series one, two, three, and four.* There are materials that supplement and can be used in conjunction with a version. These are the *Single Topic Inquiry Films,* the *Inquiry Slide Sets,* the *Pamphlet Series,* and the *Laboratory Blocks.* The latter are unique in that they can be used as the basis for a full year course as well as in conjunction with other materials. Tests and test manuals have been developed as supplements to versions, lab blocks, and the second course. Other types of publications are in the process of being developed. Each of these publications and types of materials is described in this chapter.

THE BLUE VERSION

Overview of the BSCS Blue Version

The major subject matter organization of the BSCS Blue Version, *Biological science: molecules to man,* 2nd ed., Boston: Houghton Mifflin, 1968, is built around the theme of levels of biological complexity: molecules, cells, tissues, organ systems, multicellular organisms, species, groups, societies, and communities.

This type of content organization is rather unique in the teaching of biology. The time-honored system of treating biology as only slightly more than a merger of botany and zoology needs serious reconsideration. The advantages of treating the content of biology as interrelated levels of increasing complexity in biological organization will become clear as the overview is developed.

But biology is much more than its content and subject matter. Science is a search for explanation; an interaction of facts and ideas. The nature of this search, or enquiry, should be made clear to the student through his participation in laboratory investigations. Furthermore, the content of biology should not be presented as a "rhetoric of conclusions" as though the textbook were a type of biological almanac. The BSCS Blue Version attempts to treat biology as a narrative of enquiry in which the search for explanation is revealed through the development of the gene theory, cell theory, and evolution theory.

Consequently, the BSCS Blue Version is *not* a course in molecular biology. The molecular level is only one level in the hierarchy of biological organization and, furthermore, the authors have always made the obvious assumption that not very much "molecular biology" can be taught to young people who, generally, have not even had a chemistry course.

The book contains 29 chapters grouped into 8 units. There are 66 laboratory investigations integrated into the text chapters and an additional 20 supplementary investigations at the end of the book. Ten appendices covering 25 pages provide additional enrichment material for a more detailed coverage of such topics as genetic code, fermentation, photosynthesis, Krebs cycle, respiratory chain, and a more extensive classification of living things. There is an annotated teacher's edition containing, in addition to the annotations, about 50 pages of notes with extra information and teaching suggestions for each chapter. A separate booklet, the *Answer key and teacher's guide to investigations,* provides answers to all end-of-chapter questions and also supplements the annotations relative to guidelines for the proper use of the laboratory investigations. The "Teacher's Edition Notes" contains four different teaching schedules to help the teacher plan a program which is most suitable to his teaching situation.

The overview of the text's content can most easily be understood through a brief summary and rationale of each unit.

Unit I. Biology: The Interaction of Facts and Ideas. The main goal of the first chapter is to emphasize that science is a process as well as a product. The activities of problem construction, observation, hypothesis formation, prediction, and testing are illustrated through laboratory investigations and through an example using Darwin's Coral-Island-Subsidence Hypothesis. The notion that science proceeds by recognizing problems for which hypotheses are then formed and tested is exemplified through the use of a chapter on the variety of living things, which is used as a problem, and a chapter on evolution-natural selection as a reasonable hypothesis to explain the problem of variation.

In this way, evolution theory is introduced early in the book. Evolution is probably the major organizing theory in biology and it certainly is not reasonable to wait until the end of the book before this major theory is introduced. This does not mean that the early chapter on evolution should attempt to cover the entire subject. Instead, only the major postulates of the theory are presented, and it is then pointed out that the explanatory scope of this great theory will be seen as it applies to the chapters that follow.

If living things have evolved, then an obvious question must follow: How did it all start? The origin of living things is discussed through the use of the spontaneous generation-biogenesis controversy. The solution to this controversy leads easily into the Oparin-Haldane heterotroph hypothesis on the origin of life.

The heterotroph hypothesis serves the following important functions in the Blue Version.

1. This hypothesis is "current history" and therefore the teacher is not placed in the position of using an old, and perhaps uninteresting, "case history" to illustrate the nature of science. The heterotroph hypothesis is presented as a series of interconnected assumptions to illustrate the structure of an hypothesis. These assumptions are then used to explain certain observations about the nature of life and of the cell. The tentative nature of the assumptions and the fact that they are "subject to change without notice" illustrate the interaction of facts and ideas in science.

2. The inclusion of chemisry into the high school biology course has always been a problem. Most tenth-grade students have had little exposure to chemistry by the time they take biology, although the new curriculum studies for the elementary and junior high schools may soon change this situation. The authors of the Blue Version seemed to feel that they could best present the minimum amount of chemistry needed in a high school biology course by folding it in *where needed* and in a biological context. In this way the chemistry would be seen as a necessary tool to understand the biology and not as a separate entity. The heterotroph hypothesis, building as it does on assumptions about simple inorganic materials and then progressing to amino acids and other

organic compounds, is a perfect teaching vehicle for making chemistry meaningful and logical within a biological framework.

3. The heterotroph hypothesis is also a logical beginning for a study of the levels of organization referred to at the beginning of the overview. This hypothesis suggests that something must have preceded cells, just as we are convinced that unicellular organisms must have preceded multicellular organisms. It should be pointed out, however, that the heterotroph hypothesis is not developed until the fifth chapter. This delay seemed necessary in order to start the book at a more observational level of biology where the student could deal with the living things with which he is generally acquainted. In other words, the problem-hypothesis sequence that is introduced in the first chapter is exemplified through the coral island problem and the variety-evolution relationship before the heterotroph hypothesis is presented.

4. Finally, the origin of life is always a fascinating topic to students of all ages—and the secondary school student is no exception. A topic of this nature, which can hold the attention of the student and also serve as an effective teaching vehicle for the proper introduction of contextual chemistry and scientific methodology, plays a significant role.

Unit II. Evolution of Life Processes. This unit builds on the energy concepts of fermentation, photosynthesis, and respiration. Life in its simplest manifestation requires a constant flow of energy to keep the flame alive. Although reproduction involves the survival of the species, the acquisition of energy is the immediate problem of all individual cells and is the necessary (though not sufficient) requirement for reproduction. The assumptions of the heteratroph hypothesis provide a basis for treating fermentation, photosynthesis, and respiration in this logical order.

Unit III. The Evolution of the Cell. The modern cell undoubtedly is a highly evolved organism. Without reproduction there is no evolution, and therefore the evolved cell carries the genetic materials that represent the link between old and new generations. The presentation of the genetic code and the DNA-RNA-Protein sequence in this unit prepares the way for an easier understanding of heredity and genetics at the multicellular level, which is covered in a later unit. This is one of the great advantages of the levels-of-organization approach, in that each level builds on the level below it and the study of the sequential levels makes for a more logical and interconnected approach to the subject matter content. This unit closes with a restatement of the cell theory to the effect that the cell unites the acellular and cellular levels of biological activity. It is also pointed out that the evolution theory, cell theory, and gene theory are the major unifying ideas in biology and that they merge into one broad conceptual structure in which the three theories become interdependent and mutually supportable.

Unit IV. Multicellular Organisms: New Individuals. The evolution of higher levels of organization bring new advantages; but new problems also arise. Cell division was complicated enough, but now a new individual with millions of cells must be produced with the same careful exactness. It is pointed out that the problem of cellular differentiation is one of today's major focuses of biological research. Again, the information in the early chapters is of great help in understanding these higher levels of organization.

Unit V. Multicellular Organisms: Genetic Continuity. The work of Mendel, as it applies to multicellular organisms, is eventually related to cell heredity so that the student sees the connection and interrelatedness of cell reproduction and multicellular reproduction. The discussions of population genetics reveal further the relationship between evolution, cell theory, and gene theory. As was pointed out earlier, Mendel's work becomes much easier to understand with the knowledge of cell genetics (Unit III) as a lower level of organization for background.

Unit VI. Multicellular Organisms: Energy Utilization: It now is well understood that the life of the multicellular organism resides in its cells. A person may be more than his cells, but he is at least the life of his cells. Insofar as energy utilization in the multicellular organization is concerned, the main problem arises in supplying the minimum requirements to maintain the life of the cell. The significance of the systems (higher levels of organization) then becomes obvious. The transport, respiratory, digestive, and excretory systems are levels of complexity between the cells and tissues, on the one hand, and the individual multicellular organism, on the other.

Unit VII. Multicellular Organisms: Unifying Systems. Certain systems within the multicellular organism function to unify other systems as well as to organize the individual. The regulatory, nervous, muscle, and skeletal systems seem to carry on this function. The digestive system, for example, is coordinated and related to the whole organism through the function of the regulatory and nervous systems.

Unit VIII. Higher Levels of Organisms. Consistent with the levels-of-organization approach, the final unit of the book now delves into some of the classic areas of ecology. The levels of the community and ecosystem are the most complex of the levels of organization and consequently require an understanding of the lower levels of which they are composed. The biological problems of a community are immediately dependent on the problems of the lower levels of organization. The pollution problem, for example, has as its major dimension the chemical and physiological relationship between all levels of the biological community.

The title, *Biological science: molecules to man,* indicates the scope and organization of the BSCS Blue Version. Molecules are probably the lowest level of biological organization and certainly mankind, with his various powers of abstraction and creativity, represents a level of complexity even beyond the community concept.

Suggestions for Teacher Preparation

Today's university students who are planning a career as secondary school teachers in biology are undoubtedly well prepared to teach any of the BSCS versions. However, a few suggestions might be helpful in the event that any of the following areas have not been adequately treated as part of the teacher's background.

1. **Philosophy of Science.** A good treatment of this area will help the student teacher to understand the relationship between the biological and physical sciences. This study undoubtedly will examine the structure and function of theory so that its importance can be conveyed eventually to the high school student. To understand science is to understand the importance of a theoretical foundation for experiments and observations. It is through a philosophy of a science course, properly taught, that the role of perception and the structure of scientific knowledge begins to have an impact. Sometimes it is here that the student begins to understand the role of of mathematics in science. The science student often is too close to his subject to understand what it is all about. This is not always the case if, fortunately, the philosophical structure and implications of biology are interwoven by the biology professor within the framework of biology itself. But, just to make sure, a good philosophy of science course will provide an excellent background for the BSCS Blue Version.

2. **Dialogue Teaching.** The Blue Version authors like to think of their book as an adventure in ideas. These ideas have been created as men sought explanations for problems. Creative people are not necessarily the most intelligent people. Consequently, the process of Dialogue Teaching, using materials like the Invitations to Enquiry, permits creativity to spring forth in a variety of circumstances. An original hypothesis, as a solution to some minor problem, can be a great

source of accomplishment to even a "mediocre" student. Great teachers have learned that "involvement" is the key to communication and that, consequently, a dialogue must substitute for the monologue. It is well to remember that the prefix "mono" not only means "single" but also means "alone."

3. Physical Science. Certainly, today's biology rests on a foundation of physical science. But this does not mean that biology is nothing more than physical science. It is, indeed, true that many of the recent advances in biology have been made by biologists who have had good backgrounds in biochemistry, physics, and mathematics. Genetics is a case in point. Classical Mendelian genetics involving some of the elementary crosses and their interpretation is extremely interesting. Furthermore, the ability to apply some mathematics to these crosses and to predict some consequences of population genetics adds another interesting and broadening dimension to the gene concept. If one can go another step and look at biochemical genetics in the search for explantation, one begins to comprehend the full significance of the gene theory. The application of the physical sciences to biology should be viewed as a great additional tool in our search to understand our century-old problems relating to evolution, the cell, and the gene.

Implementation in the Schools

The most successful teachers of the BSCS Blue Version are those who have used the book with average and above-average students. Gone are the days, we hope, when the lowest common denominator of the class determined the level at which the class would function. This type of discrimination, which operates against the average and high achievers, is just as evil as that which operates to the detriment of the low achievers. The BSCS *Special materials, biological science: patterns and processes,* New York: Holt, Rinehart and Winston, 1966, has been tailor-made for the "underachievers" and it is possible that some system of double tracking is the best solution. Schools that continue to limp along with a single book that, as advertized, covers the whole range of students are simply taking the easy way out and are not meeting their responsibility to challenge each student to work at his own meaningful current level of achievement.

It is no more necessary to attend an in-service institute for the BSCS Blue Version than it would be for any other program. This does not mean that a novice cannot obtain good suggestions and techniques from any experienced teacher. Some school systems have conducted special sessions to run through the laboratory investigations—not because they are difficult, but because they are new and, we hope, challenging.

No special "exotic" equipment is needed to teach the investigations in the Blue Version. There are a total of 86 investigations available, and most teachers will do about one half that many during their first year of teaching. If some investigations appear to be more extensive than your situation allows, skip it and look at the next one. No single investigation is mandatory!

As mentioned earlier, there are four different plans available in the Teacher's Edition Notes (pages 4–7) which might help the new teacher estimate the time required to satisfy the situation at his particular school.

The annotated teacher's edition should be very helpful to the new teacher. We have found that a separate teacher's guide is often lost or, at least, not available when needed. The page-by-page annotations and notes bound in at the beginning of the book provide considerable help in planning and interrelating many portions of the book.

THE GREEN VERSION

"The word "ecology" was proposed by Ernst Haeckel in 1870 to cover what he called "outer physiology." It is the point of view in biology that takes the individual organism as the primary unit of study and is concerned with how these individuals are organized into populations, species, and communities; with what organisms do and how they do it.

This contrasts with "inner physiology," the study of how the individual is constructed and how the parts work. Obviously the inside and outside of the organism are completely interdependent, and one cannot be understood without constant reference to the other. The division is arbitrary, but so are all of the ways in which biological subject matter might be split. We stress the outside rather than the inside on the assumption that this is more familiar and more easily understood. We believe, too, that it is more important for the citizen, who must participate in decisions about urban development, flood control, public health, and conservation—always as a voter and sometimes as a member of the town council or state legislature.

For disorders of inner physiology, the citizen should consult his physician. But there is no specialist for outer physiology, for disorders of the human biological community. Here each citizen shares responsibility, and biological knowledge is greatly needed for some kinds of decisions."

Thus wrote Dr. Marston Bates, the first supervisor of the BSCS team entrusted with the development of the high school biology course that was to become known as the Green Version, *High school biology, BSCS green version,* 2nd ed., Chicago: Rand McNally, 1968. The statement appeared in the first experimental edition (1960). Sociological, educational, and biological events of the years that followed have given the statement even greater import.

Aims

Within the framework of the BSCS, the Green Version has been developed on the basis of the following facts: (a) the great majority of high school students take biology; (b) a large number of these students will take no more science in school; (c) very few will become research biologists, and only a slightly larger proportion will enter the biological professions; and (d) all are potential voting citizens.

To the revision team of 1968, these facts mean essentially the same things that they meant to our predecessors: that the high school biology course should encourage a scientific viewpoint in the student; that it should provide him with a background in biology that is as advanced as he is able to assimilate; and that subject matter should be selected to increase his effectiveness as a future citizen as well as to help him assert himself in his universe. A course that does these things will serve the interests of all.

Clearly, the writers believe that secondary-school science should be presented as an aspect of the humanities. If, in some cases, a secondary-school course also arouses the student to pursue further biological studies, fine—but this should be presented as an incidental rather than a primary aim. The high school is not the place to begin the training of biological scientists.

These general aims, logically derived from four simple and undeniable facts, are the foundation of the Green Version. They explain what is included and what is excluded. They explain the manner as well as the matter. Whoever undertakes to use the Green Version in his teaching needs to understand this position.

Some Points of View

In addition to the distinctive aims presented above, the successive members of the Green Version team have carried out their writing and revising with a number of special viewpoints constantly in mind. In part, these derive from attitudes that have pervaded BSCS Steering

Committee deliberations, and they derive partly from sources within the team.

Level. The course was planned for the middle 60 percent (in interest and achievement) of tenth-grade students. For the lower range, the BSCS has developed special materials (*Biological science: patterns and processes*. New York: Holt, Rinehart and Winston, 1966). No classroom course will completely satisfy the needs of the upper range of students. However, by wise use of the "Problems" and "Suggested Readings" appended to each chapter and of the suggestions "For Further Investigation" following many of the investigations, the teacher of the Green Version may carry such students beyond the classroom and into other BSCS materials. (See, for example, *Research problems in biology: investigations for students*. 4 Vols. New York: Doubleday, 1963–1965.)

Experience with previous editions has shown that the level sought has been attained, and careful attention was devoted to maintaining this level in the 1968 revision. In the tenth grade we are not teaching children, and we are not teaching adults. Teachers who regard tenth-graders as children may be appalled by the degree of mental sophistication demanded in this course, by the lack of clear-cut definition, by the emphasis on chiaroscuro rather than on the black-and-white thought patterns whose beautiful simplicity we adults find so difficult to shake from our Greco-Hebraic heritage. On the other hand, teachers who ignore the difference between the adolescent and the adult may be scornful of the control of running vocabulary, the attention to development of ideas, and the paucity of esoteric detail.

The title of the student's book simply and adequately describes it: one version of a biology course specifically designed for *high school* students.

Scope. The Green Version is not intended to provide an encyclopedic account of biology. Many topics are omitted; many are treated cursorily. But the topics that have seemed to the authors to best fit the aims discussed above have been developed in depth. For example, the ecological concept of infectious disease seems to be more important to a thinking citizenry than theories concerning the mechanisms of immunity. The ecosystem concept likewise seems more important than the electron microscopy of cells. So also, the concept of speciation against the decipherment of protein structure. In each case, topics that are beautiful and exciting to the biological pioneers are slighted in favor of meat for the supporting troops. Yet, room has been left for the imaginative and adept teacher.

Laboratory. Paraphrasing an earlier statement by Dr. H. Bentley

Glass, Chairman (1959–1966) of the BSCS Steering Committee, the Preface "For the Student" clearly states the viewpoint of the Green Version writers. Laboratory work is *sine qua non* in this course.

A year might be devoted entirely to laboratory work. But the result would be either a myopic view of a narrow segment of biology or an episodic view such as might be obtained by glancing out of a car window at a landscape every 50 miles. In the Green Version the text is used to provide continuity and perspective. Laboratory investigations are placed at points in the course where first-hand experience is most pertinent, most feasible, and most efficient in the utilization of student time.

"Laboratory" has been interpreted broadly. The laboratory is where the work of the scientist is done; it need not be bounded by four walls. Moreover, some of the investigations involve much reasoning and little or no doing. These also may be legitimately regarded as laboratory work.

Continuity. The course intentionally avoids the designation of its sections as "units"—paradoxically, it does so in order to unify. In practice, the unit organization tends to compartmentalize—although this was certainly not the original aim. In practice, the student takes the unit test, breathes a sigh of relief, and murmurs, "Now I can forget all that; tomorrow we start a new unit." And too often he has been quite correct.

The writers of the Green Version intend to give the student no relief at any time from the healthy tension of learning. The course is designed to build up ideas from beginning to end. Everywhere, an effort is made to relate what is immediately in front of the student with what has preceded. At the end, the course returns to the beginning—like the Chinese dragon with its tail in its mouth—and relates all aspects of the course to a biological world view.

Mensural units. Long antedating the Iron Curtain, a "Mensural Screen" has divided the nations of the world into two camps. Even within nations, this screen has divided scientist from engineer. For one camp the tide is running out, slowly but inexorably; economics, not science, is the moon in this affair.

In high school science, it has been customary to *talk* about the metric system, to spend much time converting units from system to system, and to express quantities in metric units during some laboratory procedures. But consistent use of the metric system has usually been lacking, and students have gained the impression that it is of ritualistic importance only—on a par with school-dress codes and academic gowns.

In the Green Version the metric system is used throughout. For explicatory purposes, equivalents are sometimes given; for designating a few pieces of equipment, sizes are expressed in British units. However, no exercises on conversion of units are offered. In this, a modern prin ciple of language-teaching has been followed: "The word to the object, not to another word." We want the student to use centimeter measurements so frequently that he will have a mental image of 10 centimeters, not a recollection of "about 4 inches."

The Student's Book: An Overview

Before specific suggestions for teaching can be meaningful, the teacher must be acquainted with the organization of the course, with variations in organization that experience has shown to be feasible, and with the teaching aids supplied to teacher and student. The following sections are intended to provide such general background.

Organization

In the student's book, sections and chapters are of unequal lengths, but the number of pages devoted to a topic is not necessarily an indication of that topic's relative importance or of the amount of time to be spent upon it. In general, the first three sections are discursive, with a rather low density of ideas. The last three are more compact, with an increasing load of ideas per page.

Section One: The World of Life: The Biosphere. It is whole individual organisms with which the student has had experience. He is one himself. Therefore the course begins with biology at the level of the individual and treats the ways in which such biological units interact.

Chapter 1, "The Web of Life," is designed chiefly to lay some groundwork, to establish the direction of the course. The interdependence of organisms in the transfer of energy and in the cycling of matter, the interdependence of the living system and the physical environment—these things will never be stated again quite so explicitly, but they will persist in the background throughout the course. "Scientific method" is not preached; instead, the laboratory work is relied on to introduce such basic matters as observation, measurement, experimentation, and instrumentation.

The units of ecological study form a series from the individual (the most concrete) to the ecosystem (the most abstract). Chapter 2, "Individuals and Populations," deals with individuals and the various groupings of individuals that can be called populations. The concept of a population keeps turning up: species populations form the basis of classification; populations interact in communities; contemporary genetic and evolutionary theories turn largely on population studies.

The actual field study of a community—even the community in a crack in a city sidewalk— is essential to Chapter 3, "Communities and Ecosystems." Since communities available for study differ greatly among schools, a brief description of a community is provided as a basis for comparison. The kinds of ecological relationships in communities are considered and, finally, the concept of an ecosystem is introduced.

Section Two: Diversity Among Living Things. The student should have some idea of the diversity of organisms and of their classification before going further into the patterns of ecological organization. But there is no need for a "type study" of living things. Emphasis is on the variety of forms in which life can occur and on the aspects of form and function that are relevant to a useful and meaningful organization of such diversity.

On the principle of starting with the familiar, Chapter 4, "Animals," begins with mammals. This, of course, causes difficulties but, at this point, there is no basis for a meaningful discussion of phylogeny. Instead, diversity of form within the animal kingdom is stressed. But this diversity is not endless; patterns are discernible. And evolution is offered as a possible explanation for the apparent order within the general diversity.

The concept of classification does not need repetition in Chapter 5, "Plants," but another abstract idea is developed—that of nomenclature. Historically, nomenclature and classification have developed together, but this is not a sufficient pedagogical reason for presenting them to the student together. Experience has suggested that these concepts are better understood when they are presented to the student separately rather than as a single large block of abstract material.

In Chapter 6, "Protists," a third kingdom is introduced. The teacher may not agree with the system of kingdoms that has been used in the text or with the way that particular groups have been assigned to these kingdoms. But there is no classification on which all biologists agree. Pointing out the reasons for disagreement ought to give students an idea of the nature of the problems of classification.

Section Three: Patterns in the Biosphere. There are three bases on which patterns of biotic distribution within the biosphere may be constructed: ecological, historical, and biogeographical. Each of these is dealt with in this section.

A number of laboratory investigations involving microorganisms have been started during the work on Chapter 6. Therefore, to facilitate continuity of laboratory procedures, Chapter 7, "Patterns of Life in the

Microscopic World," deals with ecological groupings of microorganisms. Two of these groupings, both of great import to man's existence, are treated in some detail: soil organisms and the microorganisms involved in disease.

The theme in Chapter 8, "Patterns of Life on Land," is the distribution of macroscopic terrestrial organisms. The relation of physiological tolerances to the global distribution of abiotic environmental factors leads to the description of biomes. But ecological conditions do not wholly explain distributions. Explanation is then sought in evidence of past distribution and artificial distribution by man.

Chapter 9, "Patterns of Life in the Water," extends the principles of ecological distribution to aquatic environments. Ponds are probably more easily visualized as ecological systems than any other part of the biosphere. Running waters are treated briefly. Finally, it seems desirable that the student obtain some understanding of marine life—which is certain to become increasingly important as a resource for man.

In Chapter 10, "Patterns of Life in the Past," attention is given chiefly to the nature of the evidence in paleontology, to the kinds of work that paleontologists do, and to some principles of paleontological reasoning. Emphasis on ecosystems is maintained, and the temporal continuity of the biosphere is thereby stressed. The principle of evolution is not discussed here but is simply assumed as the most reasonable basis for interpreting the evidence.

Section Four: Within the Individual Organism. Having spent nearly half of its duration on the supra-individual levels of biological organization, the course now turns to the infra-individual levels. Some acquaintance with "inner physiology" is essential—not only for appreciation of some rapidly developing areas of modern biology but also as background for topics (such as genetics and evolution) that are important parts of the biological understanding required by the functioning citizen.

The objective of Chapter 11, "The Cell," is to provide the student with sufficient understanding of cellular structure, of some cell physiology, and of cell duplication to enable him to interpret subsequent chapters. Only the cell structures that have relevance to later discussion are treated, and the physiology deals principally with the relations of a cell to its environment. Two important topics—differentiation and aging—are presented primarily as problems undergoing current investigation.

Energy-flow in living systems has been a fundamental idea from the beginning of the course. In Chapter 12, "Bioenergetics," attention is focused on energy-storage and energy-release in cells. Here the student is exposed to some of the biochemical aspects of modern biology.

Chapter 13, "The Functioning Plant," is chiefly concerned with the structure and function of the plants with which the student comes in contact most frequently—the vascular plants.

The theme of Chapter 14, "The Functioning Animal," is the variety of ways in which the necessary functions of an animal body are carried out in different animal groups. In each case, the comparative physiology uses man as a principal example.

In Chapter 15, "Behavior," the reactions of organisms to external environment are considered as stemming from internal mechanisms. From the vast field of behavioral biology, topics have been chosen that are related to other parts of the course and that have proved stimulating and fairly comprehensible to tenth-grade students: learning, periodicity, territoriality, and social behavior.

Section Five: Continuity of the Biosphere. This may be considered the heart of the course. Perhaps the most fundamental thing that can be said about life is that it goes on. In this section, much of the matter from the preceding four sections is directed toward an understanding of this basic idea.

Reproduction is considered as a life process unimportant to an individual's continuity but—because of individual death—essential to the continuity of populations and all high levels of biological organization. Chapter 16, "Reproduction," continues the comparative method employed in the two previous chapters and again (as in Chapter 14) uses man as a principal example.

It is difficult to overestimate the importance of genetics in contemporary biology, but balance in the biology course demands that genetics should not be allowed to get out of hand. In Chapter 17, "Heredity," the topic is developed historically and, from this development, some ideas concerning the logic of evidence are derived. Mathematics is not shunned; it can be either minimized or maximized by the teacher.

The entire course—indeed, any modern biology course—can be regarded as a summary of the evidence for evolution. The main objective of Chapter 18, "Evolution," then, is not to give the *evidence* for evolution; that has already been done in several ways, both implicit and explicit. Instead, the chief aim is to give the student some idea of the mechanism of evolution. Darwin is presented as one who provided an explanation of how evolution operates, not as the originator of evolution as a concept.

Section Six: Man and the Biosphere. We are human, and biology in the high school can be justified only on humanistic grounds. After putting man in perspective with the rest of nature, the course here, at its end, focuses explicitly on him.

In Chapter 19, "The Human Animal," some ways in which man differs anatomically and physiologically from fellow organisms are discussed. Since much of man's distinctiveness is behavioral rather than physiological or anatomical, the chapter inevitably becomes involved in borderline areas between anthropology and biology. Then the paleontological evidence for the origin of man is examined. Finally, racial variation within the human species is viewed biologically.

The authors hope that they have woven the whole course together in Chapter 20, "Man in the Web of Life." The student is confronted with topics that will concern him in the future as a citizen—topics for which biological information has some relevance. These topics, of course, all go beyond biology; the primary aim is to provoke the student into continuing to think about them.

THE YELLOW VERSION

The educational challenges to the teachers and students using the BSCS Yellow Version, *Biological science: an inquiry into life,* 2nd ed., New York: Harcourt, Brace & World, 1968, stem from the factors below.

1. This will be the only laboratory science course for many high school students.
2. For even more students, this will be their only contact with biology.
3. The rapid upgrading of high school biology has led many universities to shorten or eliminate their long-standing introductory course in general biology. Thus, for many students, this will be their only course attempting a general treatment of the field of biology.
4. Biology has long occupied an important place in the science curriculum in its own right but, today, it is becoming even more important in helping us to solve some of man's more urgent problems.

Thus it is extremely important that a biology course be outstanding. In the comments that follow, suggestions are given for using the Yellow Version more effectively. These comments ignore the fact that there is much similarity of content and approach, not only among the various BSCS versions but also among various non-BSCS books. After all, if a student is using the Yellow Version the chances are that this is not only his sole biology text but also his sole science text.

The Yellow Version first became generally available for use during the 1963–1964 academic year. It was widely used in schools representing a broad spectrum of geographic, social, economic, and political situations. A comprehensive program of evaluation by students, teachers, and professional biologists was conducted for two academic years, and the criticisms and suggestions of these groups were used to prepare the second edition (1968). The remarks that follow pertain specifically to the second edition but, to a considerable extent, they apply to the first edition as well.

The contents are grouped into four major parts—unity, diversity, continuity, and interactions—each emphasizing a way of looking at biological problems and their solutions. Teachers and students will find it useful to refer repeatedly to the précis of each unit and chapter, which appear in the Contents (pages ix–xviii). These will help to place each unit in its proper context and to suggest the points of emphasis.

The eight chapters of the first part, "Unity," are themselves an introduction to general biological procedures and principles. An attempt is made in them to give the historical and intellectual basis of biology.

In Chapter 1 the usual platitudes are kept to a minimum and the student is presented immediately with an example of what biology is, who does it, how its questions are asked, how its data are obtained, and its relation to other fields of knowledge. The example chosen is malaria since, with this topic, one can deal with almost all major aspects of biology and can show the relation of this science to others.

Chapter 2 sets the stage for considering one of life's most important characteristics: its self-origin. With the prevailing conditions on the earth's crust today, it is true beyond a reasonable doubt that life arises only from life. This is a new idea, considering the totality of man's intellectual history, and the description of the slow process of coming to accept this idea introduces the student to elementary procedures of discovery and verification in science.

In Chapter 3 there is a somewhat more advanced treatment of scientific procedures built around a discussion of the historical and intellectual steps that led to the realization that all life shows a fundamental structural unity in being cellular. As this topic is developed, there are discussions of the nature of science, how science differs from technology, the role of instruments in scientific discovery, how scientific knowledge is spread, how hypotheses are made and tested, and how conceptual schemes are formulated.

For many students, elementary chemistry is something to be memorized in terror. Hopefully, in Chapter 4, we have provided a gentle and meaningful introduction; that is, the student should be able to understand much of the elementary chemistry that is necessary for an understanding of elementary biology. Once again, by using the historical approach, ideas are presented slowly and without undue detail. After all, the basic idea of this chapter is that the events that occur in living cells are, to the best of our knowledge, chemical and physical events and nothing more.

The topics considered in elementary detail in Chapter 4 are amplified, somewhat, in Chapter 5, with special emphasis on the chemical compounds and interactions that occur in living cells. It is important here for the teacher to reassure the student, who otherwise may assume that he is expected to achieve such awesome feats of memorization as learning the number and distribution of the electrons of the elements (pages 82 and 85), the structural formulas of various organic compounds (page 94), or even the sequence of amino acids in insulin (page 92). It is most certainly not our feeling that he should. The illustration of the insulin molecule, for example, is intended to give the student some feeling of the complexity of organic molecules and that this organic molecule is made up of amino acids linked together in specific ways (but not *which* specific way).

By now the student has probably discovered Chapter 39 but, if not, he should be encouraged to examine it from time to time. This chapter will give him some appreciation of levels of organization, as well as of the diversity of the objects of nature. Gradually this catalogue of nature will become part of his information, and it will serve as a convenient framework in which he can place the various animals and plants that he encounters.

In Chapter 6 the chemical facts and principles discussed in Chapters 4 and 5, plus the material on cell structure presented in Chapter 3 (and amplified here), are combined to show how cells live. The teacher will probably have to devote some effort to explaining "flow diagrams," which are introduced in this chapter (Figures 6–8 and 6–9) and are used frequently thereafter. Flow diagrams contain a large amount of information presented in a manner that is easily understood—that is, once their construction has been explained to the student.

Chapter 7 considers a broad and fundamental biological problem: reproduction. The point is developed that reproduction occurs at many levels, and the examples stressed are: chromosomes, cells, and individuals (reproduction at the molecular level is considered in Chapter 8). Experience has shown that it is somewhat easier for students to understand the *process* of mitosis if it is not obscured by a terminology that has a subtle static connotation. For this reason we have not used such familiar terms as prophase, metaphase, etc. The material on pages 138–145 is somewhat difficult and, in all probability, its full significance will not be realized without considerable guidance from the teacher. But this is important material and it forms a necessary basis for Chapter 8.

Chapter 8 carries the problem of reproduction to the molecular level. Once again, this is a demanding chapter for the student and, if he is to fully appreciate the material, considerable intellectual effort is required. If he is willing to make the effort, he will have the satisfaction of having "discovered" the Watson-Crick model of DNA for himself.

Something more general should be said about Chapters 7 and 8. Many will consider them difficult, possibly too difficult, but hopefully they will be in a minority. The amount of technical information—"facts to be learned"—in these chapters is small. Yet the student must stretch his mind if he is to attain his maximum level of understanding. The level reached will depend on the student's ability and motivation. There is something in these chapters for all. Some students will "get the general point" and little more. Most students will probably gain a fairly sophisticated insight into some of the most profound biological problems. The truly gifted and highly motivated student may find material in these chapters that will stimulate him greatly. Here and elsewhere in the Yellow Version we have attempted to write at a level that will allow the average student to master most of the material. Hopefully the below-average student will not be lost, and the evidence available to us indicates that they can perform satisfactorily. We have, however, tried not to short-change the truly gifted student. From time to time the argument is carried to levels that may seem to be for him alone but, in so doing, we hope the level attained by the average student will be enhanced.

Part 2, "Diversity," begins the systematic treatment of organisms as individuals and groups of individuals, not simply as bags of biological problems. The onset of diversity begins with the origin of life, which is used to start Chapter 9. This topic suggests an analysis of the simplest forms of life: the viruses. Viruses are intrinsically interesting; they are of obvious importance to man; and some aspects of their behavior are used to extend the account of reproduction. The problem of the origin of life raises the question of how one studies the past, and this matter is discussed. In this chapter, as elsewhere, the topics will have greater significance for the student if he is reminded of what he already knows. For example, the origin of life in the past should be related to Chapter 2 where spontaneous generation was discussed in detail. The student should also be asked to recall the experiments in Chapter 8 where viruses were used (pages 152–158). The section on genes and enzymes should be related to the discussion of enzymes in Chapter 6 (pages 105–108).

Chapter 10 and 11 deal with bacteria. Bacteria represent a comparatively simple stage of biological organization, which alone makes them worthy of study, but they are of interest for many other reasons. Experiments with them have contributed notably to our understanding of inheritance, especially at the molecular level. Two new aspects of inheritance are discussed: transduction and transformation. Bacteria are important to man and other organisms in many ways—disease, spoilage of food, various industrial processes—and the student should be alerted to their role in the cycles of nature (which is discussed in Chapter 36).

The plants are treated in Chapters 12–17. After considering the somewhat aberrant fungi, the green plants are discussed in relation to their increasing complexity, which is a reflection of their evolution. (Students should refer to Chapter 39.) Special topics are outlined against this background of evolutionary change: variations in the life cycle, relation to the environment, increase in morphological complexity, photosynthesis, plant development, and plant hormones.

The reader will notice here, as elsewhere, that the emphasis is on general principles and that the wealth of details frequently associated with discussions of groups of organisms is largely avoided.

Chapters 18–21 deal with animals. There is first a discussion, in Chapter 18, of the animal way of life and then, in Chapter 19, problems relating to classification and the evolutionary interrelations of the major animal phyla are considered. Chapter 39 should be used as a supplement. In the subsequent chapters the material is organized around the solutions of basic physiological problems by a variety of animals. For the most part, each of these physiological chapters first states the problem and then discusses how it is solved by Hydra, planaria, earthworm, grasshopper, and man. Chapters 27 and 28 are concerned with development, first in terms of the normal development of the salamander embryo, and then an attempt is made to understand the underlying mechanisms by experimental investigations.

Part 3, "Continuity," has two main divisions: the short-term aspect, or genetics, and the long-term aspect, or evolution.

Genetics was introduced in force in Chapters 7–9. It was referred to frequently thereafter (for example, pages 205–212, 250–251, and 515–517). Chapters 29–30 cover the more classical aspects beginning with Mendel's work and continuing with Drosophila. In addition, there are discussions of chromosomal abnormalities, genetics of man, and the action of genes. It is hoped that this method of discussing genetics throughout the book will lead to deeper understandings of this vital area of biology.

Evolution is another pervasive theme that is developed throughout the book, but its main treatment comes in Chapter 31–34. The student must have a good background in genetics, as well as some knowledge of the diversity of animals, before evolution can have its full impact. Chapters 33 and 34 deal specifically with the physical and cultural evolution of man.

Part 4, "Interaction," considers the most complex problems of biology. These are problems that relate not only to the interactions of individuals and groups of individuals but, to a marked degree, there must be an interaction of many different fields of biology for the analysis of the problems. The teacher must be very careful to rigidly follow a schedule that will leave time for these final chapters. They may well be the most important in the book for an understanding of some of the difficulties currently facing our species.

The complex and difficult nature of these areas in biology can be seen in Chapter 35 on animal behavior. This is a rapidly developing and fascinating field of biology, but the student will soon observe that the level of understanding is much lower than it is in genetics, for example.

Ecology, in its varied aspects, is treated in Chapters 36–38. Chapter 36 considers basic ecological relations, and the student should be asked to recall much that he already knows about microorganisms, plants, animals, and physiology. Chapter 37 deals with the various sorts of biological communities of the earth's crust and then, in Chapter 38, some of man's more pressing ecological problems are considered.

Chapter 39, as noted before, considers the vast spectrum of organization of nonliving and living objects. It is a modern "scale of nature" that is intended for reference, not for memorization.

With this synopsis of the Yellow Version completed, more general remarks on teaching might be added. Some teachers seem to feel that teaching BSCS biology is very different from teaching conventional biology. Not infrequently the question is asked, "How should a teacher prepare himself to use a BSCS version?", and one has the feeling that what was really meant is "Aren't the BSCS versions so different from traditional biology that very special preparation must be demanded of the teacher?" Most probably this is not true. In fact, it should be somewhat easier for a young teacher to teach a course based on a BSCS version than a course that emphasizes the details of morphology and systematics. Probably the only essential prerequisites for teaching BSCS biology are the willingness to use one's mind as an analytical tool and a reasonable amount of humility. After all, we expect a student with little or no preparation in biology to "learn" BSCS biology. It is not too much to expect that a person who has been educated to a level where he can be appointed a teacher of biology is fully equipped to use a BSCS version in his classes.

"But what about all that chemistry?" Possibly we should explain the philosophy of the BSCS writers. It has always been our hope that our materials would be used in courses where understanding of principles was far more important than the memorization of details. One can understand the general principles of the use of energy in cells without knowing the details of metabolic pathways or the details of molecular structure. Certainly, few of many writers who have worked on the BSCS versions could pass an examination stressing such details. Similarly, one can understand the general principles of evolution without knowing the details of morphology, classification, and population genetics. And so it goes. There are probably very few, if any, professional biologists who "know" all there is in any one of the versions, or any biology text-book for that matter. Should more be demanded of a beginning teacher?

But the two prerequisites mentioned are not always easy to obtain. By and large, our school system does not stress the use of one's brain as an analytical tool—but surely there are many trends in this direction. Basically we deal with facts and hope that the deeper understandings will emerge. In the BSCS versions we have been trying to reverse this procedure, hoping to give the student a broad framework of theoretical biology into which the details will find a logical place.

Not many students are likely to approach biology, or any other course, in this manner. It is for the teacher to give them this perspective and to train them in thinking in terms of broad concepts. Biology should be approached as an intellectual exercise, not as a recipe book ("how to make an earthworm by assembling the 493 parts, now to be listed"). And if this done, the end result will probably be that the student will know more facts—but they will be meaningful facts, not facts for facts' sakes.

Now about that other prerequisite—humility. The most effective teachers are those who participate with their students in the enterprise of learning. A teacher should be at least as interested in learning new facts, new relationships, new ways of approaching problems, and gaining new insights as his students. Such a teacher functions as a somewhat more experienced student—not as an oracle willing and able to give final pronouncements on all questions. The inexperienced teacher may regard this as a dangerous game—can one really control students unless they believe the teacher knows it all? In the beginning of one's career as a teacher, it may take considerable humility and courage to admit not knowing the answers to some of the student's questions. But to do this can become an enormously effective teaching device.

In summary, the problem is not what it takes to teach a BSCS version, it is what it takes to be a good teacher. Once that problem is well on the way to solution, the subproblem of being a good biology teacher will take care of itself.

BSCS SPECIAL MATERIALS

After the introduction of the original three versions of BSCS biology and their widespread use over a period of years, it became possible to identify and come to grips with the problem presented by the students using these materials who were unable to achieve a satisfactory degree of academic success. These students, throughout much of their school career, have exhibited a pattern of failure in their school work. Their characteristics are quite diverse in most respects, but they show a common pattern of being academically unsuccessful.

The few books then in existence that aimed at the needs of this group differed from the standard textbooks only in vocabulary level and depth of content. There had been no attempt to reach these students in a way that was different from the way in which they had already experienced failure, a way that involved reading a textbook, memorizing facts, and being quizzed on each chapter. Even if this methodology were abandoned in the face of its massive failure, the retreat was often marked by a return to the old general science course or, sometimes, just hygiene.

The BSCS Special Materials Committee, which was charged with the study of this problem, developed three basic assumptions on which the Committee's work was based: (1) all students can learn more than they have thus far learned in school, since the difficulty is not so much with the students or the nature of the material to be learned as with the teaching approach; (2) every high school student should have the opportunity to learn the concepts and enquiry processes of modern biology; and (3) a variety of teaching approaches, including laboratory activities, group discussions, films, and readings, are necessary to help the students learn.

With these basic assumptions in mind, the Special Materials Committee undertook to develop a new approach. The same basic concepts, themes, and objectives of the original three BSCS versions were incorporated into a new format: the reading material was held to a minimum; there was no "big book" to intimidate the student from the beginning; the activities in the laboratory and field were expanded considerably; part of the material was programed; the overall pace was slackened and individual differences could be accommodated; and a comprehensive, step-by-step Teacher's Handbook was prepared. The latter provides detailed examples for structuring learning situations. This structure is designed to build a rapport between the teacher and the students as well as to place the student into direct interaction with the biological subject matter to be learned.

The result of this effort was *Biological science: patterns and processes,* published in 1966 (New York: Holt, Rinehart and Winston) after two experimental editions. It has been widely accepted in the schools, far above expectations, and many teachers have provided BSCS with "feedback" that is both gratifying and challenging. The challenge, of course, is to revise the materials on the basis of this expanded experience in the classroom and to produce a new edition that will even more effectively meet the needs of a group of students that has heretofore largely been ignored. Revision was begun in the summer of 1968, and a revised edition should be available in 1970. (See *BSCS Newsletter* 36.)

The Students

Much research has been aimed at defining the characteristics of academically unsuccessful students. These studies range from finding correlations with socioeconomic factors to identifying characteristics of the students that are interpreted as a part of their "nature" which sets them apart from other students. One example of the latter, which has often been described, is "short attention span." Teachers report, however, that if students are engaged in interesting, meaningful activities, these students often show more patience and a *longer* attention span than students who ordinarily succeed in school. This is but one example of much evidence from direct classroom experience which has led the writers of *Biological science: patterns and processes* to view the student in terms of *what he can do now* rather than in terms of his past history. From this starting point, it is then possible to look for ways that will enable the student to learn, since it is expected that he will learn if his current achievement level is accepted.

What we are suggesting, then, is a new set of assumptions about the student that enable us to take a new approach to teaching and learning. The teachers who have been willing to adopt these assumptions as working hypotheses report that they are pleasantly surprised, if not amazed, at the changes that occur in the students. These changes often take the form of increased interest in learning and attending school as well as an unexpected achievement in learning biological science. BSCS Special Publication No. 4, "The Teacher and BSCS Special Materials" discusses this point of view at greater length. (Also see below, Implementation, and Chapter 7, "Evaluation.") Other suggested readings that elaborate this new view of the student are the following.

Rosenthal, Robert, and Lenore Jacobson, 1968. *Pygmalion in the classroom.* Holt, Rinehart and Winston.
Combs, Arthur W., et al., 1962. *Perceiving, behaving, becoming.* Association for Supervision and Curriculum Development. (See, especially, pp. 75-82.)
Maslow, Abraham, 1968. "Some Educational Implications of the Humanistic Psychologies." *Harvard Educational Review,* Vol. 38, No. 4 (fall, 1968) pp. 685-696.

The Student's Book

The student's book is bound as a paperback and has been developed as a set of individual and group activities. The activity sheets are removed from the book and are used by the student when his class activities have prepared him for their use. These sheets are then placed in the student's notebook. Interspersed with these activity sheets are material summarizing class discussions, data from laboratory activities, copies of tests, and other products of class work. The result is that the student develops his own textbook during the year.

The range of reading achievement of academically underdeveloped students is great; hence, writing for a specific grade level is not advisable. It is more important for reading material to be relatively short, to have an interesting style, to be used only when needed, and to be used in a meaningful context. With these techniques, the authors' aim is to develop the student's interest in reading and to improve his reading ability.

Five major areas of biological science are included in *Biological science: patterns and processes.* These are "ecological relationships," "cell energy processes," "reproduction and development," "genetic continuity," and "organic evolution." Although these are organized as separate units, many of the major biological concepts developed in one are referred to or used in others so that there is an overall unity throughout the course.

The Teacher's Book

In the Teacher's Handbook the entire sequence of the course is presented in detail, topic by topic. The teacher is given suggested patterns of discussion questions and possible student responses that are designed, for example, to develop the understanding of a concept or an aspect of enquiry, to prepare the students for a forthcoming laboratory activity, to summarize a laboratory activity just completed, or to provide a situation for formulation of a biological problem. The Handbook also provides suggestions for helping students with reading, evaluating the student's progress, leading class discussions, utilizing the laboratory, and introducing programed instruction. The Handbook outlines the quantities of materials needed for each laboratory activity and gives directions for preparing special solutions and media. In addition to the suggested sequence of activities, many options and alternative activities are described. In short, the Handbook is designed to help the teacher be successful in teaching the academically underdeveloped student.

Teacher Preparation

Many times when we speak of teacher preparation we think only in terms of formal preparation—class meetings, reading of textbooks, and writing of papers. Although this formal training is important in teacher preparation, it is essential to realize that teacher preparation can occur on an informal, self-training basis.

The introduction to the Teacher's Handbook presents many items of value to the teacher who undertakes a self-training program for *Biological science: patterns and processes*. It provides a rationale for the course and reviews the teaching strategies used in the Handbook.

Any self-training program should include a complete reading of the first topic, "ecological relationships." This reading will provide the teacher not only with an overview of the biological concepts presented but also with statements from the authors telling why this particular organization of the topic was chosen. The initial reading will also help to familiarize the teacher with the format of the Teacher's Handbook. A second reading should be undertaken but, this time, for the purpose of organizing the topic into a teaching plan. During the second reading, questions such as the following should be answered. How many weeks of classroom work will I devote to this topic? Will we do the

suggested activities, or should I substitute some of the alternate suggestions? How do I organize the class—groups of two, groups of four,
or in some other way? Will I mix the special solutions for the students,
or should they be allowed to mix them? Are there any materials that
must be ordered in advance? When the teacher has completed this
phase of his self-training program he should have a teaching plan that
will allow him to present the first topic, allowing for minor changes
that class progress or school activities dictate.

The suggestions for a self-training program for teachers in connection
with the first topic should be repeated for each of the topics in the
course. It is essential that the teacher be familiar with each topic prior
to teaching it in the classroom. Then and only then can he concentrate
on the teaching and learning situations of the classroom rather than on
the subject matter and worrying about proper sequence.

The third phase of the self-training program should include a reading
of BSCS Special Publication No. 4, *The teacher and BSCS special materials*. This publication contains short articles, written by members
of the BSCS Teacher Preparation and Special Materials Committees,
designed to provide the teacher with a background for understanding
the students for whom *Biological science: patterns and processes* was
prepared.

Implementation: The Teacher and the Administrator

As more has been discovered about the learning process and a
greater effort has been made to offer optimum educational experiences
to all students, our schools have become more complex. In spite of this
trend, the fundamental ingredients of education—the student, the subject matter, and the teacher—remain of primary importance. All other
aspects of school administration and organization are peripheral and
secondary to these fundamental ingredients. What is the role of the
teacher in this dynamic situation? What qualities make it possible for
a teacher to provide an exciting and profitable learning situation?

Important characteristics of the teacher of the up-to-now academically
unsuccessful student are appropriate attitudes toward and appreciations
of the learner. These include an accepting attitude, patience, and a
belief that students who have not profited from the conventional school
program are as important as other students. Other attitudes and appreciations include the following.

1. Respect for every child or youth and faith in his ability to learn.
2. Awareness of the emotional components of all aspects of life and
learning.

3. Concern for and interest in all students.

4. Ability and willingness to see things from the eyes of the students.

5. Affection for all students—rejection of offenses but not offenders.

6. Full acceptance of the authority role—firmness without cruelty or hostility.

7. Consistency.

8. Absence of fear or shock at language and behavior foreign to his own experience.

9. Avoidance of ridicule, embarrassment, or humiliation of students.

10. Avoidance of impatience, disparagement, or comparisons of students.

11. Maintenance of a democratic atmosphere.

12. Warm, outgoing, friendly communication with parents; seeks opportunities for such communication.

Ideally, these characteristics should be true of all teachers; however, they are essential for the teacher of the academically underdeveloped student.

There are essentially three characteristics, other than those in the realm of attitudes and appreciations, that the teacher should have: (1) an understanding of the learning process and the various channels and media of learning; (2) an understanding of the logic of the subject matter developed within the topics of *Biological science: patterns and processes;* and (3) an understanding of and ability to use the pedagogic techniques which bring the first two understandings together.

Some specific principles of learning that are prominent in the current literature and that have been used in the development of *Biological science: patterns and processes* are the following.

1. Learning becomes meaningful when students relate specific facts into conceptual wholes. Learning is discovery of meaning, not merely a change in behavior.

2. Students learn better when they have success and know that they are successful.

3. An atmosphere of respect and acceptance, friendliness, and helpfulness is conducive to learning.

4. Start with skills and understandings the students have and build on these.

5. Learning is more effective when there is active mental participation by the students. Problem solving and situations of interest to the students are likely to encourage active participation.

6. Reinforcement aids learning, particularly when multiple media

are used (for example, laboratory, discussions, and readings) and all pertain to the same idea or information.

7. Students learn better when they have some responsibility for determining what they learn.

The most important single function of administrators in this program is the selection of students to use *Biological science: patterns and processes*. The methods incorporated into the materials are valid for use with all students; however, these materials have been designed specifically for the student who has not profited from the conventional school program. It is difficult to determine exactly how many students fit this category. Some investigators estimate that these students constitute one-third to one-half of the national high school population, yet within a given school, or school district, the proportion may range from 1 percent to more than 70 percent. Therefore, only the local school, or school district, can establish realistic procedures for selecting students who will use *Biological science: patterns and processes*.

In establishing groups, it must be remembered that no single criterion can serve as a valid source of grouping information. To base all grouping on the results of one standardized test would not take into account all of the factors influencing a youngster's success or nonsuccess in school. Any instrument used for the measurement of a student's current achievement in standard school tasks should contain components other than a language or verbal component. Numerical skills or abstract reasoning are other areas that might be included. In addition to such information, some measure of the student's reading achievement level and the recommendations of previous teachers should also be available for possible use.

Administrators can support education in the classroom by selecting and grouping students for the program. However, grouping must remain flexible. A reduction of class size will allow the student to develop adequate skills in the manipulation of laboratory equipment and strengthen their learning through laboratory activities. In addition, the general school climate must be appropriate to maximize the kind of learning that occurs in the classroom.

We can sum up the critical points of the school climate by quoting Eugene McCreary:

"We ought to be concerned about any aspects of our schools' programs which lead to passivity and boredom or to compulsive conformity, fear, shame, or guilt. We ought to be sensitive to practices that break down youths' sense of identity and individuality. We ought to be critical of circumstances that detract from students' feeling of their own worth and significance. We ought to look sharply at programs that set student against student and undermine the essential social ties of youth. We ought to be sure that our schools are places where all students of every class and ethnic background are able to find rewarding, meaningful tasks that they perform successfully in association with others. We ought to be sure that students are not afraid to think differently or behave differently from others within the limits of law and safety. We ought to be sure that all students encounter ideals that can be meaningful to them." (Eugene McCreary, as quoted in BSCS Special Publication No. 4, p. 42.)

THE BSCS LABORATORY BLOCKS

It has been suggested that the biology teacher using BSCS materials can teach a course tailor-made to the interests and needs of his students, school, community, and himself. This possibility is realizable because of the variety of materials developed and the variety of teaching approaches that can be used with these materials.

The Laboratory Blocks are unique among the materials developed in the BSCS Program. They offer a design for student experiences that can enable the students to investigate a series of questions in the biology laboratory following a pattern similar to that of a professional biologist. In keeping with the BSCS rationale, emphasis in the use of the Laboratory Blocks is on student participation—both as individuals and in groups. Furthermore, the nature of this participation is enquiry.

The Nature of Laboratory Blocks

The laboratory blocks have been developed as individual books. Each provides a pattern for a group of laboratory investigations by the student of a specific biological subject. A six-week "block" of time is required for these investigations—hence the name.

Laboratory teaching has two purposes: *illustrative* and *investigative*. Much of our laboratory teaching serves the illustrative rather than the investigative purpose. However, if the student is really to understand the nature of the scientific process, active participation in that process is required. The student must investigate some problems for which he does not know the answers.

The Laboratory Blocks are designed to serve the investigative purpose of laboratory teaching. They guide the student to the making of discoveries for himself. Some questions or problems are posed, but no answers are given in advance. Thus, the student is really working with an unknown. In making discoveries for himself, the student will generally follow in the footsteps of scientists who have preceded him. On some occasions, however, he may follow an entirely new approach. The student should be encouraged to keep in mind the possibility of alternative procedures. He should be on the lookout for new problems, should design new experiments to investigate them and, as much as is practicable, should investigate his own new problems.

The student is expected to acquire the background information needed to investigate the problems at hand. Some of this information is provided in the Laboratory Block and in other books and journals. Other information may be provided by the teacher *at appropriate times.* However, for the most part, the student will study a given group of problems by making observations and obtaining data relative to them. Then he will be expected to interpret his observations and data and arrive at conclusions that not only will answer specific questions relating to the problems he has investigated but also will increase his understanding of certain biological concepts that underlie the individual Laboratory Blocks.

The Laboratory Blocks are intended to help students to realize the need for constructing reasonable hypotheses and planning experimental designs with appropriate controls. Students should learn to distinguish different kinds of data, selecting those that are relevant, identifying possible sources of error, and accurately presenting data in various ways (graphs, symbols, quantified, verbal) while employing suitable procedures for analysis of the data. As a research biologist must do, they should learn to base their interpretations or generalizations on the evidence at hand, realizing the extent and limitations of these generalizations and the frequent lack of clear-cut "yes" or "no" answers to questions.

It is intended that in working through a specific Block the student will actually experience, on a somewhat reduced scale, the various processes used by scientists in biological research. They should also gain an understanding of various laboratory and research techniques. For example, they may become quite skillful in the use of such equipment as a temperature gradient box or a dissection microscope. They may develop skill in preparing culture media, counting bacteria, pithing frogs, incubating eggs, running a centrifuge, or determining the pH of a sample. But these techniques are incidental to the objectives of each Block and are *not* emphasized as primary skills which should be developed by each student. Certainly the Laboratory Blocks are not designed to develop and train laboratory technicians. Instead, each Block is designed to help a student understand more about some facet of the biological sciences and some of the characteristics of science and scientists.

The Laboratory Blocks, at least to some extent, should bring the student to the frontier of science for a particular subject. He should learn some facts and some techniques. But—what is more important— he should gain invaluable practice in working like a scientist. Thus, his image of a scientist will come from his own experiences. The carry-over of these experiences is valuable to any individual in modern life, regardless of whether he intends to become a professional scientist.

The Development of Laboratory Blocks

In the development of the Laboratory Blocks an active research scientist was requested to provide the initial material for a given Block. The material was then turned over to a staff of experienced high school teachers, project associates, who tried out the materials in a special Research and Development Laboratory set up by the BSCS for this purpose. The project associates did the experiments suggested by the research scientist for the Laboratory Block under various conditions and with such modifications as they thought might make them practicable for use in the high school laboratory. Recommendations were then given to the scientist who provided the original material, and revisions were made as needed. Each Laboratory Block was then "pretested" in a special classroom situation under a regular classroom teacher and again it was revised as needed. Finally, each Block was tried out by teachers in a limited number of BSCS testing centers and was again revised as indicated by the evaluation and feedback from teachers in the program.

This somewhat elaborate procedure was followed in order to provide accurate and up-to-date biology content and to have it developed in such a way as to make its use practicable in actual high school classes.

A Teacher's Supplement has been written for each of the Laboratory Blocks. In the Supplement, specific information concerning each individual investigation in the Block is given, as well as suggestions for conducting the study as a whole. Although answers are provided in the Supplement for most of the questions raised in the Block, the teacher is strongly urged to elicit answers from the students rather than give them to the students directly. If the Laboratory Block is to serve its most useful purpose, specific answers should be given to the student only as a last resort. Students should be encouraged to seek their own answers, and answers volunteered by students should be carefully discussed in terms of their relevance and limitations. At times, there is no one "right" answer, but it may be decided that following one line of thought can be more productive than another, in terms of the further work in the Block.

In addition to answers to questions, samples of data obtained during preliminary trials of the Laboratory Block work are given for many of the investigations. These are included only to give a teacher using the Block for the first time an indication of the kind of results that may be obtained. The teacher should not expect results identical with those in the Teacher's Supplement, nor should he necessarily interpret different results as a failure. Differences in data obtained may be accounted for by differences in geographical location, temperature, light, and many other factors, including variations in manipulative skill and technique. If the results should be very different from the samples given in the Supplement, it may be advisable to explore with the class possible reasons for such differences, but results or answers should not be given to the students in advance of the investigation.

Each Teacher's Supplement also contans an example of a suggested Daily Schedule for the Block and a list of special supplies and equipment that may be needed.

Teaching a Laboratory Block

Teachers using BSCS Laboratory Blocks for the first time will find themselves in a situation similar, at least to some extent, to their first teaching position or the first year that they moved from one teaching position to another. For many teachers, the use of Laboratory Blocks will be a new teaching approach, and its success will depend largely on their understanding of and commitment to this approach.

The nature and objectives of the Blocks should be recognized and understood. One objective is to provide students with the opportunity to learn biology—but with a Block, only a selected topic, not all of biology, is the objective. Furthermore, the student is to learn the selected topic in depth. *However, an even more important objective, insofar as teaching the Laboratory Block is concerned, is for the learning opportunity to be organized in such a way that the student will actually carry out a series of investigations on the given topic that will enable him to develop an understanding of the nature of research in the biological sciences, as well as of the scientific enterprise as a whole.* In other words, the student is to learn biology as a biologist learns it.

Many teachers pay lip service to this idea—but most secondary school teachers, whether the amount of their training in biology is adequate or not, have had little or no experience in actual research. A good understanding of research is obtained best, if not exclusively, by experience. However, it is reasonable to expect that teaching a Laboratory Block may help to correct any lack that might exist in the teacher's understanding of the processes of biological research.

But in terms of student behavioral objectives the important question is: How does the Laboratory Block help the student to understand the process of research or investigation in the field of biology? It does so by providing him with an opportunity to investigate a specific biological topic, following the pattern a scientist might employ to investigate the same problem if he were starting at the same place as the student. On the other hand, it should be understood that some decisions regarding organization, techniques, and equipment to be used in the investigations are made for the student. Admittedly, this procedure is somewhat different from that followed by a research scientist, but it is necessary because of time and schedule limitations in the total high school program.

A teacher who chooses to include a Laboratory Block as a part of the biology course will find it necessary to make careful plans for its use in relation to the other part of the course. It is particularly important to make such plans in advance, including the timing of the Block in relation to other parts of the course. Since the teacher will have only 30 weeks to cover the biological topics that otherwise would be covered in 36 weeks, some reduction will need to be made in the amount of time given to topics in that portion of the work; it may be necessary to leave out some topics, even though such an omission is sometimes a traumatic experience for the conscientious teacher.

It should be pointed out that a lot of biology will be taught in one of the Laboratory Blocks and, in the long run, it is believed that there will be a net gain rather than a loss in using a Block, even if some topics are condensed or cut from the course as a whole. It should be kept in mind that in any biology course some important topics will have to be omitted: a teacher simply cannot teach all of biology in one year.

Most of the Laboratory Blocks can be scheduled at any time during the school year. However, it is usually advisable to start the year's work with the introductory units of the text and laboratory materials. This procedure will introduce the student to the subject of biology and to much of the basic laboratory equipment (microscope, balances, etc.). Because of the continuing nature of many experiments in some of the Laboratory Blocks, it may be advisable to avoid interruption resulting from extended vacation periods.[1] The selection by the teacher of the specific Block to use ideally takes into account his own training, background, and interest, as well as the particular school situation.

Notice that Laboratory Blocks can be used with any textbook under a wide variety of conditions. The Blocks were designed originally for use with tenth-grade classes but have been used successfully with some ninth-grade classes and with eleventh- and twelfth-grade classes. It is recommended that only one Block be used in a regular nine-month course. One exception to this suggestion is the use of a sequence of Blocks in an advanced biology class if the students have already had a ninth- or tenth-grade biology course and have not previously had the Blocks. In such a situation, it may be necessary either to institute some modifications of the Blocks in order to avoid the overlapping of techniques or to develop appropriate instructional materials to connect one Block with another during the course.

The use of a Laboratory Block also requires careful planning to make sure that appropriate amounts and kinds of equipment, as well as adequate storage and other facilities, are available. Teachers and administrators frequently are concerned about the possible cost of doing a Laboratory Block and the extent to which special equipment and supplies are required. It is true that in some instances some additional equipment will be needed, but not on any large scale. One must recognize that the modern teaching approach in all of the sciences at the secondary school level involves greater emphasis on laboratory work, on individual student participation, and on investigation. If a biology laboratory is equipped for modern teaching, little if any extra equipment and facilities will be needed to teach a Laboratory Block.

Summary of Content of Each Laboratory Block

Laboratory Blocks have been developed on a variety of subjects. The following list gives the titles, authors, and a brief description of the nature of the investigations designed for each of thirteen Blocks [2] developed by the BSCS.

[1] For sample schedules made up for each Laboratory Block to be used with each of the three versions of BSCS textbooks as well as other useful information in using Laboratory Blocks, see BSCS Special Publication No. 5, *Laboratory blocks in teaching biology*, edited by Addison E. Lee, David L. Lehman, and Glen E. Peterson. Published by the Biological Science Curriculum Study, P. O. Box 930, University of Colorado, Boulder, Col. 80302, 1967.

[2] All of the BSCS Laboratory Blocks are published by D. C. Heath and Company, Boston, Mass.

ANIMAL GROWTH AND DEVELOPMENT
Florence Moog, Washington University

Students observe animal development through the series of events by which a fertilized egg converts itself into the organized array of differentiated structures that make up a complex animal. Working with living, developing eggs of frogs and chickens, students have their attention directed to the evolutionary changes in the reproductive mechanism that helped make it possible for vertebrates to become land-living animals. In addition, a series of experiments is presented concerning the influence on development of such agents as temperature, hormones, and enzymes. Optional sections are provided to investigate regeneration and growth in planaria and hydra.

THE COMPLEMENTARITY OF STRUCTURE AND FUNCTION
A. Glenn Richards, University of Minnesota

Investigations begin with an attempt to gain the concept of mechanical advantage of levers with the study of levers of a frog skeleton and the muscular forces that move the levers. The physical concepts learned are then related to the degrees and kinds of motion that are possible with levers in higher animals. Muscular movements without levers are observed in the crawling of earthworms and the pulsation of blood vessels. Movement by pressure changes is studied in the breathing movements of the mammalian lung and by the extension of the proboscis of a housefly. Nonmuscular movements are investigated using representative cases such as the movement of epithelial cilia and the locomotion of single cells. Finally, the student investigates the minute details of the complementarity of structure and function of the muscle itself.

PLANT GROWTH AND DEVELOPMENT
Addison E. Lee, The University of Texas at Austin

This Block includes studies of germination, increase in size, internal organization, metabolism, and regulation, from embryos to mature plants. With seed germination, the student determines seed viability, volume change during imbibition, and order of emergence of various parts from the seed. Changes in the external features of seedlings during early growth and development are observed. The effects of temperature, oxygen supply, nutrient supply, and light are investigated by conducting various experiments. The cellular basis for growth is observed in cell size, number, and specialization. Attention is then turned to metabolism through experiments with enzyme activity, respiration, and photosynthesis. Finally, the student investigates regulation of growth and development.

FIELD ECOLOGY
Edwin A. Phillips, Pomona College

A field area is mapped to establish permanent boundaries and a map of the area is drawn to scale by the class. Measurements are made of factors in the field area including temperature, moisture, light, and wind. Students, working in pairs, select an animal and a plant for observation during the six-week study. Sampling techniques are used to determine frequency, abundance, and cover of both animal and plant types. Correlations are determined between the structure of vegetation and environment. Measurements are made and students prepare pictorial diagrams of the vegetation cover. Experimentation is conducted in physiological ecology including transpiration, photosynthesis, respiration, and temperature of plants. In a final study, evidence is sought for changes which indicate a succession in the ecosystem of the field area.

MICROBES: THEIR GROWTH, NUTRITION, AND INTERACTION
Alfred S. Sussman, University of Michigan

Ubiquity and versatility of microbes are investigated through studies of samples of pond water and with selective enrichment cultures. Studies of growth are conducted through determination of various measurement and growth rates with generation time analyzed as an exponential function and growth rate plotted on semilogarithmic graph paper. Nutritional requirements of microorganisms are analyzed, and assays for a specific requirement are conducted. Finally, the interactions of microorganisms are investigated with studies of positive interactions and negative reactions among microbial organisms.

REGULATION IN PLANTS BY HORMONES: A STUDY IN EXPERIMENTAL DESIGN
William P. Jacobs, Princeton University
Clifford E. LaMotte, Boston University

Hormone regulation in plants is studied with a conscious focus on the problems of experimental design. Regulation of leaf abscission in the *Coleus* plant is investigated through a series of experiments involving: (1) the effect of auxin on leaf fall, (2) the relationship of individual leaf fall to the rest of the plant, and (3) the effect of the apical bud on the rate of leaf fall. Data are analyzed statistically. Excised root tips of tomato seedlings are cultured under sterile conditions to determine whether special substances are required for growth in culture. On completion of an evaluation of their own experimental work, the students make a study of the technical literature concerned with investigations by research scientists of the same problems. The students are asked to reevaluate their conclusions in light of these extensive investigations.

LIFE IN THE SOIL
David Pramer, Rutgers, The State University

The student first analyzes the chemical and physical properties of soil. Soil organisms are studied by searching for arthropods, nematodes, protozoa, algae, fungi, bacteria, actinomycetes, and viruses in the soil. The abundance of kinds of organisms is determined with statistical methods used to estimate the precision of the sampling. A buried-slide technique is used to observe certain qualitative characteristics of the soil population in relation to its environment. Techniques are provided for hunting and observing nematode-trapping fungi as well as mycoparasites of the soil. A last section considers such activities of soil organisms as cellulose decomposition, nitrification, the Fusarium wilt of tomato plants, and the production of antibiotics by soil actinomycetes.

ANIMAL BEHAVIOR
Harper Follansbee, Phillips Academy

The nature of animal behavior and the approaches and techniques used in its study are considered. Basic types of unlearned behavior are studied through investigations of a conditioned reflex and of habits. Three films are used to demonstrate techniques for exploring the processes of simple learning. Students perform a series of investigations concerned with responses of *Daphnia* to environmental changes. Finally, the students make a series of observations of the American chameleon.

GENETIC CONTINUITY
Bentley Glass, Johns Hopkins University

Biogenesis is investigated by repeating early classic experiments. Following this introduction, the student looks closely at the nature of cell division and learns how chance and probability work in the production of a fertilized egg. Data are collected on various human traits, and an effort is made to determine whether a given human trait is hereditary. The relative importance of hereditary and environmental differences are explored; the study is extended to examine genetic difference at the levels of enzymes, biochemical products, and morphological differences. The gene pool of the present population is investigated. In considering mutations, the student engages in an actual mutation hunt, within a population of *Escherichia coli,* for the rate of mutation of a specific gene.

MOLECULAR BASIS OF METABOLISM
Peter Albersheim, University of Colorado
John Dowling, Johns Hopkins University, Medical School
Johns Hopkins III, Harvard University

The first section provides background information regarding molecular structure, bonding, and simple organic chemistry. Investigations begin with a study of macromolecules and their subunits. Proteins, nucleic acids, and glycogen are extracted from yeast cells; the three groups of macromolecules are separated, following which each type of macromolecule is hydrolyzed into its subunits for analysis. The relationship of energy to molecular structure and to chemical reactions is investigated through the role of ATP in the luminescent system of fireflies. The dependence of a biological system on enzymatic activity is considered; respiration and photosynthesis are studied with volumetric systems. Photosynthesis is demonstrated by means of radioautography. In conclusion, the student explores the use of energy for muscular contraction with the addition of ATP to glycerinated rabbit muscle fibers.

PHYSIOLOGICAL ADAPTATION
Earl Segal, San Fernando Valley State College
Warren J. Gross, University of California at Riverside

This Laboratory Block is concerned with physiological adaptation performed by individuals in response to a particular environment. The sections of the Block are called Levels with each level representing a different time course of a response. Level I discusses the simplest kind of responses that come about as an immediate reaction to an environmental factor. It includes a study of the reactions of different animals to light, to contact (touch or pressure) and to the forces of gravity. The kind of responses a simple animal, an isopod, makes to different percentages of humidity are then explored in greater depth. In Level II a type of physiological adaptation that generally is longer-lasting than that of Level I is studied. The interaction between the external environment and the internal environment of an organism and how the internal environment of an organism is controlled to a constant or steady state, known as homeostasis, is emphasized. Level III includes a type of physiological adaptation that may run a time course of years in some animals, depending upon environmental conditions. An investigation of three weeks' duration is performed regarding temperature acclimation in *Porcellio,* a land crustacean (sometimes called pill bug, sowbug, or woodlouse). In Level IV the most lasting type of adaptation is investigated—a genetic adaptation, produced in this case by a mutation.

EVOLUTION
S. David Webb, University of Florida

The major projects include direct investigation of five different groups of organisms: fruit flies, bacteria, crickets, horses (fossil teeth), and cactuslike plants. Each series of investigations emphasizes somewhat different aspects of evolution, although similarities in the process are stressed throughout. In working with bacterial colonies the process of mutation is emphasized, while in the sexually reproducing forms the importance of natural selection is stressed. Investigations with bacteria and fruit flies demonstrate dynamic aspects of evolution in a short-term, direct way. They are complemented by projects with crickets which suggest how species first diverge. Observations on fossil horse teeth illustrate how evolution operates over a span of several million years to product distinct species as well as genera.

RADIATION AND ITS BIOLOGICAL EFFECTS
William V. Mayer, University of Colorado

The Block opens with a general discussion of radiation with an illustration of the various forms of radiant energy as we know them today. From this introduction the student is led into a discussion of the generalized effects of radiation. The purpose of this section is to instill in the student respect but not fear, care but not avoidance of radioactive material. The next sequence of investigations includes the properties of radioactive material, such as the effects of distance, concentration, and time (particularly half-life). Then the student learns the techniques of handling radioisotopes, preparing autoradiographs, and using a Geiger counter. Here the investigations involve uptake of radioactive phosphorous by a plant and by various animal organs, the lethal effects of radiation on microorganisms, and the relation of tissue metabolism to radiation injury in planaria.

The Use of Laboratory Blocks in Special Programs

Although it is obvious that the Laboratory Blocks were conceived and developed for use in a regular high school biology course, the potential for their use extends beyond the high school course. In the following paragraphs, other ways of using the Laboratory Blocks are suggested.

Uses in Special High School Programs. A very promising use of Laboratory Blocks with high school students is in various summer enrichment programs. These programs are generally of two types: some are sponsored locally and others are sponsored by a particular college or an outside agency (such as the National Science Foundation, the National Institutes of Health, the Jackson Memorial Laboratory, and the Worchester Foundation for Experimental Biology). The use of one or more Laboratory Blocks, with the freedom of such noncredit courses, for highly motivated youngsters offers them an excellent opportunity to learn about biology and research through a summer of laboratory investigation.

A teacher might also have some gifted students for whom a Laboratory Block would provide an ideal opportunity to explore, on their own time, a specific topic in greater depth (after school or at home if facilities and equipment were available) through laboratory investigations. In BSCS Bulletin No. 2—*Teaching high school biology: a guide to working with potential biologists* [3]—just such a suggestion is made, as well as other possible uses of Laboratory Blocks with gifted students.

The BSCS second-level course, *Biological science: interaction of experiments and ideas,*[4] was designed primarily for students who have taken a first course in biology and who desire to further their studies in biology while still at the high school level. The initial step in the development of this course was to combine four, then three Laboratory Blocks in a single, unified, second-year program. Later it was decided to use a modified Block approach for the entire course, but not to try to combine a series of individual Blocks. However, much of the material in the present version of the second-level course has been adapted from various Laboratory Blocks. On the other hand, it is still possible that an experienced teacher whose students have not previously had a Laboratory Block might choose, for example, to take three Laboratory Blocks and, with supplementary material of his own, weave them into his own second-year biology course.

[3] Brandwein, Paul F., Jerome Metzner, Evelyn Moreholt, Anne Roe, and Walter Rosen, *Teaching high school biology: a guide to working with potential biologists,* Biological Sciences Curriculum Study Bulletin No. 2, American Institute of Biological Sciences, 2000 P Street, N.W., Washington, D. C., 1962.

[4] Biological Sciences Curriculum Study, *Biological science: interaction of experiments and ideas,* Prentice-Hall, Englewood Cliffs, N. J., 1965.

Uses in College Programs. The impact of Laboratory Blocks on college biology programs is difficult to separate from the impact of the total BSCS effort. Certainly the impact of the total effort has been considerable in providing new up-to-date materials and in suggesting ways of teaching science as enquiry. But in addition to these general results, the Laboratory Blocks have influenced the college programs in at least three ways:

1. They have provided sources of material for college laboratory teaching. In some instances, the materials have been used directly, and in other instances the materials have been modified appropriately for the college course.

2. They have been used intact in a number of college programs, primarily in the preservice or in-service training of biology teachers. For example, a number of NSF-supported summer institutes and academic year institutes use one or more of the Blocks directly in their programs.

3. They have been used in a number of college and university educational research programs.

Uses in Community Education Programs. Laboratory Blocks can be used in such community-sponsored science programs as those offered through a local zoo or museum. Natural science centers, summer recreation programs of nature study, and summer camp programs—all could use one or more carefully selected Blocks successfully, even with upper-grade-school youngsters, although the Blocks would require some modification. For example, the *Field Ecology* Block would provide youngsters with an excellent opportunity to spend a summer exploring, in an organized manner, the living organisms to be found in their own community. A local zoo might select the Block on *Animal Behavior* to teach as a part of a Saturday morning community education program. The opportunities for using short blocks of time for biological education seem limitless.

Use in International Activities: Although the total BSCS influence in foreign countries has been considerable, the use of Laboratory Blocks to date has been limited. However, use is not entirely lacking, and greater use of the Blocks in the BSCS International Cooperation Programs is expected. Laboratory Blocks have been taught in Brazil, Japan, Venezuela, and Italy. Representatives from the following countries have come to the United States and have received special training in the use of Laboratory Blocks: Japan, Panama, Rhodesia, Philippine Islands, and Mexico. In addition, visitors to the BSCS Research and Development Laboratory have come from England, India, Ethiopia, Formosa, and Thailand.

Thus, it is apparent that, although the Laboratory Blocks were designed primarily for use with high school students in a regular introductory biology course, numerous other uses have been made of them. No doubt other innovations in the use of Laboratory Blocks or modified forms of them are possible, and this possibility is in keeping with the BSCS philosophy of trying to provide a wide variety of materials and programs for use in biology education.

Special Resource Books to Use with Laboratory Blocks

As previously indicated, development of the Laboratory Blocks involved considerable experimental effort in the BSCS Research and Development Laboratory and extensive tryouts in the schools. As a byproduct of these efforts, two special resource books were prepared.

Equipment and Techniques Resource Book. In the development of the Laboratory Blocks, not only have the problems and needs of the teacher for a given Block been kept in mind but attention has been devoted also to general matters concerning the biology teaching laboratory, its facilities, equipment, and use. These matters should be of interest not only to biology teachers but to school administrators and other policy-making individuals as well. Moreover, it has been necessary to investigate and solve certain equipment problems along with the development of the laboratory work. As a result, some innovations in equipment have been developed, some new equipment has been designed and built, useful modifications of existing equipment have been made, and unexpected uses have been found for equipment not normally employed in the high school laboratory. As an outgrowth of these efforts, *Innovations in equipment and techniques for the biology teaching laboratory* [5] was published in 1964 as a resource book for science teachers. This book discusses techniques and equipment in the following areas: the biology teaching laboratory, equipment and techniques for culture of microorganisms, laboratory animals and their housing, equipment and techniques for the study of plant growth, equipment and techniques for physiological experiments, temperature and heat control equipment, light-control equipment, and construction and use of some models and special equipment.

[5] Barthelemy, Richard, James R. Dawson, Jr., and Addison E. Lee, *Innovations in equipment and techniques for the biology teaching laboratory*, D. C. Heath, Boston, 1964.

The Test and Evaluation Resource Book. A given high school teacher can teach a given Laboratory Block. The teacher may react favorably to it and his students also may be pleased. However, all of the reports notwithstanding, teachers are faced with the problem of testing and evaluating student achievement in the Block program.

Students, parents, and teachers are still living in an academic world in which grades are an expected and accepted (although not always respected) measure of success. Grades are usually determined by teachers as a result of scores made by students on examinations. Examination scores are not always the only criteria used by teachers to determine grades, but they still remain one of the most common, if not the most popular, means of assigning grades to students in high school classes.

To help meet the demand from teachers using the Laboratory Blocks for help in testing and evaluating the work, a resource book on *Testing and evaluating student success with BSCS laboratory blocks* [6] has been developed. It serves as a companion to *Innovations in equipment and techniques for the biology teaching laboratory* to aid biology teachers in their continuing effort to improve biological education. The test and evaluation resource book has a pool of questions for each Laboratory Block as well as other materials useful for this phase of biology teaching.

SINGLE TOPIC INQUIRY FILMS

The BSCS Single Topic Inquiry Films are radically different from the usual classroom films. They are designed to arouse and encourage an attitude of enquiry. The films pose questions, raise problems, and present experimental data in such a way as to promote active student participation. Teachers and students must interact during the film presentation. The film does not dominate the development of the topic. The role of the teacher is to guide student discussion and interaction. The Super 8 mm projection equipment and cartridged loop films are uniquely suited for this purpose.

The films are designed to be held on certain frames or stopped at critical points to allow student reactions. The students are expected to supply hypotheses, interpret data, and suggest implications of experimental results, or are invited to offer ideas for additional experiments. The design often calls for the lights to be turned on and the

[6] Lee, Addison E., *Testing and evaluating student success with BSCS laboratory blocks,* D. C. Heath, Boston, 1969.

projector turned off for extensive discussion before proceeding with the film. Although each film is only four minutes in length, full class periods are used for their presentation.

Teachers are cautioned to avoid giving answers to problems or questions in order to encourage students to develop concepts by analyzing the situations given in the film. Discussions among small groups are suggested in the Teacher's Guide in order to promote participation by all students. Written responses are sometimes required before verbal discussion so that each student may react to key questions or problems posed by the film.

The usual 16 mm sound film presentation involves students only as passive observers and is often solely fact-oriented. Such films serve to reinforce the authoritarian aspects of science and to eliminate possibilities for differing opinions among students. The BSCS Single Topic Film format differs in requiring students to assume an active role and enhances opportunities for the teacher to present science as enquiry.

Although the basic approach in each film is the same, each film differs not only in subject matter but also in level of difficulty and in emphasis on the various elements of scientific enquiry. Some films stress the role of critical observation; others involve the interpretation of data, the design of experiments, the formulation of hypotheses, or the identification of problems. The films are structured to help the teacher guide the students as active participants, discovering and developing topics as they unfold in the film presentation.

The Teacher Has the Key Role

The basic philosophy underlying the development of BSCS materials is exemplified in the use of BSCS Single Topic Films. The teaching methodology incorporated in the films illustrates what the BSCS has found to be good practice in science instruction. Past experience with other BSCS programs has shown that the proper implementation of the BSCS Single Topic Film depends upon the teachers who use them.

The BSCS and other NSF-supported curriculum projects appear to agree on the need to teach science as an active process involving enquiry rather than relying on exposition of content alone. Unfortunately the methodology built into the materials produced by the BSCS and others is not always put into practice. Teachers often short-circuit the design of the materials and revert to the "tell about science" approach. They do not always recognize that enquiry is part of the content and design of the materials they are using.

The films and Teacher's Guides are structured so that the teacher is placed in the role of promoting student thinking and discussion. Neither the teacher nor the film represent authority but, instead, the students are encouraged to interpret and analyze situations presented to them by the teacher and film. The teacher and film are data sources for the students.

The films place the teacher in a nonauthoritarian role that promotes attitudes and skills that have carry-over value. Perhaps teachers will reconsider the way they present discussions, laboratory exercises, and lectures on the basis of their experiences with the films. Classroom visits by BSCS Consultants and interviews with teachers who have successfully used the films indicate that this does occur. Teachers have described the incorporation of newly acquired insights into their teaching after using the films.

The Teachers Guide—Key to Implementation

Each BSCS Single Topic Inquiry Film is accompanied by an extensive teacher's guide. The film serves as a data source and the guide offers the teacher a program of teaching style and strategy. The following, which appears as an introduction to each Teacher's Guide, illustrates its importance in helping teachers to use the films successfully.

HOW TO USE THIS FILM

BSCS Single Topic Films are designed to encourage an attitude of enquiry. They pose questions, raise problems, and present experimental data that promote student participation. The design of each film is such that students and teacher are engaged in active discussion during its presentation. The teacher's role is to guide student discussion and interaction and not to dominate the presentation. The film should not involve students as passive spectators.

1. STUDY THE FILM AND TEACHER'S GUIDE BEFORE CLASS USE.

It is virtually impossible to use this film successfully without an understanding of its basic design and rationale and that of the teacher's guide. Study both carefully before presenting the film to the students. Many teachers have found that one or more practice sessions with a small group of students or colleagues greatly enhances the class presentation.

You will note that the "Guide to Film Presentation" is divided into three sections. The first of these provides visual cues for the teacher, which include photographs of each scene as well as notes on when to start or turn off the projector or to hold the picture.

The second section gives information and questions that the teacher may wish to read or paraphrase to the students.

The third section includes information to supplement the teacher's background information, notes on how a specific part of the film should be handled, and the kinds of responses that the teacher can expect from the students. Typical student responses to questions are indicated by the special background color. They do not however, include all of the possible answers or comments which students will make.

2. TO ENCOURAGE STUDENT ACTIVITY, STOP OR HOLD THE FILM ON SELECTED FRAMES AS INDICATED IN THIS GUIDE.

Problems and questions are presented by the teacher and the film. The students are expected to supply hypotheses, interpret data, or suggest implications of experimental results.

The following directions are used in the guide:

HOLD FILM

At certain key places in the film a small square will appear in the upper right-hand corner of the screen. The square has been placed on the film to aid you in selecting the most opportune point to hold on a given scene.

The motion picture projector is to be stopped in order to project a "still" of a frame for the purposes of:

eliciting class discussion

posing further problems

inviting ideas for experiments

recording data

permitting detailed study of a selected frame.

HOLD, THEN TURN OFF PROJECTOR

The film is to be held on the selected scene for further student observation; then the projector is to be turned off and the classroom lights turned on for more extensive discussion.

TURN OFF PROJECTOR

The projector is to be stopped and the classroom lights turned on. This procedure is followed where major questions or problems are posed.

START FILM

Room lights are to be turned off or dimmed and the next section of the film begun.

3. AVOID GIVING ANSWERS TO PROBLEMS OR QUESTIONS.

Encourage students to develop concepts by having *them* analyze situations presented in the film and think through problems raised.

4. ENCOURAGE ALL STUDENTS TO PARTICIPATE.

Do not allow a small group of students to dominate discussion. Reticent students often make unique and excellent contributions to classroom discussion if given an opportunity to do so. When key questions or problems are posed, you may wish to have each student respond in writing before you ask for verbal responses. You can encourage greater student participation by having small groups discuss the problems and questions.

TITLES OF SINGLE TOPIC INQUIRY FILMS

The first series of twenty Single Topic Films was completed in 1963. Titles and brief descriptions of the films in this series follows.

SOCIAL BEHAVIOR IN CHICKENS

Observe and describe the interactions in a flock of chickens. What are the implications of the behavior?

WATER AND DESERT ANIMALS

How could some animals survive without drinking water? Suggest experiments that would show which of two animal species is best able to conserve water.

PHOTOTROPISM

Design experiments to determine why certain plants bend toward the light. Evaluate your hypotheses. Why is it unsatisfactory to say, "Plants bend toward light because they need it?"

LIFE IN THE INTERTIDAL REGION

How do tides affect intertidal organisms? What special adaptations would you expect to find?

AN INQUIRY—THE IMPORTANCE OF THE NUCLEUS

What is the role of the nucleus? Can an amoeba survive without a nucleus? How would you test your ideas?

PREDATION AND PROTECTION IN THE OCEAN

What protective adaptations are evident in marine organisms? Relate these to possible predators. How might warnings be communicated between individuals of the same species?

GROUSE—A SPECIES PROBLEM

Observe two closely related populations of birds. How could you determine whether or not they are separate species?

CONVERGENCE

Account for the similar appearance of unrelated organisms in different parts of the world. What traits do you think are most important for determining genetic relationships between groups of organisms?

PRAIRIES AND DECIDUOUS FORESTS

What ecological factors might explain why forests and prairies are found in definite regions? Rainfall alone does not seem to account for the distribution of forests and prairies—what other variables are important?

THE KIDNEY AND HOMEOSTASIS

What are the functions of the kidney? Based on the data presented, describe the kidney as a homeostatic organ.

MOUNTAIN TREES—AN ECOLOGICAL STUDY

Based on your observations, suggest why trees are distributed unevenly on mountains. How do the physical factors affecting mountain trees interrelate?

AN EXAMPLE OF THE BIOLOGICAL SIGNIFICANCE OF COLOR

Design an experiment to determine the role of color in food selection by tortoises. What are some ways that color and color pattern may be significant in the life of animals?

THE PEPPERED MOTH: A POPULATION STUDY

Why might dark moths outnumber light moths in certain areas? What hypotheses do the data support?

THE INTERTIDAL REGION

Observe the distribution of organisms in the intertidal region. Having seen the distribution of organisms on a rock face and a wharf piling, would you expect to find a similar distribution on a sandy beach?

MATING BEHAVIOR IN THE COCKROACH

Describe the behavior of both males and females. What is the significance of the events you observed? How do the experiments affect your hypotheses?

AUSTRALIAN MARSUPIALS

Suggest reasons for the uneven distribution of marsupials on the continents of the world. Account for the diversity of Australian marsupials.

TEMPERATURE AND ACTIVITY IN REPTILES

What effect does the change in season have on the activity periods of reptiles? What does the graphed data suggest? How is behavior associated with temperature regulation in reptiles?

MITOSIS

Observe and describe the changes you notice in the nucleus of the cell. What might be the significance of these changes?

WATER AND DESERT PLANTS

What adaptations would you expect to find in desert plants? Relate these adaptations to the environment. Design a model desert plant.

MIMICRY

Organisms are frequently encountered which, though distantly related, are similar in appearance, behavior, or other characteristics. What factors might account for these similarities?

A second series of Single Topic Inquiry Films is in production. Below is a list of titles correlated with BSCS themes that will be focused on in each film.

1. A Study of Oak Populations (Evolution, Genetics, Environment)
2. Regeneration in Acetabularia (Structure and Function, Genetics, Regulation)
3. Chemical Communication (Behavior, Regulation)
4. Nerves and Heartbeat Rate (Structure and Function, Regulation)
5. Prey Detection in the Rattlesnake (Behavior, Structure and Function)
6. Structure, Function, and Feeding Behavior in Herons (Diversity)
7. A Study of Frog Development (Genetics)
8. Imprinting (Behavior)
9. Engelmann's Inquiry into Photosynthesis (Environment, Behavior)
10. The Behavior of a Purple Bacterium (Regulation, Behavior)
11. Gene Flow in a California Salamander (Genetics, Evolution)
12. Temporal Patterns of Animal Activity (Regulation, Behavior)
13. Photoreception and Flowering (Structure and Function, Regulation)
14. Flowering (Environment, Regulation)
15. Feeding Mechanisms of Oyster Drills (Structure and Function, Behavior)
16. Planarian Behavior (Genetics, Behavior)
17. Genetics of Bacterial Nutrition (Genetics)
18. Feeding Behavior of Hydra (Regulation, Behavior)
19. Fossil Interpretation (Genetic Continuity, Evolution)
20. Locomotion in an Amoeba (Structure and Function)

THE BSCS INQUIRY SLIDE SETS

The BSCS Inquiry Slide Sets represent yet another innovative departure in the design and development of instructional materials and media. Each set contains several 35 mm slides that present a sequence of visual data for observation. Along with the visuals are projected carefully selected and designed questions that involve students in loosely structured patterns of enquiry and discussion. The slides are accompanied by a teacher's guide that defines the behavioral objectives of each slide and offers suggested and fully tested strategies for the teacher.

A unique feature of the slides is a processing technique that allows them to be projected in a reasonably well-lighted room onto the chalkboard rather than on a screen. The only light that needs to be avoided is artificial and natural light that shines directly onto the chalkboard. Under these conditions, the slide presents a sharp, clear image easily visible throughout the room. In addition, the teacher or students may use white or colored chalk to modify, add to, or highlight some aspect of the image. On the image, chalk takes on a luminescent glow that is immediately striking and effective.

Some of the slide sets follow the general pattern of the Invitations to Enquiry. Others present a variety of content material in ways designed to increase pupil opportunity and motivation to enquire, as well as to aid teachers in the conduct of enquiry-oriented discussion.

Teachers using the BSCS Inquiry Slides will find them an exciting and different kind of instructional aid.

THE BSCS SECOND COURSE

The BSCS Second Course, *Biological science: interaction of experiments and ideas,* was developed to meet an increasing demand for a second course in biology that would build upon the major BSCS themes. The course was developed over a period of three years of classroom testing and was first published commercially by Prentice-Hall in 1965. The book was prepared primarily for students who have taken a first course and who have found that they enjoyed the study of biology, particularly in its investigative aspects, and who desire to continue their study in greater depth.

The Second Course differs considerably from the kind of biology program usually found at the senior high school or beginning college level. As implied in the title, *Biological science: interaction of experiments and ideas,* the course is laboratory-oriented. A primary goal is to provide experiences that simulate biological research so that students will gain an understanding of science from direct experience rather than from being told about the processes of science. Yeast metabolism, population dynamics, microbial genetics, plant and animal growth, regulation and development, and animal behavior provide the major subject areas for laboratory study. There are 38 scheduled laboratory investigations and numerous related investigations for further study to allow the teachers a wide latitude in working with students of diverse interests. Stress is placed on the application of elementary statistics to evaluation of data and on the role of constructive controversy as an important process of science. The investigations involve a parallel study of original research literature in biology, both comtemporary and classical. The authors do not believe that a textbook alone can provide the entire substance of a good course in biology. It is hoped that *Biological science: interaction of experiments and ideas* will be regarded as a guide to learning rather than a complete instructional manual.

The teacher's edition is an essential component of the Second Course. It contains photographed pages of the student edition together with annotated comments and appropriate discussions of various ways that the student material might best be presented.

The Second Course is an integrated laboratory text in which students are given problems to solve but are not provided with answers. Wherever possible, a single kind of organism is studied to allow for depth in laboratory investigation. For example, the first eight experiments deal with the common yeast, S. *cerevisiae*. Each investigation is designed to become increasingly complex so that students will have an opportunity to experience both the frustration and elation that comes with understanding biology through laboratory investigation.

It is recommended that the Second Course *not* be a substitute for chemistry or physics and that teachers be given an opportunity to select those students who indicate an interest in a further study of biology and who evidence the capability required to carry on intensive study with a minimum of teacher supervision.

Approximately 12,000 students used the Second Course during its first year of commercial release. Reports from teachers indicate that the course is challenging both to students and to teachers who must work cooperatively in carrying out the investigations. Many of the investigations border on the frontier of what is known about the particular organism being studied. Reports further indicate that students of average achievement, who have a strong desire to learn, succeed well in the Second Course indicating that high scores on aptitude tests should not be the only criterion used in the selection of students. Motivation and enthusiasm are also important.

Teacher Preparation

Ideally, the prospective teacher of *Biological science: interaction of experiments and ideas* should have had extensive graduate training in each of these specialties: microbiology, genetics, population dynamics, plant and animal growth, regulation and development, animal behavior, and biostatistics. Also he should be a highly skilled research biologist; he should possess a thorough knowledge of how teenagers learn; and he should be a highly creative and imaginative teacher. It is doubtful whether any biology teacher possesses all of these qualifications and, even if he did, he might experience difficulty in teaching the program.

What kind of preparation *is* needed to teach the Second Course? Actually, any biology teacher with an *active* interest in teaching a research-oriented laboratory program can achieve success. The Teacher's Edition is, in itself, a course on how to teach the program and should be studied carefully and intensively. The teacher should also have a personal library that contains at least one good reference text for each of the major subject areas given above. The reference texts should be referred to in conjunction with a study of the Teacher's Edition of the Second Course.

Perhaps the most difficult aspect of teaching the Second Course arises from the way the course is designed. *Biological science: interaction of experiments and ideas* is not a textbook in the usual sense. Instead, it is a series of laboratory investigations that concentrate, in depth, on specific biological problems. Students are expected to read widely in various publications as the need arises after performing laboratory investigations. Perhaps in this way the Second Course differs from most advanced biology programs. The major stress is on laboratory investigation and a study of related literature rather than on reading in a single text. Laboratory work with living things involves many unpredictable factors including the risk of failure to achieve the desired results. The following comment from the Teacher's Edition expands this point:

"If the nature of experimental biology is not thoroughly understood, the cooperative nature of organisms in failing to do what is expected of them can provide varying degrees of frustration to the teacher and sometimes disillusionment to students. Carefully planted seeds may rot rather than germinate, unexpected contamination may ruin a pure culture of a microorganism, and animals may fail to behave in the manner desired by the investigator. The failure of an organism to respond should not imply that the experiment has failed, at least in terms of understanding the nature of science. Seeking the reason for "failure" may provide a greater challenge and subsequent learning experience than would have occurred had the experiment proceeded as scheduled. (The only way to avoid the occasional failure of an experiment is to talk about experimental method, and let someone else do the experimenting.) The student's natural desire to solve real problems offers the best promise for exercise of his potential intellectual ability. Many problems will arise for which immediate answers are not known.

Because of the uncertainties accompanying each investigation, both the teacher and the student must be prepared to face a unique discipline."

Another problem may arise from the extensive emphasis on teaching science as enquiry. Few of us were taught by an enquiry method, and it is natural to teach as we were taught. This problem can best be illustrated by a passage from the Second Course Teacher's Edition:

"Teaching science as a process of inquiry rather than as a collection of facts is both difficult and rewarding. It is difficult for many reasons. When a student asks for the 'correct' conclusion to a laboratory investigation the easiest thing to do is to give him an answer. In this way our own urge to tell is relieved and the student usually feels satisfied. Unfortunately, the conclusion given is probably not entirely correct, and a false image of science as a rhetoric of conclusions is reinforced. The student also has little chance to sharpen his mental facility and experience a sense of creativity.

"The reward to the teacher who has patience enough *not* to supply the answer comes from the response of students who, if encouraged to think, find they enjoy the process."

Teachers experience more difficulty in using the Second Course the first year than in subsequent years. There is probably no way to overcome the first-year frustrations. It would be valuable to attend a good summer institute that focuses on the Second Course prior to teaching the program. But, today, very few summer institutes of this type exist. Courses in the subject areas mentioned in the first paragraph of this section provide valuable background.

The lack of formal institute training need not deter a teacher from teaching the Second Course. A study of the Teacher's Edition should disclose areas in which the teacher feels a need for additional knowledge and thus should guide his selection of reference texts.

Many teachers have not had college courses dealing with statistical interpretation of data. Yet the use of a null hypothesis and t and chi-square tests of significance enable students to understand the conditional nature of scientific "truth" and is a valuable part of the program. No attempt is made to derive these tests of significance. They are used only as tools to give students a better understanding of how to interpret data resulting from laboratory investigation. Some of the mathematical formulas may seem difficult, but if a teacher will work through the sample problems, the difficulty becomes routine.

Teaching the BSCS Second Course can be a highly rewarding experience for both student and teacher. The best preparation is a desire to teach a laboratory-oriented program and a willingness to spend considerable time in study.

BSCS RESEARCH PROBLEMS IN BIOLOGY: INVESTIGATIONS FOR STUDENTS

The series of four volumes of *Research problems in biology* was initiated by the BSCS Special Student Committee, headed by Paul F. Brandwein. Each volume consists of 40 separate investigations, all prepared by active researchers in the area of the investigation. These represent problems that have some promise of yielding results that are publishable, that are genuine contributions to science. Obviously they are directed toward only the highly motivated, creative high school student, and require persistent effort, often more than a year's time, to carry out. Somewhat unexpectedly, the Research Problems have found wide use in colleges and universities, as well as in the high schools. Volumes 1 and 2 were published in 1963 and Volumes 3 and 4 in 1965 by Doubleday as paperback Anchor Books.

The bulk of the instructional material developed for the potential biologist consists of prospectuses designed to help the student carry forward individual research. In so doing, the student is led to *seek* his own problems, to *design* his own experiments, and altogether to engage in the creative work of the biologist. The student, in short, is *confronted* with the need to enquire into a field in which knowledge is not available, and to engage in the research required to secure the knowledge. He is brought face to face with a blank wall.

Biologists in the several fields have contributed these prospectuses; hence they reflect the cutting edge of the research laboratory. A subcommittee of the Special Student Committee, under the chairmanship of Jerome Metzner,[7] has edited these prospectuses for inclusion in the volumes published by Doubleday and Company under the title *Research Problems in Biology for the Schools*. Each volume contains forty prospectuses ranging over the field of biology.

Several examples of these prospectuses are included here to indicate the nature of the investigations with which the student is faced. Note especially the suggestions made to the student; note as well that certain "pitfalls" are highlighted. The prospectuses, it should be emphasized,

[7]Members of the subcommittee are Philip Goldstein, Robert Hull, Irving Reich, and Walter Rosen.

include investigations in botany, zoology, physiology, embryology, anatomy, genetics, ecology, and microbiology.

Four examples of prospectuses follow, each under the name of the biologist who prepared the original. The examples are offered as types of prospectuses rather than as indicators of the range of difficulty. The prospectuses do range in difficulty; some are "easier" to get at than others. They also have a range in the kinds of skills and arts required of the young investigator.

Nutrition of Excised Plant Tissues and Organs[8]

Background. Many plant tissues can be grown in sterile cultures in the laboratory, completely isolated from the plants from which they were derived. Some tissues and organs will grow on relatively simple media consisting of inorganic salts, a few vitamins, and amino acids. Others will not grow at all, or will gradually cease growing, unless they are provided with additional vitamins, trace elements, plant growth hormones, or unidentified factors found in natural products such as coconut milk.

Some plant tissues cultured in the laboratory will grow "normally," forming organized new tissue. Others will form only undifferentiated growths called callus. Some tissues cultured in the laboratory will form new tissue, or callus, or both, depending on the composition of the culture medium.

Suggested Problems. The study of plant tissue cultures, therefore, provides a means of learning about the nutritional requirements of plant tissues and the factors that govern various types of tissue development.

Single ingredients may be omitted, one at a time, from the basic nutrient. A fairly clear idea may then be gained of the importance of each ingredient. The investigation may then be expanded by varying the concentrations of individual components, singly, to establish optimal concentrations. Once a notion of the complex nature of nutritional balance is obtained, other problems may be undertaken. An attempt may be made to find adequate nutrients for tissues such as tree cambiums, whose specific requirements are still unknown. This program can be varied by using embryos removed from seeds or stem tips excised from seedlings as the experimental material.

Materials Needed. It is not generally realized that experimentation with small, isolated parts of plants can be carried on with a minimum of equipment. The following project involves the use of three types

[8] By Philip R. White, Roscoe B. Jackson Memorial Laboratory, Bar Harbor, Maine.

of plant material: tomato roots; carrot callus; and callus from a maple, beech, poplar, or pine tree. The first two are easily cultured; the third will provide more challenge.

The materials needed for routine maintenance cultures are the following:

1. Two hundred small containers such as Falcon disposable 60-mm petri dishes (cost about 10 cents each), small wide-mouthed medicine bottles, 4-oz French squares, or 6-oz prescription bottles.
2. Pressure cooker (a 5-qt canner is best, but any cooker will do) or an autoclave.
3. Supply of Difco Bacto-agar #B140 or #B142; Difco T. C. Medium, White, Sugar #0782, #0783, or #0784.
4. Supply of salts, vitamins, and other supplements, listed in the Difco T. C. Manual, needed to make up the solutions for nutritional experiments.
5. Set of cork borers.
6. Set of scalpels, preferably Bard-Parker replaceable blade type.
7. Set of assorted forceps.
8. Alcohol lamp.
9. Double boiler for sterilizing instruments.
10. Operating shield made of a piece of Plexiglass held in a wooden frame about 10 in. above the table.
11. Ripe field-grown tomatoes (greenhouse tomatoes often do not have viable seeds).
12. Firm, smooth carrots.

Suggested Approach

Tomato Roots. Make up two sets of nutrients by preparing suitable dilutions of Difco stocks, one set without agar, the second containing 1.5% agar. Place 15 ml of the nutrient material containing agar in some of the containers. Do *not* close the containers tightly. Sterilize the containers in the cooker; then set them aside to cool. Wash a tomato and break it open, being careful not to touch the seeds with the knife used in cutting the surface. With sterile forceps remove the seeds and place them on the agar. When they have sprouted and produced roots an inch long, cut off the root tips with a sterile scalpel and transfer them to containers with liquid nutrients. If cut back and transferred once a week, the tips should grow indefinitely. One strain grown by the author is in its twenty-seventh year.

Carrot Callus. Wash a carrot and break it across, being careful not to touch the interior. With a sterile ¼-in. or smaller cork borer remove a series of cores at such a position that they are traversed by the cambium (the line between the deep orange core and the lighter cortex). Transfer these cores to a sterile petri dish and slice them into disks about 1–1½ mm thick. Transfer these disks to agar nutrient. Callus will grow from the cambium. At the end of a month, this callus should be cut away and transferred to fresh agar. The older tissue is discarded. If cut up once a month, this callus also should grow indefinitely. Some individual carrots will produce good strains of cultures, others will not. All will grow freely if about 1 part in 10,000,000 of the common weed killer 2,4-D is added to the nutrient. Dissolve 10 mg of 2,4-D in a liter of water and add 10 ml to each liter of nutrient before sterilizing. One strain grown by the author is in its twenty-third year.

Wood Callus. Under the guidance and with the assistance of an experienced botanist or forester, select a tree in the woods, and with a saw and chisel mark out a 6-in. square on the bark. With the chisel remove a block, being sure to split it out at least ½ in. to 1 in. below the cambium. Bring the block into the laboratory and shave off all the dead bark until living phloem is exposed. Then wash the surface with 95% alcohol. Using extreme care, burn off the excess alcohol. Continue to remove the phloem with a sterile chisel down to within 1–2 mm of the cambium. Then, with a sterile scalpel, lay out a series of blocks 5 × 10 mm, penetrating 1 mm below the cambium. Remove these small blocks and with sterile, strong forceps push them on end into an agar nutrient so that about two-thirds of each piece protrudes above the surface. Callus will develop from the cambium. At the end of a month, this callus can be excised and removed to fresh agar. The older tissue may be discarded. The callus will probably *not* continue to grow on the basic medium. Each different tree will require special study. The commonly recognized supplements include the auxins 2,4-D and IAA at concentrations varying from 10^{-10} to 10^{-5} M; vitamins such as biotin, pantothenic acid, inositol, and choline; reducing agents such as cysteine, glutathione, tyrosine; and additional metal catalysts such as cobalt.

All these experiments should be performed in a reasonably cool laboratory. The cultures are placed on a shelf or in a cupboard. They do not require light and are not injured by diffuse light. They do not require humidity control except to reduce the drying up of the nutrient. They do not require microscopic examination. Measurements can be made with a celluloid rule. They *do* require cleanliness and precision of manipulation. There are ample opportunities for observation and for ingenuity in planning experiments.

REFERENCES

Bonner, J., and A. W. Galston, 1952. *Principles of plant physiology*. W. H. Freeman and Co., San Francisco. (Especially Chapters 3, 15, 16, and 19.)
Brookhaven Symposium Series. *Abnormal and pathological plant growth*.
Gautheret, R. J., 1958. *La culture des tissus vegetaux*. Masson et Cie, Paris.
White, P. R., 1943. *A handbook of plant tissue culture*. The Ronald Press, New York.
——, 1954. *The cultivation of animal and plant cells*. The Ronald Press, New York.

Communication by Trail Laying in Ants[9]

Background. Members of ant colonies communicate with each other in large part by chemical substances conveniently referred to as *pheromones* (to distinguish them from the purely internal hormones). The pheromones are produced by glands that open to the outside of the body and are perceived chiefly by the antennae. Despite the great popular attention that has been devoted in the past to the subject of ant communication, it is only within the last several years that the glandular sources of some of the pheromones have been discovered. The chemical communication system is both more complex and more efficient than previously believed and is a fitting subject for imaginative original research.

The pheromone most accessible and conveniently studied is the trail substance. Each ant species thus far studied produces a substance peculiar to itself. The trail pheromones are powerful attractants produced in tiny quantities in various of the abdominal glands. They are sufficient, when drawn out in trail form, to recruit workers from a nest; no other signals, such as antennal stroking, appear to be necessary.

Suggested Procedure. In warm weather, place baits composed variously of sugar water, bread crumbs, and freshly killed insects in selected habitats. The ant species attracted to them can be quickly classified into two classes: those that forage solitarily, and those that recruit by trails. Creighton's book (see the references), can be used to classify any or all of these species, and it is possible at this point to launch a study into the little-explored subject of other ecological differences between the two classes of species. For instance, do they differ in habitat

[9] By Edward O. Wilson, Harvard University, Cambridge, Massachusetts.

choice, colony size, or other characteristics? Perhaps twenty or more species need to be studied in this regard to make your conclusions sound.

You may prefer to select one trail-laying species and examine the properties of the trail pheromone. Track the workers to their nest, excavate the colony, and set it up for laboratory study as described in Wheeler (Appendix A). Allow the workers to forage into a spacious area, such as the bottom of a terrarium or a section of the floor of a room surrounded by a plaster-of-Paris wall.

The abdominal glands can be dissected in distilled water under a dissecting microscope using fine forceps and thin needles set in match-stick holders (this takes time and patience but with ants to spare it can be developed into an art). Figures of internal anatomy of the abdomen are given by Wheeler (Chapter 3); first note the subfamily that the experimental species belongs to. With the abdominal glands separated, it is easy to smear them out from the nest in artificial trails, following the method of Wilson (1959) and of Wilson and Pavan (1959). Positive responses are immediate and often startlingly clear-cut. The chances are that the species you select, and perhaps the genus to which it belongs as well, has never been studied in this way, since the glandular source of trail was unknown for any ant species prior to 1959.

Suggested Problems. Once you have located the source gland of your species, your imagination can lead you in many directions in behavioral research. For instance, you may ask, Will the artificial trail work in other trail-laying species? Will the queen follow it? How long does a trail persist, and what consequences does this have on the efficiency of communication? If you were designing a communication system for your ant species, could you conceive of something plausible that improves on the system in existence? At this point you will be on your own in one interesting sector of behavioral research.

REFERENCES

General

Creighton, W. S., 1950. The ants of North America. *Bulletin of the Museum of Comparative Zoology, Harvard,* Vol. 103. Harvard University Press, Cambridge, Massachusetts. 585 pages.

Wheeler, W. M., 1910. *Ants.* Columbia University Press, New York. 663 pages.

Specific

Wilson, E. O., 1959. Source and possible nature of the odor trail of fire ants. *Science,* Vol. 129, 643-644.

Wilson, E. O., and M. Pavan, 1959. Glandular sources and specificity of some chemical releases of social behavior in dolichoderine ants. *Psyche* (Harvard Univ.), Vol. 66, 70-76.

Are Traces of Water Necessary for the Survival of Plant Cells?[7]

Background. Seeds, spores, and even the vegetative parts of plants may survive for long periods of time in the "air-dry" state. Such tissues retain about 10% water. They may be maintained for as long as 6 months *in vacuo* over sulfuric acid. Under these conditions, however, very small amounts of water may remain in the tissues. Some workers claim that if this water were removed, life would cease. Others (Harrington and Crocker) have reduced water contents to as low as 0.1% without injury.

Suggested Problem. Will the seeds retain their ability to germinate after the last traces of water are removed? How will the seeds of various species react? What is the effect of complete desiccation on the spores of ferns, mosses, and fungi; on bacteria; and on pollen grains?

Suggested Approach. Wheat seeds may be used since they maintain a practically unaltered germination percentage for years. The per cent germination of the seed sample is first determined. The water content is also determined by weighing several samples, then transferring them in their weighing bottles to an oven at about 105°C, and then drying until a constant weight is reached. About a thousand seeds from the same lot are placed in a vacuum desiccator over either phosphorus pentoxide or magnesium perchlorate rather than sulfuric acid.

One disadvantage of using H_2SO_4 is that it makes it difficult to maintain a constant relative humidity. As soon as it absorbs some water, the relative humidity of the air above it rises slightly. Phosphorus pentoxide and magnesium perchlorate do not have this disadvantage. They combine chemically with the water and therefore maintain a relative humidity of essentially zero per cent.

The desiccator is then evacuated to as low a pressure as possible and left for about 6 months. If available, three desiccators may be set up in order to obtain statistically significant results. At monthly intervals two ten-seed samples are removed and the water content and percent germination are determined. When no further loss in weight occurs at room temperature, and if germination is still normal, the desiccator is transferred to an oven at 40°C. Monthly weighing and germination tests are again made until weight equilibrium is reached. If germination still remains normal, the temperature is then raised to

[7] By J. Levitt, Department of Botany, University of Missouri, Columbia, Missouri.

50°C and the test is repeated. Since dry seeds have been shown to survive as high as 140°C (dry heat), this procedure may be repeated with successive 10°C increases in temperature without heat injury. At each temperature there will be a slight additional loss of water.

Pitfalls. Phosphorus pentoxide and magnesium perchlorate should be handled with care. Observe the precautions printed on the reagent bottle or consult a chemist or a handbook of chemistry for full instructions on their use.

If the seeds decline in germination rate as the temperature is increased, the investigator must decide if the decline has resulted from the decrease in water content, the increase in temperature, or both. He should be sure to include adequate controls in his experiments.

The rate at which dried materials become remoistened may be critical in determining their capacity to resume active growth. Consult the references to determine the proper conditions for germination tests. If no germination is obtained by the usual methods, try remoistening the seed in two styles: first by allowing them to come to equilibrium with an atmosphere of 95–100% relative humidity, then by adding water in the usual way.

REFERENCES

General

Bonner, J. F., and A. W. Galston, 1952. *Principles of plant physiology.* W. H. Freeman, San Francisco.

Harrington, G. T., and W. Crocker, 1918. Resistance of seeds to desiccation. *J. Agr. Research,* Vol. 14, 525-532.

Levitt, J., 1956. *The hardiness of plants.* Academic Press, New York.

Wilson, C., and W. Loomis, 1957. *Botany.* Dryden Press, New York.

Detection of the Effects of Genes Carried in Heterozygous Condition [8]

Background. Most known mutations in man, *Drosophila,* corn, and other well-studied organisms are recessive to the wild type. This means that they must be present in double dose (homozygous) for their effect to be observed. An organism which carries a recessive mutation in single dose, combined with a single dose of the wild-type gene (heterozygous), appears normal.

However, there are many examples where the presence of a recessive mutation in single dose with the wild-type gene *does* alter the outward appearance of the organism. This is true in cases where blending prevails, as for example when pink color shows in a plant

[8] By R. C. Lewontin, University of Rochester, Rochester, New York.

bearing allelic genes for red and white. As a matter of fact, recent evidence has accumulated to show that so-called "recessive" genes may not be completely recessive at all. Rather, the effects of these genes in single dose are so subtle that they cannot be easily observed or detected. With special techniques the presence of a recessive gene may be detected in some cases, even when its effect is not outwardly visible. The detection of differences between normal homozygotes and heterozygotes which outwardly look alike is important for two reasons.

First, it will help us gather evidence on the nature of the action of these genes. Even though a single dose of a recessive gene has no apparent effect on the final phenotype, there must be some difference between homozygotes and heterozygotes at the end of primary gene action. Does the mutant gene make a different substance than does the wild-type gene? Or does the mutant gene make the same substance in greater or lesser quantity? Does the primary action of the mutant gene interact with that of the wild-type gene?

Second, in human genetics and in the genetics of economically important plants and animals, detection of heterozygotes carrying recessive genes is of great practical value. Genes, the effect of which is visible only in homozygous condition, are very difficult to select against, especially if the particular gene is rare. For example, if a mutant gene is present in a population at a frequency of one per thousand, then 99.8% of these mutant genes are in heterozygotes, while only 0.2% are in homozygotes. Since only these 0.2% can be identified by appearance, any selection must completely miss the other 99.8%. However, if we did have methods of detecting the heterozygotes, selection against these genes would become tremendously effective.

Suggested Approach. A very useful organism for studies of this sort is the fruit fly, *Drosophila melanogaster*. Many recessive mutants are known in this animal, and mutant stocks are easy to obtain from university laboratories. The fly is easily cultured, and it produces large numbers of offspring. The best mutants to work with are those known to cause simple biochemical changes. Among these are mutants affecting eye color and body color. The general idea would be to attempt to detect differences in pigment composition or in intensity of pigmentation between homozygous normal flies and flies heterozygous for the recessive mutation. Of course, this *cannot* be done by simple inspection, because if there were a visible difference the mutant gene would not have been classed as recessive.

One approach for eye-color mutations would be to crush the heads of a number of flies on filter paper and run the pigments in chromatographic analysis. Side-by-side comparison of the chromatographs of homozygous wild type, heterozygous mutant, and homozygous mutant flies might easily show differences in pigment composition. Chemical reactions for some of the pigments are known so that direct chemical assay of the concentration of pigments is possible. One such method is discussed in the paper by Green.

For mutations not concerned with pigment formation, an amino acid analysis of whole flies is one possible approach. Again, chromatographic analysis is a good way of attacking the problem.

Possible Pitfalls. The differences for which you are searching are quite subtle, and it is possible that observed differences in pigment composition or intensity may be due to genes other than the one under investigation. For this reason it is important that the mutant heterozygotes and the wild-type homozygotes be as much alike in their other genes as possible. This can be accomplished by crossing wild-type stocks with mutant stocks and allowing the offspring to produce a third, fourth, fifth, and sixth generation. From the sixth generation, pure breeding stocks of wild-type and mutant-type flies can be isolated for future experiments. These stocks will carry about the same genetic backgrounds, yet they will differ in the specific character under investigation.

REFERENCES

General

Block, R. J., E. L. Durrum, and G. Zweig, 1958. *A manual of paper chromatography and paper electrophoresis.* Academic Press, New York.

Demerec, M., and B. P. Kaufman, 1961. Drosophila *guide.* Carnegie Institute of Washington, Cold Spring Harbor, New York.

Green, M. M., 1959. The discrimination of wild-type isoalleles at the white locus of *Drosophila melanogaster. Proc. Natl. Acad. Sci.,* Vol. 45, 549-553.

Lederer, E., and M. Lederer, 1957. *Chromatography.* D. Van Nostrand, New York.

Sinnott, E. W., L. C. Dunn, and T. Dobzhansky, 1958. *Principles of genetics.* McGraw-Hill, New York.

Materials for the teacher are published in *A Guide to Work with Potential Biologists,* Bulletin No. 2, Biological Sciences Curriculum Study, AIBS, 1963.

The *Guide* is a statement of purposes, principles, and practices hopefully useful to teachers who desire to work with the very able biology student. The book contains material on the characteristics of the "gifted" student (with particular reference to science); practices in stimulating the development of an art of investigation; promising practices in the teaching of students of high ability in biology as observed in the classrooms of the United States; an introduction to use of the library and a fairly complete bibliography on "giftedness."

TESTS

An often unsung aspect of curriculum development (but a highly important one) is that of preparing tests to be used with the new curriculum materials. Over and over again, studies have shown that the educational objectives of a course can be vitiated by instruments that test for different, or at least not completely sympathetic, objectives. Students do study and learn what they think they will be asked on a test, or what their teachers want them to be able to answer on a test.

Much of the early effort of BSCS was in developing tests to evaluate the students' success in achieving the objectives of the new BSCS high school biology courses. A series of two parallel quarterly examinations for each of the three versions was prepared, tested in classrooms, revised, tested again, and finally published. Alternate Comprehensive Final Examinations for use with one of the three versions were likewise developed. All of these tests, a total of 26 different instruments, were eventually normed (a process for determining what score may be expected from the average student), and the results were made available in the test manuals and published in the *BSCS Newsletter*.

Similar procedures were followed in developing test instruments for the BSCS Second Course, *Biological science: interaction of experiments and ideas,* and for the Special Materials, *Biological science: patterns and processes.* A pretest, first called the *Impact Test* and later the *Processes of Science Test* (POST), was also developed to gauge a student's grasp of the methods of scientific investigation before he is exposed to a BSCS course.

The next job of test construction was to develop a book of test items for teachers of the versions to use in preparing their own tests. This was done in response to the need that many teachers expressed for a flexible selection of test items that could either be used when their students had finished work on a particular section of the book or when the regular quarterly tests were not appropriate for reasons of timing or coverage. Three Test Booklets, one for each version, were prepared and published in experimental editions that were distributed free of charge to teachers who requested them from the version publishers.

As the Laboratory Blocks reached the final stage of development, a need became evident for instruments to measure student achievement of the objectives in the Blocks. A Resource Book for Teachers, *Testing and evaluating student success with laboratory blocks,* has been prepared for this purpose. As with all BSCS tests, the Laboratory Block Resource Book was developed cooperatively by the Test Construction Committee and the writers of the material for which the tests were prepared—in this instance, the authors of the Blocks and the staff of the BSCS Research and Development Laboratory at the University of Texas.

As the BSCS versions are revised, tests that accompany them are also revised. This involves revising, testing, and rewriting the 26 instruments that were developed for the first editions of the versions.

As other curricular materials are tested or revised, other evaluation instruments will be developed.

OTHER PUBLICATIONS

In addition to the BSCS publications described in detail in the preceding pages, BSCS has contributed to the development of other materials, some of which are now in the planning or prepublication stage. Only brief descriptions will be provided here; additional information will be available in future BSCS Newsletters.

The Bulletin Series consists of monographs on topics related to biological education. Titles in the series to date are as follows.

Hurd, Paul DeHart, 1961. *Biological education in American secondary schools, 1890-1960.*

Brandwein, Paul, et al., 1962. *Teaching high school biology: a guide to working with potential biologists.*

Grobman, Arnold B., et al., 1964. *BSCS biology: implementation in the schools.*

Grobman, Arnold B., 1969. *The changing classroom: the role of the biological sciences curriculum study.*

Special Publications. As programs for teacher preparation developed, the need emerged for specific guidelines to assist teacher training faculties in the development of their programs for preparation of teachers to teach the BSCS courses. This series was initiated by the Teacher Preparation Committee to meet this need. Six booklets have been developed thus far in the series:

No. 1: 1962, *BSCS biology guidelines for preparation of in-service teachers.*
No. 2: 1963, *Patterns for the preparation of BSCS biology teachers.*
No. 3: 1964, *BSCS materials for preparation of in-service teachers of biology.*

No. 4: 1966, *The teacher and BSCS special materials.*
No. 5: 1967, *Laboratory blocks in teaching biology.*
No. 6: 1969, *New materials and techniques in the preparation of high school biology teachers.*
No. 7: 1969, *Guidelines for junior high school science.*

The Pamphlet Series consists of 36-page booklets, each written by one or two professional biologists who have done creative research in their field of specialization. The Pamphlets are directed toward the teacher who wishes to broaden his knowledge in a given area, the especially interested student, and the layman with an interest in biology. These sometimes highly original, although simply written contributions, each on a single topic in biology, have also found use at all levels in the universities and in research laboratories.

Titles, commercially published by D. C. Heath, are as follows.

1. *Guideposts of animal navigation,* Archie Carr.
2. *Biological clocks,* Frank A. Brown, Jr.
3. *Courtship in animals,* Andrew J. Meyerriecks.
4. *Bioelectricity,* E. E. Suckling.
5. *Biomechanics of the body,* E. Lloyd Du Brul.
6. *Present problems about the past,* Walter Auffenberg.
7. *Metabolites of the sea,* Ross F. Nigrelli.
8. *Blood cell physiology,* Albert S. Gordon.
9. *Homeostatic regulation,* Thomas G. Overmire.
10. *Biology of coral atolls,* Richard A. Boolootian.
11. *Early evolution of life,* Richard S. Young and Cyril Ponnamperuma.
12. *Population genetics,* Bruce Wallace.
13. *Slime molds and research,* C. J. Alexopoulos and James Koevenig.
14. *Cell division,* Daniel Mazia.
15. *Photoperiodism in animals,* S. Farmer.
16. *Growth and age,* Lorus J. Milne and Margery Milne.
17. *Biology of termites,* E. Morton Miller.
18. *Biogeography,* Wilfred T. Neill.
19. *Hibernation,* William V. Mayer.
20. *Animal language,* Nicholas E. Collias.
21. *Ecology of the African elephant,* Horace Quick.
22. *Cellulose in animal nutrition,* R. E. Hungate.
23. *Plan systematics,* Peter H. Raven and Thomas R. Mertems.
24. *Photosynthesis,* Hans Gaffron.

Pegasus Series. To meet the need for scientific books that may be read and understood by the layman, the BSCS has developed (in cooperation with Pegasus, a division of Western Publishing Company) ideas for a series of 40 paperbacks on interesting and controversial biological topics and issues facing every individual in our society today. Books on such subjects as birth control, air and water pollution, mental health, and the meaning of evolution will be commissioned from knowledgeable authors.

As with other BSCS materials, this series will stimulate the reader to ask questions and will challenge his imagination to explore further. It will give the layman a foundation that will enable him to interpret and evaluate the biological information he encounters in newspapers, magazines, and on television. Students at the secondary and undergraduate college levels should find these books informative. Although the BSCS does not plan for these books to be formal text supplements, students and teachers may find them valuable for their studies.

A partial list of authors and tentative titles in the series follows: Adrian M. Wenner, *The bee language controversy;* James T. Enright, *Animal timekeeping;* E. Peter Volpe, *Human genetics and congenital defects;* James V. McConnell, *Psychotherapy and human behavior;* Thomas C. Cheng, *Symbiosis;* Garrett Hardin, *Birth control;* Andrew J. Meyerriecks, *Man and birds;* and David M. Prescott, *Cell biology and cancer.*

Patterns of Life Series. A new series of paperback books on biological topics will soon be published under the BSCS imprimatur by Rand McNally. The series will be part of the Rand McNally "Patterns of Life" series—an arrangement that will effect economies in printing and distribution—but each of the books prepared by the BSCS will be clearly identified as such on the book itself and in all promotional material about the series.

At present, eight manuscripts are in press. They are *Antibiotics,* David Perlman; *Behavior of tortoises,* Walter Auffenberg; *Bird migration,* Albert Wolfson; *Defensive secretions of arthropods,* Thomas Eisner and Rosalind Alsop; *Energy transfer in ecological systems,* Richard G. Wiegert; *Ionizing radiation and life,* Thomas G. Overmire; *Island life,* C. J. McCoy; and *Plant morphogenesis,* Frits W. Went.

CUEBS-BSCS Biomethods Committee. For several years the Commission on Undergraduate Education in the Biological Sciences (CUEBS) and the Biological Sciences Curriculum Study have cooperated in a common concern for both the in-service and preservice education of biology teachers. The interests of both CUEBS and BSCS are obvious; CUEBS, because it is concerned with the college curriculum that prepares the biology teachers; BSCS, because it has developed a substantial number of new teaching materials for teachers of secondary biology that require special knowledge and skills. To identify the critical areas in the education of teachers, a joint working committee has been formed. Efforts thus far have been directed toward producing several models of teaching situations for use in a biology methods course. Material resulting from the work of this committee was published in 1969.

McREL-BSCS Cooperative Program. Since 1966, the Midcontinent Regional Educational Laboratory (McREL), located in Kansas City, Missouri, and the Biological Sciences Curriculum Study have been cooperating in a program directed at identifying those aspects of the BSCS program concerned with self-directed learning by students. It is the view of both McREL and BSCS that an important contribution to developing the self-directed learner may be achieved through enquiry processes.

The long-range goals of the program include identifying teacher behaviors that encourage the enquiry processes on the part of the student. Once these have been identified, then plans are to determine ways to help teachers develop and use these behaviors. The major work of the program to date has been centered on identifying measurable student and teacher behaviors in both cognitive and affective domains. Another phase of the program involves the video taping of biology teachers in action in the classroom. The aim here is to obtain tapes that may be analyzed for different kinds of teacher behaviors, including nonverbal behavior, and that seem to be characteristic of teachers who are actually encouraging the enquiry processes among their students.

Work completed as of July 1969 by the committtee, which consists of personnel associated with BSCS or McREL, is included in Chapter 2, "Objectives for Teaching Enquiry Processes." Other materials produced by the committee were published later in 1969.

Techniques Films. Early in the history of BSCS, several films were developed to aid in teacher preparation. They are mainly concerned with laboratory techniques with which teachers were not always familiar. They were produced by Thorne Films and are available through them either as 16 mm sound or 8 mm loop silent films. Titles are "Bacteriological Techniques"; "Culturing Slime Mold Plasmodium"; "Genetics: Techniques of Handling Drosophila"; "Measuring Techniques"; "Neurospora Techniques"; "Paper Chromatography"; "Removing Frog Pituitary"; "Smear and Squash Techniques"; and "Weighing Techniques".

International Cooperation. With support from the Ford Foundation and The Agency for International Development (through the National Science Foundation), the International Cooperation Program is moving along at a brisk pace. Forty adaptation projects in 20 languages are in some stage of development. These projects involve biologists from more than 50 countries working with groups headquartered in 33 different countries.

Initial adaptations have been completed in Argentina, Australia, Brazil, Canada (French language), Ceylon, India, Israel, Italy, Japan, Korea, Mexico, New Zealand, Peru, the Philippines, Scotland, Thailand, Taiwan, Turkey, and the USSR. Other adaptations are underway in Afghanistan, Czechoslovakia, Colombia, Denmark, Finland, Great Britain, East Pakistan, Hong Kong, Indonesia, Singapore, Sweden, Uruguay, Venezuela, and Yugoslavia. In addition, the BSCS (through a special grant) cooperated with the UNESCO Biology Pilot Project for 15 English-speaking countries in Africa.

Participation in the BSCS International Cooperation program has been intellectually rewarding for all concerned, but funds are still limited for these "biologist-to-biologist" efforts toward improved biological education. Thus, each of the persons in these programs has had to make substantial personal contributions of time and talent toward the present development of international programs. As far as these biologists are concerned, however, their contributions have been justified because of the common good that can result from this effective cooperation for biological education. They believe it is important to the future of all citizens of the world to understand the biological basis of behavior, health, race, population growth, food supplies, pollution, and many other matters of great personal and international concern. The program will certainly continue and expand as biology becomes more and more important to the lives of all of the citizens of the world.

Some BSCS Laboratory Activities.

Anesthetization and examination of fruit flies.

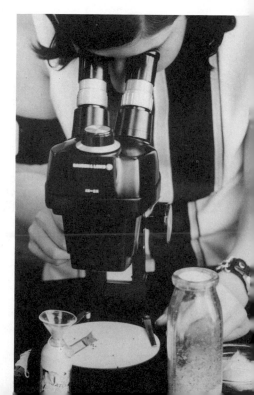

Determinants of blood type.

Photos by Jeff Smith

Observation of plant growth and development.

Photos by Jeff Smith

SECTION TWO

Invitations to Enquiry

Index

Note. Four additional Invitations are included in Chapter 5. Two of these were developed by teachers, and two were developed by the writers of the Australian adaptation of BSCS biology.

4

The Nature and Use of Invitations to Enquiry [1]

REASONS FOR TEACHING SCIENCE AS ENQUIRY

We have remarked that teaching science merely as authoritative facts and dogma has had an extremely bad effect on American attitudes toward science and scientists. Such methods of teaching science divorce the conclusions of science from the data and the conceptual frames that give conclusions their meaning. As a consequence, the student often learns a lesson we never intended to teach. He learns that science is unreliable and unrelated to reality.

If this statement appears strange, consider a student who has garnered the impression that science consists of unalterable truths. Five or ten years after graduation, he discovers that many of the matters taught him are no longer used as knowledge or no longer considered sound. They have become obsolete and have been replaced. Unprepared for such a change and unaware of what produced it, the former student can do no better than to doubt the soundness of his textbooks and his teachers. In a great many cases, this doubt of teacher and textbook becomes a doubt of science itself, and of professional competence in general. The former student has no recourse but to fall into a dangerous relativism or cynicism.

Consider, too, the student who has never been taught to discriminate the conceptual from the physical. For such a student, a change

[1] The name "Invitations to Enquiry," their format, and all the statements in the Invitations and the Interim Summaries that pertain to enquiry in science and biology are drawn from the following, with permission of the author: *The teaching of science as enquiry*, Harvard University Press, Cambridge, Mass., 1962; *The practical teaching of science as enquiry* (to be published by John Wiley and Sons), both by Joseph J. Schwab, University of Chicago. This material may not be reproduced in any form without the permission of the author.

in a formulation of scientific knowledge which he once learned is not only an instance of unexpected change in what he thinks is fixed and certain, it is also an incomprehensible change. Since he does not know that some of the "things" taught him are not literal facts but facts embodied and organized in ideas, he cannot account for their sudden disappearance, and he is still further confused by the new body of statements that refer to "things" of which he had learned nothing at all. The effect of this experience can only reinforce the impression that science is whimsical or mysterious, and without relevance to the everyday realities that count in life.

These are some of the consequences of dogmatic teaching which lead to the need to teach science as enquiry. To teach science as enquiry means, first, to show students how knowledge arises from the interpretation of data. It means, second, to show students that the interpretation of data—indeed, even the search for data—proceeds on the basis of concepts and assumptions that change as our knowledge grows. It means, third, to show students that because these principles and concepts change, knowledge changes too. It means, fourth, to show students that, though knowledge changes, it changes for good reason—because we know better and know more than we knew before. The converse of this point also needs stress: The possibility that present knowledge may be revised in the future does *not* mean that present knowledge is false. Present knowledge in science is based on the best-tested facts and concepts we presently possess. It is the most reliable, rational knowledge of which man is capable.

Note that we have said, *"show* the student that knowledge arises from data"; *"show* the student that knowledge changes." The point is that the mere *telling* of students about such things is not effective. The need is to exhibit science in operation, not to talk *about* science except as a summing up of what has been shown.

The Invitations to Enquiry are one useful means to such ends. They are teaching units that bring before the student small samples of the operation of enquiry, samples graded to his competence and knowledge.

Each sample is incomplete. There is a blank, an omission, which the student is invited to fill. This omission may be the plan of an experiment, or a way to control one factor in an experiment. It may be the conclusion to be drawn from given data. It may be an hypothesis to account for data given.

Whatever the blank place in the enquiry may be, it is marked out by what is provided in the Invitation. Its outlines are well (but not too well) sketched. Thus the student is enabled to fill the hole, contribute to the enquiry.

The Invitation to Enquiry teaches enquiry in two ways. First, it poses example after example of the process itself. Second, it *engages the participation* of the student in the process. Thus, for the less able student there is one channel toward understanding—the Invitations as examples of enquiry. For the more able student there are two channels—the Invitations as examples of enquiry and his *own* contributions toward solving the problem each one poses.

An additional word should be said about the aims of such teaching. The primary aim is an understanding of enquiry. It is mainly for the sake of this aim that the active participation of the student is invoked. Both practical experience and experimental study indicate that concepts are understood best and retained longest when the student contributes to his own understanding.

Two further objectives, however, are served by active student participation. First, the student can discover through the kinds of problems posed, and through his ability to contribute toward their solution, that science is something more than merely learning from others what others already know. He finds that science is also an activity of the mind, a challenge to the imagination, and a place where thought and invention are rewarded. Second, through the process of participating in such activities, the quality of the activity itself may be shaped and improved. The student may develop skill in the interpretation of data and the understanding of scientific knowledge.

PRACTICAL CONSIDERATIONS

Ease and Difficulty

Each of the earlier Invitations poses a very simple problem, and problems are posed one at a time. Then, as each group of Invitations proceeds, the isolated bits and pieces of problems are put together. For example, the first Invitation asks only that a student think of the bare outline of an experiment. A later Invitation introduces the idea of a *controlled* experiment. A still later Invitation asks for both a general plan and certain control features.

In the same way, an early Invitation will ask the student to draw a conclusion from very simple, all-or-nothing data. Another will introduce him to the fact that data are rarely all-or-nothing. Subsequent Invitations will then use realistic data as a matter of course.

A somewhat different grading of difficulty is used for the biological content of the Invitations. Each Invitation provides the specific biological information needed to understand the problem it presents. Necessarily, however, some degree of *general* understanding of biology is called for. It is this general understanding which is graded from little to more as the Invitations proceed. The first Invitations require very little understanding. Later Invitations assume that the student has had a week, two weeks, or a month or more of biology, but they do not assume that he knows a particular body of information.

By this kind of grading of biological content, the Invitations are freed from dependence on the particular textbook the teacher may be using.

Methods of Use of Invitations

The Invitations in this book are set up for use primarily in classroom or laboratory period discussions. (Some of them may also be used in another way, which will be dealt with shortly.)

We suggest that the teacher follow the procedure we shall sketch here. The teacher presents the background information of the Invitations orally and, when necessary, at the blackboard. He then poses the problem of the Invitation and invites student reaction. Thereafter, he deals with student responses as they arise, asking diagnostic questions which help students see what is wrong with poorer answers, reviewing the logic that justifies good responses. Sound responses to early problems of an Invitation then lead naturally into the next problem in the Invitation, and the procedure of diagnostic and analytical questioning continues.

When you turn to the Invitations themselves on the following pages, you will note that each one has two components, set off by different arrangements on the page. There is a component for the student. These are always marked "To the student." The other component, set off by brackets and smaller type, consists of comments and guide materials for the teacher. These materials indicate the purpose and content of the Invitation, give examples of sound answers to the problem, and provide forewords on what is coming next.

This format is indicative of the fact that the Invitations have been planned for use primarily in classroom and laboratory. That is, each Invitation has a psychological ordering or "structuring" designed to make use of the kinds of communication and interaction that occur within a student group. For example, at certain points, in some Invitations, the materials will be so structured as to elicit and make constructive use of a certain degree of competitiveness among students. At other points, the structure will invite complementary contributions of differing kinds from students with different abilities, thus fostering and using a certain kind of cooperation among students. Still other points in an Invitation may be structured to encourage a kind of dialogue or back-and-forth contribution between students and teacher.

The succession of problems, suggestions, and explanations in each Invitation are intended as aids to the teacher in the mapping and conduct of each *enquiring discussion*. These aids and their particular arrangement have been settled on after tests conducted by some seventy teachers in as many classrooms with students of differing personalities and abilities. We should bear in mind, however, that it is impossible to harness the human mind or to predict its behavior exactly. Some students will need guidance and aid beyond that provided. In such cases, the teacher is the only person who can supply the needed assistance, for it is only the teacher, alert to the moment's unique difficulty, in a particular classroom and with a particular student or group of students, who can detect what is needed.

In still other cases (especially in the earlier and "easier" Invitations), the student may race ahead of the parts and steps as presented in the Invitations. In such cases, the student should not only be encouraged but applauded. Where this occurs with only one or a few students in a class, these students can then become invaluable aides in conveying the gist of the Invitation to less able students.

After Invitations have been used in the context of laboratory or classroom discussion for some time, it is sometimes profitable to treat them as private challenges, to be done by students singly, at home or in study period. For use in this service, the "To the student" units are identified and separately duplicated for use as homework material.

Invitations designed especially for such "private" use have been tested and found effective. These experimental "private" Invitations, however, were constructed with an eye to replacing the motivations and rewards inherent in the discussion context with rewards and impulsions appropriate to the "private" context. The "To the student" units of the Invitations provided in this volume do not contain this special structure. Hence the teacher should not expect the same effectiveness in private use as in the discussion context.

It should also be noted that the second and successive "To the student" units in many Invitations incorporate the *answers* to previous problems. This is a necessary reflection of the fact that these Invitations deal with a continuing problem, not isolated parts of the problems. In such cases, the successive "To the student" units may be used as homework only if the teacher can provide copies of them on successive days or periods.

Place and Time of Use

Invitations can be used in any of three different time-place contexts. First, they may be used as an independent series of learning experiences. That is, one period each week, or some fraction of a period, can be set aside as the regular time of use. This method takes maximum advantage of the fact that the Invitations are arranged to form a progressive series.

Second, Invitations may be used in conjunction with laboratory exercises. There are two possibilities here. Invitations may be selected for the biological subject matter they discuss and used in conjunction with laboratory experiences dealing with the same biological material. Second, they may be selected for their relevance to the aspect of enquiry that a laboratory experience illustrates. To facilitate such selection, each group of Invitations is preceded by an index that lists both the biological subject and the topic of enquiry treated by each Invitation.

Third, Invitations may be used as in the second alternative given, but in conjunction with classroom work. For this use, they would be selected to match the biological subject under study at a given time. When they are used in this way, there will be reciprocal reinforcement in some cases. Not only will the biological material under discussion be of aid in work on the Invitation, but the Invitation may throw additional light on the textbook treatment.

Guidelines for Using Invitations

1. Look through each group of Invitations before using the first one. Give special attention to the *interim summaries,* which are included in each group. These summaries recapitulate the conceptions that the Invitations develop. They also contain material you may wish to incorporate in your presentation of the Invitations.

2. Read and reread each Invitation before starting it with your students.

3. Note whether it will require preparation of duplicated materials for the students, require graphs on the blackboard, and so on, or whether it can be best presented orally.

4. Help the student to see that these are graded samples of enquiry. The first are necessarily short and simple (even to the point of insult for the semisophisticated). Help him to understand that these are

models used to illustrate the Invitations and that soon he will have more to chew on.

5. Remember that the aim of an Invitation is not to obtain the "right" answer immediately. Rather, the aim is to invite the student to use *his* information and intelligence in an effort to *find* the answer. Each answer given by a youngster should be "honored" by discussing with him what may have been an incorrect response (not whether it was right or wrong but where his reasoning may have gone wrong in producing it).

6. Remember the built-in time trap. The Invitation is a model problem, not a subject matter dissertation, review, or introduction to your favorite subject. If you try to "tell" the story, you will fail to "show" science as enquiry. Excessive inclusions unnecessarily prolong and may even destroy the value of the Invitation.

7. Know at what time in the presentation you can include some of the bracketed materials or interim summary material to illustrate the Invitation well without giving answers prematurely.

8. Keep in mind the possibility of developing new invitations of your own. Those presented here barely sample the content of the BSCS text you are using. The Supervisor, Joseph J. Schwab, would appreciate copies of any you may develop. Address them to him care of The University of Chicago, 5835 Kimbark Avenue, Chicago 37, Illinois.

INVITATIONS TO ENQUIRY, GROUP I

SIMPLE ENQUIRY: The Role and Nature of General Knowledge, Data, Experiment, Control, Hypothesis, and Problem in Scientific Investigation

INVITATION	SUBJECT	TOPIC
1	The cell nucleus	Interpretation of simple data
2	The cell nucleus	Interpretation of variable data
3	Seed germination	Misinterpretation of data
4	Plant physiology	Interpretation of complex data

Interim Summary 1, Knowledge and Data

5	Measurement in general	Systematic and random error
6	Plant nutrition	Planning of experiment
7	Plant nutrition	Control of experiment
8	Predator-prey; natural populations	"Second-best" data

INVITATION	SUBJECT	TOPIC
9	Population growth	The problem of sampling
10	Environment and disease	The idea of hypothesis
11	Light and plant growth	Construction of hypotheses
12	Vitamin deficiency	"If . . . , then . . ." analysis
13	Natural selection	Practice in hypothesis

Interim Summary 2, The Role of Hypothesis

14	Auxins and plant movement	Hypothesis; interpretation of abnormality
15	Neurohormones of the heart	Origin of scientific problems
16	Discovery of penicillin	Accident in enquiry
16A	Discovery of anaphylaxis	Accident in enquiry

INVITATION 1

SUBJECT: *The cell nucleus*
TOPIC: *Interpretation of simple data*

[This is an extremely simple introduction to the Invitations to Enquiry. You may wish to commence Invitation 2 immediately after completing Invitation 1, completing both within one class period.]

To the student: You may already know that animals and plants are made of one or more very small living units called cells. You may also know that most kinds of cells contain a still smaller part called the nucleus. A biologists wanted to know whether the nucleus of cells is necessary for life, or whether cells could live without a nucleus. He found a way to break cells into two pieces so that one piece contained the nucleus and the other piece did not. He performed this experiment on a number of different kinds of cells.

Suppose the results of this experiment were that all the cell pieces without the nuclei died. All the cell pieces that had nuclei soon re-formed into smaller but normal-looking cells, grew then to their former size, and in all other ways continued to behave exactly as do cells that have not been broken. What interpretation would you make of this experiment and its results?

[The evidence, as stated, will probably impel most of the students to conclude that the nucleus is necessary for the continued normal life of the cell. Even in this highly idealized Invitation, however, in which all the data are artificially perfect, there are, of course, doubts that could be raised and almost endless qualifications that could be added to the interpretations to make it "safer." We may suspect, for example, that the nucleus is necessary only for the repair of injury—hence the death of the non-nucleated fragments. Or, ignoring that, we may doubt whether what was found to hold for the kinds of cells tested necessarily holds for all cells. Or we could take pains to point out that the conclusions should be restricted to cells which normally have nuclei, and so on.

[Doubts such as these lurk behind almost all the general statements our textbooks ordinarily assert to be scientific "truths." Nevertheless, doubt can be cast on almost any statement that goes beyond the immediate and particular (such as, "That *one* automobile in front of *this* house is black

138

in *this* light to *my* eye.")   Hence it is not the business of science to be infinitely cautious as a condition for being "right." Over-caution is just as much a handicap to the growth of dependable knowledge as reckless over-generalization.   Scientific knowledge increases and becomes more dependable only as we *do* draw conclusions, interpret our data, and go on to further problems that these interpretations in turn suggest—to further experiments designed to solve these new problems, to new data, and to new interpretations of the whole body of data.   The life blood of science is not indefinite caution and indecisiveness, but ongoing enquiry, enquiry that refines earlier conclusions, makes them more precise, and extends their scope.   Perhaps that is why scientific work is called *re*search.

[The present Invitation and those following it develop along similar lines. Later Invitations will successively exemplify one and then another of the common sources of error in interpreting experimental data, and different ways of refining and expanding scientific knowledge.   Hence, for the moment, it is desirable that the students be permitted to have confidence in the unqualified conclusion to which the idealized data of this Invitation will probably impel them—that the nucleus *is* necessary to the continued life of a cell.   Indeed, if some students have a habit of over-caution instilled into them, it might be desirable to point out some of the ideas we have stated.]

INVITATION 2

SUBJECT: *The cell nucleus*

TOPIC: *Interpretation of variable data*

To the student: (a) Let us recall our earlier enquiry into the cell nucleus. We were told then that all the cell fragments which lacked a nucleus soon died. We were also told that all the cell fragments in the experiment which retained a nucleus recovered and went on to live normally.

Now, in the experience of biologists, such neat, either/or results occur rarely, if ever. Rather, results are more likely to look like these:

Number of *non-nucleated* fragments studied:		100
Number of *non-nucleated* fragments surviving:	1 day	81
	2 days	62
	3 days	20
	4 days	0
Number of *nucleated* fragments studied:		100
Number of *nucleated* fragments surviving:	1 day	79
	2 days	78
	3 days	77
	4 days	74
	10 days	67
	30 days	65

If you *had* to make a positive interpretation of the data about the importance of the nucleus, what would you say? What is there about the data which could make you feel uneasy about the interpretation you made?

> [The purpose of these questions at this point in the Invitation is to ensure that the students examine the data carefully before going on. Their answers may anticipate to some extent what is to follow.]

To the student: (b) Let us assume that the scientists who did the work stated their interpretation as follows: "Our experiment was judged

140

satisfactory and was terminated at the thirtieth day. Our data indicate that the nucleus is normally necessary for the continued life of the cell."

Now as we can see, the conclusion drawn by the scientist "goes beyond the data,"

["Data" are the recorded facts from which we try to derive an answer to a question, a solution to a problem.]

for we see that not only did some non-nucleated fragments live for three days, but that a number of nucleated fragments died, some after one day, more after two days, still more in ten days, and so on. It is entirely possible that if the experiment had been extended for sixty days or more, many more nucleated fragments might have died.

How, then, can we defend the statement with which the scientist concluded his piece of research? The general answer is that *one* experiment can rarely, if ever, *prove* a scientific statement beyond any shadow of doubt. There is always some doubt, and successive experiments try to remove one doubt after another.

In short, it is a little bit misleading to say that a scientist draws a *conclusion* from his data. For his summing up is rarely a *conclusion,* an *end* to research on that subject. Rather, it is a beginning or a continuation. It is more informative to say that a scientist *interprets* his data. He expects other scientists to check him, refine his interpretation, and extend it. For science it a social, cooperative enterprise.

Let us go back to our new data about the number of nucleated and non-nucleated fragments which survive. A more cautious scientist complained to our biologist that his interpretation of his data was unacceptable until he gave at least one reasonable explanation for the death of so many nucleated fragments. Come to the defense of our biologist. Provide him with such an explanation.

Hint 1: Remember how the nucleated fragments were obtained.

Hint 2. Think of what probably would happen if you tried to nurture 10 baby chicks or 100 baby kittens to maturity.

[The teacher may or may not need to assist students to see that the experiment itself probably involved fatal damage to some nucleated fragments; and that some fraction of any population of live things is likely to die in a given period of time from many common causes. In any case, it is important to stress the point that in the present instance data will be less than perfect because of many factors which are virtually incontrollable—that in almost any case variability will characterize results as contrasted to the uniformity contained in interpretations. A later Invitation will return to the matter of variability and exemplify the idea of *experimental error* as another source of variability.

[Obviously, a better experiment could be designed. If your students feel that the Invitation is thus far an easy one, you may wish to invite them to improve it. Problems of adequate sampling and control arise, however. These points are raised in later Invitations.]

To the student: (c) There are cells which *normally contain no nucleus* and yet live a long time. Suppose for the moment that our experimenter is right—that in cells *with* a nucleus, there is something in the nucleus indispensable to survival. How would you explain the healthy long lives of cells that do not have nuclei?

[The most extreme possibility would be that cells *without* nuclei have greatly different organizations, which do not require whatever it is in the nucleus that seems to be required for survival of cells not so organized. A more conservative possibility would be that the "something" within the nucleus of the nuclated cell is not located within the confines of a discrete nucleus in non-nucleated cells.]

To the student: (d) One interesting example of non-nucleated cells is the mature human red blood cell. The nucleus degenerates during the development of the cell. Even that "something" in the nucleus degenerates. In view of this information, what would you expect to be the length of life of such cells?

[They have a limited survival time, on the average around three to four months.]

INVITATION 3

SUBJECT: *Seed germination*

TOPIC: *Misinterpretation of data*

[It is one thing to take a calculated risk in interpreting data. It is another thing to propose an interpretation for which there is *no* evidence—whether based on misreading of the available data or indifference to evidence. The material in this Invitation is intended to illustrate one of the most obvious misinterpretations. It also introduces the role of a clearly formulated *problem* in controlling interpretation of the data from experiments to which the problem leads.]

To the student: (a) An investigator was interested in the conditions under which seeds would best germinate. He placed several grains of corn on moist blotting paper in each of two glass dishes. He then placed one of these dishes in a room from which light was excluded. The other was placed in a well-lighted room. Both rooms were kept at the same temperature. After four days the investigator examined the grains. He found that all the seeds in both dishes had germinated.

What interpretation would you make of the data from this experiment? Do not include facts that you may have obtained elsewhere, but restrict your interpretation to those from *this experiment alone.*

[Of course, the experiment is designed to test the light factor. The Invitation is intended, however, to give the inadequately logical students a chance to say that the experiment suggests that moisture is necessary for the sprouting of grains. Others may say it shows that a warm temperature is necessary. If such suggestions do not arise, introduce one as a possibility. Do so with an attitude that will encourage the expression of unwarranted interpretation, if such exists among the students.

[If such an interpretation is forthcoming, you can suggest its weakness by asking the students if the data suggest that corn grains require a glass dish in order to germinate. Probably none of your students will accept this. You should have little difficulty in showing them that the data some of them thought were evidence for the necessity of moisture or warmth are no different from the data available about glass dishes. In neither case are the data evidence for such a conclusion.]

143

To the student: (b) What factor was clearly *different* in the surroundings of the two dishes? In view of your answer, remembering that this was a deliberately planned experiment, state as precisely as you can the specific problem that led to this particular plan of experiment.

[If it has not come out long before this, it should be apparent now that the experiment was designed to test the necessity of light as a factor in germination. As to the statement of the problem, the Invitation began with a very general question: "Under what conditions do seeds germinate best?" This is not the most useful way to state a problem for scientific enquiry, because it does not indicate where and how to look for an answer. Only when the "question" is made specific enough to suggest what data are needed to answer it does it become an immediate useful scientific problem. For example, "Will seeds germinate better with or without light?" is a question pointing clearly to what data are required. A comparsion of germination in the light with germination in the dark is needed. So we can say that a general "wonderment" is converted into an immediately useful problem when the question is made sufficiently specific to suggest an experiment to be performed or specific data to be sought. We do not mean to suggest that general "wonderments" are bad. On the contrary, they are indispensable. The point is only that they must lead to something else—a solvable problem.]

To the student: (c) In view of the problem you have stated, look at the data again. What interpretation are we led to?

[It should now be clear that the evidence indicates that light is *not* necessary for the germination of *some* seeds. You may wish to point out that light is necessary for some other seeds (for example, Grand Rapids lettuce) and may inhibit the germination of others (for instance, some varieties of onion).

[*N.B.:* This Invitation continues to deal with the ideas of data, evidence, and interpretation. It also touches on the new point dealt with under paragraph *(b)*, the idea of *a problem*. It exemplifies the fact that general curiosity must be converted into a specific problem.

[It also indicates that the problem posed in an enquiry has more than one function. First, it leads to the design of the experiment. It converts a wonder into a plan of attack. It also guides us in interpreting data. This is indicated in *(c)*, where it is so much easier to make a sound interpretation than it is in *(a)*, where we are proceeding without a clear idea of what problem led to the particular body of data being dealt with.

[If your students have found this Invitation easy or especially stimulating, you may wish to carry the discussion further and anticipate to some extent the topic of Invitation 6 (planning an experiment). The following additions are designed for such use.]

To the student: (d) Granting that light is not *necessary* for the germination of corn seeds, different amounts of light may speed up or slow down germination. How might the experiment check on this possibility?

[Counting the number of germinated seeds per day in lighted and non-lighted dishes would provide some evidence on this point.]

To the student: (e) Now that you have learned to test for the role that light plays in seed germination, plan an experiment in which you test the effect of temperature on seed germination.

[This sort of experiment, of course, should involve setting up the same moisture and light conditions, but varying the temperature for different containers of seeds]

INVITATION 4

SUBJECT: *Plant physiology*

TOPIC: *Interpretation of complex data*

[This is a further exercise in the interpretation of data; no new ideas about enquiry are introduced. No information is needed by the student except that included here. Tell the students that they should confine their interpretations to the evidence given. The situation presented is not factually correct, as you will note. Plants in a sealed bell jar are alleged to die in a short time. Actually they survive for a very considerable time.]

To the student: (a) Suppose you place a plant in a sealed bell jar, and place a similar plant beside it in the open air. Suppose that after a few days the plant in the bell jar dies, while the plant in the open air continues to flouish. Suppose, further, that when the experiment is repeated a number of times the same results are obtained.

What explanations can you suggest for the death of plants in sealed bell jars?

[There are two relevant, possible answers. One is that plants need something from the air which is eventually depleted if there is a limited amount of air available. The other is that plants produce one or more substances which pollute a small amount of air. From the evidence presented, we cannot determine which of these two possibilities is correct.]

To the student: (b) If you were to experiment with pairs of mice as you did with the plants, the results would be similar. The mice in the sealed bell jar would die; the mice in the open air would flourish. What interpretation should we make of the results of this experiment?

[The same two answers are possible for mice as for plants.]

To the student: (c) If you were so unwise as to assume that all living things were alike, what unwarranted interpretation might you have drawn from the two experiments? *Hint:* The *observed* results were the same in the two experiments.

146

[The unwarranted interpretation is that plants and mice need the same "something" from the air, or pollute the air in the same way.]

To the student: *(d)* Now suppose you were to place a healthy specimen of the same kind of plant and a healthy mouse of the same species together underneath the same sealed bell jar. You would find that both mouse and plant would live considerably longer together under the same sealed bell jar than when kept under separate sealed bell jars. (Hopefully, you would have had the foresight to shield the plant from the mouse with a screen of some sort.) Repetition of this experiment should give similar results each time. What interpretation would you give of this view of the results of previous experiments?

[One interpretation is that there is some kind of relation between the requirements of plants and animals, which is such that each produces something the other needs.

[Another interpretation is that each organism removes the "polluting material" produced by the other.

[At the appropriate time, the teacher may wish to point out that most plants can live a long time in a sealed bell jar, but that mice (and other animals) cannot. When the students have learned about photosynthesis and respiration, the class can return to this point and understand why plants can survive in these circumstances whereas mice cannot.]

INTERIM SUMMARY 1

Knowledge and Data

THE IDEA OF GENERAL KNOWLEDGE

With Invitation 4 we have completed our introduction to the student of simple enquiry. In this introduction, we have tried to show him the operation of two ideas basic to the scientific enterprise. Let us review these ideas briefly and at the same time note certain relations among them.

One idea is central to all others in defining the scientific enterprise. It is the idea of *general* knowledge, knowledge which is not about one particular thing in one particular time and place but knowledge which embraces whole sets of particulars. This idea sets science apart from almost all the other forms of knowledge and enquiry man has invented. History and biography, for example, are dedicated to the particular. Generally, they try to recover and recount specific, unique events, each occurring at a specified time and place—in a certain city on a certain day, or in the mind of a certain leader or thinker in the midst of a specified, single moment of thought or decision. Rarely, indeed, does the historian try to discover general laws among his particulars. And when he does, he is quite ready, even proud, to say that what he is trying to write is *scientific* history.

The idea of general truths about particulars also sets science apart from mathematics—although in an entirely different way from that which distinguishes science from history. Mathematics is as general as science; perhaps, indeed, much more general, even universal. What geometry has to say about "triangle," for example, holds for all triangles and has bearing on all other figures as well.

Nevertheless, the generality of mathematics does not make it the same as science, for the general knowledge characteristic of science is general knowledge of concrete, existing, observable, measurable particulars. It is about the world of things and events. The knowledge of mathematics is not derived from the world of things and events. The general or universal statements of the mathematician have their origin in something else—perhaps in ideas in the mind of the mathematician, ideas which he studies and whose component ideas become the content of his theorems. Men are divided, in fact, about the nature and origin of mathematics, but one thing is clear; whatever the origin of mathematics may be, it is not an origin in concrete, existing, observable, measurable particulars. The Pythagorean theorem, in its mathematical form, did not arise *from* measurements of the sides of many triangles.

The essence of science, then, is that it is, paradoxically, *general* knowledge from *particulars,* from existing, observable, measurable particulars belonging to the world of things and events.

THE IDEA OF DATA

Science, then, begins with particulars, but it does not rest in them. Particulars are indispensable to science, but only as raw materials. Further, they are indispensable raw materials of a certain limited kind. They are like ore to the refiner of metals. Some of the ore is unwanted material. Only a part of it is the desired gold. So, too, with particulars as the ore of science. Only *some* aspects and items of a group of particulars are relevant and useful to a scientist in search of a general truth. The remainder of the manifold aspects of the particulars are irrelevant to his search.

Only *some* aspects, then, of a group of particulars serve as raw materials for any one general truth. It is this condition that defines what we mean by scientific *data.* Data are the gold extracted from the ore. They represent the facts about particulars that the scientist selects from all the available facts because he thinks they will best serve the aim of science— lead him to the most revealing truths about the particulars he is studying. *Data,* then, are *selected facts.*

Our Invitations (1 and 2) begin with this point. A scientist wants to know about the role of nuclei in cells. He selects the facts which are relevant to this question. He examines these selected facts, these data, and tries to draw from them *(interprets them)* a general truth about cells and nuclei. Note how the major points we have mentioned are involved in these beginning hints about enquiry. First, the data are "real"; they represent the facts. Second, the data represent selected facts; they are only those facts that will answer the question about the nucleus which the scientist has formulated as the central portion of the problem. Third, the data are only raw materials, starting points. What the scientist can *make* of his data is the climax of his enterprise.

INVITATION 5

SUBJECT: *Measurement in general*

TOPIC: *Systematic and random error*

[This Invitation introduces the notion of experimental error. It can be omitted if your regular laboratory projects cover the following points:

1. Experimental error is inevitable; it can be reduced but not eliminated.

2. Since error is inevitable, data are practically always equivocal. That is, the defensible interpretation is almost always "cleaner" than the data, for what we are reluctant to include in our interpretation we can often ascribe to experimental error.

3. One man's "experimental error" may be the food of the next man's research. That is, the variation in the data may *not* have been due entirely to experimental error. The variation may indicate a factor overlooked in the design of the experiment.]

To the student: (a) Over a period of a week, four students made five measurements each of a metal bar which was kept in their laboratory. They were given the impression that each was measuring a different bar each time. The record of their measurements (in millimeters) is shown in Table 1.

TABLE 1

Measurement	Student 1	Student 2	Student 3	Student 4
1	500.0	500.0	499.8	500.1
2	499.9	500.0	499.9	500.1
3	500.0	500.0	500.1	500.2
4	500.1	500.0	500.2	500.3
5	500.2	500.0	500.2	500.3

Notice the difference between the measurements reported by student 2 and those reported by students 1, 3, and 4. How would you describe the difference? What do you think explains it?

[The difference between the report of student 2 and the others lies in the uniformity of student 2's measurements. Some of your students—those least

150

experienced in the laboratory—may explain this as due to the greater accuracy of student 2. The more sophisticated of your students may suggest the contrary—that student 2 let his first measurement influence his reading of later measurements, or even that he is cheating in an extremely silly way. You can then appeal to their own experience to indicate that the first interpretation is the *less* likely of the two, that uniformity is *not* characteristic of actual measurements and is always suspect. This point should end by introducing the term "experimental error," meaning unavoidable inaccuracy and inconsistency of measurement.]

To the student: (*b*) Now compare the measurements reported by student 4 with the other reports. What overall difference is there? How would you explain this difference? *Hint:* Think of the measurement you would obtain if you read the right end of the measuring stick while standing at its left end, with the stick on top of the measured bar. Then think of what the measurements might seem if you stood in line with the right end of the measuring stick.

[Student 4's measurements are uniformly higher than most. The hint is intended to lead the student to see that conditions of measurement may vary from one investigator to another, and thus lead to different data. You may wish to have these variations actually experienced. If so, obtain a measuring stick and have one student stand well to the left, another well to the right, while each estimates some measurement.

[If you feel your class is ready for it, you can drive the idea of experimental error home by clarifying the difference between *random* error and *systematic* error. Student 4 represents a case of systematic error. The entire group of measurements, irrespective of the students making each measurement, would come close to exemplifying random error. The most important reason for making the distinction between random error and systematic error is that our common ways of making use of numerous measurements (by calculating their arithmetic mean, their mode, or their median, for example) are ways of correcting *random* error. We have no readily available way of detecting or correcting systematic error except by watching our habits. Hence, systematic error is potentially a source of danger, whereas a reasonable amount of random error is not.]

To the student: (*c*) Suppose a later experiment required an estimate of the length of the bar to be used in *many* arithmetic calculations. In that case, what estimate would you choose as being "good" enough and also convenient?

If the future experiment required the *best* estimate from these data of the length of the bar, what number would you choose to use, and how would you go about calculating it?

[The two questions are put together so that the students will have a chance to contrast the idea of a "convenient" measure with the idea of a "best" measure. The most convenient measure would, of course, be 500.0. One "best" estimate, making no further allowances for differences between early and later measurements and measurements done by different students, would be a simple arithmetic mean, an average. This is computed, of course, by summing all the measures and dividing by the number of measurements.]

To the student: (d) What could we do to make this "best" estimate even more reliable?

[The answer, of course, is to secure more measurements.]

To the student: (e) There are two ways to secure additional measurements. One would be to have the same students do the job. Another would be to call on additional students to contribute. What kind of error, systematic or random, would be reduced more by the second method than by the first?

[Method 2 would reduce the overall systematic error more than would method 1, if you are lucky or wise in your selection of students.

[Thus far we have treated the problem of measurement only from the viewpoint of error introduced by the investigator. In the remaining portion of this Invitation we deal with error introduced by unnoticed changes in the thing being measured. This is, of course, one of the kinds of problems to be met in controlling an experiment.]

To the student: (f) Now inspect each column of measurements from the top down. What *trend* is noticeable? What—besides a new sort of systematic error—might explain this trend? *Hint:* Remember that the measurements were made over a period of a week. Remember, too, that the bar being measured was made of metal, and that metals are subject to changes due to environmental factors.

[There is a clearly noticeable trend (ignoring student 2) toward larger measurements on the later days. The second hint is intended to turn the student's attention to the possibility that the bar actually became longer as the days grew warmer. You will probably need to guide the students into seeing and understanding this possibility.

[In actuality, no readily available metal has a coefficient of expansion as great as is indicated by our fictitious measurements.]

To the student: (g) Suppose that the trend just discussed had been overlooked or treated as experimental error. Further, suppose that the measurements were used to support a certain theory according to which the rod ought to be nearer 500.1 mm long than to 500.0. Unless this experiment were repeated by others at different times and places, what might happen to the field of study that defended this theory?

[It might well have adopted the theory in question. Then, only when later inconsistencies appeared would the theory have been called into question. Only then might the experiment have been repeated by someone and the error discovered.

[This exemplifies the point made in the introduction to this Invitation— what one man ascribes to experimental error may be the basis, through research, of other and more useful investigations.]

To the student: (h) In view of your answers to *(e)* and *(g)* explain what is meant by saying that science is a *social* enterprise.

[The idea here is that for some problems many heads are better than one; that science depends on debates and differences, on alternative approaches to problems.

[By now, your students are, hopefully, on the way to regarding data as only approximations of what is, and to seeing that what is is ever elusive, although we may refine our attempts to define it more accurately.]

INVITATION 6

SUBJECT: *Plant nutrition*

TOPIC: *Planning of experiment*

[So far we have dealt only with some aspects of the interpretation of data. Now we introduce a second major factor—the planning of an experiment. Emphasis here is on two points: (1) the planning of a simple experiment, omitting consideration (for the moment) of the not-so-simple matter of adequate control; (2) the anticipation of what may result from an experiment, done for the sake of interpreting the resulting data. The latter *uses* the idea of hypothesis in the form of a series of "If . . . , then . . ." logical analyses, but the term "hypothesis" is not introduced. It will be introduced only in a later Invitation, since we wish to minimize talking about such things until examples of them are well in hand.]

To the student: (a) We know that as living things grow they increase in weight. They do so by taking from *elsewhere t*he materials needed for their growth. Suppose you wanted to know what percentage of the raw materials for the growth of a plant comes from the soil in which it is planted. Suppose you also wanted to know what part of these raw materials comes from the air that surrounds the plant. Plan a simple experiment that would begin to solve this problem. Begin by stating a more specific form of the problem. If you have forgotten what this means, remember Invitation 3 about seed germination.

[Because of the way in which the question is phrased, some students may begin by trying to deal with *both* possible sources—soil and air as well as the plant. It is for them that we suggest starting by reformulating the problem. In addition, we want to repeat the lesson about problem formulation.

[The general problem can be made more specific by reformulating it in terms of *either* air or soil. We do not need both. For example: To what extent is weight gain by the plant accounted for by weight loss by the soil? The same plan, substituting air for soil, would embody the second way in which the problem could be more specific. An extremely able and quantitatively minded student might phrase the problem as follows: Is weight loss by the soil equal to or exceeded by weight gain by the plant? The same two

154

formats, with "air" substituted for "soil," would provide the second way in which the problem could be specified. Both possibilities should be elicited from the class. Do not burden them at this time with the problem of water in soil and air.

[Now it is time to move toward the actual plan of experiment. Concentrate on the soil version of the problem since it is easier to work with. One simple procedure is to pot a growing plant of known weight in a known weight of soil. After a period of growth, the plant and soil are each weighed.

[It may be of considerable importance here to point out to students what *experiment* means, or it may be postponed to Invitation 10. All too often, students have the impression that any looking, measuring, or noting that occurs in a laboratory room is an "experiment." This, of course, is not the case. An experiment can be defined as follows: It is the instituting (setting up, creation) of a planned situation; the situation is so planned as to yield certain specified, wanted data.

[In many experiments a planned *alteration* of the ordinary or "normal" course of events is contrasted to the normal. However, *alteration* of the ordinary or normal is not always characteristic of experiments; *contrast* is; either a contrast of before to after or a contrast between two set-ups differing in some specific, planned, and, therefore, known way.]

To the student: (b) What results would you expect from this experiment if the plant took all raw materials for growth from the soil?

[Weight increase of plant equals weight decrease of soil.]

To the student: (c) If the plant derived none of its raw materials from the soil, what results would you expect?

[Weight decrease of soil equals 0.]

To the student: (d) If the plant derived half its raw materials from the soil, what results would you expect?

[Weight increase of plant equals twice the weight decrease of soil.]

To the student: (e) If the plant derived some but not all of its raw materials from the soil, what results would you expect?

[Weight increase of plant is greater than weight decrease of oil.

[The experiment is obviously without adequate control. This problem is taken up in the next Invitation.]

INVITATION 7

SUBJECT: *Plant nutrition*

TOPIC: *Control of experiment*

[This Invitation takes up the problem ignored by the previous one—the problem of adequate control in the planning of an experiment. In these two experiments we have chosen a situation in which the major uncontrolled factor is fairly easy to notice, very easy to understand when pointed out, but extremely difficult to bring under control. Keep in mind that though the Invitation calls for "insight" from the student, its main purpose is to introduce students to the *idea* of control, and to the great difficulty in achieving such control. However, we feel it important at this point not to talk about control as such. The term and the concept are best postponed until the idea is implanted as an aspect of enquiry. Later (Invitation 11) the concept and the term will be introduced.

[You will note that we also introduce here in a similarly inconspicuous way the terms "assume" and "assumption." These, too, will be treated more formally in a later Invitation.]

To the student: (a) Suppose that we published the experiment planned in Invitation 6, together with its results. Immediately another scientist writes us that the entire experiment is worthless because of something we have overlooked.

You will remember that what we did was to put a plant of known weight into soil of known weight, permitting the plant to grow there. We then weighed plant and soil a second time. Let us imagine that the soil lost 0.2 mg, while the plant gained 2.0 mg. From these data we conclude that one-tenth of the plant's growth had taken place at the expense of the soil.

We are now told that our conclusion is valueless because the experiment was badly planned. Our critic does not complain about any possible inaccuracy in our weighing. He says only that the experiment was badly *planned*.

Now visualize that flowerpot sitting on a desk day after day while the plant grows. Remember that we took it for granted (assumed) that

156

the weight lost by the soil was gained by the plant. What is wrong with our making such an assumption?

[This, of course, is an exercise requiring insight, the putting of two and two together. The class may or may not perceive the possibility that, with the soil open to the air, it could be losing materials not only to the plant but to the surrounding air. An actual growing plant to look at may facilitate the perception. If it does not, then a further leading suggestion can be given. For instance, have someone come up and feel the *moist* soil. Then ask again if they see what was wrong. Or remind them of what happens to wet hands even though they are not towelled. At some juncture the point may be noted, or you may need to tell the students explicitly. Then it is time to move on to the following.]

To the student: (b) We see, then, that it is necessary to guard against any loss or gain by the soil except to the plant. Let us take only one part of this problem by assuming that the soil gains nothing from the surrounding air and that it is only *water* which the soil might lose to the surrounding air. Then our problem is to ensure that the only water the soil may lose is water taken up and retained by the plant. How might we go about doing that?

[Begin by inviting a simple attempt at this control. One might be to put the pot containing the weighed soil in a pan containing a weighed amount of water and then to put pot, plant, soil, pan, water, and all under a large bell jar.

[If this is suggested, go on to the next point. Again, if the class does not suggest it, provide it yourself. In either case, then invite criticism of the control.]

To the student: (c) Now let us see if this bell jar arrangement really takes care of our whole problem. For example, if you actually did this experiment, you might find that after some time beads of moisture accumulated on the inside of the bell jar. This would not occur if the jar had not contained our plant-soil-water set-up. What does this suggest might be a weakness in our control?

[The point here is that some water would be lost to the air within the bell jar unless the jar had been put over the experimental set-up in an atmosphere already saturated with water. And, of course, temperature changes would affect this even further. When this weakness is noted and the correction (saturation at start of experiment plus temperature control) made, you can point out still another weakness, as follows:]

To the student: (d) Suppose now that our bell jar arrangement does, in fact, control water loss. Remember, however, that our large problem was to find out how much of the plant's requirements came from the soil and how much from the air. This problem arose because, let us pretend, we already knew that the plant required something from each. If this is the case, our new experiment is just as bad as, or worse than, the first one. Try to see what is wrong with the modified experiment.

[The point here is that the limited amount of air within the bell jar may well permit so little growth that our "after" data will show no measured increase in plant weight or decrease in soil weight. This has already been suggested in Invitation 4, and you may find it useful to remind students of it.

[By class discussion or by the teacher's description, the class may now be led to understanding controls as elaborate as those described next.

[We measure the *dry* weight of the soil and the weight of the plant. We place into the soil a tube through which known quantities of water can be added. We enclose the pot so that the soil is moisture-sealed, and place the plant in a bell jar through which air can be circulated. An arrangement is added whereby any water which is in the air going to or coming from the bell jar can be absorbed and measured. With this arrangement, almost all the water added to the soil and not used by the plant should either remain in the soil or should find its way into the air of the bell jar. At the end of the experiment, moisture remaining in the soil is evaporated and measured, and this amount is added to the amount absorbed from the air leaving the bell jar. The difference between the amount of water added during the experiment and the amount not used by the plant can then be calculated. The terminal dry weight of the soil and the weight of the plant are measured as in the first experiment.]

INVITATION 8

SUBJECT: *Predator-prey; natural populations*

TOPIC: *"Second-best" data*

[In this Invitation we again provide occasion for the interpretation of data and the planning of experiments. We also introduce two new points. First, the Invitation begins in an inconspicuous way to stress the fact that scientific problems are conceived or made by the investigator, not objectively "given" by the subject matter. This point, however, is passed over lightly. The second new point is more emphatically treated. It is the point that the data which would best serve to throw light on a problem are not always obtainable. Often this inaccessibility is due to limitations in our technical knowledge or competence. At other times, we refrain from eliciting very useful data because the steps necessary to collect them will ruin the possibility of later investigations, which we deem more important. It is this latter possibility that is used here. The "best" data are rejected in favor of less definitive data, and the shortcomings of the "second-best" data are then examined.]

To the student: (a) A zoologist was studying the size of the populations of small rodents and owls in a certain geographical area. He was interested in finding out what factors tended to control the size of the populations.

Because of a complex series of factors which need not concern us, the rodent population was greatly reduced during the period of the study. It was known that adult owls normally fed on small rodents, but when the numbers of these rodents were reduced, the *adult* owl population—approximately fourteen—remained unchanged. On the other hand, the number of newly hatched owls was found to be greatly reduced by comparison with previous counts. What might explain the fact that the reduced food supply resulted in fewer young owls but no reduction in the number of adult owls?

[The first answer may be the vague one that the reduced food supply caused a decrease in the population. If so, point out that the decrease has been *differential,* and press for possible ways in which the decreased food supply might have acted to produce this *differential* decrease.

[Some of the possibilities are that (1) the underfed owlets are more susceptible to disease; (2) the adult owls may be eating some of their young; (3) underfed female owls are producing fewer eggs; (4) a lower percentage of laid eggs hatch.

[A sufficient variety of these or similar answers should be elicited or, if necessary, supplied, to make possible the following question.]

To the student: (b) The possibilities we have developed seem to be of two different kinds. That is, there seem to be two different routes by which the population of young could be made smaller. What are they?

[The answer is that some of the suggested possibilities increase the mortality among the young owls. Others act by decreasing the natality. Students may not use the words "natality" and "mortality," of course. If not, this is a good context in which to introduce them.]

To the student: (c) The reduction in owlet population may be due, then, to an increased infant mortality or to a decreased natality rate. How might we find out which of these two possibilities was operating?

[The possibility of increased infant mortality would require either observing the owls in their nests or finding other evidence that the young owls were dying or being killed by their parents or predators. Time-lapse infrared photography could perhaps produce such evidence. It would be easier to check on lowered natality by making counts of laid and newly hatched eggs.]

To the student: (d) Suppose it was found that there was, indeed, a lowered natality. We have already seen two ways in which this reduction might occur: Fewer eggs produced; fewer laid eggs hatched. In what different ways might the food affect the adult owls so as to bring these effects about?

[There are a number of possibilities: Poor metabolism affecting ovaries so that fewer eggs are laid; similar effects on testes leading to a decrease in the percentage of fertilized eggs among eggs laid; no effect on eggs or sperm but reduction in frequency of mating.]

To the student: (e) What should we study in order to obtain data (information, evidence) that would help us decide which of these possibilities are, in fact, the case? Note that all three of them, or any two of them, may contribute to the reduced size of the new generation of owls.

[The question asks only what we should study and not what data to obtain. This is intentional, because this is an early exercise and students may not yet know enough biology to speculate responsibly about the precise data required. In any case, obvious and incomplete answers are as follows:

HYPOTHESIS	SUBJECT REQUIRING STUDY
Poor development of germ cells	Ovaries and testes of well-fed owls, as against owls of this population
Change in mating frequency	Frequency of mating in the local owl population as against owls in a similar environment but amply supplied with food
Percentage of eggs hatching	Comparison of number of eggs incubated to number hatched]

To the student: (f) Remember that the biologist we are talking about is interested in studying what happened to the *natural* population of owls. Look back at the evidence you have in hand and see if you can determine why he might decide that he must *not* gather the data needed for further information about our first possibility.

[The point here is that the owl population is so small that to kill a few for study of the gonads would make a large artificial change in the population.]

To the student: (g) How might the biologist use the *laboratory* to get useful information on the germ cell production possibility?

[Putting *caged* owls (from other sources) on good and poor diets, and so on.]

To the student: (h) Show that there is at least one serious objection to drawing a conclusion about the natural owl population from data obtained by a study of the gonads of caged owls.

[There is the possibility that caging could, itself, lead to a significant change in germ cell production.]

To the student: (i) Since the first possibility is not easily tested, let us consider the second one—that there has been a change in mating behavior. Suppose a study of the owl's behavior showed that they mated less often on the reduced rodent diet than on the previously plentiful food supply. Suppose, further, that a biologist was rash enough to decide from these data that the cause of the reduced size of the new generation was the reduction in frequency of mating. Show the biologist why he is rash in drawing such a conclusion.

[The evidence indicates that mating behavior may be *one* factor in reducing the size of the new generation. The evidence does *not* rule out the possibility that the other factors also contribute.

[This exercise has been designed to introduce the student to some very specific aspects of biological research and to some of the names we give to the steps and materials of scientific enquiry.

[In order to bring these steps, aspects, and terms into sharp focus, it is desirable at this poin tto look back on, diagram, and thus review what has happened. A structure for such a review is given next.

WHAT WENT ON IN THIS ENQUIRY	SOME TERMS
1. A study of a stable population was under way when the population was violently changed by a disease.	A problem
2. The investigator saw the opportunity to study another problem [the cause of the drop in population of owls—*(a)* and *(b)*].	Cause
3. He first considered some *possible* causes *(b)*.	
4. He then tried to see what data might throw light on whether these possible causes were actually operating.	Data, evidence
5. He found that obtaining the most desirable data about one possibility would spoil the population for further study.	
6. Substitute (less desirable) evidence was considered [*(f)* and *(g)*].	
7. Another hypothesis was tested and found to operate.	Test
8. An unjustifiably sweeping interpretation of the evidence was made *(i)*.	Interpretation of data
9. This interpretation was criticized.	Multiple causal factors

[Certain aspects of this suggested structure for review call for emphasis.

[In the first place, the word "hypothesis" has been purposely omitted. This is because we intend to deal with it separately in the following Invitations. There are two reasons for this separation. In the first place, the idea of "hypothesis" has sometimes been given a very special and narrow meaning borrowed from certain methods of enquiry used more often in the physical than in the biological sciences. (These are treated in the following Invitation.) In the second place, we need to put special emphasis in the case of the present Invitation on another idea of great significance in biology: the idea of study of a *natural* situation as against an experimental one. We have touched on this matter in the present Invitation by the queries on the unwisdom of removing owls from the natural population in order to study their gonads and germ cells. Students need to begin to see that, for all the knowledge we owe to experimental manipulation, the careful study of undisturbed ("natural") situations also has an important place in biological science.

[The second aspect of Invitation 8 needing emphasis is the fact that we say nothing about the so-called "four or five steps" of *the* scientific method. We feel that it is important *not* to formulate scientific enquiry as following some such pattern, for these reasons:

1. Scientific research uses many different patterns.

2. The pattern a scientist follows depends on what his problem is. The problem, in turn, derives from preceding enquiries and from the knowledge of the subject accepted at that time. "Method" depends to a very great extent on "content." In short, there is no one abstract pattern or method in science which can be understood in its true significance unless it is seen in specific, on-going researches.

[For these reasons, we feel that it is wise to postpone any seemingly complete story about scientific method. Instead, it would be most helpful to limit explanation to what went on in *this* research without generalizing from the one case to all research. Thus, certain data were sought. In the light of them, problems were formulated. As a guide toward solution of these problems some "might-be's" or possibilities were thought of. (*Note:* the colloquialism "might be" or the more formal "possibility" are good substitutes at this point for the technical word "hypothesis.") With the possibilities envisaged, data that would test them were reasoned out. For one such test, one useful kind of data was rejected because collecting it would spoil the situation for further study. Less useful data were substituted because they would throw *some* light on the question yet not spoil further enquiry. Finally, the data were over-interpreted, and the over-interpretation was criticized.

[Note, especially, that when the Invitation ends the problem is still open. There is no complete settlement once and for all. This is typical of most lines of enquiry in the sciences.]

INVITATION 9

SUBJECT: *Population growth*

TOPIC: *The problem of sampling*

[In this Invitation we raise a further problem involved in obtaining wanted data: the problem of *adequate sampling*. The idea of *sample* is developed, and methods of sampling a mouse population are examined. The uncertainty of data based on sampling is stressed, and the possibility that sampling may alter the population under study is raised.]

To the student: (a) This Invitation is based on investigations conducted by Dr. John Emlen and his associates into the food and space requirements of a colony of mice.

In one of these experiments a colony of mice in an old building was provided with 250 g of food per day. A daily study of the size of the mouse population was then carried out. According to the investigators' report, the colony grew rapidly until there were as many mice as the food would support. Then, it was reported, mice began to leave the colony and continued to do so at about the same rate as young mice were born into the colony.

How could data be obtained on which to base such a report? To see the difficulty of this problem, remember the following. The study is intended to find out how a *natural* population grows under *natural* conditions. Therefore, data are needed about population size in the old building at frequent intervals in the period of the population growth. Let us attack this problem in two steps. First, if all we wanted to know was the size of the population at some *one* time, what would we need to do?

[The answer, of course, is that we would need to capture all the mice in the building.]

To the student: (b) Assume that by such a mass capture we now have a good estimate of the size of the population at the beginning of our study. Hereafter we wish to know only whether the population increases from day to day and, if so, by what proportion. We no longer need absolute numbers but only percentages. How might we go on to obtain data that would give us some idea of the *change* in size of the population, yet not involve us in the attempt to trap all the mice?

[The solution to this problem is the technique of *sampling*. We set a fixed number of traps in the same places each day, baited in the same way, for example. We take the number trapped on the first day as representative of the population whose size is known from the first trapping, just described. Thereafter we take the fluctuating size of the trapped sample as an indicator of fluctuations in the size of the whole population.]

To the student: (c) Suppose that live traps (traps that do not kill the captured mice) are used for such a sampling technique. After each morning's count the captured mice are turned loose. Now name at least three factors that could cause this sampling technique to give us misleading information.

[Some of the possibilities are as follows: (1) Mice too young to leave the nest will not have been counted. (2) As time goes on, more and more mice may learn to avoid the trap. (3) A greater percentage of the mice may be attracted to the trap when the population becomes large compared with the food supply than when the population was smaller. (4) Many mice may have areas of activity that do not bring them near the traps.]

To the student: (d) Now let us name some ways in which the effect of these sources of sampling error could be reduced.

[(4) can be minimized by a wide distribution of the traps used and by increasing the number used. (3) can be reduced by providing no food except that provided in the traps. (2) could be avoided by using kill traps instead of live traps. (1) could at least be taken account of by assuming that the measured increases in size of population lag a little behind the actual increase.]

To the student: (e) Show that the "corrective" suggested for (2) might well ruin the entire experiment.

[The point, of course, is that kill trapping is an interference with the very thing under study—a "natural" population and its pattern of growth. We have no way of knowing exactly what effect we are having by killing a certain percentage of the population each day, but it is entirely too probable that *some* unwanted warping of the normal course of events is taking place.]

INVITATION 10

SUBJECT: *Environment and disease*

TOPIC: *The idea of hypothesis*

[In this Invitation the ideas of *hypothesis* and of *controlled experiment* are introduced by name. It is worth pointing out in advance that there are two kinds of hypotheses commonly encountered in science.

[The most common kind of hypothesis currently used in biology is one we shall call the "glass-box" hypothesis. We use such a name to emphasize the point that what is hypothesized is something that, at least in good part, can be *seen,* that is, it is open to direct examination. Suppose, for example, that we make the hypothesis that nerves are necessary for control of some process. This is a typical glass-box hypothesis, since nerves are things accessible to observation and so is their absence. We may not be able to observe the act of control, but we can observe the presence of nerves, remove them, and note the consequent existence or disappearance of control.

[The other kind of hypothesis we may call the "black-box" type. In black-box investigations we conceive of an underlying mechanism inaccessible to direct observation to account for—give coherence to—a wide variety of visible behaviors. Mendel's genetic determiners and the mechanism for their segregation and recombination constituted such a hypothesis at the time he made it. Sutton, of course, by pointing out the parallel between the behavior of Mendel's hypothetical units and the visible behavior of chromosomes in meiosis and fertilization, took it out of the class of black boxes, but this does not alter the fact that it *was* a black-box affair when Mendel created it.

[Unlike the first kind of hypothesis, the black-box variety cannot, by definition, be verified "directly." Instead, we seek necessary *implications* of the black-box mechanism which can be looked for. From Mendel's mechanism, for example, we should get a 1:2:1 ratio when some F_1 organisms are bred to one another. Again, if the hypothesis is a good one, we expect that two independent characteristics, both with a dominant alternate, will show in the ratio of 9:3:3:1, and so on.

[Note that we said if the hypothesis is a *good* one some of the consequences necessarily implicated in it will be found. We did not say that discovery of implicated consequences *proves* the hypothesis to be true. As

166

many logicians have pointed out, there is no complete "verification" of a black-box hypothesis—only a demonstration of how much it will account for and how well it does it. These—the how much and how well—are the measures of the "goodness" or usefulness of a hypothesis. It is usefulness, not truth, which we demand of black-box hypotheses. Thus in physics we have, for example, the Bohr model of the atom, which accounts for a great many phenomena. We also have a more recent and more complicated model that accounts for still more. The scientist interested in the "still more" will use the more comprehensive model. But a scientist interested only in the phenomena embraced by the Bohr atom need feel no self-consciousness about using it on his problems (as long as he knows what he is and is not using).

[For both the black-box and the glass-box kinds of hypotheses, the same logical form precedes and guides the experimental test of its value. This is the "If . . . , then . . ." logical linkage. For example, "*If* nervous connections are necessary for the normal function of tissue A, *then* severing the nerves to it should result in failure of function." Or, a black-box example: "*If* every chemical element occurs as numerous indivisible particles, *then* every combination of some two elements to form a specific compound should involve definite proportions of the two elements"; and: "*If* two elements combine in differing proportions to form two different compounds, *then* the proportion in one case should be a whole number multiple of the other."

[In either the glass-box or the black-box case, then, the "If . . . , then . . ." logic tells us what data to look for. In this invitation we shall introduce the student formally to this logic, but only for the case of glass-box hypotheses.]

To the student: (a) Let us look back for a moment at Invitation 8. There we asked you several questions of the same kind. We asked, for example, "What *might* explain the fact that reduced food supply resulted in fewer young owls but no change in the adult owl population?" Later we asked, "State two ways in which food lack *might* lead to lowered natality."

In both these questions—and some earlier ones too—we were asking for "might-be's." Such "might-be's" are extremely useful sorts of educated guesses for the scientist. They help him see what data to look for and what meaning to give to them when he finds them. For example, we saw that lowered natality *might be* due to a change in breeding habit. This led us to a plan for studying breeding behavior. We saw that lowered natality *might be* due to a smaller proportion of fertilized eggs reaching the hatching stage. This would have led us to study the nests of owls to see how many eggs were incubated, how many were hatched, and how many of the unhatched eggs showed signs of having been fertilized.

Such guiding "might-be's" have a definite name. They are called *hypotheses*. We can define "hypothesis" thus: A hypothesis is a "might-be," a possibility, which we intend to test. (There are other *"might-be's"* which scientists use in a given piece of research which they *don't* intend to test just then. They use them as if they were true as far as that piece of research is concerned. These *"might-be's,"* the taken-for-granted ones, are called *assumptions.* But more about these later. For now, we are interested mainly in hypotheses—*"might-be's"* intended for test.

Let us practice constructing some hypotheses. Let us assume that you have just moved to Boulder, Colorado, on the slope of the Rocky Mountains. Before your family moved you lived in Chicago, Illinois, on the shore of Lake Michigan. Soon after arrival in Boulder, you and your twin brother develop the same illness. You both have runny noses, little appetite, and are sick to your stomachs (have nausea) most of the time. Then you both return to Chicago for a short visit; overnight, your illness disappears. When you go back to your new home in Boulder, however, the illness again appears. So you .*assume* without further evidence that there is something about Boulder, Colorado, that makes you and your twin brother ill.

Now let us make at least six *"might-be's,"* six educated guesses, six testable hypotheses about what *might be* the cause of this illness. Remember, Boulder is in a different part of the country from Chicago and is different in many ways.

> [The list of possible causes can be very long: the water; new foods; pollen allergies to plants that do not grow near Chicago; different strains of bacteria; homesickness; rising much earlier because of the earlier sunlight; altitude; heavier doses of ultraviolet rays because of the cleaner air; different kinds of dusts.]

To the student: (b) We have already pointed out that one of the values of a hypothesis consists in the fact that it can be made to point like an arrow toward data and experiments which may solve our problem. Let us see how this works.

We make a hypothesis serve as an arrow by thinking of a situation which will be one sort if the hypothesis is a good one and of another sort if the hypothesis is a bad one. For example, suppose we take the hypothesis that there is something in the Boulder water supply which makes you and your twin ill. We say to ourselves, *"If* the water is responsible for our illness, *then* while drinking the water, we shall be ill; and while drinking pure distilled water instead, we should recover in time. Contraiwise, *if* the water is not responsible, *then* drinking pure distilled water as against Boulder water should make no difference."

Notice the form which our reasoning takes in thinking of such plans. It is a form of thought that is called, "If . . . , then. . . ." We say, *if* the hypothesis is a good one, *then* such-and-such should happen. *If* the hypothesis is a poor one, *then* another such-and-such should happen.

Notice that in both our "If . . . , then . . .'s" the *if* is followed by the hypothesis. The *then* is followed by the consequences we expect, though a different consequence in each of the two "If . . . , then . . .'s." In the first case, *then* is followed by what we expect to happen if the hypothesis is a *good* one. In the second case, *then* is followed by what we expect if the hypothesis is a poor one.

Now let us try constructing "If . . . , then . . .'s" for some of our other hypotheses.

> [The dust hypothesis could be tested by wearing a mask; the sunlight by staying indoors or wearing covering clothing; and so on. Note that we have yet to convert the "If . . . , then . . ." logic into an actual experiment with reasonably adequate control. This is done next.]

To the student: (c) Suppose, now, that we use our "If . . . , then . . ." about water to plan an experiment. Suppose we do so as follows: You and your brother use nothing but pure, distilled bottled water for a week. At the end of the week you and your twin are no longer ill. You compare this consequence with your illness of the week before and decide that your hypothesis was a good one.

But have you really shown this? Suppose this week had been rainy. If it had, what unanswered question would remain?

> [Of course, the unresolved problem is that we do not know whether it was the absence of bright sun or the absence of Boulder water which led to recovery.]

To the student: (d) Let us see what led us into such an inadequate experiment. Our logic originally read, roughly speaking: *If* Boulder water, *then* illness; *if* no Boulder water, *then* no illness. But, of course, it was also saying something else. It was saying, "Of course, we mean change the water and *only* the water because, if we change *two* things, we can't know which of the two changes (or, maybe, both) brought about the change in us from illness to health."

This point, that we must arrange things in our experiment so as to be reasonably sure that we have changed *only* the factor under test, is called *controlling* our experiment. It is very hard to be sure that we have a *perfectly* controlled experiment, but at least we can try to control the factors we have recognized in our other hypothesis. We make a mental list of them and check the list off to be sure that in our experiment they remain as they were while only the water is changed.

A way to come closer to complete control—even to control of unknown factors, unrecognized hypotheses—is to change the hypothetical factor in one case and not change it in another. For example, we might have *you* drink bottled water while your *twin* continues on Boulder water. (It would be even better if there were several of you and several twins).

With an experiment that uses this pattern, we would expect the *then* (recovering from the illness) to occur in one case (you), but not in the other (your twin). If both of you recover, something other than the water change must have been responsible. If neither of you recovers, then the water alone is not to be blamed for the illness. Only if you recover and your twin does not would we have some basis for deciding that the water was responsible for the illness.

Now let us plan controlled experiments to test your other hypotheses.

INVITATION 11

SUBJECT: *Light and plant growth*

TOPIC: *Construction of hypotheses*

[This Invitation continues the problems of constructing hypotheses and controlled experiments. One new idea is added, the idea that the results of an experiment may be a surprise, that they may not follow from what seems to be the logical expectation from the hypotheses we propose.]

To the student: (a) What effect do you think light has on growth in height of plants? Consider at least two possibilities.

[Without sound information, some of your students will make the commonsense suggestion that the more light there is, the taller and healthier plants will be. It is useful to elicit all three of the obvious possibilities: increased growth, inhibited growth, and no effect.]

To the student: (b) Now, let us convert one of our hypotheses—that light promotes growth—into an "If . . . , then . . ." which points toward an appropriate experiment.

["*If* light promotes growth, *then* the more light (up to a point) the greater the height of the growing plants. *If* light does not promote growth, and so on." If Invitation 10 was not completed very recently, you may wish to remind your students of the meaning of "If . . . , then . . ."]

To the student: (c) Now let us use our "If . . . , then . . ." as a guide in planning a *controlled* experiment.

[In this case it is desirable to introduce a hint of the problem involved in sampling, as well as the conventional "experimental-control" pair. That is, several pairs of pots should be planned for, using different kinds of seeds; corn and bean, for example.]

To the student: (d) Now, keeping in mind our "If . . . , then . . ." hypothesis, indicate what interpretation you will make of the possible results from such an experiment. Use the following form:

If plants grown in the light are taller than those grown in darkness, we conclude that _____.

171

[The students should construct such a statement for the possibility "If plants grown in the light are shorter than those grown in darkness, we conclude that . . . ," and "If plants grown in the light are the same height as those grown in darkness, we conclude that . . ."]

To the student: (e) If you performed this experiment, you would see these results:

Corn plants grown in *light:*   Stems not visible; leaves large, spread out; plants shorter

Corn plants grown in *darkness:*   Stems not visible, leaves small and not spread out, plants taller

Bean plants grown in *light:*   Stems thick and short

Bean plants grown in *darkness:*   Stems tall and slender

What does this show about the usefulness of making predictions from hypotheses about the result of experiments?

[The general point is, of course, that our predictions are always based on previous enquiries—previous knowledge. Such enquiries are often incomplete. Hence, our actual results may be more complex than our predictions, or very different.

[The specific point is that our predictions assumed that all seed plants—at least those involved in the experiment—would behave in the *same* way. We learn that we cannot be sure, before trial, how far similarities extend among things we have put into the same class.]

INVITATION 12

SUBJECT: *Vitamin deficiency*

TOPIC: *"If . . . , then . . ." analysis*

[This Invitation extends the making of hypothesis, "If . . . , then . . ." analyses, the planning of experiments, and so on, to animal physiology. You will find a special emphasis on the difficulty of achieving control. This special emphasis leads to the point that we can gain confidence in scientific conclusions in spite of the difficulty of perfect control. We do so, not only by improving control from enquiry to enquiry, but also by formulating and working on a number of *different* problems that converge on the point in question. See *(e)*.]

To the student: (a) A scientist noted that chickens fed mainly on polished rice (rice grains from which the outer layers have been removed) developed a disease. Their muscular coordination was affected. Chickens fed a varied diet of table scraps, other grains, and some polished rice remained healthy.

In view of these data construct two hypotheses concerning the cause of the nerve-muscle disease.

[A dietary deficiency in the polished rice diet. Some factor in rice which causes the disease.]

To the student: (b) A scientist named Eijkman, using further information at his disposal, was inclined to the hypothesis that the disease was due to lack of a dietary factor which was present in the outer layers of grain but not in the inner rice kernel. Construct an "If . . . , then . . ." to guide you from this hypothesis toward an experiment that would test it.

[*If* a necessary dietary factor is present in the rice husk but deficient in the remaining grain, *then* chickens fed on polished rice will develop the disease while chickens fed on the whole grain will remain healthy.]

To the student: (c) Now plan in careful detail a controlled experiment that would test this hypothesis.

173

[A population of uniform young chickens divided into two groups; one group fed whole grain and water, the other group fed polished grain and water, and so on. *Note:* To this point, the Invitations have not used the terms "control group" and "experimental group," although the ideas back of the words have, of course, been used repeatedly. With this Invitation the terms themselves should be introduced.]

To the student: (d) From this experiment it was found that almost all the animals fed polished rice developed the disease during the period of the experiment while all chicks fed the whole grain remained healthy. Assume that the experimenter interpreted these data to mean that a deficiency of a necessary dietary factor caused the disease in the experimental group.

Pretend that at this time another extremely skeptical scientist visited the laboratory and noted that the control group had been housed in one room and the experimental group in another. He immediately rejected the interpretation of the first scientist on grounds of inadequate control.

What uncontrolled possibility might have been involved in the separate housing of the control and experimental groups?

How would you do the experiment so as to control this possibility?

[The experimental group might, coincidentally, have been subject to an infectious disease that did not spread to the control group housed in another room. One obvious control of this factor would be to mark experimental and control birds so that they could be housed together except at feeding time.]

To the student: (e) Even with this particular possibility now controlled, we cannot be sure that our experiment is perfect; for we do not know what still *unrecognized* factors may still be uncontrolled. Since this possibility always exists, scientists use two different ways of making their interpretations more and more reliable. First, they try to recognize uncontrolled factors and replan their experiment to take care of them. Second, they try to see new and different problems which are connected with the first problem.

For example, here we have a situation in which one approach—change of diet—suggests that a necessary dietary factor is responsible for a nervous-muscular disease, while criticism suggests that the disease may be due to an infection—a virus, for example. Now note some of the separate ideas present here:

1. The idea of a necessary dietary factor.

2. Normal and abnormal behavior of nerve and muscle.

3. The possibility of an infectious agent.

Using one or more of these separate ideas, try to think of new and different problems that might help us to decide for or against the dietary-deficiency hypothesis. This is not an easy task, for there is no "method" we can learn which will guide us to the new problems. Thinking of new problems is an example of "creative thinking." Let us try our hand at it.

[Some "new" problems are:

1. An attempt to isolate the dietary factor from rice husks and use it to "cure" the disease.

2. An attempt to find and isolate the infectious agent and show that it produces the disease.

3. A study of nerve-muscle action in diseased and healthy birds to determine what has gone wrong in the sick birds.]

INVITATION 13

SUBJECT: *Natural selection*

TOPIC: *Practice in hypothesis*

[Many of your students may know too much about the subject of this Invitation to find it challenging. If so, you may wish to omit it. However, another way to deal with this problem is to point out that reports we read in popular publications are not always complete and accurate. Hence it might be worth their while to approach this Invitation with open minds about the outcome.]

To the student: (a) A group of men were working with dairy cattle at an agricultural experiment station. The population of flies in the stables where the cattle lived was so large that the animals' health was affected. So the men sprayed the barn and the cattle with a solution of DDT, an insecticide. They found that this killed nearly all the flies.

A week or so later, however, the number of flies was again large. The workers again sprayed with DDT. The result was similar to that of the first spraying. Most of the flies were killed.

Again the population of flies increased, and again DDT spray was used. This sequence of events was repeated several times. After four or five sprayings, it became apparent that the DDT was becoming less and less effective in killing the flies, until finally spraying with DDT appeared to be useless. Construct several different hypotheses to account for these facts.

[These may include the following:
1. Decomposition of DDT with age.
2. DDT effective only under certain environmental conditions, which changed in the course of the work; for example, temperature, humidity.
3. And the crucial hypothesis—selective killing of flies genetically most susceptible to DDT.
(3) should not be elicited at this point nor developed if suggested.]

To the student: (b) One of the men noted that one large batch of DDT solution has been made and used in all the sprayings. He there-

176

fore suggested the possibility that DDT solution decomposed with age. Suggest at least two *different* approaches toward testing this hypothesis.

[The emphasis on *different* approaches continues the new point made at the end of Invitation 12, that investigation of several different problems may contribute to the reliability of the conclusions drawn from each. In the present instance, one approach would be to use sprays of different ages on different barn populations of flies. A quite different approach would consist simply of chemical analysis of fresh and old solutions to determine if changes had occurred. Neither approach is perfect, of course. But each has a contribution to make.]

To the student: (c) A fresh batch of DDT was made up. It was used instead of the old batch on the renewed fly population at the experiment station barn. Nevertheless, despite the freshness of the solution, only a few of the flies died.

The same batch of DDT was then tried on a fly population at another barn several miles from the experiment station. In this case the results were like those originally seen at the experiment station: Most of the flies were killed. Here were two quite different results with a fresh batch of DDT. Moreover, the weather conditions at the time of the effective spraying of the distant barn were the same as when the spray was used wtihout success at the experiment station.

Now let us take our problem situation apart and try to list its major components. They might run as follows:

1. Something used (the DDT).
2. The conditions under which the "something" was used.
3. The way in which the "something" was used.
4. Something *on which* the DDT was used (the flies).

Thus far, all our hypotheses have had to do with just a few of these components. Which ones?

[The hypotheses so far have concerned only items 1 and 2.]

To the student: (d) The advantage of analyzing a problem situation, as we have done in our lists, consists in the fact that it lets us see what possibilities we have not made use of, what "blinders" we may have unconsciously had around our mind's eye.

What possibilities in the list have we not made use of in forming our hypotheses?

[Items 3 and 4. Item 3 may be pursued as a further exercise if the teacher so wishes. However, emphasis should go to the major possibility contained in item 4. This is developed next.]

To the student: (e) Let us see if we can make use of the fourth entry in our list. From your knowledge of biology, see if you can think of something that might have happened within the fly population that would account for the decreasing effectiveness of DDT as an insecticide.

[The students are very likely to need help here, even if they have learned something about evolution and natural selection. One way help can be given follows:

[Ask the students to remember that after the first spraying most, but not all, of the flies were killed. Ask them where the new population of flies came from, that is, who were their parents? Were the parents among the more susceptible or the more resistant as far as the effects of DDT are concerned? Then remind them that the barn was sprayed again. If there are differences in the population in the susceptibility to DDT, which individuals would be more likely to survive this spraying? Remind them that dead flies do not produce offspring—living ones may. The students may thus be led to see that natural selection, in this case in an imposed environment (the presence of DDT), might have resulted in the survival of only those individuals best adapted to live in the new environment (one with DDT).]

INTERIM SUMMARY 2

The Role of Hypothesis

EXPERIMENT AND CONTROL

Invitations 1 to 4 emphasized only the ideas of *data* and *general knowledge.* Their relations were summed up in Interim Summary 1. Data (selected facts) are the raw materials of science. What the scientist can *make* of his data is the climax of the enterprise.

In Invitation 2 we began to hint at some of the difficulties the scientist faces in this enterprise. It was pointed out there that data rarely come "clean"; they always involve variability, which the scientist must somehow cut through in order to arrive at general knowledge. On the whole, however, these early Invitations were confined to exemplifying the use of data in arriving at general knowledge.

The theme of Invitations 5 through 13, on the other hand, is the problem of obtaining wanted data. Let us look back briefly at the order in which different aspects of this problem were raised. Invitation 5 points out the existence of *error* and the fact that it can be reduced but not eliminated. Invitation 6 then introduces *experiment* as the means, *par excellence,* for obtaining our chosen data. Experiment is defined as a situation so planned as to yield specified, wanted data. However, the difficulties in the way of *carrying out* the plan of an experiment are not dealt with here but carried over to Invitation 7. Here again the emphasis is on the difficulty of obtaining wanted data, the particular difficulty being that of ensuring adequate *control.*

OPERATIONAL UNCERTAINTY IN SCIENCE

Invitation 8 raises a quite different aspect of the problem of obtaining wanted data. In an ecological setting we see the possibility

179

that the operations required to obtain one set of wanted data may so warp the situation under investigation that little further useful data can be obtained. This kind of handicap to the progress of enquiry is of considerable importance. It arises whenever we are investigating a sequence of events in time. In such a case, we usually require data about the progress of the sequence—what it is like from moment to moment. This requirement often leads, in turn, to the need for intruding ourselves or our instruments into the process under investigation. By this act we may so alter the process that its further behavior is no longer appropriate to our investigation.

One of the most extreme examples of this problem occurs in those investigations in the physical sciences where the objects under investigation are very small. In such cases, the most refined methods of measurement must use devices of much the same size as the objects investigated. How are we, for example, to measure the velocity of an electron without using something *on* the electron that is as big or bigger than it is, and that therefore alters the velocity to a degree which may make the measurement worthless for our purposes?

The most extreme examples of this handicap may occur only in small-particle physics, but it is, nevertheless, a continuing problem in biology, too. It arises in the experimental study of the developing embryo, in the experimental study of growing and waning populations, in the experimental study of population genetics and evolution, and so on. It is illustrated once again among our enquiries in the case of Invitation 9, where the problem of sampling a mouse population is not merely a problem of obtaining a fair sample but also a problem of minimizing the effect of sampling on the population itself.

THE ROLE OF HYPOTHESES IN SCIENCE

Invitations 10 and 11 continue to treat the problem of obtaining wanted data. They introduce the idea of *hypothesis* in close connection with the use of "If . . . , then . . ." reasoning.

In these Invitations the hypothesis is treated in large part as if its role in science was primarily to be established or disestablished, "proved" or "disproved." In fact, however, hypotheses play another important role. They serve as guides pointing the way to new discoveries.

We can make the significance of this role clearer by re-examining the "If . . . , then . . ." logic associated with the use of hypotheses. When we look at such a logical chain from one point of view, it seems to be purely and simply a major step toward "proving" the "If." "If *A*, then

B." We seek for B. If we find it, our hypothesis, A, is rendered possible. If we fail to find B, we are in doubt. If we show that B does not exist, we have disproved the hypothesis, A.

Now let us look at an "If A, then B" from another point of view. The A, the hypothesis, is a new idea, or a familiar idea in an untried and unfamiliar context. Therefore, the B is also something new or something familiar but located in a context as yet unsearched; so we proceed to search for the new thing or in the new context. In this new search *something* new may be discovered. Whether the something new corresponds to our hypothesis or not, it sets off a stream of fresh enquiries likely to lead to fresh and important knowledge. This is the point suggested in Invitation 11 and repeated in Invitations 12 and 13.

There are two important ways to emphasize this role of hypotheses. One way is to point out that science is not one process but two. It is a process of discovery as well as a process of "proof."

A second important way to present the role of hypotheses in the process of discovery is as follows. The problem of obtaining wanted data is twofold. First, it is the problem of obtaining the data when we know what data we want. The planned and controlled experiment is the best solution to this problem.

In addition, the problem of obtaining wanted data is the problem of *knowing what data to want*. The conceiving of fresh hypotheses is the way we solve this problem.

The two problems are, of course, connected. Hence the devices that solve them are connected too. Although it is the experiment that leads us to our data, it is the hypothesis that leads us to our experiment. And the fresh hypothesis leads us also to fresh problems.

INVITATION 14

SUBJECT: *Auxins and plant movement*

TOPIC: *Hypothesis; interpretation of abnormality*

[In this Invitation we drop the planning of experiments to concentrate on the formation of hypotheses and interpretation of data. This is done in order to emphasize two major new points. The first of these concerns the vexing question of "teleology." One unacceptable form of "teleological" interpretation is contrasted to the acceptable "functional" interpretation.

[The second point concerns the significance of ablation experiments. The Invitation emphasizes the fact that most biological enquiry is focused on the *normal* organism. Data obtained by removal of an organ, for example, are often intended to help us know what the organ contributes when in its normal place and condition.]

To the student: (a) As some of you know, plants tend to bend toward the light. This usually results in an exposure of more leaf surface to the available light. It is mainly within the leaves of such plants that a process occurs by which the plant "captures" light energy. Now, in view of this information formulate an answer to the following question. Why do plants bend toward the light?

[In an average class, at least one student will reply, in effect, "They bend so as to get more light, which they need in order to grow."

[From such an answer discussion should move toward two points. First, help the students see that such a formulation suggests—or even intends to assert—something like *intelligent* behavior, behavior that *consciously* anticipates future needs. With this clarification made, discussion continues by developing the point that we have no reasonably sure data to indicate that such is the case. Further, the information we do have about the structure and organization in *us* which permits such conscious, intelligent behavior goes a long way to suggest that plants are not likely to have this competence. (In the case of some animals, there is basis for supposing conscious, intelligent behavior; there is certainly plenty of evidence of intelligent behavior.)

[The second main point of the ensuing discussion should show that although *"teleological"* interpretations which assign conscious purpose to the individual plant may be unsound, *functional* interpretations are sound and useful. That is, there is a contribution to our understanding of living things when we link the actions of a given part of an organism to what is achieved by this action as far as the organism as a whole is concerned. Plant movements may, indeed, be reasonably interpreted to have *survival value* or *adaptive significance*. The plant may well "profit" from its movements, carry on some other activities more effectively with such movements than without them. Other examples (human) of adaptive behavior may be cited: learning behavior, even in infants; the pupillary reflex; various protective withdrawal reflexes; inflammation in areas of local infection.

[It is worth pointing out what the term "teleological" means, strictly and technically. Strictly speaking, a teleological explanation is nothing more than an explanation in terms of *outcomes*. For example, suppose we interpret a series of events in embryonic development by pointing out how they lead to a mature structure. This would be a teleological explanation.

[Perhaps pointing out the contrary kind of interpretation would also help to clarify what is meant by "teleological" in its strict sense. The contrary kind of explanation would be one in terms of *origins*. For example, we might interpret the misbehavior of a child in terms of events which occurred in its earlier home life.

[As these examples may indicate, there is nothing illogical or factually wrong with "teleological" interpretations in their strictest form. Indeed, phenomena in the physical sciences are often interpreted in terms of *outcome*. Teleological interpretations fell into disrepute in biology because they were so often made without justifying data and by assigning consciousness and intelligence where there was little ground for doing so. As is so often the case, we tended to throw out the baby with the bath, and sweepingly rejected all teleological explanations. It is to correct this over-correction that the present Invitation distinguishes between teleological and functional explanations.]

To the student: (b) In view of the lack of evidence that plants display purpose in their "actions," how could you improve on the phrasing used by many students, "They bend so as to get more light?". . .

[One possible rephrasing is: "Plants respond to light by bending toward it; this results in more efficient utilization of the available light."

[A close reading will show only a little difference in meaning between this formulation and the preceding one. Only if our students are ready to understand adaptive behavior as arising through adaptive evolution can we thoroughly take care of what is undesirable in teleological formulations.]

To the student: (c) About eighty years ago, the biologist Charles Darwin and his son Francis began an investigation of plants bending toward the light. They had noticed that canary grass seedlings which received light mainly from one side bent toward the light. Most of the bending took place near the bottom of the seedling, just above the soil. Other observations, however, suggested to the Darwins that the seedling *tip* might have something to do with the bending. Therefore they performed the following experiment. With a razor they cut off the tips of seven young seedlings. These, together with seven uncut seedlings, were then exposed to light from one side only. Those with intact tips bent sharply toward the light. The seedlings without tips did not bend.

Explain this experimental result. What may be back of the failure of the tipless grasses to bend?

[Some students may suggest that only the tip of the plant can "perceive" light. Others may suggest that the act of cutting, rather than the absence of a tip, may have interfered with the plant's bending. Over-cautious students may go no further than to say that the absence of tip somehow prevented bending. These students should be urged to make a more positive contribution. You should point out that the aim of such experiments is not to find out what *absence* of the tip leads to, but what the tip contributed to the bending process *when it was present*. Absence, in short, may be a lead, a sign, evidence, of presence. If you wish, you may refer to Invitation 12, in which the absence of something in rice husks was seen as conceivable in two ways. The absence of the factor could be thought of as a cause of the disease; or the whole experimental result—absence of the factor resulting in certain symptoms—could be thought of as evidence about what the missing factor contributed.

[The word "perceive" as used here can provide a second way to emphasize the difference between the teleological and the functional. The word is generally used to convey the idea of meaningful sensation. Thus, we show that we have perceived when we say: "There is a light down there." On the other hand, we are said to sense when, for example, our eyelids close in response to a visual experience, before we are aware of the vision.

[This distinction can be made for your class, if you wish, by asking whether they think plants perceive light in the same way we do. The ensuing discussion can be brought to a head by pointing out that engineers often refer to a thermostat as a heat sensor, to a hygrometer as a moisture sensor, but do not refer to such devices as heat or moisture "perceivers."]

To the student: (d) The Darwins then took more seedlings and covered the tips of half of them with blackened paper. The other half were left uncovered. Both sets were exposed to light from one side. The uncovered seedlings bent toward the light, but the seedlings with covered tips did not bend.

In view of the possibilities uncovered by the previous experiment (cutting the tips), what was this experiment designed to do?

[You may need to remind the student that the previous experiment left two mutually exclusive possibilities—loss of some perception-like tip function and injury of the remainder of the plant due to the clipping. The second experiment and its results eliminate the latter. Once this is made clear, move on to the next item.]

To the student: (e) Remember that curvature occurs at the *base* of seedling. In view of the results of both experiments, what hypothesis can we now propose concerning the contribution of the tip of the seedling to bending?

[This question is designed to expand and define the notion of a plant tip "perceiving." Some students may say again that only the tip reacts to light. If so, encourage them to make constructive guesses about how the tip sensitivity results in base curvature. Many possibilities exist here, depending on how far your students are in their knowledge of biology. They may suggest that light may lead to production of some chemical in the tip which reaches and affects the base, or an electric signal between tip and base, or nerves, and so on. All such suggestions that fit the data should be treated as valid possibilities. The Darwins' suggestion was: ". . . when seedlings are freely exposed to lateral light, some influence is transmitted from the upper to the lower part causing the latter to bend."]

INVITATION 15

SUBJECT: *Neurohormones of the heart*

TOPIC: *Origin of scientific problems*

[This Invitation reviews previous Invitations by putting all the major ideas and processes we have exemplified so far into connection with one another in one coherent enquiry. The ideas and processes in question are the traditional constituents of simple enquiry:

> Derivation and statement of a problem
> Formulation of hypotheses
> Design and performance of experiments
> Interpretation of the forthcoming data

We also take pains to illustrate, through usage, the sense of some other useful terms; for example, *assumption, generalization,* and *inference.*

[The Invitation is not, however, only a review. There are certain significant additions to the idea of problem formulation—the point that problems arise from contradictions, inconsistencies, incoherencies, or blank places created by bringing our principles of enquiry and other assumptions into connection with already garnered knowledge. This renewed emphasis on the problem may be considered the major emphasis of the Invitation.]

To the student: (a) Over many years, many investigators concentrated their attention on the transmission of impulses from one part of the body to another through nerves.

Nerves, although small and simple in appearance, turned out to be exceedingly difficult to understand. This difficulty was so great that investigators realized they could make little headway in their study if they insisted on waiting until they were completely certain about one thing before going on to another. Rather, they decided that enquiry would proceed more satisfactorily if they took some conclusions for granted, even though the evidence was far from complete. This would enable them to move on to other problems, and the exploration of these other problems would then provide a check on the items that had been taken for granted.

186

It was as if a doctor had said to himself, "This person's symptoms suggest that he is suffering from disease *A*, though it is by no means certain that it is disease *A*. Nevertheless, let us assume for the moment that it is, and try a few doses of the drug which is known to cure disease *A*. If our diagnosis is right, the treatment will help the patient. If our diagnosis is mistaken, the treatment will not help but it will also not do harm. Meanwhile, it will give us a check on our tentative diagnosis."

One of the most important of the tentative generalizations about nerves which the investigators took for granted was: The impulses in all nerves are essentially the same. The impulse along one nerve may lead to increased activity of an organ. The impulse along another nerve may lead to decreased activity of that same organ. Nevertheless, the impulses along these nerves are much the same.

The *facts* in the statement are listed with the generalization:

FACTS	GENERALIZATION
There are two nerves carrying impulses to organ *A*. The activity of organ *A* is increased by an impulse from one of these nerves and decreased by an impulse from the other.	The nerve impulses in each of these two nerves are essentially the same.

Between these facts on the one hand and the generalization on the other, you will see what seems to be an inconsistency, something like a contradiction. You will also see that if we insist on keeping both the facts and the general statement we have a problem for research. The problem is to discover additional facts that will make the inconsistency disappear. Now try to state the problem for investigation suggested by the seeming inconsistency between our facts and our generalization.

[The reasoning that makes the problem clear may be summarized for the class after recognition of the main features of inconsistency has been achieved in discussion: "If the *nerve impulse* is the same in all nerves, then some factor other than the nerve impulse must account for different nerves having different effects on the same organ." In other words, "Assuming that differences in the nerve impulse in different nerves do not exist, then the opposite effects of two distinct nerves, both leading to the same organ, must be accounted for in another way."]

To the student: (b) You will notice that one of the two statements leading to the problem is an *assumption*—namely that no important differences in nerve impulses exist. Remember that such assumptions are commonly used in science to arrive at problems.

Assuming that all nerve impulses *are* the same, construct an hypothesis to account for the fact that some nerve impulses increase the activity of an organ, whereas other nerve impulses to the same organ decrease its activity.

> [The important hypothesis is that the two nerves enter the organ at different places. Hence impulses from each nerve may activate different processes within the organ.]

To the student: (c) It is possible for the investigator to make different assumptions; he has *alternative* ways for formulating a problem. Instead of proceeding on the assumption we have stated, that differences in nerve impulses do not exist, he may reject it, maintaining that differences do exist, although he has not found them. Then his problem changes from looking for differences of effects of impulses to looking for differences in impulses, or for some differences in the nerves.

In short, we have here another case where analysis of the overall situation leads us to see a second possibility, one that is opposite to the one brought out in part *(b)*. In this situation there is a nerve and there is the organ in which the nerve terminates. The difference we are seeking may lie in either of the two places, the nerve or the organ.

We are now in a position to order our assumptions, problems, and hypotheses. Once we do this, we shall be able to see the situation more clearly.

In view of the discussion, see how well you can work out appropriate statements of the problems and the hypotheses that accompany the two assumptions we have made.

ASSUMPTIONS ABOUT THE NERVES	THE PROBLEMS	THE HYPOTHESES
1. The nerves and the nerve impulses are the same for increase and decrease channels.	1. What factor accounts for the different nerves having different effects on the same organ, if both nerves and nerve impulses are the same?	1. Different processes within the organ cause it to react differently to identical nerve impulses.
2. The nerves or the the nerve impulses are different.	2. If the nerve impulses are different, what is the difference?	2. The organ reacts differently to different kinds of nerve impulses, or to something else differing between the two kinds of nerves.

[These parallels also indicate that what hypothesis we favor and feel is worth testing first is also determined in part by our view of the situation and our formulation of the problem.]

To the student: (d) The heart contains two different nerve pathways with different effects. There is an "accelerator" nerve, which was given this name because its impulses increase the heartbeat rate. There is also a "vagus" nerve which decreases heartbeat rate. Here, then, is a concrete case of the problem we are attacking.

Now note the following new fact about the heartbeat. As the heart as a whole contracts and relaxes, *all* its different muscle fibers contract and relax together. There may be some laggards and leaders among the fibers, but they are few.

In the light of this new fact, which of our two hypotheses seems to be the more likely?

[This new fact tends to disfavor the idea of different processes within the heart, and to favor the possibility of some difference between the nerves.]

To the student: (e) To test a similar hypothesis, an investigator named Otto Loewi performed the following experiment: He removed the hearts from two frogs and cut all the nerves from one heart (heart *B*). He left both the *vagus* and the *accelerator* nerves attached to the other heart (heart *A*). Both hearts continued to beat at the same rate. He then connected these two hearts by means of a small glass tube in such a way that a salt solution from the heart with attached nerves was pumped through the tube into the heart from which the nerves had been removed (heart *B*).

He found that when he stimulated the vagus nerve leading to heart *A*, the rate of the beat in heart *A* decreased, and soon thereafter the rate of beat in heart *B* also slowed. Similarly, when he stimulated the accelerator nerve leading to heart *A*, the rates of both hearts increased.

Now state in *detail* and as specifically as you can the hypothesis which this experiment was designed to test.

[Emphasis on the fact that the only connection between the two hearts is by way of the salt solution within the tube can lead the class to recognize that the investigator had in mind the possibility that some substances might be released in response to nerve impulses to heart *A* and that these substances might be carried to heart *B* by the salt solution.

[When this much has been made clear, a more complete statement of the specific hypothesis should be made by the teacher: "The vagus and accelerator nerves have different effects on the rate of heartbeat because each nerve releases a different substance within the heart muscle—the vagus substance slows down the contraction of all fibers, the accelerator substance speeds up the contraction of all fibers.]

To the student: (f) In the hypothesis just stated, point out the one major term which the experimental results do not throw any light on. State the interpretations the results do justify.

[Different substances mediate the different effects of vagus and accelerator impulses, but there is as yet no evidence that these substances originate from the nerves. Now read to the class these excerpts from an autobiographical sketch by the man who actually performed the investigations described here.]

THE IDEA OF CHEMICAL TRANSMISSION OF NERVOUS IMPULSE [1]

Now I have to turn to the best known of my scientific achievements, the establishment in 1921 of the chemical theory of the transmission of the nervous impulse. Until 1921 it was generally assumed that transmission was due to the

Assumption direct spreading of the electrical wave accompanying the propagated nervous impulse from the nerve terminal to the effector organ. Since the character of that potential is everywhere the same, such an assumption would not explain the

Known fact well known fact that the stimulation of certain nerves increases the function of one organ and decreases the function of an-

Possible problem other. A different mode of transmission had, therefore, to be considered.

As far back as 1903, I discussed with Walter M. Fletcher from Cambridge, England, then an associate in Marburg, the fact that certain drugs mimic the augmentary as well as

Suggestive analogy the inhibitory effects of the stimulation of sympathetic and/or parasympathetic nerves on their effector organs. During this discussion, the idea occurred to me that the terminals of those nerves might contain chemicals, that stimulation might liber-

Hypothesis ate them from the nerve terminals, and that these chemicals might in turn transmit the nervous impulse to their respective effector organs. At that time I did not see a way to prove the correctness of this hunch, and it entirely slipped my conscious memory until it emerged again in 1920.

The night before Easter Sunday of that year I awoke, turned on the light, and jotted down a few notes on a tiny slip of thin paper. Then I fell asleep again. It occurred to me at six o'clock in the morning that during the night I had

[1] From an autobiographic sketch by Otto Loewi, *Perspectives in biology and medicine,* Vol. IV, No. 1, autumn 1960.

written down something most important, but I was unable to decipher the scrawl. The next night, at three o'clock, the idea returned. It was the design of an experiment to determine whether or not the hypothesis of chemical transmission that I had uttered seventeen years ago was correct. I got up immediately, went to the laboratory, and performed a simple experiment on a frog heart according to the nocturnal design. I have to describe briefly this experiment since its results became the foundation of the theory of chemical transmission of the nervous impulse.

Design of
experiment

The hearts of two frogs were isolated, the first with its nerves, the second without. Both hearts were attached to Straub canulas filled with a little Ringer solution. The vagus nerve of the first heart was stimulated for a few minutes. Then the Ringer solution that had been in the first heart during the stimulation of the vagus was transferred to the second heart. It slowed and its beats diminished just as if its vagus had been stimulated. Similarly, when the accelerator nerve was stimulated and the Ringer from this period transferred, the second heart speeded up and its beats increased. These results unequivocally proved that the nerves do not influence the heart directly but liberate from their terminals specific chemical substances which, in their turn, cause the well known modification of the function of the heart characteristic of the stimulation of its nerves.

Results;
data

Interpretation

The story of this discovery shows that an idea may sleep for decades in the unconscious mind and then suddenly return. Further, it indicates that we should sometimes trust a sudden intuition without too much skepticism. If carefully considered in the daytime, I would undoubtedly have rejected the kind of experiment I performed. It would have seemed likely that any transmitting agent released by a nervous impulse would be in an amount just sufficient to influence the effector organ. It would seem improbable that an excess that could be detected would escape into the fluid which filled the heart. It was good fortune that at the moment of the hunch I did not think but acted immediately.

For many years this nocturnal emergence of the design of the crucial experiment to check the validity of a hypothesis uttered seventeen years before was a complete mystery. My interest in that problem was revived about five years ago by a discussion with the late Ernest Kris, a leading psychoanalyst. A short time later I had to write my bibliography, and glanced over all the papers published from my laboratory. I came across two studies made about two years before the arrival of the nocturnal design in which, also in search of a substance given off from the heart, I had applied the technique used in 1920. This experience, in my opinion, was an essential preparation for the idea of the finished design. In fact, the nocturnal concept represented a sudden association of the hypothesis of 1903 with the method tested not long before in other experiments. Most so-called "intuitive" discoveries are such associations suddenly made in the unconscious mind.

Many questions connected with and raised by the discovery of chemical transmission were studied in laboratories all over the world as well as in my laboratory. We found, for instance, that the effect of the *Vagusstoff* on the heart quickly fades because it is inactivated by an ester-splitting enzyme (cholinesterase). The *vagusstoff* was soon identified as acetylcholine. Also, it was proved that the activity of cholinesterase is prevented by the alkaloid physostigmine.

This was the first identification of the point of attack of an alkaloid and, to my knowledge, the first elucidation of the mechanism underlying all the effects of an alkaloid. Not until 1936 could I identify the transmitter liberated by stimulation of the accelerator nerve as epinephrine.

[During the reading, contrast the story as presented in this Invitation with the author's own anecdotal account. The following points should be stressed:

1. The problem was much as originally stated at the beginning of this exercise.

2. The hypothesis appears to have arisen in a free speculation about a possible analogy with certain drug effects.

3. The hypothesis was not submitted to any sort of test for seventeen years after its formulation; it occurred to the investigator long before any means of testing it could be found.

4. The crucial experiment used a technique which the investigator had already put to other uses.

5. The investigator *states* that he considers it unlikely that he would have thought the experiment sufficiently promising to be worth even a try had he considered the matter carefully!

6. Experiments to identify the neurosecretions followed, but epinephrine was not identified until 1936, thirty-three years after its existence was first suggested.]

INVITATION 16

SUBJECT: *Discovery of penicillin*

TOPIC: *Accident in enquiry*

[The discovery of pencillin has been so much popularized that this Invitation may fail to "invite." If you think that this may be the case, you will find an alternate in Invitation 16A.

[Invitation 11 made the point that the results of a planned experiment may exceed the plan—be unexpected. Invitation 15 begins the task of showing that problems are not objectively given by phenomena but are "seen" or "made" when the scientist juxtaposes already garnered knowledge with the principles and assumptions which underlie the garnered knowledge. In this Invitation we bring these two points together.

[Apparently minor unexpected results are obtained in an experiment. They *might* have been due to a mere slip in technique. But the investigator is alert enough and responsible enough to think twice before discarding the apparently minor and apparently technical slip in results. Instead, he pauses to consider the possibility that the unexpected result may be due to something new, something not included in existing knowledge or presently used principles and assumptions. The result of this concern of the scientist with something apparently trivial is the discovery of antibiotics!]

To the student: (a) A bacteriologist in St. Mary's Hospital, London, was working with a variety of strains of staphylococcus bacteria, trying to identify the one that was causing an outbreak of infections in the hospital. A number of culture plates containing colonies of bacteria were on the laboratory bench and were opened from time to time and examined. One plate was found to contain a contaminating mold.

The contaminating mold appeared as a white, fluffy mass growing near the center of the plates. Immediately surrounding the mold was a clear zone in which no bacteria grew. This zone was surrounded by a flourishing colony of the bacteria (as seen in Figure 1). What might account for the clear zone?

[The clear zone might have lacked some essential element in the nutrient medium due to incomplete mixing of the agar when originally prepared.

193

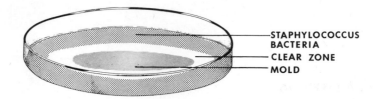

STAPHYLOCOCCUS
BACTERIA
CLEAR ZONE
MOLD

Figure 1

The staphylococcus might have started to grow in the middle of the dish
and the accumulation of waste products of the aging culture killed the
colony. A product of the mold may have diffused out and destroyed the
bacteria. A substance diffusing from the mold and another substance diffusing from the bacteria might have united in the midzone to form a product
inhibitory to both mold and bacteria.]

To the student: (b) The bacteriologist who found the culture plate
with the clear zone surrounding the mold growth soon discovered that
when he transferred and grew the mold in nutrient broth in which
bacteria grew well, the nutrient broth acquired the ability to destroy
several types of bacteria. What new problem thus developed from the
investigator's effort to *grow* bacteria?

[The problem that originally involved *maintenance* of the growth of
bacteria has become a problem involving *inhibition* of the growth of
bacteria.]

To the student: (c) What further lines of investigation would you suggest?

[A new line of investigation might be concerned with the chemical analysis
of the inhibitory product, the best way to make such a product, determination of the nature of the product—virus or nonliving molecule—or determination of the effect on the bacteria.]

To the student: (d) This Invitation is based on the discovery of the
antibiotic penicillin by Alexander Fleming in England, the pursuit of
the problem in Italy by Fleming's student, Florey, and the production
of penicillin by American scientists working for pharmaceutical com-
panies.

INVITATION 16A

SUBJECT: *Discovery of anaphylaxis*

TOPIC: *Accident in enquiry*

[Invitation 11 made the point that the results of a planned experiment may exceed the plan—be unexpected. Invitation 15 begins the task of showing that problems are not objectively given by phenomena but are "seen" or "made" when the scientist juxtaposes already garnered knowledge with the principles and assumptions which underlie it. In this Invitation we bring these two points together.

[Apparently unexpected results are obtained in an experiment. They *might* have been caused by a mere slip in technique. But the investigator is alert enough and responsible enough to think twice before discarding the apparently incorrect results. Instead, he pauses to consider the possibility that the unexpected result may be due to something new, something not included in existing knowledge or presently used principles and assumptions.]

To the student: (a) A physiologist was interested in studying the effects of poisons secreted by certain marine organisms. He decided to extract the poison and inject it into some dogs. Most of the dogs died, but a few survived and appeared to recover fully within a few weeks.

Subsequently, the investigator decided to repeat the experiment with the survivors of the first experiment. He again injected the dogs with an extract of the poison from the marine organisms. On the basis of his knowledge of how animals react to *infections,* the investigator expected either of two possible results. What would you say the possibilities were?

[If he assumed that the dogs recovered from the first injection without any permanent internal effect, he would predict that these dogs would react as before with the same severity of symptoms for about the same length of time. He might have assumed, however, on analogy with the effects of some infections that the recovered dogs had developed an immunity to the poison. In that case, the dogs would not become sick at all, or would display milder symptoms of poisoning for perhaps a shorter period than before.]

To the student: (b) Neither of these two possibilities occurred. Rather than either, the dogs reacted more severely than they had to the first injection. In fact, a very small dose produced such a violent response that the dogs met an early death.

Inasmuch as the experimenter had clearly expected either one of two results and instead got a third, unexpected result, what course of action would you suggest he follow?

[Clearly he should now repeat his entire first and second experiments to check the correctness of his techniques, records, and observations.]

To the student: (c) This is, in fact, what the experimenter did do. Although he did not, at first, believe that the dogs' second reaction had anything to do with the first dose of poison, he repeated his experiment a number of times. Each time the results were the same.

In view of the previous understanding of how animals recover from diseases, and in view of the results from the many experiments, what position would you say the investigator ought now to take?

[By now the investigator would have to acknowledge that he had encountered something new. He would have to modify a previously held belief, held firmly not only by himself, but by his colleagues and by the general public as well.]

To the student: (d) The dogs had become sensitized to the poison. This process of first being exposed to some foreign material, standing up well to the exposure, then responding more violently to subsequent exposures (even to smaller doses) is called anaphylaxis.

You know, perhaps, that some persons respond to injections of penicillin, or other drugs, in much the same fashion as the dogs did to the poison from the marine organisms. The same is true of exposure to poison ivy and to insect and snake venom.

[Anaphylaxis was originally discovered by Charles Richet. The essential parts of his discovery form the subject of this Invitation. Richet received the 1931 Nobel Prize in Physiology for his work.]

INVITATIONS TO ENQUIRY, GROUP II

THE CONCEPTION OF CAUSE IN BIOLOGICAL ENQUIRY:
Causal Factors, Multiple Causes, Time Sequences, Negative Causation, Feedback

INVITATION	SUBJECT	TOPIC
17	Thyroid action	Unit causes
18	Disease and treatment	Unit causes
19	Photosynthesis	Serial causation
20	Several examples of sequential analysis	Serial causation
21	Parathyroid action	Multiple causation
22	Control of pancreas	Diverse causation
23	Control of pancreas, continued	Diverse effects of diverse causes
24	Control of thyroid secretion	Inhibitory causes
25	Pituitary-gonad mechanism	Feedback mechanisms

Interim Summary 3, The Concept of Causal Lines

INVITATION 17

SUBJECT: *Thyroid action*

TOPIC: *Unit causes*

[Many of the Invitations up to this point have involved analysis of "cause" into "unit causes." In Invitation 3, for example, the overall condition of seed germination was analyzed into several "unit" factors—moisture, temperature, and so on—and a beginning was made at a one-at-a-time test of each factor.

[Nevertheless, the idea of cause has been used without explanation, and so far no attention has been given to the complications that arise in doing casual research in biology. With this Invitation we begin to emphasize a few of these complications.

[One of the easiest of complications to understand is the fact that we can always try to break big causes into little ones. A whole gland, such as the adrenal, for example, can be treated as a cause. We remove it and note the consequences of removal. Then we interpret these consequences as indications of the normal effects of the adrenal gland. There is nothing erroneous about such an experiment and interpretation, but it is certainly incomplete, for we can go on to divide the gland into parts, such as medulla and cortex, and redo our experiment using one such part at a time. Later, we may analyze still further. Perhaps we would proceed by distinguishing different kinds of cells in the cortex and trying to remove one kind of cell at a time—and so on down to molecules.

[Back of this progressive analysis of causes into finer and finer parts is the idea that somewhere we will arrive at irreducible unit causes, or casual *elements*. However, the possibility of such elemental causes does not mean that grosser, composite causes, such as whole organs or tissue components of organs, are "wrong" or useless. On the contrary, they may be very useful, not only in applied biology, as in medicine, but also in research.

[Hence, what we want to convey to students is not that one level of analysis is better or more "scientific" than another, but only that there are many levels.]

To the student: (a) There are two masses of tissue—often referred to as the thyroid glands—which lie in the throat on either side of the "windpipe." A physiologist named X was interested in finding out what these tissue masses did. He performed the usual experiment for this kind of research. He removed the glands from a number of animals of species A. For control, he performed similar surgery on another group of animals of the same species but left the thyroid tissue intact and in place. He then looked for the differences between the two groups of animals. He found that the following symptoms appeared in the animals without thyroid tissue:

1. Muscular tremors, followed by severe muscular cramping. These symptoms disappeared after about 20 days, but the following additional symptoms remained.

2. Lowered body temperature.

3. Thickening of the skin and the appearance of much soft, jellylike material throughout the body.

4. Great lethargy: Although the normal animals were playful and active, the animals without thyroid sat huddled in one place most of the time.

Scientist X then studied these symptoms closely to see if he could understand what might underlie them. He decided that all were signs that the animal lacked some sort of control of its ability to utilize food. He interpreted symptom 1 as indicating malnutrition of nerves and the other symptoms as indicating reduction of the ability to utilize food for energy in the body generally. He therefore tentatively interpreted his data to mean that thyroid tissue controlled ability to utilize food materials.

Another scientist, Y, then announced the results of what he considered the *same* experiment performed on *another* but similar species of animal. He found that only symptoms 2, 3, and 4 appeared. There was no sign of the muscular tremors and spasms found by scientist X.

Let us begin the job of trying to see what might account for this difference in result. In Invitation 13 we saw that it was helpful in trying to understand something, to break it up—in our minds—into a few major parts. For instance, in analyzing the problem raised when the same batch of DDT had different effects on two fly populations, we listed the following components of the problem situation: (1) something used (DDT); (2) the conditions under which the "something" was used; (3) the way in which the "something" was used; and (4) something on which the DDT was used.

Now back off and take a bird's-eye view of our present situation: two scientists performing a removal experiment. What are the major parts of the situation at which we might look closer to find a possible explanation for the difference in results? Try for no more than three major aspects.

> [In this particular case, an extremely convenient and sensible analysis of the situation follows from characterizing the situation thus: Somebody removed something from something. This gives us a division into three places to look for what we are seeking: the investigator (his technique); the tissue removed; the organism used as the experimental animal.]

To the student: (b) Now let us identify at least some possible sources of the differences in experimental results—one source in *each* of the three parts of the situation.

> [In the case of the experimenter, it is possible that one of them inflicted damage on other structures in the course of removing the thyroid. In the case of the tissue removed, one possibility is that the tissue masses are only superficially alike. There may be physiological differences in spite of the anatomical similarity. In the case of the experimental animals in general, they may react differently to the normal output of thyroid tissue, even though the tissue itself is much the same in the two animals.]

To the student: (c) Still a third scientist, Z, working on still a third species, noticed two very small bits of tissue, one on each thyroid mass, which were somewhat darker than the larger thyroid masses. He removed the thyroids from a few animals, but carefully left the small, dark bits of tissue in place. His operated animals then showed symptoms 2, 3, and 4 of our original list, but not symptom 1.

In view of the results obtained by all three experimenters, what tentative explanation might we now select that would account for the differences in results reported by scientists X and Y?

> [The obvious possibility is that X removed both large and small tissues, whereas Y removed only the large tissues.]

To the student: (d) Scientist X, on hearing about the results of Y and Z, made a careful study of tissues in the thyroid region of his experimental animal. He found no such darker bits. Scientist Y made a similar study and did find such darker tissues. However, in his species of experimental animal they were located well to one side of the thyroids and had escaped removal for this reason. Incidentally, scientist Y gave a name to these smaller and darker bits of tissue, a name to suggest their location. He called them parathyroids.

Now let us make two assumptions based on these results. Let us assume, first, that scientist X is correct in reporting no visible dark masses on the thyroid of his experimental animals. Second, let us assume that all three species of experimental animals are so closely related that all of them have the same major organs and that these organs do much the same things in each animals. Granting these assumptions and keeping all three experiments in mind, what might we advise scientist X to look for to explain his unusual results?

[Given the assumptions, it follows that the animals used by X have parathyroids; and given the diversity of position of these glands reported by Y and Z, it is quite likely that they are in still a third position in X's animals. We could account for X's failure to see them on re-examination by the possibility that they are *embedded within* the thyroid masses. Hence we might advise X to look for identifiably different tissue masses *within* the thyroid.]

To the student: (e) Let us suppose that these anatomical mysteries are all cleared up. All three scientists find the newly discovered small bits (called parathyroids) in their animals—though differing in location in each species.

There is still some unfinished business, however, for X still has four sets of symptoms which appeared from his surgery, while Y and Z have but three of them.

1. What experiment should X now perform?

2. What other experiment should Y and Z now perform?

3. If we have been right in our assumptions so far, what results should X discover? What results should Y and Z discover?

[X should now try removal of thyroid alone and should find symptoms 2, 3, 4, but not symptom 1. Y and Z (and X, too, for that matter) should try removal of parathyroids alone and should discover the nerve-muscle symptoms without symptoms 2, 3, and 4.]

To the student: (f) In closing, let us try to make a general rule to cover the experience of scientists X, Y, and Z. One point, to be sure, is that different species, though very similar, may differ sufficiently from one another to make for complications when we try to combine results of experiments performed on them.

There is another very important rule for us to derive by examining the work done by scientist X on the animal which had parathyroid tissue inconspicuously buried within the thyroid tissue.

You will recall that removal of the thyroids by scientist X resulted in removal of the parathyroids as well. What was thought to be an operation removing *one* kind of organ was really an operation removing *two* kinds of organs, and the symptoms noted were attributed partly to the removal of the thyroids and partly to removal of the parathyroids.

The rule for us to derive deals with cause and effect. The idea is that anything that happens (the effect) is brought about by something else (the cause). Everything would be so much simpler if there were always a one-to-one correspondence between causes and effects. The trouble is, however, that when we remove what we think is one possible cause and note the major effect, it may be that the "one cause" is, in fact, —————. And again, what appears to be "one effect" is really —————.

INVITATION 18

SUBJECT: *Disease and treatment*

TOPIC: *Unit causes*

[In this Invitation we return to the concluding point made in Invitation 17 in order to reinforce its learning. The situations used are much simpler ones.]

To the student: (a) Sometimes when we "'catch a cold" we have a runny nose, a sore throat, and a cough. At other times when we "catch a cold" we have a runny nose and a sore throat but no cough. With still other colds, we may have only a cough, or only a runny nose.

You read in the newspaper the following story: "Professor John Smith of our local university has been given a grant of $100,000 to find the cause of the common cold."

Suppose you met the reporter who wrote the story. What might you point out to him to help him do more accurate scientific reporting in the future?

[The point is, of course, that "*the* common cold" may be many different combinations of effects, and that therefore there may be several relatively independent causes at work: combinations of different viruses, combinations of a virus and a bacterium, and so on.]

To the student: (b) A drug chemist sets out to construct a drug that will prevent airsickness. After many trials he prepares a very complex chemical compound which he has reason to think will work.

A supply of the new compound is turned over to other investigators for public trial. They report as follows: "The new compound is effective. It acts to prevent airsickness in almost all cases. However, it also causes and unpleasantly dry mouth and considerable sleepiness."

When a vice-president new to the drug business receives this report, he sends the following memo to the chemist who had prepared the new compound. "Your new compound prevents airsickness but it also causes dry mouth and sleepiness. Therefore we have decided not to market this drug and herewith request you to start over, trying for an entirely different compound."

If the vice-president had known what you discovered at the end of Invitation 17, what different request might he have made to the chemist?

[He would have requested an effort to find what *part* of the compound was responsible for the unwanted effects and for an effort at a similar compound that lacked these parts.]

INVITATION 19

SUBJECT: *Photosynthesis*

TOPIC: *Serial causation*

[The second complication involved in causal analysis is very much like the one treated in Invitations 17 and 18. There we were concerned with reducing a composite causal factor to simpler components. Here, we are again concerned with analysis of composite causes. This time, however, we are concerned with the time dimension instead of the space dimension. That is, we wish to show students that a causal chain, such as *A* leads to *B* leads to *C* leads to *D*, can be analyzed more finely so as to read: *A'* leads to *A''* leads to *B'* leads to *B''*, and so on.

[Scientific enquiry usually proceeds to investigate causal chains by fiirst looking for a few large links. Then these links are subdivided again and again. In this analysis of *sequences* of causes into smaller and smaller events, there is the idea, as in the case of spatial analysis, that we will arrive at elemental events requiring no further analysis. In this analysis too, however, the possibility that smaller and more numerous events can be identified in a sequence does not mean that previous analyses were "wrong" or useless. On the contrary, to know the beginning, middle, and end of a sequence may well be all we need to know in many cases. In other cases, of course, we may find it very useful to know about further links between the beginning and the middle, the middle and the end.]

To the student: (a) Textbooks often say that the green leaf uses carbon dioxide and water to produce starch and oxygen. They also sometimes say that light is necessary for the process to go on. Suppose we want to test the statement that light is necessary for production of starch.

The experiment we need would be like a number of those you have seen or invented in earlier Invitations. We would begin with leaves kept in darkness, then expose them to light and see what difference we find in the two cases.

Suppose you performed the following experiment. You took two groups of coleus plants. Both groups were placed in the dark for 24 hours. You then placed one group in the light and left the second group in the dark. After another 24 hours you took several leaves from plants in each group. Using a reliable test for starch, you found that the leaves from the plants in the light contained starch while those in the dark contained none.

What is the reasonable conclusion from this evidence?

> [Probably students will say, "Light is necessary for the production of starch." This is a reasonable conclusion from the evidence at hand if we add, "in normal coleus leaves."]

To the student: (b) Your conclusion that light is necessary for starch production is sound as far as the evidence goes. But now add one more piece of evidence. Potato tubers grow underground where it is quite dark. Furthermore, starch, a compound insoluble in water, cannot be removed from cell to cell. Yet potato tubers are loaded with starch. Now what must we do about our previous conclusion?

> [A knowledgeable or imaginative student may note here that perhaps the explanation is that light is necessary for some precursor of starch, a precursor which is soluble and can move from cell to cell. This is an unlikely response at this point and also unnecessary as far as this exercise is concerned. It is enough to see that the two pieces of evidence seem to point in different directions and that more study is necessary in order to resolve the apparent contradiction.]

To the student: (c) Suppose that you took several leaves from coleus plants which have been in the dark for 24 hours. You placed the petioles of half of these leaves in water in a petri dish. The petioles of the other half were placed in a 5% sugar solution in a similar dish. Both sets of leaves were then covered with a light-tight box and left for 24 hours. You then tested leaves from each set for starch. You found that the leaves whose petioles were in the sugar solution contained starch; those in water did not. Now what can we conclude from this experiment alone?

> [It should not be long before a student will point out that coleus leaves can produce starch in darkness *if they have a source of sugar.*]

To the student: (d) You now have three bits of data: one from the first experiment, another about tubers, and, third, the data from the petiole experiment. What interpretation can you now make, taking account of all of them?

> [It appears that light is necessary for the production of sugar (or sugar phosphate) in the green leaf and that this in turn is necessary for the production of starch.]

To the student: (e) Assume that what we have found in coleus leaves is generally the case for photosynthesis in green plants. Then, were we wrong when we said at the beginning: "Light is necessary for the process by which green leaves produce starch"? Were we wrong, at the end, when we said: "Light is necessary for the production of sugar and sugar is necessary for the production of starch"?

> [Both statements are equally sound. Error would have arisen only if we had interpreted our earlier evidence to mean: "Light is necessary for the production of starch in green leaves." For then we would not so much have been making large links in a casual chain as ignoring the possibility of an intermediate link.]

To the student: (f) How can we understand that two such different statements can both be sound?. *Hint:* Suppose John tossed a ball to Paul and Paul tossed it to Peter. Would it be wrong of Peter to say that he got the ball from John? From Paul?

To the student: (g) Let us put our answer to *(f)* in a more general form. Suppose we have evidence that there is a chain of causes and effects as follows: (Read the symbol → as "causes.")

$$A \rightarrow C \rightarrow E$$

In the same way, diagram the possibility that corresponds to what we have seen about light → sugar → starch.

$$[A \rightarrow \mathbf{B} \rightarrow C \rightarrow \mathbf{D} \rightarrow E]$$

INVITATION 20

SUBJECT: *Several examples of sequential analysis*

TOPIC: *Serial causation*

[This Invitation is designed to reinforce the idea of finer analysis of causal sequences brought out in Invitation 19.]

To the student: (a) In Invitation 19 we saw that a sequence of cause and effect, such as light → starch, could be further broken down into finer and more numerous links, such as light → sugar → starch.

Let us practice the understanding of this idea by applying it in a number of different situations.

Suppose you heard a convicted criminal say, "The police put me in prison." Using your knowledge of social studies, analyze the big link between police and prison and find two or three finer links.

["The police collected evidence. The evidence convinced a grand jury. The grand jury sent me to trial before a court. The court convicted me. The conviction required the police to send me to prison."]

To the student: (b) Do the same thing to the following two-link chain: "The mainspring in my watch moves the hands."

[All we need here is to visualize and refer to the several different gear wheels which intervene between the mainspring and the hands.]

INVITATION 21

SUBJECT: *Parathyroid action*

TOPIC: *Multiple causation*

[We now go on to a third complication of causal analysis: that a given effect may be evoked by what appear to be several *different* causes. This is sometimes referred to as *multiple causation,* but the term is something of a misnomer. It suggests that several genuinely different things or events can cause precisely the same effect. This *could* be the case in nature. However, the concept of causation underlying much biological research into causes includes the idea that, ultimately, if several instances of an effect are identical, the cause will also be found to be the same.

[It follows from this concept that when we think we have a case of *different* causes leading to the same effect, we proceed to make finer analyses (either spatial or sequential) in order to find what is common to the apparently different causes.

[Thus, multiple causation is a complication of enquiry at two levels. First, it involves the simple possibility that an event may have "several causes." Second, it involves the more abstruse notion that several causes may really be one—though, of course, the common factor, the "true" cause may be located in several places or react in several different circumstances, or be involved in several different sequences.

[In this Invitation we deal with "multiple causation" in its simplest form: the idea of the same causal source lying in several different locations. We do not introduce the complication of an underlying common cause acting in different circumstances or through different sequences.]

To the student: (a) Now we go back to the problem raised in Invitation 17. There we witnessed the discovery by scientist Z of small bodies called parathyroid glands. He found two of them, one located on the top surface of each thyroid gland.

We also saw in Invitation 17 the discovery of the function of the parathyroids. Experimenter Z found that with their removal the animal suffered severely from some sort of nerve-muscle illness. First tremors developed, then severe muscular cramping and jerking. These severe symptoms usually resulted in death.

209

Soon after this research a young graduate student discovered a new complication. He removed the parathyroid from the top surface of each thyroid gland in 50 animals of the species used by scientist Z. He was very careful to remove *all* of the two parathyroids without removing any substantial amount of thyroid tissue.

He expected only to find a check on the results obtained by scientist Z, using a larger number of animals. Instead, however, he got very mixed results, as follows:

Thirty of his animals developed the tremors, cramps (tetanus), and jerks (convulsions) Z had reported. Further, these 30 animals died, again as Z had reported. Ten of the experimental animals, however, developed only mild symptoms. There were tremors, a few brief convulsions, and mild tetanus in the early days after the experimental surgery, but these symptoms slowly disappeared until the animals seemed to be normal in all respects and continued to live in health. The remaining 10 animals, much to the surprise of the experimenter, recovered from the surgery and showed none of the nerve-muscle symptoms whatever.

What might account for these mixed results? The investigator, being very modest and doubtful about his surgical technique, suspected that perhaps he had failed to remove all of the two bits of parathyroid tissue from the 20 animals that survived. He therefore asked another and more experienced investigator to check his work. The experienced investigator did so and found that the operations on these 20 animals had gone according to plan. There was no sign of parathyroid tissue on the tops of the thyroids in any of the 20 surviving animals.

The puzzled graduate student was most reluctant to challenge the results reported by scientist Z, despite the positive check on the student's surgical technique. So he went back to the scientific journals and reread the reports published by scientists X and Y as well as Z, as we saw them summarized in Invitation 17. From these results he suddenly had an idea that would explain his results and the difference between them and the results of scientist Z. He quickly made a new examination of the thyroid region of the animals that had not developed the full set of symptoms. Then he checked on those that had developed the symptoms and died of them. He found that his idea was correct.

What this young student did you can probably do too. Go back to Invitation 17. Reread part *(d)* and the first paragraph of part *(e)* where the observations of X and Y are reported. What possibility is foreshadowed there? What difference might there be among different

animals, even of the same species, which would result in different reactions to the experimental surgery performed by our graduate student?

[The possibility is, of course, that there may be more parathyroids than
the two located conspicuously on the outer surfaces of the thyroids. Many
individual rabbits have only these two. Others have an additional pair
located on the inner surface of the thyroid bodies. Still others may have
only one additional parathyroid body. Two of the four are usually sufficient
to maintain normal muscular function. We have, in addition, inserted the
not uncommon possibility that one such body may be only partially adequate
but that the one may increase in efficiency on removal of the conspicuous
two, or that the animal may in other ways adjust to the deficiency.]

To the student: (b) In Invitation 17 we made a general rule to cover
what we learned about causes there. It read: "When we remove what
we think is a *single* cause and note a *single* effect, we had better be
careful about drawing too firm a conclusion. For what we think is a
single cause may be a combination of several *different* causes. And
what we noted as a single effect may in turn be a combination of effects."

We could also have made a rule to cover what we learned about
cause in Invitation 19. It would read: "When we find that cause *A*
leads to *B* and *B* to *C,* we had better be careful about drawing too
firm a conclusion that these steps are all there are. For there may be
other steps between *A* and *B* and still others between *B* and *C.*"

Let us draw up still a third rule about cause to cover what we have
seen exemplified here in the number and location of parathyroid glands.
It would read: "The cause of a given effect may not have a single
location."

We could say this a little more quickly and simply if we keep in
mind exactly what we mean; we could say: "There may be more than
one cause for an effect." This way of saying it is a little misleading,
however. It carries with it the implication that *one kind* of effect may
result from causes which are of *different* kinds. Now this was not the
case for our parathyroid glands. There were several of them in different
places but the available data show that they were all of the *same* kind.
They were all made of parathyroid tissue and all manufactured the
same chemical material.

We have yet to find out whether very *different* kinds of things can
have the *same* effect. Later Invitaitons will go into this point.

[In Invitation 19 we led up to the general rule stated here but omitted
its formulation. This was intentional. It enables us to ask for recall and
new formulation and to state three general rules about cause in close and

reasonable juxtaposition. These devices help the learning process.

[The alternate formulations of the third rule, together with the remarks about them, also are designed to serve a teaching purpose. In a small way, they say something about *precision* of thought and formulation. (It also enables us to use the word "implication" properly.)]

INVITATION 22

SUBJECT: *Control of pancreas*

TOPIC: *Diverse causation*

[In Invitation 21 we raised the problem of multiple causation in its most primitive form: the case of mere repetition (in different locations) of the same causal factor, parathyroid tissue. Now we examine a case that goes well beyond mere repetition, a case of what seems to be two quite different sequences, which nevertheless terminate in the same end effect. As we develop the problem in Invitation 22, however, we shall see that the idea of entirely different causes having precisely the same effect is of questionable value as a principle for guiding scientific enquiry. It is a case of both nervous and humeral control of rate of secretion.]

To the student: (a) Let us review some necessary background information. Very important steps in the digestion of foods take place in the upper part of the small intestine, a region called the duodenum. Much of this digestion is carried out by enzymes that are poured into the duodenum by a large, separate, glandular organ called the pancreas. The pancreatic juice containing the enzymes reaches the duodenum from the pancreas by way of a short duct.

There appears to be a very neat and economical control of pancreatic secretion, because, although there is some secretion by this organ all the time, secretion in large amounts occurs only when food enters the duodenum from the stomach. It is the control of pancreatic secretion that will concern us in this Invitation. We want to know how it is carried out.

First, let us construct a number of hypothetical mechanisms by which pancreatic control might be carried out. Remember these facts:

1. The pancreas is a short distance away from the duodenum.
2. Pancreatic juice reaches the duodenum by way of a duct.
3. Both duodenum and pancreas are well supplied with blood vessels.
4. Nerves can be seen attached to both organs.

213

5. Copious secretion occurs only when food reaches the duodenum from the stomach.

Remember, too, what you have learned about control (integrating) mechanisms in general. Now let us construct at least five hypotheses, differing from one another in both large and small ways.

[The repetition and emphasis on detail are intended to lead the student as quickly as possible to the realization that two major possibilities exist: nervous control and some kind of chemical (perhaps humeral) control. This, in turn, is done so as to make the point that such merely categorical suggestions are not enough, that we want detailed hypotheses. Some possibilities are as follows:

1. Nervous stimulation by the physical presence of food in the duodenum.
2. Nervous stimulation by one or more chemical constituents of the entering material.
   (a) Juices from the stomach.
   (b) Food materials themselves entering from the stomach.
3. Chemical stimulation by way of the pancreatic duct. The subpossibilities are the same as those given for 2.
4. Chemical stimulation by way of the blood stream,
   (a) by one or more components of the material from the stomach;
   (b) by a hormone produced by the duodenal wall, in turn stimulated by one or more of the materials from the stomach—or by physical stimulation.

[Most of these possibilities should be brought out—for two reasons. First, except for possibility 3, all these possibilities will be involved in this and later Invitations. Second, the effort to construct these hypotheses in detail affords a freshened review of physiological mechanisms in general.

[It might be helpful at this point, from the point of view of motivation, to appoint one member of the class as "Recorder of Protocol." It would be the duty of this person to keep a record of all suggestions made and adopted, together with the names of those who make the proposals.]

To the student: (b) Now let us turn to the problem of planning tests of some of these hypotheses. Let us begin with the possibility of nervous control. Describe two different ways in which we might obtain some information on this point.

[The two clear alternatives are (1) to stimulate nerves leading to the pancreas, noting whether increased secretion results; (2) to cut all nerves leading to the pancreas and determine whether copious secretion on entry of food to the duodenum is missing.]

To the student: (c) Investigators paid little attention to the possibility of chemical control by way of the pancreatic duct. Mentally review the information we have given you and indicate why this possibility seems unlikely compared with the other two major possibilities.

[Since there is some secretion from pancreas to intestine at all times, it appears unlikely that there could simultaneously be a flow in the reverse direction which could occur quickly enough to provide the control we have identified.]

To the student: (d) An investigator began by stimulating nerves leading to the pancreas. He found that increased secretion occurred although no food had entered the intestine. What conclusion about pancreatic secretion appears to be called for?

[There is no doubt that there are nerves *to* the pancreas whose stimulation leads to increased secretion. As yet, however, we have no evidence that these nerves are stimulated by events in the intestine, much less any information as to what events may be the stimulus. There is always the possibility that the efferent innervation of the pancreas is stimulated by events remote from the intestine—the sight or smell of food, for example, or swallowing. Some of these reasons for *not* drawing a firm conclusion from the stimulation experiment should be gone into, but not at length, for that would distract from the main point of the Invitation—the existence of alternate routes of control.]

To the student: (e) To test their first tentative conclusion, investigators then proceeded to the second experimental test you suggested. They cut all the nerves leading to the pancreas. They found, much to their surprise, that this did *not* stop pancreatic control. When food entered the duodenum of these experimental animals, the pancreas proceeded to secrete copious amounts of fluid, apparently just as it did in intact animals.

How can we interpret the results of this experiment—*together with the results of nerve stimulation?*

[An easy evasion of this problem is to suspect that some nerve connections were not cut. If this possibility is raised, agree that it is an important matter but that it was thoroughly investigated and found not to be the case. The important alternative and the point of this Invitation is that there may be *two* modes of control. In addition to nervous control, there seems to be another.]

To the student: (f) To make very sure that they knew what was occurring, the investigators proceeded as follows. They cut the pancreatic nerves in a number of experimental animals. From other animals they took food that was in process of digestion in the stomach and introduced small quantities of this food directly into the duodenum of the animals whose pancreatic nerves had been cut. They found that copious secretion of pancreatic juice took place.

In view, then, of all these experimental results, what tentative conclusion about pancreatic control seems called for?

> [That there appear to be two quite different routes of control, one by nerves, one by some other route—presumably by way of the bloodstream.]

To the student: (g) You will recall that we have lately discovered certain complications in the investigation of biological causal factors. In Invitation 17 we saw that what seems in one context to be one cause may turn out to be a combination of several. In Invitation 19 we saw that a sequence of causes may be analyzed into finer and more numerous links. In Invitation 21 we saw that one factor may exist in several different places.

Now, in the case of control of pancreatic secretion we have what may be still another complication, for we appear to have evidence that there is both nervous and chemical control of this organ. What general rule about biological causal factors might we construct to cover this case?

> ["In living organisms, the same effect may be brought about by different causes." Or, "Different causes may bring about the same effect."]

To the student: (h) Be warned that the rule we have just made is based on only *one* investigation, and that one is incomplete. There are many loose ends. Let us not, then, be too sure that our rule tells the whole truth and nothing but the truth.

INVITATION 23

SUBJECT: *Control of pancreas, continued*

TOPIC: *Diverse effects of diverse causes*

[Invitation 22 gives us merely the following information: There is nervous stimulation of the pancreas; there is also another route, involving the presence of food in the duodenum. It is not yet clear how the presence of food in the *intestine* can stimulate the *pancreas*—although the fact that all nerves to the pancreas are cut leaves it highly probable that a hormone mechanism is involved. There is such a mechanism—the production of secretin in the intestinal wall in response to acid in the entering food material.

[If this were the whole story, there would be no doubt that the general rule formulated at the end of Invitation 22 is a sound one; we would have a case of two quite different "causes" of one and the same effect. This is by no means the whole story, however. It is found that the juice secreted by the pancreas in response to secretin is apparently quite different from the composition of the juice when other control routes operate. The secretin route seems to lead to a pancreatic fluid that is plentiful and alkaline but contains comparatively little of the three major enzymes produced by the pancreas. Enzyme secretion, on the other hand, seems to depend on the nervous route of control and perhaps on a hormone route different from secretin stimulation.

[In short, our rule about different causes having the same effect needs modification. We probably need to say something like this: There may be quite different routes (sequences of causes and effects) by which a biological effect is made to occur, but if the effect in each case is *identical,* we ought to suspect a common factor operating in the alternative routes. On the other hand, if the routes appear to be different in all respects, we ought to suspect that the effects are different too.

[In this Invitation we finish the story about secretin and then introduce the discovery of differences in the composition of pancreatic juice with different stimulations. This will enable us to introduce the idea of essential connection between cause and effect—that, if causes are different, we ought to suspect different effects too.]

To the student: (a) Let us recall where our story of Invitation 22 ended.

217

We found that there could be nervous stimulation of the pancreas. We also found that the *pancreas* was stimulated when digesting food entered the *duodenum*—even though all the nerves to the pancreas were cut.

Let us forget about nervous control for the time being and concentrate on the other situation—on the fact that with all nerves cut (and ruling out the possibility of material going from duodenum to pancreas by way of the pancreatic duct) the entry of food into the duodenum from the stomach is quickly followed by pancreatic secretion. State as precisely as you can one problem this situation suggests.

[How a stimulus acting in the *duodenum* is transmitted to the *pancreas*.]

To the student: (b) There is still another problem here. Remember that all we know so far about the stimulus to the duodenum is that it is in some way the material which enters from the stomach. Remember what a typical meal may contain and what happens in the stomach. What is the other problem?

[What specific substance or factor in the mixture from the stomach is the effective stimulus? We have partially digested foods of the three major kinds, plus HCl and stomach enzymes. Any one or some combination of these may be the stimulus.]

To the student: (c) The puzzle of pancreatic control interested a number of scientists. One investigator, aiming mainly at our second problem, introduced some food materials that had not gone through the stomach into the duodenum. He found *no* increased secretion of pancreatic juice.

What positive inference should we make from the data of this experiment concerning the original stimulus of pancreatic secretion?

[Since food that has passed through the stomach is a stimulus and food that has not passed through the stomach is not, we infer that a product of stomach digestion or one of the stomach secretions is the operating stimulus.]

To the student: (d) In view of the results of this exepriment, Professors Bayliss and Starling at the University of London tried introducing the various components of stomach material one at a time. They soon found that HCl from the stomach was alone sufficient to lead to pancreatic secretion. Knowing this, they proceeded immediately to attack the other problem we have formulated—how a stimulus that acts in the duodenum is transmitted, in the absence of nerves, to the pancreas.

First, they tried injecting HCl into the bloodstream and also into the pancreatic duct. No increased pancreatic secretion occurred. Noting, then, that acid in the intestine resulted in increased secretion but that the acid itself did not stimulate the pancreas, they arrived at a hypothesis in which they had great confidence. What was it?

[That an intermediate substance was produced by the intestine in response to acid—the intermediate being carried by the blood to the pancreas, where it acted as the stimulus. Of course, we have here a case where our second rule applies—there may be a chain of several intermediates—but this is unimportant at this point.]

To the student: (e) To test this hypothesis, a very pretty experiment was performed by investigators at Northwestern University in Chicago. A bit of intestinal tissue was transplanted to a place just under the skin. A bit of pancreatic tissue was transplanted nearby. When the transplants were nicely healed, a bit of dilute acid was placed in contact with the duodenal tissue. Very quickly the *pancreatic* transplant was seen to secrete.

Clearly, *something* had passed from the duodenal tissue to the pancreatic tissue which stimulated the pancreas to secrete. There were no nerve connections, and, from a previous experiment, it was clear that it could not be the acid. So it was concluded that it was a substance produced by the duodenal tissue.

As you know from other studies, there are chemical "messengers" in the body that are transmitted from place of origin to place of work by the bloodstream. Such "messengers" are called hormones. The hormone produced by the intestine that stimulated the pancreas was called secretin.

To the student: (f) Let us stop for breath and see where we are. We know that there is nervous stimulation of the pancreas. We also now know that there is stimulation by way of a hormone—secretin. The secretin arises in the duodenum in response to acid from stomach contents, passes by way of the blood to the pancreas, and there stimulates the pancreas to copious secretion.

So now the story seems to be complete and the rule we formulated tentatively in Invitation 22 seems to be justified. We do seem to have two causes—nervous stimulation and hormone stimulation—for the same effect. But—do we? Is the story really complete or have we ignored something important?

Let us check up in the way we found useful several times before. Let us break our overall situation into a few major pieces and note whether there is or is not a major piece which we have ignored.

First, try for a very general description of the situation we are and have been working on. Make it so general that we don't even mention duodenum, nerves, hormones, pancreas, or secretion. And remember that we want a description of the *problem situation* we are working on, not of what we have found out about it. Now try for such a description; a single sentence should be enough.

> [An analysis of the requisite generality would be as follows: Something stimulates something to do something.]

To the student: (g) Very well. There are three major parts of the situation:

1. A stimulus.
2. Something stimulated.
3. A response to the situation.

Now which of these is our "cause" and which is our "effect"? Have we paid attention to both?

> [The stimulus is our "cause." The response is our "effect." Very obviously, we have taken the sameness of our "effect" for granted.]

To the student: (h) Biologists now know that the pancreatic juice produced in response to secretin is mainly water and inorganic material. The concentration of digestive enzymes in it is very low. By contrast, the pancreatic juice produced after nerve stimulation contains a very high concentration of the enzymes which attack protein, carbohydrate, and fat food products.

As far as our general rule about causes and effects is concerned, how can we restate it so that it will reflect this knowledge? Here is a more appropriate version:

In living things we often find different causes of *similar* effects. If the causes are different, however, we ought to search for possible differences in the effect.

Conversely, if the effects do appear to be identical, we ought to look for a common factor operating in the alternate causes.

Notice that we do not state a law about nature here. We do not say that the effects *must* have like causes or that unlike causes must produce unlike effects. We say only that we ought to keep these possibilities in mind as guides to our research. Such ideas, assumed for the purpose of guiding research, we call *principles of enquiry*.

To the student: (i) You may be interested in some recent work on this subject. Professors Greengard, Grossman, and Ivy, working at Northwestern University, have found that pancreatic secretion is *adaptively* controlled. When there is a large amount of carbohydrate to be digested, there is a high concentration of the carbohydrate-digesting enzyme in the pancreatic juice. When there is a large amount of protein to be digested, there is an increase in the concentration of protein-digesting enzyme in the secreted pancreatic juice.

It is not yet at all certain just how this adaptive control of proportions of enzymes takes place. Using what you have learned from Invitations 22 and 23, construct a hypothetical mechanism that would accomplish this adaptive control.

> [Different nerve pathways stimulated by the different food materials; different hormones similarly triggered; or a combination of the two.
>
> [Note the deliberate open-endedness of this Invitation—the pointed conclusion on an inconclusive note.]

INVITATION 24

SUBJECT: *Control of thyroid secretion*

TOPIC: *Inhibitory causes*

[Our studies of causal factors so far have treated causes as always positive—leading to the *appearance* or the *increase* of something. We now illustrate the fact that a causal factor may be *negative,* lead to *inhibition* of something.]

To the student: (a) By now you probably know that the thyroid gland produces the hormone thyroxine, which increases the rate of oxidation of food materials in animal tissues. If a person's thyroid produces too little thyroxine, he develops a disease called *myxedema* or *hypo*thyroidism. If his thyroid is over-active, he develops another disease called *hyper*thyroidism. Most humans are not troubled by these diseases, indicating that there is some mechanism by which the secretion of thyroxine is kept at a fairly constant level.

A hint about the nature of this control mechanism came when experimenters found that removal of the pituitary body (located at the base of the brain) led to symptoms of *hypo*thyroidism. Later examination showed that the thyroids in most of these cases had degenerated.

What possibility is suggested by these results? By what experiment could we get further and *different* evidence on this point?

[That something from the pituitary is necessary for normal functioning of the thyroid. Further evidence would be obtained by injecting pituitary extract into animals whose pituitaries had been removed.]

To the student: (b) It was soon found that one substance produced by one part of the pituitary (the anterior lobe) when regularly injected into animals without pituitary glands would maintain the normal action of the thyroid gland. It was also found, however, that larger injections of this pituitary hormone led to symptoms of *hyper*thyrodism! What is our problem now, in view of the fact that we are looking for thyroid *control,* not merely thyroid stimulation?

222

[The pituitary may stimulate the thyroid, but how is the pituitary so controlled that it does not over-stimulate?]

To the student: (c) Note that the pituitary hormone must be injected regularly and repeatedly in order to maintain normal thyroid function in animals without pituitaries. What system for control of thyroid secretion by a pituitary hormone does this suggest?

[That the stimulating material from the pituitary is normally used up or destroyed at a rate which keeps *its* concentration at the level required to give stimulation but not over-stimulation of the thyroid.]

To the student: (d) The idea that pituitary substance just happens to be destroyed or used up at the appropriate rate is a hypothesis that depends too much on coincidence: It leaves unanswered the question of how the secretion of pituitary material is kept at just the right rate. Let us try to design another and a more complete hypothetical mechanism. Let us try for one that is as neat and economical as possible, a system of control which uses no parts besides those discussed so far: a stimulating substance from the pituitary, plus the thyroid gland and its secretion, thyroxine. In short, let us assign a role to thyroxine that in cooperation with the pituitary substance would give us a control system.

[The solution to the problem is a stimulation-inhibition cycle. Let the pituitary substance stimulate the thyroid. Let thyroxine be an *inhibitor* of the pituitary. Then, as thyroxine concentration rises in response to pituitary stimulation, this very rise will reduce secretion of the stimulant by the pituitary. Then, since the pituitary substance is used up or destroyed in a short time, the stimulation of the thyroid will fall off.

[In the remainder of the cycle the converse occurs. As thyroid secretion falls off and thyroxine is used up, the inhibition of the pituitary is relaxed, its increased secretion restimulates the thyroid, and so the cycle repeats.

[If the possibility of an inhibitory action does not occur to your students, you may choose to read them a stimulant to their thinking: "In a sense, the thermostats in our houses stimulate the furnaces to produce heat. What does the heat from the furnaces do to the thermostats?"

[The question put to the students as phrased in *(d)* does not call for the detailed tracing of the steps in control we have outlined. This must be done in discussion, by leading students from their first insights to other details of the process. The first step must be conceiving an inhibiting action of thyroxine. This *seems* to solve the problem. When the consequences of such a mechanism are traced out, however, it soon becomes clear that we must also assume the relatively rapid disappearance (because of use or instability) of both the stimulating hormone and of thyroxine. Discussion can close by telling students that experimental evidence indicates that this hypothetical mechanism does exist. Finally, the general principle exemplified by the Invitation should be pointed out: Causal factors may act to inhibit as well as to stimulate.]

INVITATION 25

SUBJECT: *Pituitary-gonad mechanism*

TOPIC: *Feedback mechanisms*

[This Invitation has two major points. First, it exemplifies the fact that experimental tests of hypotheses *cannot*, as a rule, verify them. As a rule we can demonstrate only that the hypothesis is not possible or that it *might* be the case. Only when *all* the possible alternative hypotheses are known can experimental test, by eliminating all but one alternative, verify that one. This situation is sometimes referred to as "the falsifiability, but not the verifiability of hypotheses." The point is that we are rarely in a position to say with certainty that we know all the possible alternatives.

[The second function of the Invitation is to add a last item to our understanding of cause-effect enquiry: the existence of complex *interactions* of causes.

[The Invitation is based on a real case—the sexual cycle in the mammalian female. However, it refers to organs and materials only by code letters, *A, B, C,* and so on. This is done for two reasons. First, it makes it possible to use the Invitation at any time, since background information is not required. Second, it permits us to put our emphasis on the general pattern of interaction, rather than on the specific case.

The real equivalents of the code symbols are as follows.
A = pituitary gland;
Hormone A = FSH and LH from the pituitary;
B = ovary;
Hormone B = estrogen and progesterone;
C = uterus;
C-1 = estrus uterine phase in rats (comparable to proliferation phase of endometrium in human) ;
C-2 = diestrus uterine phase in rats (comparable to secretory endometrial phase in human) .]

To the student: (a) Organ A, a ductless gland, is removed from a number of adult, female rats. By comparison with control animals, it was found that organ B (also a ductless gland) and organ C ceased functioning in the animals deprived of organ A. Clearly, A is a necessary factor in the functioning of organs B and C. Describe and diagram (using an arrow to mean "controls" or "stimulates") three quite different routes by which the control of B and C by A may take place.

[The three possibilities are (1) A controls B and also controls C, with no control relation between B and C. (2) A controls B, and B in turn controls C. (3) A controls C, and C in turn controls B. Diagrammatically,

(1)   (2)   (3)

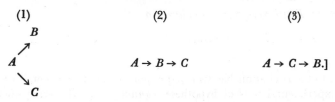

$A \to B \to C$ $A \to C \to B$.]

To the student: (b) Design an experiment to test hypothesis 2. State the "If . . . , then . . ." reasoning that leads from the hypothesis to the plan of experiment.

["*If* the control of C by A is mediated through organ B, *then,* removal of organ B, leaving organs A and C intact, should result in stoppage of organ C." Obviously, a student is unlikely to use the word, "mediated." It is entirely satisfactory if he says, "If A affects B, and B in turn affects C, then . . .". However, "mediated" is a useful word for talking about causal series. It could be introduced at this point.]

To the student: (c) The experiment of removing B, leaving A and C intact, was done. It was found that organ C ceased to function. But does this expected result, alone, verify ("prove") our hypothesis? Let us check this closely. Certainly, it is the result expected if the hypothesis is correct. But perhaps the same result would be expected on the basis of other hypotheses. Hence, we must check all the alternative hypotheses by "If . . . , then . . ." as carefully as possible. First, show by "If . . . , then . . ." reasoning that alternative 1 is eliminated.

[If A affects C directly and not through an effect on B, then removal of B should make no change in the control of C by A. However, we do find an effect of the removal, which is evidence that B is involved.]

To the student: (d) Note that what we were able to say here was that B is *involved in* the control of C. We cannot say that B is involved in the *particular* way indicated by the hypothesis we thought we were testing, unless we are sure that there is no alternative way in which the A and B control of C could occur. However, we have an additional piece of evidence—the result of our original experiment, in which we found that removal of A led to stoppage of both B and C. If we include these results, can we say that the control pathway is surely $A \to B \to C$? Again, let us look closely before committing ourselves.

What do we know from the first experiment?

[That B and C are both dependent on A.]

What do we know from the second experiment?

[That C is dependent on B.]

Putting the two together, what do we know about the control of C?

[That C is dependent on both A and B.]

There *is* a relation between A, B, and C other than the one we thought we were testing, which would give us these same results. What is it?

[B That is, both A and B are necessary for C, and A is
    ↗ ↘   necessary for B.]
   $A \to C$

To the student: (e) There is a very important lesson to be learned from this experience. It is as follows: When we find what we expect from an "If . . . , then . . ." chain of reasoning, we *do* thereby show that our hypothesis is *possible*. But we do *not* show that it *must* be the case.

We can "prove" a hypothesis only if we can list *all* possible alternatives and show by experiments that one is possible and the others impossible. The ever-present limitation here is that we are rarely, if ever, sure that we have conceived of all the possible alternatives.

To the student: (f) Our experiments so far have left us with two of our conceived alternatives, as follows:

$$B$$
$$\nearrow \searrow$$
$$A \to C \qquad A \to B \to C$$

Let us now devise an experiment that will eliminate one of these. Note, first, the one major difference between the two possibilities. What is it?

[The direct dependence of C on A in the first case but not in the second.]

To the student: (g) What experimental plan would check on whether there is a direct action of A on C? Remember that A and B are ductless glands. Use an "If . . . , then . . ." to show the logic of your plan.

[In the absence of organ A, inject an extract from B. If A and B are *both* necessary for the action of C, the injection of B alone will have no effect on C. If A is necessary only for B (and B for C) then the injection of B will have an effect on C.

[What follows now is a major simplification of the mammalian reproductive cycle as so far known. The simplification is designed to emphasize one further point—that our linear sequence A → B → C is found, on further study, to be a circular sequence.]

To the student: (h) In the normal female rat, organ C regularly goes through two stages, C-1 and C-2. For a few days it shows condition C-1. This is then transformed into C-2, which lasts a few days, then C-1 reappears, and so on.

Add to this new datum what we already know: that A and B are ductless glands, and that the control of C is from A by way of B.

With these facts in hand, the next problem faced by investigators was to discover how the sequence $A \rightarrow B$ could lead to the C-1 → C-2 → C-1 → C-2. It was easy to see how the sequence $A \rightarrow B$ could lead, say to a change in C from C-1 to C-2 (or vice versa), but it was very hard to conceive of any way in which the simple sequence $A \rightarrow B$ could lead to the continual cyclic change from C-1 to C-2 and back again to C-1.

First, the investigators pinned down which effect (C-1 or C-2) was brought about by B. They removed both A and B and found that in their absence organ C remained steadily in condition C-1. Clearly then (since there are only two alternatives in this case), condition C-2 is the effect of hormone B. As soon as this was clarified the investigators saw a possible way in which, despite the presence of organs A and B, there could be a cyclic return to condition C-1 followed by reappearance of condition C-2. What is this possibility? *Hint:* Remember what we saw in Invitation 22.

[The possibility is that above a certain concentration hormone B *inhibits* organ A while stimulating organ C. There is also, of course, the possibility that the inhibition is due to something from organ C when in condition C-2. It is also possible that it is organ B rather than organ A which is *inhibited*.

$$A \longrightarrow B \longrightarrow \left\{ \begin{array}{c} C\text{-}1 \\ \uparrow \quad \downarrow \\ C\text{-}2 \end{array} \right.$$

However, these are all variations on the same major theme.

[This conception of a stimulation-inhibition control of a cycle may be a difficult conception for students, in spite of the lead given in the previous Invitation. If it proves too difficult, the hypothesis should be given them—but stated simply, as in the first sentence of the last paragraph.]

To the student: (i) Let us be sure we understand this hypothetical mechanism by tracing out each step in it. Assume that organ *C* is in condition *C*-1. Now state, step by step, the sequence of events that will carry it through condition *C*-2 and back to *C*-1.

[Organ *A* secretes hormone *A*. Hormone *A* stimulates organ *B* to produce hormone *B*. Hormone *B* stimulates organ *C* to pass over to condition *C*-2. At some point during the latter process, hormone *B* reaches a level of concentration at which it inhibits organ *A*. Therefore organ *A* ceases to stimulate *B*. Hence, *B* ceases to secrete. Hence, organ *C* reverts to condition *C*-1. At the same time, as *C* reverts to *C*-1, the inhibition of *A* by *B* falls off and the cycle begins again.]

To the student: (j) There is one condition not so far stated, which is necessary for this sequence of events to occur. Imagine the situation in which *C* is in condition *C*-2 and organ A has just been inhibited. Why should not this stage of affairs continue indefinitely? After all, hormone *B* is present in concentrations sufficient to have achieved its two effects. The question is clearer put the other way around. What must happen in order to cut off both the inhibition of *A* and the maintenance of *C*-2 so that the cycle can move on?

[The answer is that hormone *B* must either be used up in the process of stimulating *C* and inhibiting *A,* or decomposed, or in some other way taken out of circulation.

[This Invitation can be used in close conjunction with the texts to make their treatment of the mammalian reproductive cycle more easily understood. If the Invitation *precedes* study of the chapter, the students will already have some understanding of the basic pattern of action and reaction involved. Then the added complications, which may be presented in your text, such as two hormones from the pituitary (organ *A*), two hormones from the ovary (organ *B*) and so on, will only be variations on the basic theme of cyclic control through stimulation, inhibition, and consumption.

[This basic theme, incidentally, like the basic pattern described in Invitation 24, is nowadays often referred to as a *negative feedback* mechanism. All that is meant by the idea of negative feedback is that the output of something controls the rate at which the something produces its output. In the present case, the feedback is the output of organ *B* "fed back" to organ *A* in such a way as eventually to prevent continuing production of the output of hormone *B*. In Invitation 24 there was a similar negative (inhibiting) feedback of thyroid material to the pituitary.]

[It is worth noting that the possibilities referred to by the phrase "negative feedback mechanism" were known to biologists a good many years before the name was invented and dramatized.]

INTERIM SUMMARY 3

The Concept of Causal Lines

The Invitations up to this point *seem* to present a reasonably complete conception of scientific research. We have dealt with the ideas of data and their interpretation, the planning of experiments, the construction of hypotheses, the formation of problems, and the search for causal factors.

In fact, however, these Invitations represent only one pattern of enquiry used in biological research, a pattern based on the conception of independent causal lines. In later Invitations, we deal with other patterns—the pattern based on the idea of *function* and the pattern based on the idea of *regulation*. Before embarking on these, and in order to begin to clarify the differences among these lines of biological enquiry, let us look back at the idea of separate causal lines.

The idea of separate causal lines is one of the simplest but also one of the most useful principles of enquiry used in science. Basically, the idea is that we treat any given whole *as if* it consists of many separate parts, each of which operates as a separate, *independent* entity. That is, we plan experiments and interpret data as if what each part *is* and *does*, it *is* and *does* regardless of the company it keeps and regardless of whether it keeps company or not. Each part is considered independent in the sense that we treat it as if its behavior and effect are always the same whether the other parts act or not.

The advantage of such a principle of enquiry is that it lets us study a given subject matter by studying each of its parts without worry about connections between the studied part and other parts. If we assume that each part operates as if it is independent, we can ignore the company it keeps or even remove it from the complicated company it keeps and study its properties and behavior when alone—which is much

231

simpler and less confusing than studying it when it is working as a part in the larger whole.

Conversely, the assumption of independent parts and separate causal lines lets us remove a part from the whole and treat the remainder of the whole as if none of the remaining parts had been affected by the removal of the one part in question. As so many of our previous Invitations have indicated, the assumption permits us to interpret differences between an animal without a part and a complete animal as if these differences were due entirely to the absence of the part removed.

The most useful biological version of the conception of separate and independent parts is that of a chain of antecedents and consequents—causes and effects. A leads to B leads to C. Meanwhile L leads to K leads to J, and X leads to Y leads to Z. But $A \rightarrow B \rightarrow C$ is independent of, unaffected by $L \rightarrow K \rightarrow J$, and each of these, in turn, is independent of $X \rightarrow Y \rightarrow Z$.

The usefulness and simplicity of research based on such a principle are seen with special clarity in simple versions of genetics and equally simple versions of physiology. In simple genetics based on the idea of separate parts and independent causal lines, we talk of gene A as the determiner, let us say, of brown eye color; of a the determiner of blue. In similar fashion we may speak of B as a determiner of height, C of curly hair, and so on, assuming as we speak that no one of these will modify the effect of the other. That is, we speak as if an animal whose genetic composition is $AABBCC$ will have precisely the same eye pigment as one whose composition is $AAbbcc$. As long as A is thought of as *the* gene for brown eye pigment, then whether it is in the company of B or b, C or c, there will be no influence on the effect A has on eye pigment.

In simple physiology, similarly, we have experiments and conclusions based on the assumption of separate causal lines. We speak of organ or tissue A as having effect a; we speak of organ B as having effect b, and so on. Here, as in the case of genetics, there is the underlying basic assumption that organ A will produce effect a whether organ B or C or D is present or absent. Again, each chain of causes and effects, each causal line goes its way independent of other causal lines.

It is this conception of independent causal lines, independent chains of antecedent-consequent connection, that leads to the pattern of enquiry illustrated in our Invitations so far. Let us take a problem in the area of physiology to exemplify the major steps in this pattern. The basic steps are as follows. First, we must identfy the members of a set of antecedents. Second, we must identify the mass of consequents made up of the many different consequents flowing separately from each antecedent. With our identified set of antecedents in hand, together with the mass of consequents, the next task is to perform an experiment which will identify the consequent of each given antecedent. Let us see how this works out in the case of physiology.

Let us take the cat as our case in point. We take all the separate organs identified by anatomical study of the cat as our set of identified antecedents. We then treat the whole physiological condition of the normal cat as made up of the many consequents of all these antecedents. Then our experimental identification or linking up of consequent to antecedent proceeds as follows. In a number of cats we remove *one* particular organ. When the animals have presumably recovered from the effects of the surgery, we scrutinize the physiology of these experimental animals in search of differences from intact animals.

When these differences are noted we then treat them as signs or evidence of the *absence* of that physiological consequent which normally follows from the removed organ when that organ is present. We then try to interpret these signs of absence so as to be able to say what the "presence" is; that is, what the organ does when it is present.

Thus, if we find that removal of the thyroid gland leads to lower temperature, placidity, obesity, lower respiratory rate, reduced consumption of oxygen, and so on, we interpret these as absences, as indicators of something called the "basal metabolism" which is normally maintained by the thyroid gland. This maintenance of basal metabolism is then the discovered consequent of thyroid activity taken as antecedent.

The vast quantity of biological knowledge that takes the form of antecedent-consequent connections bears ample witness to the strength and usefulness of the conception of separate causal lines. It has a central weakness, however, which is, in fact, the basic idea behind the causal line—the idea of its independence from every other line. Biologists know that "causes" or "antecedents" in living organisms may *not* be independent. They do not necessarily act the same way regardless of the company they keep. On the contrary, a living organism is so much an integrated whole that the opposite is almost the rule: Every "cause" is likely to have a different effect, depending on what other causes are operating at the same time.

This massive interaction of causes can arise from either of two different behaviors. On the one hand, if cause *A* is the cause we are interested in, we may think of cause *B* as affecting cause *A* in a certain way. Then, of course, the effect of cause *A* will differ depending on whether cause B is acting or not, since A in these two cases will, in effect, be two different causes. On the other hand, cause B may have an effect on the same end organ as does cause A. In that case, again, the apparent effect of cause A will differ depending on whether cause B is acting or not.

Physiology offers us an excellent example of this kind of interaction. If we remove islet tissue of the pancreas from a number of experimental animals, we find that absence of this tissue leads to violent upset in the utilization of sugar and fats. If in another group of experimental animals we removed the pituitary gland (anterior lobe), we find cessation of bone growth and of sexual maturation and function. If we now remove *both* organs from the *same* animals we discover consequences very different from the expectation based on the notion of separate causal lines. Animals with both these tissues removed do not exhibit merely the same failure of sugar and fat metabolism together with the same failure of bone growth and sexual maturation seen in the cases of separate removal. On the contrary, failure of regulation of sugar metabolism is much *less* severe when both organs are removed than it is when the islet tissue alone is removed. In short, the concept of an organism as consisting of separate and independent causal lines is a highly simplified model of the organism.

The fact that the concept is a highly simplified model does not mean, however, that it is one to be discarded. On the contrary, it has given us and will continue to give us a vast quantity of knowledge about the organism. The fact that the model is simpler than the reality means only that we need other models leading to other kinds of experiments, other kinds of data, and, therefore, other kinds of knowledge of the organism.

Specifically, we need a concept which brings back into the picture what the idea of separate causal lines so conspicuously leaves out—the organism as a whole. The idea of separate causal lines says, in effect, that there is no whole in any functional sense. Instead, the concept treats the organism simply as a collection, not an organization, of causes. Each causal line, taken separately, is object of investigation. The web formed by these lines is not investigated.

The concept of *function* is one of the principles of enquiry that does bring back the organism as a whole. This concept and the simple way it operates are illustrated in the Invitations of Group IV.

INVITATIONS TO ENQUIRY, GROUP III

QUANTITATIVE RELATIONS IN BIOLOGY: Linear Relations, Exponential Relations, Rate, Change of Rate, Units, and Constants

INVITATION	SUBJECT	TOPIC
26	Oxygen and carbon dioxide in respiration	Linear relation
27	Light intensity and photosynthesis	Linearity; limiting factors
28	Rate of fermentation	Change of rate; complicated variables
29	Growth regulation in leaf	Nonlinear polynomial of degree > 1
30	Light and auxin formation	Nonlinear polynomial of degree < 1
31	Population growth in bacteria	Exponential functions; exponent > 1

INVITATION 26

SUBJECT: *Oxygen and carbon dioxide in respiration*
TOPIC: *Linear relation*

To the student: (a) A scientist was interested in the respiratory processes in germinating seeds. He wanted to know if the processes were the same in different species. He began his study with experiments to determine the ratio between the amount of carbon dioxide produced and the amount of oxygen used. (This is called the respiratory ratio or quotient.) First, he placed germinating wheat grains in a suitable chamber equipped with inlet and outlet tubes through which air could be circulated. Using standard methods, he measured the amount of oxygen and carbon dioxide in the air that entered the chamber and in the air that left the chamber. From this information he could calculate the amount of oxygen used and the amount of carbon dioxide produced by the germinating seeds. Some of his data are presented in Table 2.

TABLE 2

	CO_2 Produced (ml)	O_2 Used (ml)
Series 1	10.0	9.8
	12.0	12.1
	25.4	25.2
Series 2	4.1	4.4
	25.4	22.1
	25.4	26.2
Series 3	9.1	9.8
	14.8	17.0
	15.5	15.0
	16.2	13.9
	17.0	17.5
	20.3	19.8
	21.5	21.0

Plot these data on simple graph paper with the volume of produced carbon dioxide as the abscissa (horizontal axis of the paper) and the volume of oxygen used as the ordinate (vertical axis). Do all the points fall on a straight line? What reasons can you give for their failure to do so?

[If this exercise is used after the students have done Invitation 5 or have had experience in conducting experiments, many of them will recognize the results of *experimental error*.]

To the student: (b) Draw the straight line that most nearly fits the points on your graph of these data. *(Teacher:* See Figure 2.)

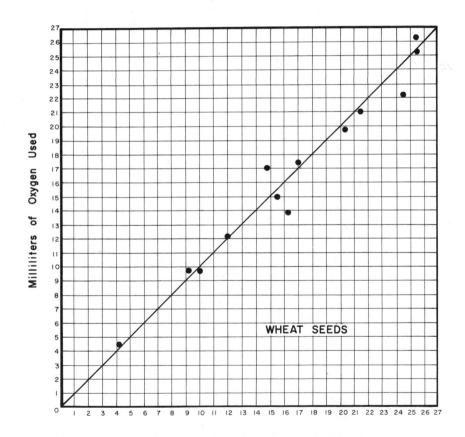

Milliliters of Carbon Dioxide Produced

Figure 2

You may remember from some of your work in mathematics that we can write an equation by which we can determine other points on this line. Usually, we denote the values on the horizontal axis (abscissa) as x, those on the vertical axis (ordinate) as y. In this case you can see that according to this straight line 1 ml of carbon dioxide was liberated for each milliliter of oxygen used. That is , x/y—the respiratory quotient—is 1. Let us factor $x/y = 1$ into $x = y$.

Because, when all the points are plotted, they lie on a straight line, or nearly so, we say that the relation here is "linear." Many relations are linear, however, in which y does not equal x, but equals some multiple or fraction of x.

Next, the scientist collected similar data from germinating castor bean seeds. The data for this series of experiments are presented in Table 3.

TABLE 3

CO$_2$ Produced (ml)	O$_2$ Used (ml)	CO$_2$ Produced (ml)	O$_2$ Used (ml)
5.1	7.5	5.0	6.8
4.0	5.0	11.0	15.0
13.0	17.5	15.0	22.2
9.0	13.3	19.0	27.0
2.5	4.0		

Plot these values on graph paper as you did the data for the earlier experiments. (*Teacher:* See Figure 4.) Again the points will not all fall on a straight line, but a straight line can be drawn which will come reasonably close to all the points.

Now compute the respiratory quotients for the pairs of data for the castor bean. What is the average of these ratios?

[The mean ratio is approximately 0.7.]

To the student: (c) We plotted the amount of carbon dioxide on the x axis and the amount of oxygen on the y axis. We now have an equation in the form of $x/y = $ _____. Replace this equation by the two equivalent equations $x = $ _____ and $y = $ _____.

[$x/y = 0.7$; $x = 0.7y$; $y = 1.43x$.]

To the student: (d) When we have a linear relation between x and y as in the data for these two series of experiments, we can say that $y = kx$, where k is a constant. In the data for germinating wheat grains,

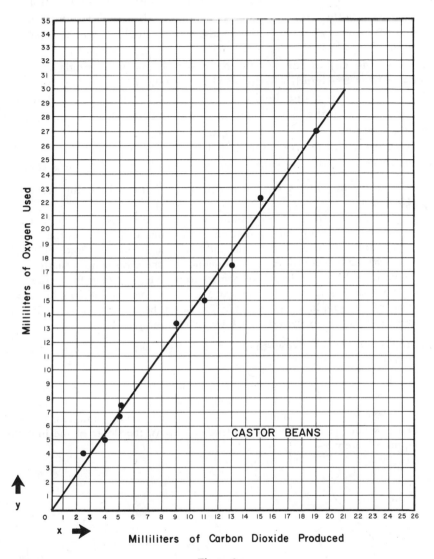

Figure 3

the value of k was 1; therefore we could write $y = x$ without mentioning the k. For the data for the germinating castor bean seeds, the value of k is 1.43. In this sort of relation, we say that k is a measure of the *slope* of the line, for it measures how steeply the line slopes above the x axis. Thus, in the case of wheat grains, for every unit extending the line from left to right along the x axis, there is a corresponding rise of 1 unit along the y axis. In the castor bean case, there is a rise of 1.43 units in the value of y for every unit that the x value of the line is extended from left to right.

Draw the curves for wheat and castor bean on a single sheet of paper. Do it free-hand, just to see the different slopes of the two lines. Now draw another line beginning in the lower left corner, but just halfway between the wheat line and the bottom axis. Would the k for this line's equation $y = kx$ be greater or less than 1?

Now draw still another line halfway between the wheat line and the vertical axis. Would the value k for this line be greater or less than 1?

[For the first line k will be less than 1; for the second line k will be greater than 1.

[For the students who are familiar with the basic trigonometric functions it may be useful to point out that the slope is often expressed as the tangent of the angle between the line and the base line (abscissa). Since in a right triangle the tangent of an angle is defined as the ratio of the length of the opposite side to the length of the adjacent side, the tangent in this case is y/x. See Figure 4. Calculation of this ratio from the data, or by inspection of the graph, shows that y/x for castor beans is approximately 1.43 which, of course, is the value of k we have found.]

To the student: (e) Now that the scientist had plotted his data, of what use are the plots?

[They show him that under the conditions of the experiments the ratio of the amount of oxygen used to the amount of carbon dioxide produced is linear for the two kinds of seeds. They also vividly show him that the ratio is not the same for the two species; that is, the slopes of the lines are quite different.]

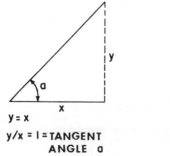

$y = x$

$y/x = 1 =$ TANGENT
ANGLE a

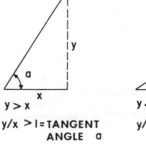

$y > x$

$y/x > 1 =$ TANGENT
ANGLE a

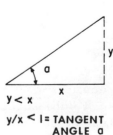

$y < x$

$y/x < 1 =$ TANGENT
ANGLE a

Figure 4

To the student: (f) We know that oxidation here consists of union of oxygen with some foodstuff. In view of this fact, what factor might account for the different respiratory quotients in the two species?

[The answer lies in the possibility that the foodstuff may differ. In fact, wheat seeds are composed very largely of starch, whereas castor beans contain a large percentage of fat. The significance of this difference for the respiratory quotient is developed next.]

To the student: (g) If you do not know, look up the chemical composition of starch and of fat. Assume that each kind of seed was able to oxidize its food material completely into carbon dioxide and water. Which kind of food, starch or fat, will require the most oxygen to produce a given amount of carbon dioxide?

[The formula for starch is often written $(C_6H_{10}O_5)_n \cdot H_2O$. In complete oxidation it will be converted to CO_2 and H_2O. The students should have little difficulty in seeing that no additional oxygen will be needed to oxidize the hydrogen. It is already present in the ratio of two hydrogens for each oxygen. But it will take two atoms of oxygen (one O_2 molecule) to oxidize each carbon atom to CO_2. Since this will produce one molecule of CO_2 for each O_2 used, and since a given number of molecules of one gas occupies the same amount of space as the same number of molecules of any other gas, the volume of O_2 used will be equal to the volume of CO_2 produced.

[The formula for fat, by contrast, contains much less oxygen in relation to the amount of carbon and hydrogen. We might write the formula of castor oil as $C_{57}H_{104}O_8$. Now the student should be able to see why so much more oxygen is required to oxidize the fat to carbon dioxide and water than was the case for starch.

$$C_{57}H_{104}O_8 + 79O_2 \rightarrow 57CO_2 + 52H_2O$$

The ratio CO_2/O_2 then becomes 57/79, which is approximately 0.7, when the gases are measured by volume.

[Measurements of this sort are often used to give us knowledge about the type of food being used in respiration.]

To the student: (h) Suppose that the scientist had determined the *weight* of the carbon dioxide produced and the *weight* of the oxygen used instead of determining the *volumes,* and that he obtained the data in Table 4 for the germinating wheat seeds.

Plot these data on graph paper, again using the amount of carbon dioxide on the x axis and the amount of oxygen on the y axis. You will notice that again you obtain a straight line. (See Figure 5.) Examination of the data or the graph will tell you, however, that y is not equal to x. If we divide the number of milligrams of oxygen used by the number of milligrams of carbon dioxide produced, (y/x), we obtain the values shown in the third column of the chart. The mean of these values is approximately 0.74. Hence we may write

$$y = 0.74x$$

TABLE 4

Carbon Dioxide Produced (mg) x	Oxygen Used (mg) y	Oxygen/Carbon Dioxide (mg) y/x
19.6	14.0	0.71
23.5	17.3	0.74
49.8	36.0	0.72
8.0	6.3	0.79
49.8	37.5	0.75
39.8	28.3	0.76
17.8	14.0	0.79
30.4	21.4	0.70
33.3	25.0	0.75
42.1	30.0	0.71

[It may be useful to point out to the students once more that this equation measures the *slope* of the line, and that it says that, for every increase of one unit of x, y increases 0.74 units.

[The slope of the line, however, depends on the units we have used. In (a) of this Invitation, the amounts of carbon dioxide and oxygen were measured in milliliters. In that case the slope was 1.

[If time permits, ask the students to plot the weight of oxygen used on the x axis, and the weight of carbon dioxide produced on the y axis. If this is done they may see that the slope is changed to one represented by the equation $y = 1.36x$.]

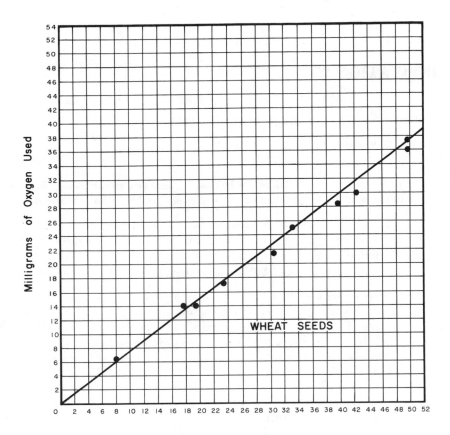

Milligrams of Carbon Dioxide Produced

Figure 5

INVITATION 27

SUBJECT: *Light intensity and photosynthesis*

TOPIC: *Linearity; limiting factors*

[This Invitation provides additional experience in handling data in a linear relation. It presents as a new concept the idea of a limiting factor.]

To the student: (a) An investigator wished to determine the quantitative relation between light intensity and the rate of photosynthesis. He recognized that *many* factors other than the amount of available light might also affect this rate. What other factors should he consider?

[Temperature and the availability of raw materials are the more obvious factors. Less obvious ones include the quality of light, that is, the wave length or distribution of wave lengths, the availability of minerals not needed directly as raw materials but nonetheless essential for the process, for example, iron, magnesium; and the kind of plant or plant tissue. It is not important that an exhaustive list be compiled. The point of the listing is to help students realize that photosynthesis is a complex process and that changes in any one of many factors might bring about changes in the photosynthetic rate.]

To the student: (b) If only the effect of light intensity on photosynthetic rate is to be studied, what must be done about the other factors?

[This query is intended merely to recall in a fresh context the need for experimental control.]

To the student: (c) What would be the result of controlling, that is, standardizing or holding constant, all the component factors in such a way that some of them, carbon dioxide for instance, are in short though constant supply?

[The class should recognize that if photosynthesis is to proceed at a rate influenced by light alone, *none* of the raw materials can be left in short supply. All the other conditions must permit the process to go on at any rate, however high, that the light might promote. Realization of this brings out the important fact that control, or simple standardization, of

244

conditions is not enough. It is further necessary that none of the component factors be limiting except the one being studied—in this case light intensity. The concept of a *limiting factor* may be a new one in the experience of the class. Bring out clearly the notice that in separating out one component factor for study it is necessary that none of the others be limiting. This is an example of what is often referred to as the *law of the minimum*.]

To the student: (d) An experiment was performed in which care was taken that all the known factors were carefully controlled and none of the raw materials was in short supply. As a measure of photosynthetic rate the investigator chose to take the amount of carbon dioxide absorbed by a given area of leaf surface per minute. He made measurements of carbon dioxide uptake per minute at eight different light intensities and repeated his series of measurements three times. His results are summarized in Table 5.

[The teacher might put this table on the blackboard for the students' inspection.]

What factors could account for the differences in the measured amounts of carbon dioxide absorbed at given light intensities among the three series of determinations?

[*Experimental error* is, of course, the appropriate blanket suggestion (see Invitation 5). But this should be broken down into a few specific kinds of possible experimental error: Inaccuracy in measuring carbon dioxide uptake, inadequate control of the amount of illumination, failure to bring under control other contributing factors. Be sure that the students realize that there is no way of distinguishing these factors by examining these data

TABLE 5

Light Intensity (ft-c)	Microliters of Carbon Dioxide Absorbed (per Minute)			
	Series 1	Series 2	Series 3	Mean
x	y_1	y_2	y_3	y
100	15	17	16	16
200	34	36	38	36
300	52	49	49	50
400	67	69	68	68
500	88	85	85	86
600	101	101	101	101
700	122	123	124	123
800	133	136	136	135

alone—the source of the observed variation cannot be pinpointed without further experimental work, if at all.

[It may be important to point out once more that the existence of variation in experimental measurements is inevitable. It is always desirable that an investigator reduce it as far as possible, but its presence does not mean a bad experiment.]

To the student: (e) Using the data given, make a graphic representation in the form shown in Figure 6.

Plot the points, but do not connect them. From inspection of this plot and the imaginary line that fits them, what can you say about the relation of light intensity to carbon dioxide uptake that you could not say at once from examination of the table? In other words, what understanding have you gained by making this plot?

[If Invitation 26 has been studied, the class will again recognize that the construction of a plot has made the nature of the relation clearer—the points fit closely to a straight line. (See Figure 7.)

[The analysis can be carried a further step to the end that the class more fully appreciates the nature of this order. Preliminary to the following question add a column, y/x, to the table of data, which would then look as shown in Table 6, omitting all values of y except the mean.]

To the student: (f) What is apparent from inspection of the series of values for y/x? What significance can you attribute to what you observed?

[The near equality, or close similarity, of the successive values of y/x should be immediately apparent. Although the value is not constant, there

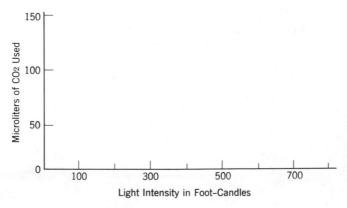

Figure 6. **CO_2 utilized per minute ($\times 10^3$).**

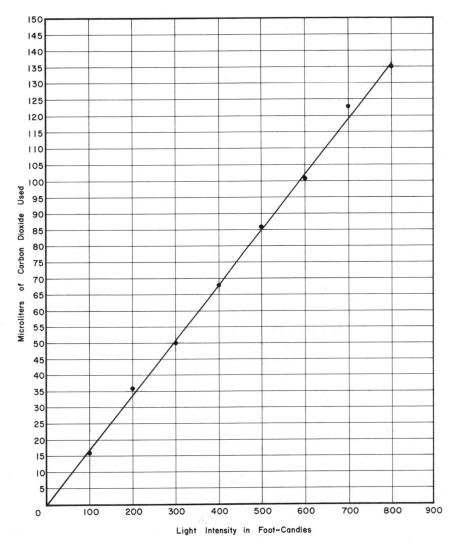

Figure 7

seems to be no system to the variation. The value seems to fluctuate randomly about an average value. Point out that the easiest way of getting an overall estimate is, in fact, to take a simple average and accept it as characterizing the relationship. The average value of y/x is 0.17.

[This relation should, at this point, be given a name. The ratio y/x is called a *direct proportionality*. *Emphasize* that recognition of this ratio

TABLE 6

Light Intensity (ft-c)	CO₂ Used, Average (ml)	Ratio
x	y	y/x
100	16	0.160
200	36	0.180
300	50	0.166
400	68	0.170
500	86	0.172
600	101	0.168
700	123	0.176
800	135	0.168

provides *another way* of expressing the relation between two variables that fall into a straight line when presented in the form of a graph.]

To the student: (g) How would you go about determining the amount of carbon doixide that would be utilized at light intensities for which no measurements were actually made? At 150 or 250 ft-c, for example.

[Some students will very likely suggest referring to the graph and estimating the value of y by inspection. Others will, it is hoped, recognize that the constancy of the ratio y/x is another key to answering such questions. Point out, if necessary, that since the ratio of observed x-y pairs is nearly constant the ratio of intermediate values is likely to be. This will suggest that the average of the y/x values be used as an algebraic formula, which, of course it is:

$$y/x = 0.17$$

Hence we may say

$$y = 0.17x$$

and compute. For $x = 150$ ft-c,

$$y = 25.5$$

and for $x = 250$ ft-c,

$$y = 42.5$$

[The class will have arrived at a mathematical formula which expresses the relation between the two variables. In summary, point out that the table does not bring out clearly the systematic relation between the variables. The graph does this but does not provide a concise means of expressing the relation, whereas the mathematical formula does.]

To the student: (h) The manager of a greenhouse asked the investigator to give him a reasonable estimate of photosynthetic rate for the species used at 250 ft-c. The investigator willingly did so. The manager then asked him for a prediction of the rate at 1200 ft-c and at 5 ft-c. The investigator refused to make such an estimate. He pointed out that his data did not permit such predictions. If he had not been so cautious, what values would he have given the greenhouse manager?

To the student: (i) Why would a careful investigator be more cautious about making predictions *beyond* the range of light intensities studied than about unmeasured light intensities *within* the range studied?

[For all we know from these data, the process may change radically at some point. A factor previously in generous supply may suddenly become a limiting factor. For example, the plant may not be able to bring carbon dioxide into the appropriate cell organ at a rate fast enough to utilize greater light intensities. Similarly, there may be a *threshold effect* at the lower extreme: Light may have to reach some minimum intensity to permit the process to go on at all.

[In general, *interpolation* of values between measured values is safer than *extrapolation* beyond the range of measured values. In a later Invitation we will encounter a case in which the rate is quite different for different parts of the range.]

INVITATION 28

SUBJECT: *Rate of fermentation*

TOPIC: *Change of rate; complicated variables*

[Where previous Invitations have dealt with a *rate*, this Invitation deals with a *change* in rate. The Invitation should not be used until the students well understand Invitation 26 or 27. It may be omitted entirely if you feel it is too difficult, because Invitation 29 does not depend on this one. However, this Invitation has the special value of showing how graphic and algebraic representations can give us clues to biological processes which are not shown by simple tabulation of data.]

To the student: (a) By 1900 it was known that yeast juice (an extract of ground-up yeast cells) can ferment simple sugars such as glucose and that one or more enzymes are the active agents. It was also known that the end products of fermentation are carbon dioxide and alcohol. These are produced in proportion to the amount of sugar in the fermenting mixture, and energy is released in the process. The fermentation process can be summarized as

$$C_6H_{12}O_6 \rightarrow 2C_2H_5OH + 2CO_2 + \text{energy}$$

$$\underset{\substack{\text{Simple} \\ \text{sugar}}}{} \quad \underset{\text{Alcohol}}{} \quad \underset{\substack{\text{Carbon} \\ \text{dioxide}}}{}$$

This equation shows that two molecules of carbon dioxide are produced for each molecule of sugar fermented. Therefore the amount of carbon dioxide produced can be used as an indicator of the amount of sugar fermented or of the amount of fermentation that has occurred.

A few years later it was found that addition of sodium phosphate (inorganic phosphate) increases the total amount of fermentation brought about by a given volume of yeast juice. But the reason why the addition of inorganic phosphate resulted in a greater amount of fermentation was not certain. For a time this result was explained as due to the alkalinity of the inorganic phosphate which kept the reacting mixture from getting too acid—since acidity slows down fermentation.

A scientist decided to study more closely the fermentation process in the presence of inorganic phosphate. He first determined the rate of fermentation without phosphate and later wtih phosphate added to the fermenting mixture. His first experiment was as follows.

Five grams of glucose (a large excess; remember Invitation 26) in a water solution were added to 25 ml of yeast juice. An apparatus was devised whereby the quantity of carbon dioxide produced during the fermentation could be measured, and observations were made at short intervals. The following table summarizes the data obtained from the observations.

Data for Curve A, Figure 8*

Time (min)	5	10	15	20	25	30	35	40	50	55	60	65	75	85	105
CO_2 (ml)	3	10	13	15	17	18	20	21	23	25	27	28	30	33	38

* Graphed as the hollow circles.

Plot these data, placing time on the abscissa and the volume of carbon dioxide on the ordinate.

[See curve *A* of Figure 8.]

To the student: (b) How can we describe this curve in words? Notice that the curve is not a simple line. Remember that as you follow it from left to right you are proceeding along the time axis. Keeping this point in mind, describe it as you would any change in time; that is, use the form, "First _____, then _____. This is followed by _____."

[First there is a short adjustment period. Then there is a later sharp rise. This is followed by a more gradual rise. The transition between the latter two portions is sharp.]

To the student: (c) Now "translate" your description of the curve into a description of the process it represents. What does the first part of the curve say about the rate of production of carbon dioxide as compared to the second part of the curve? What about the process might account for this difference?

[The initial sharp rise indicates that carbon dioxide is being produced at a fairly rapid rate. The second major portion of the curve has a more gradual slope, indicating that carbon dioxide is being produced at a lower rate. There is a sharp transition between the two parts of the curve. This suggests that different factors are involved in the process at different stages.]

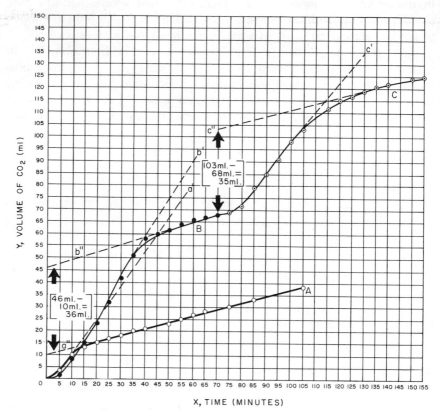

X, TIME (MINUTES)

Figure 8

To the student: (d) This curve is complicated—more complicated than the simple straight lines we have worked with before—and the algebraic formula for a complicated curve is hard to derive. Suppose we wanted to represent with reasonable, if not perfect, accuracy the data represented by this curve.

Suppose we also wanted to provide *simple* equations that would give at least an approximate representation of the data contained in our complicated curve. How might we think of this curve so that such simple, though only approximate, equations could be derived?

[The curve could be thought of as three separate curves, each one a stragiht line.

[See dotted lines *a'* and *a''* in the figure. The transition phase can also be approximated by a straight line.]

To the student: (e) Now let us make a rate calculation for each of our straight-line approximations to the original curve.

[For *a'*, suggest that figures be taken from the *extended* portion of the line so that at least a 20-minute time span can be used in the calculation. If the volumes of carbon dioxide at 45 minutes and 25 minutes are used, we find that the slope of *a'* represents a rate of 1.40 ml/min. The calculation is:

$$\frac{60 \text{ ml} - 32 \text{ ml}}{45 \text{ min} - 25 \text{ min}} = \frac{28 \text{ ml}}{20 \text{ min}} = 1.40 \text{ ml } CO_2/\text{min}$$

[For *a''*, figures can be taken from a part of the line that corresponds to the actual curve. If the volume of carbon dioxide at 75 minutes is used, we find that slope *a''* represents a rate of 0.25 ml/min. The calculation is:

$$\frac{30 \text{ ml} - 25 \text{ ml}}{75 \text{ min} - 55 \text{ min}} = \frac{5 \text{ ml}}{20 \text{ min}} = 0.25 \text{ ml/min}]$$

To the student: (f) What we did for line *a'* looks like the act of *extrapolation* we warned against in Invitation 27. Can you show that what we did for *a'* does not involve the dangerous assumptions involved in extrapolation to the extent that they were involved in Invitation 27?

[The slope of the line has been determined by actual data, and the "extrapolation" has been done only to obtain numbers convenient for calculation. We do not assume that the first phase of the process goes on to the later times.]

To the student: (g) In a second experiment the scientist added inorganic phosphate and proceeded as follows. To 5 g of glucose in a water solution and 25 ml of yeast juice (the same amounts used in the first experiment) he added a known quantity of phosphate. By means of other chemical methods he could calculate that this quantity of phosphate was "equivalent" to 35.9 ml of carbon dioxide. (By "equivalent" he meant there was a 1-to-1 ratio between the number of atoms of phosphorus added and the number of molecules of carbon dioxide released.) Using the same arrangement as before and making observations every few minutes, he obtained the data summarized in the following table. Plot these data and let us see what their curve looks like.

Data for Curve B, Figure 8*

Time (min)	5	10	15	20	25	30	35	40	45	50	55	60	65	70
CO_2 (ml)	2	9	15	23	32	42	51	58	60	62	64	66	67	68

* Graphed as black circles.

[Suggest that they plot the data on the same sheet and that they use the same coordinates as for curve A, so that data from the first and second experiments can be readily compared. See curve B of Figure 8.]

To the student: (*h*) How does the curve plotted from the data of the second experiment (curve B) compare with curve A?

[Points of comparison that may be mentioned are that (1) B is longer than A; (2) they have the same general shape (steep slope followed by a more gradual slope); (3) the steep portion of B persists for a longer period of time; (4) the later slope of B and the later slope of A seem to indicate that a steady and similar rate of fermentation is eventually established in both cases.]

To the student: (*i*) After 70 minutes, when a steady rate appeared to have been established, the scientist added a second quantity of inorganic phosphate to the same fermenting mixture. The second quantity of phosphate was equal to the first amount added. He again observed the carbon dioxide production every few minutes and obtained the results shown in the accompanying table. How should we plot these data?

*Data for Curve C, Figure 8**

Time (min)	75	80	85	90	95	100	105	115	120	125	130	135	140	150	155
CO_2 (ml)	69	73	79	85	91	98	103	112	115	117	119	121	122	124	125

* Graphed as circles with dots in the center.

[Some students may not immediately see that these data are a continuation of the data from the first part of the second experiment. Of course, if the data are plotted on the same graph it will become apparent; but in terms of understanding the procedure of the experiment it should be clear before graphing. When the plotting has been completed, the curve should correspond to curve C of Figure 8.]

To the student: (*j*) How does curve C compare with the other two curves?

[Again C is like curves B and A in that it has the same general shape of a steep slope followed by a more gradual slope.]

To the student: (*k*) Now let us simplify curves B and C and make rate calculations as we did for curve A.

[The straight lines representing curve B are b' and b''; those representing curve C are c' and c''. See Figure 8. These lines can then be represented by a rate calculation. In each case data can be taken from a portion of the line corresponding to the actual curves. The following calculations can be made:

for b': $\dfrac{51 \text{ ml} - 0 \text{ ml}}{35 \text{ min} - 0 \text{ min}} = \dfrac{51 \text{ ml}}{35 \text{ min}} = 1.47 \text{ ml/min}$

for b'': $\dfrac{69 \text{ ml} - 60 \text{ ml}}{75 \text{ min} - 45 \text{ min}} = \dfrac{9 \text{ ml}}{30 \text{ min}} = 0.30 \text{ ml/min}$

for c': $\dfrac{103 \text{ ml} - 73 \text{ ml}}{105 \text{ min} - 80 \text{ min}} = \dfrac{30 \text{ ml}}{25 \text{ min}} = 1.20 \text{ ml/min}$

for c'': $\dfrac{124 \text{ ml} - 117 \text{ ml}}{150 \text{ min} - 125 \text{ min}} = \dfrac{7 \text{ ml}}{25 \text{ min}} = 0.28 \text{ ml/min}$

The points used in the calculations were selected because they are near intersecting lines and thus can be read with greater accuracy.]

To the student: (*l*) Now examine the figures for all six rates. See if you find similarities and differences which enable you to classify the six rates in a way that might tell us something more about the fermentation process.

[It will be noted that three of the rates are of the same order of magnitude and the three other rates are of another order of magnitude. Thus we can separate the rates into two groups:

$$a' = 1.40 \quad \text{and} \quad a'' = 0.25$$
$$b' = 1.47 \quad\quad\quad b'' = 0.30$$
$$c' = 1.20 \quad\quad\quad c'' = 0.28$$

Notice that the members of each group correspond to similar portions of the three curves.]

To the student: (*m*) What further simplifying way of processing the data does this suggest?

[It suggests that we might average the rates in each group. We find that the average for a', b', and c' (the steep portion of the curves) is 1.35 ml/min. The average for a'', b'', and c'' (the more gradual portion of the curves) is 0.28 ml/min.]

To the student: (*n*) Now use these calculated rate values to produce equations for the two parts of our one derived curve.

[If we designate time as x and volume of carbon dioxide as y, then the two general formulas for the two portions of this curve are, for the steep portions, $x = 1.35y$; for the more gradual portions, $x = 0.28y$.]

To the student: (o) It was pointed out earlier that the two portions of the curves suggested that there may be two stages in the fermentation process. Has anything been done during the experiment and the processing of our data that might suggest an explanation for the difference between the two?

[If the students need a hint, it can be pointed out that a small amount of inorganic phosphate is normally present in yeast juice. This accounts for the *brief* steep slope of curve A. The steady rate represented by a'' can be interpreted as due to yeast juice alone, this small amount of phosphate having been exhausted. If they did not see it immediately, the hint should make it clear that the greater steepness of the slopes representing rapid rates of fermentation is probably due, therefore, to the presence of phosphates, and that the slower rate of fermentation is due to yeast juice alone.]

To the student: (p) Is there any additional way of analyzing the data which might strengthen the conclusion that the rapid rate of fermentation is due to the presence of phosphate?

[This question may be more difficult for the students to answer. If it is, point out that, if a'' and b'' are extended, they intersect the axis of zero time at points corresponding to 10 ml and 46 ml of carbon dioxide, respectively. Since the steady rate represented by a'' has been interpreted as being the fermentation rate due to yeast juice alone, then 10 ml can be considered the amount of carbon dioxide due to juice alone. If this is subtracted from 46 ml, we find that 36 ml of the total volume of carbon dioxide can be attributed to the added inorganic phosphate. If we extend c'' to the 70-minute time axis (the time at which the second amount of phosphate was added), we can determine the extra amount of carbon dioxide evolved in the second period. We find that 103 ml − 68 ml = 35 ml.]

To the student: (q) How do these figures compare with the calculated amount of carbon dioxide equivalent to the added inorganic phosphate?

[They correspond quite well, since the calculated amount was 35.9 ml of carbon dioxide.]

To the student: (r) What conclusion may be drawn?

[The addition of inorganic phosphate to a fermenting mixture of yeast juice and glucose causes the production of an equivalent amount of carbon dioxide.]

To the student: (s) Does this suggest a way in which the phosphate is involved in the process of fermentation?

[Since the amount of carbon dioxide produced is an indicator of the amount of glucose used in fermentation, and since carbon dioxide production is found to be proportional to the available phosphate, we infer that a definite chemical reaction occurs in which sugar and phosphate are involved. It may be added that this has been confirmed by further experiments in which it was found that at the end of the rapid phase of fermentation nearly all of the phosphate has been converted to *organic* phosphate.]

INVITATION 29

SUBJECT: *Growth regulation in leaf*

TOPIC: *Nonlinear polynomial of degree* > 1

[This Invitation should not be used until the students have a good understanding of the basic linear relations as presented in Invitations 26 and 27.]

To the student: (a) A scientist had noticed that some of the plant growth regulators such as 2,4-D (also used as a weed killer) will inhibit the growth of the trifoliate leaves of bean plants when it is applied to them. The investigator wished to test the effects of a new growth regulator. He grew a number of bean plants in the greenhouse until the first trifoliate leaves had just started to expand. He made up several concentrations of the new growth regulator, with the lowest containing 20 μg/ml. He then put an amount of the most dilute solution which would deposit 0.1 μg of the growth regulator onto the trifoliate leaf of each of four plants. He repeated this with each concentration of solution, using a different group of plants for each. He applied a similar amount of water to the leaves of another group as a control. In order to obtain the weight of the trifoliate leaves at the time of the application of the growth regulator, he cut off the leaves of still another group of plants and weighed them. He determined the average weight of the leaves and assumed that this represented the average weight of the leaves in each of the other groups at this time.

After ten days he cut off the leaf of each treated plant and weighed them separately.

From these data he calculated the amount of inhibition or decrease in growth which was caused by the growth regulator.

He obtained the data shown in Table 7.

Now plot these data on graph paper using the amount of growth regulator employed per plant as the abscissa and the inhibition of growth as the ordinate. What is the shape of the line that connects the points of the graph? How does this compare with the graphs you

258

TABLE 7

Growth Regulator Applied to Each Plant (micrograms)	Inhibition (grams)
None	0.00
0.1	0.01
0.2	0.04
0.3	0.12
0.4	0.15
0.6	0.37
1.0	0.90
1.5	2.35

made for the data in Invitations 26 and 27? Do you know what a curve with this shape indicates about the algebraic relation between the ordinate variable and the abscissa variable?

[Some students may recognize this as a logarithmic relation. (See Figure 9.) It is likely that it may *not* be recognized by any member of your class. This is handled in *(b)*. The data have been selected so that each time the concentration of the growth regulator is doubled, the inhibition is increased by *four* times, that is, the amount of inhibition increases with the square of the concentration. This is actually a sharper increase than would be obtained for most weed killers.]

To the student: (b) Prepare a table listing the amount of inhibition for concentrations of the growth regulator of 0.1, 0.2, 0.4, 0.8, 1.6 μg per plant. Some of these data are available from the table given you; the rest can be read from the graph. What relation can you now see between the figures in one column and those in the same rows in the other?

[It may be necessary to help some of the students see that while the amount of growth regulator applied was being doubled in each step, the amount of inhibition was increasing by approximately four times. Ask the students to try other positions on the curve, such as the points for 0.3, 0.6, and 1.2 μg of growth regulator per plant.]

To the student: (c) Now if we wish to express the relation of the concentration of growth regulator to the inhibition of growth in terms of an algebraic formula we can do so. You have already noticed that when x goes from 1 to 2, y goes from 1 to 4. When x goes from 1 to 4, y goes to about 16. What is the arithmetic relation for each of these pairs?

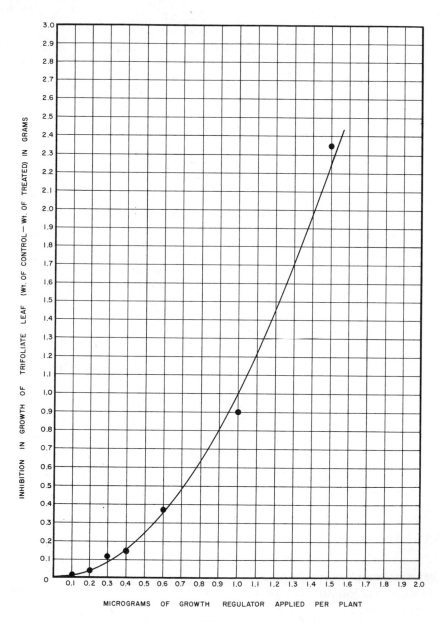

Figure 9

[A number and its square.]

To the student: (d) Using the relation we have identified, complete the equation $y =$ _____.

[$y = x^2$]

To the student: (e) Now you can plot any number of points on the curve by using the formula. Which gives the smoothest curve, the experimental data or the calculated values? Which values are most nearly correct?

[The student should be able to see that we cannot answer this question. We obtained the formula from the experimental data, therefore it cannot be more reliable than the data.

[It could be that our calculated values would be very near the values we would obtain if we used a larger number of plants for each amount of the growth regulator.]

INVITATION 30

SUBJECT: *Light and auxin formation*

TOPIC: *Nonlinear polynomial of degree < 1*

[This Invitation should be used only if the students already understand the linear relations presented earlier. Probably it should follow Invitation 29, where $y = x^2$.]

To the student: (a) A scientist had noticed that in weak light the amount of auxin produced in plants apparently increased with an increase in the light intensity to which they were exposed. In an attempt to gather more information on this relation, he performed the following experiment. He selected a number of uniform tobacco plants and placed all of them in a dark room for several days. He then divided the plants into five groups with ten plants in each. Each group was then placed under a light source of different intensity. After two weeks he determined the amount of auxin in each plant by a standard method. His average values were as listed in Table 8.

Plot these data with the light intensity as the abscissa and the amount of auxin as the ordinate. Compare this with the curve obtained

TABLE 8

Light Intensity (ft-c)	Average Amount of Auxin Found per Plant in Relative Units
5	2.2
10	3.1
20	4.5
40	6.4
80	9.2
160	12.0
400	20.5
600	23.9

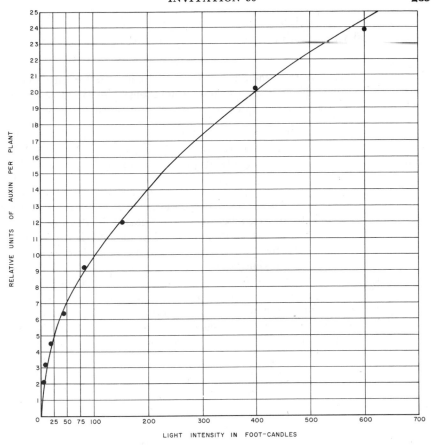

Figure 10

in Invitation 29. What does the shape of this new curve indicate about the relation between the amount of auxin found and the light intensity to which the plants were exposed?

[Again it is probable that only the students with a good mathematics background will recognize this curve as an exponential relation. In this case the exponent is less than 1 (actually $\frac{1}{2}$) so that the curve bends in the opposite direction to that noted in Invitation 29. It may be easier to explain this as a root, since most of the students will not be aware of the use of fractional exponents. To make it simpler for the student to see the relation, the data provided were "invented" so that the curve will be that for $y = \sqrt{x}$. The actual relation between light intensity and auxin yield has been shown to follow a curve similar to this, but its formula is more complicated. (See Figure 10.)]

To the student: (b) Examine the data in the table. What increase in light intensity is required in order to result in doubling the auxin yield?

[Most of the students should now be able to see that for each fourfold increase in light intensity there is only a doubling of the yield. Some will see that this is similar but opposite to the relation noted in Invitation 29.]

To the student: (c) Now we can express this relation in the form of an equation similar to that of Invitation 29. In this case, when x is increased fourfold, y is approximately doubled. Thus y increases as the square root of x. We may write $y = \sqrt{x}$.

[The students may wish to know some of the reasons for the sort of relations illustrated by this exercise. It seems probable in this case that we are dealing with at least two processes. The first of these is the formation of auxin and second is the destruction of auxin. In many, but not all cases, the *production* of auxin seems to increase with low levels of light intensity. There is also much good evidence that the *destruction* of auxin is hastened by light. The optimum light intensity for the production of auxin is perhaps lower than the optimum for auxin destruction. In the first part of the curve we see mostly the effect of light on auxin production, while in the latter part we see chiefly the effect on destruction.]

INVITATION 31

SUBJECT: *Population growth in bacteria*

TOPIC: *Exponential functions; exponent* > 1

To the student: (a) A microbiologist wanted to know how rapidly a certain kind of bacterium multiplied. He half-filled several large flasks with a culture medium he knew was favorable for the growth of this bacterium. When these flasks were plugged, sterilized, and cooled, he inoculated each one with approximately the same number of bacteria. The flasks were placed on a shaking machine so that the cells would be evenly distributed throughout the medium.

Using sterile pipettes, the investigator then removed measured quantities of the medium from each flask and returned the flasks to an incubator to keep the temperature constant. The number of bacteria in each sample was then estimated by standard counting methods. The sampling and counting were repeated at 20-minute intervals. The investigator obtained the data listed in Table 9.

Plot the means of these data using time as the x axis (abscissa). What is the shape of the line that connects the points on this graph? How does this compare with the graph you made for the data in

TABLE 9

Time After Inoculation (min)	Number of Bacteria per Milliliter of Medium				
	Flask 1	Flask 2	Flask 3	Flask 4	Mean
20	18	24	17	21	20
40	35	42	43	37	39
60	80	80	84	88	84
80	168	162	142	153	156
100	306	308	322	324	315
120	660	664	632	645	650
140	1200	1220	1260	1240	1230

Invitation 29? What does a curve with this shape indicate concerning the relation between time and the number of bacterial cells?

[This is a logarithmic (exponential) relation, in which the curve bends upward more and more sharply as x increases. By contrast, the curve of Invitation 29 has a steadily increasing rate. The difference is that the exponent that generates the curve in Invitation 29 remains the same exponent (2) throughout, whereas in this curve the exponent increases. It is possible that this curve will not be recognized by anyone in your class. It exemplifies an ideal growth curve, for cells dividing by fission (one cell producing two—which can be thought of as original and "daughter") at approximately 20-minute intervals. This kind of curve can readily be obtained for short periods. Later, such factors as depletion of the medium, formation of inhibiting substances, and so on, slow down the rate of fission.] (See Figure 12.)

To the student: (b) Note that the number of cells approximately doubles every 20 minutes. We can call this time the "generation" time for this bacterium under these conditions. So if we divide each of the values in the first column of our data chart by 20 minutes, we will have the number of generations. Let us call this number g. Note also that if the number of cells at the end of the first generation was twenty, it must have been ten at the beginning of the experiment, because each cell would have become two cells at the end of the first generation time.

Let us call this value of 10 the *concentration* at time $t = 0$. To make the numbers easier to work with, let us use the number of cells per 0.1 ml, rather than the number per 1 ml. To keep this count from being confused with that recorded on the graph in Figure 11, let us call it count c. Let us interpret the original data to mean that the number of cells exactly doubled each generation. Now we can tabulate these values (Table 10).

TABLE 10

Number of Generations (g)	Number of Cells (c) (per 0.1 ml)
1	2
2	4
3	8
4	16
5	32
6	64
7	128

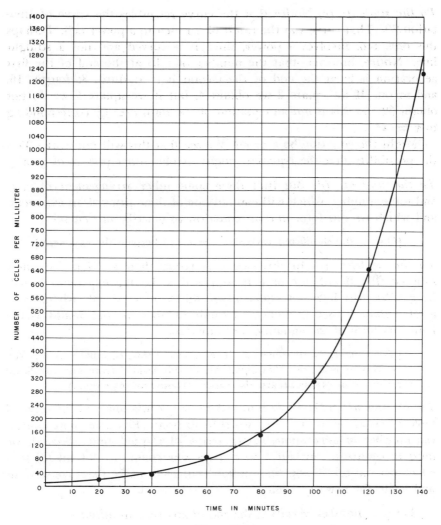

Figure 11

Find an equation that will show the relation between g and c.

[The equation is $c = 2^g$.

[Your students will probably need help. Guidance along the lines sketched may enable many of them to see how to write the equation. First, ask them to glance from each g to its corresponding c and try to detect the one relation that holds for each pair of numbers. Often this sort of direction of their attention is all that is necessary to promote "insight." If this procedure fails, proceed to part (c).]

To the student: (c) Notice that the curve resembles the curve of Invitation 29. That curve was the result of x raised to a power (x^2). Perhaps this one, too, involves a power, since it is curved and not a straight line. Notice, however, that the number of cells at the end of the first generation is 2; at the end of the second it is 4; third, 8; fourth, 16, and so on. If we square 2 we obtain 4, but if we square 5 we do not obtain 32. So this is *not* an equation of the form $c = g^2$, corresponding to $y = x^2$.

Might the equation be $g = c^2$? We see that this will not work. Test the possibility that the equation might involve other powers of g or c.

To the student: (d) But isn't there some other arrangement of c, g, and some number, which will contain one of these three as an exponent, and which we can form into an equation to describe our curve?

> [One of the significant variables c or g can become the exponent.]

To the student: (e) In this case c is larger than g. Hence we may suppose that if either c or g is supposed to be the exponent, it will be g. We will have an equation of the form $c = m^g$, where m is greater than 1. See if there is a small number m that would satisfy both this form of the equation and our data.

> [The number is, of course, 2. The equation thus becomes $c = 2^g$. You may wish to point out that equations of this form, having one significant variable as an exponent, are also called exponential equations.]

To the student: (f) If we wish to make our formula apply to the original table of data, we must make two substitutions. First, we must express the number of generations (g) in terms of elapsed time. Second, we must introduce a constant k to convert our count per 0.1 ml (c) to a count per 1.0 ml. Using the symbols indicated, make these conversions and write the new equation.

Let $y =$ number of cells at the end of any time in minutes
   $x =$ number of minutes elapsed
   $t =$ number of minutes per generation (20 in this case)
   $k =$ the constant to convert 0.1 into 1.0 ($k = 10$)

[$y = 2^{x/t} \cdot k$. Substituting our values, $y = 2^{x/20} \cdot 10$.]

INVITATIONS TO ENQUIRY, GROUP IV

THE CONCEPT OF FUNCTION: Evidences for Inferring Function, the Doubtfulness of Functional Inferences, Argument from Design versus Argument from Adaptation

INVITATION 32

[We concluded Interim Summary 3 by pointing out that the concept of causal lines has no place for the organism as a whole. Instead, the concept treats the organism simply as a collection of such causal lines, not as an organization of them. Each causal line, taken separately, is the object of investigation. The web formed by these lines is not investigated. The conception of function is one of the principles of enquiry which brings the web, the whole organism, back into the picture.

[This Invitation introduces the student to the idea of *function*. This concept involves much more than the idea of causal factor. It involves the assumption that a given part (organ, tissue, and so on) encountered in an adult organism is likely to be so well suited to the role it plays in the life of the whole organism that this role can be inferred with some confidence from observable characteristics of the part (its structure, action, and so on). As we shall indicate later, this assumption, like others in scientific research, is a *working* assumption only. We do not assume that organs are invariably perfectly adapted to their functions. We do assume that most or many of the organs in a living organism are so well adapted (because of the process of evolution) that we proceed farther in studying an organ by assuming that it is adapted to its function than by assuming that it is not.]

To the student: (a) Which of the various muscle masses of the human body would you say is the strongest?

[Students are most likely to suggest the thigh muscles, or the biceps, on the grounds that they are the largest single muscle in the body. If not, suggest the thigh muscle yourself, and defend your suggestion on grounds of size.]

To the student: (b) We decided that the thigh muscle was probably the strongest of our body muscles, using *size* as our reason for choosing it. Hence size seems to be the datum on which we base this decision. But why size, rather than color or shape? Behind our choice of size as the proper criterion, are there not data of another sort, from common

270

experience, that suggest to us that larger muscles are likely to be stronger muscles?

[In considering this question students should be shown that their recognition and acceptance of this criterion of muscle strength is derived from associations from common experience: A drop-kick sends a football farther than a forward pass, a weight lifter has bulkier musculature than a pianist, and so on.]

To the student: (c) Now a new point using no information beyond common experience. What can you say happens to a *muscle* when it contracts?

[The question here is *not* what a muscle does to other parts of the body, but what the muscle *itself* does—its change of shape in a certain way—becoming shorter, thicker, firmer by contraction. Have the students feel their arm muscles as they lift or grasp.]

To the student: (d) To the fact that the motion of a muscle is as you have found it to be, add two further facts: Many muscles are attached to some other parts of the body, and many such muscles are spindle-shaped, long, narrow, and tapering. From these data alone, what do you think muscles do?

[The motion, attachment, and shape taken together suggest that muscles in general move one or all of the other parts of the body to which they may be attached. Such inferences about function are only probable. But so are practically all inferences in science. In *(e)* and later queries, we shall make a point of the doubtful character of functional inference.]

To the student: (e) In many automobile engines there are revolving levers on the top, each of which slaps a rod at each revolution. This slapping has three results: It makes a noise, it rubs the rod against the sides of the hole in which the rod moves, and it opens a valve in one of the engine's cylinders. From these consequences of being slapped, what do you think these levers *do?*

[From the evidence as given, there is no way to decide whether the essential action is to make a noise, wear away the rod or its sleeve, or open the valve. The student who knows engines will immediately point to opening the valve as what the lever "does," or "is for." He can then be asked how he knows. (His answer will indicate that many kinds of evidence other than the visible motions and changes of the parts in question are required to make a reasonably defensive inference about function or "doing.")

[It is also possible that the student answer to this question will be premised on *design*, on what the planner of the engine intended the tappet and valve stem to do ("is for") . If so, point out that we do not know the intentions of the designer of living things, that we infer functions from other evidence. This point is unavoidable at some point in the honest discussion of the concept of function in biology.]

To the student: (f) In view of what we have noted about our inference concerning automobile tappets and valve stems, what must we now say about our inference concerning the "doing" of muscle?

[That it is questionable, to say the least.]

To the student: (g) Under a microscope we can see that a muscle consists of a bundle of very tightly packed cells that are spindle-shaped— long, narrow, and tapering toward the ends. These cells may be as long as the whole muscle.

We also find that any one of these cells, when separated from the muscle, shortens when stimulated. When you add these data about muscle *cells* to our observation of the motion and attachment of the whole muscle, what happens to your confidence in your original inference about what a muscle "does"?

[For most biologists confidence would increase, for we now have determined that our inference is not only consistent with the behavior of the gross part but also with the behavior of its finer structure.]

To the student: (h) Suppose you had also noted that as the muscle cell contracted a fluid was squeezed out. In that case, what would be your view of your original inference?

[We would have to fall back on the same state of doubt as in the case of three observed consequences of the motion of a valve stem. And we would need to seek still further evidence in order to decide whether the fluid exudation or the contraction itself was the essential action of muscle.]

To the student: (i) Remember that we now know that muscle cells are spindle-shaped and contract in the direction of their longest dimension. Add to this knowledge the following: The cells of a muscle usually lie in such a way that their longest dimension coincides with the long dimension of the whole muscle. Would this be expected, desirable, or necessary if the main function of muscular contraction were to squeeze out a fluid? Does your answer to this question increase or decrease your confidence in your original inference?

[No, and with this answer, we now have several kinds of data with which our inference is consistent. Hence confidence might well increase. These varieties of data are named next.]

To the student: (j) Here we have listed most of the data we have used so far. On the right we have listed by number some general names for such *kinds* of data. Match these two lists by writing in the blank space before each specific piece of evidence the general name that best fits it.

[2] Muscles contract

[3] Many muscles are attached to other parts of the body

[1] Attached muscles are usually spindle-shaped; in such cases, they are attached at both ends

[4] Muscle cells are long and spindle-shaped

[5] Muscle cells contract

[6] Most muscle cells lie with their long dimension in the same direction as the long dimension of the whole muscle

1. The overall shape and appearance of an organ or part
2. The observable change or motion of that part
3. The relation of that part to other parts

4. The shape and appearance of that organ's *components*
5. The observable change or motion of the *components*
6. The relation of the *components* to one another

To the student: (k) Notice that our inference about what a muscle "does" has been tested against six different kinds of evidence. But notice especially that we began with common knowledge: that we run and jump, kick and cling with a strength proportionate to muscle size. It was this common knowledge about our behavior that gave us the idea in the first place.

When we infer the "doing" of an organ from observation of the actions of a whole organism and then successfully test this inference against the kinds of evidence we listed, the tested "doing" has a special name. It is called the *function* of that organ.

INVITATION 33

SUBJECT: *Simple examples of evidence of function*

TOPIC: *Seven evidences of function*

To the student: (a) A certain part of the body of an animal is found to be short, tubular, and hollow. (See Figure 12.) It is attached at one end to an organ which is also tubular and hollow but very much bigger. It is attached at the other end to an organ which consists of a nearly solid mass of cells. It is known that all the food eaten by this animal passes through the large, hollow organ. What might be the function of the short tube?

[It very probably conveys something. The conveyance might be a secretion from the solid organ to the digestive tube. It might, however, convey digested food to the solid organ.]

To the student: (b) Name several other kinds of evidence you would require before having reasonable confidence in your answer.

[See the list of evidences in Invitation 32.]

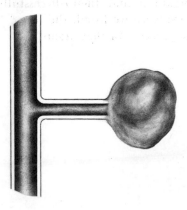

Figure 12

274

To the student: (c) In an aquatic animal an organ is found which is shaped into many fine branches, all of them in contact with the water. This animal has a circulatory system and all the blood flows through the many-branched organ at each circuit through the body. What might be the general function of this organ? What remains in complete doubt?

> [Some kind of transfer from environment to organism. There is no evidence to indicate whether the transfer is inward or outward.]

To the student: (d) Chemical analysis shows that the water surrounding the many-branched organ often contains substances in a much higher concentration than is found in the water farther from this organ. What do you now suspect this organ's function may be?

> [The added evidence suggests excretion.]

To the student: (e) A certain animal is seen to have a very large mouth cavity in proportion to its size. (See Figure 13.) Its jaws are equipped with comblike structures which meet and slightly overlap when the jaws are closed. It is noted that the animal habitually swims quickly through the water for a few yards with its jaws wide open—until its mouth is filled with water. Then it closes its jaws and squirts the water in its mouth out through the comblike structures. What might be the function of the comblike structures. Cite the evidence and logic you use to arrive at this possibility.

> [The fine structure of the comb, together with the swimming behavior and the jaw motion, suggest a filter used to retain solids in the mouth—in short, a feeding device.]

To the student: (f) A small burrowing animal is found to have three similar organs constructed as follows. Externally, there is a small, dome-shaped area of skin about a quarter-inch in diameter. This small dome

Figure 13

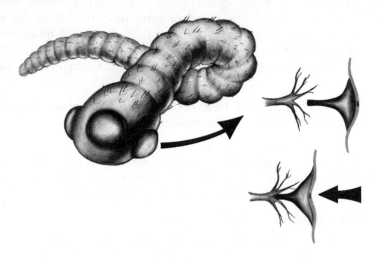

Figure 14

is elastic; if touched, it yields with ease, then immediately returns to its dome shape. The dome is connected by tiny levers to certain complex nerve ends in such a way that compression of the dome leads to pressure on the nerve ends. (See Figure 14.) One such organ is located on each side of the head. The third is located on top of the head. Study of the habits of this burrowing animal shows that it keeps these dome-shaped organs in contact with the earth of its burrow almost continually. What might be the function of these organs?

> [This animal, like the unicorn, is an imaginary one, of course. The data given make it likely that the organs in question are vibration detectors with good direction-discriminating capacity.]

To the student: (g) Now go back to the descriptions we have worked with. For each one, decide what kinds of evidence were used. The list of evidences are given again, at the right. At the left is the name of each part or animal we have investigated. Write the number of each kind of evidence used in the blank space after the name. Notice that we have added a new kind of evidence; behavior of the organism.

The small tube [1, 3, 7].

1. The overall shape and appearance of an organ or part

The branched organ of the aquatic animal [1, 3, 7].

2. The observable change or motion of that part

The comblike organ of the other aquatic animal [1, 2, 3, 7].

The dome-shaped organs of the burrowing animal [1, 2, 3, 4, 5, 6, 7].

3. The relation of that part to other parts

4. The shape and appearance of that organ's *components*

5. The observable change or motion of the *components*

6. The relation of the *components* to one another

7. The behavior of the organism

INVITATION 34

SUBJECT: *Muscle synergism and function*

TOPIC: *Function in a system*

[In Invitation 32 we began with a muscle and reached some understanding of it in terms of its *parts*. In this Invitation we move upward in biological level. We consider a muscle *system* and treat individual muscles as parts.]

To the student: (a) The arm is said to be "fully extended" when held out straight. The arm is said to be "flexed" when the finger tips are brought to the shoulder. Muscles which act to extend the arm (or any other limb) are called *extensors,* those which bring about flexion are termed *flexors.* Further, the flexors and extensors of a given part, because of their opposite action on that part, are spoken of together as a group of *antagonistic* muscles.

Grasp *lightly* with your left hand the muscles of your upper right arm in such a way that you can feel both biceps and triceps. Now from a position of full extension quickly half-*flex* your right arm. That is, bring the hand about halfway to the shoulder. What *functional* explanation can you give to the fact that you can feel extensors (the triceps) as well as flexors (the biceps) contracting during the flexion of the forearm? In answering this question, consider what might happen to your nose or ear during flexion if the extensors did not contract.

[The action of extensors during flexion permits control of this motion. If there were no tension of the triceps during contraction of the biceps, there would be no precise control of the degree of flexion.]

To the student: (b) Now go back to the list of evidences used in Invitations 32 and 33. Which *two* of these kinds of evidence were used by you to infer the function of triceps tension during flexion?

[Most conspicuously, we used motion of *parts* and arrangement of *parts*. That it was evidence about *parts* may not be clear to the students. This is treated in the next query.]

278

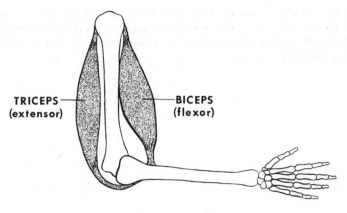

Figure 15

To the student: (c) In the previous Invitation we looked at a muscle as if it were by itself a *whole*. We then examined its parts—the muscle cells. What is the "whole" we are investigating now? What are the "parts"?

> [The extensor-flexor system is the whole. The individual muscles are the parts.]

To the student: (d) You have heard the statement that big fleas have little fleas. So do "parts" have parts, and "wholes" are often parts of larger wholes. For instance, we have treated a muscle cell as part of a muscle and a muscle, in turn, as part of a muscle system.

If the duodenum is part of the small intestine, what is the small intestine part of? And what is this still larger whole in turn a part of?

> [The small intestine is part of the digestive system. The digestive system is one of the few major systems making up the body. The point of (c) and (d) is to make clear that "part" is a relative idea and "whole" is relative also. Nothing intrinsic to a particular thing tells us whether it is a "part" or a "whole." This depends on the points of view we take. If we choose to consider a thing *A* in relation to other things *B, C,* and *D,* which are outside *A*'s boundaries, then we are considering *A* as a "part" in relation to a "whole" made up of *A, B, C,* and *D.* On the other hand, if we consider a thing *A* in relation to things *B, C,* and *D,* which are related to one another and *within A*'s boundaries, then we are treating *A* as a "whole."]

To the student: (e) Name some larger wholes of which you are a part.

> [The family, the class, the neighborhood, the community, and so on.
>
> [The word "synergism" (working together) for the interdependence of muscles in accomplishing controlled movement may be introduced in this Invitation if you wish.]

INVITATION 35

SUBJECT: *Muscle and bone*

TOPIC: *Function in a system*

[Invitation 34 worked with a system of two muscles. We now consider a system composed of muscles and their bony levers.]

To the student: (a) What significance can you see to the fact that the muscles that move the *forearm* are located in the *upper* arm?

[Discussion of this question should make the students aware of the fact that muscles of the flexor and extensor type invariably work across a joint; that is, the flexors and extensors of the forearm are in the upper arm, those of the hand are in the forearm, and so. This arrangement of parts is not only appropriate to their action in this case but, given the parts (bones, joints, and a muscle), this arrangement is *necessary* to the action of the whole. The parts, if they are to act as they do, *must* be arranged in this way as far as we know from engineering and leverage laws.]

To the student: (b) With the palm of your hand turned up, and without moving your wrist, alternately clench your fist tightly, then stretch the hand out flat. Where would you say the muscles that move the fingers into a fist are located?

[The tensing and change in shape of the foreman as the fist is made indicate that the body of muscles involved in this motion are located mainly in the forearm. Some of the much smaller muscles of the hand, are, of course, also involved, but they are of secondary importance in this movement and may be ignored here.]

To the student: (c) What significance can you give to the fact that the muscles concerned with movement of the hand are located in the forearm?

[The point to be elicited here is as follows: If the whole mass of muscles moving the fingers were located in the hand itself, the hand, if it were to be equivalent strength, would have to be several times larger than it is and

281

therefore clumsy. This point merits further discussion. The placement of some of the finger muscles (which might conceivably lie in the hand) in the forearm permits an action that facilitates the performance of certain delicate finger movements without sacrifice of strength.]

INVITATION 36

SUBJECT: *The valves of the veins*

TOPIC: *Experimental evidence of function*

[This Invitation is based on one of the famous experiments performed by William Harvey. Students should work in pairs with one member serving as the subject, the other as the manipulator. The leaner members of the class will make the best subjects for the observations.]

To the student: (a) Make a tourniquet around your upper arm by tying a handkerchief tightly around it. Turn the palm of your hand up and note the veins in your forearm. If the veins are not easily seen after tying the tourniquet, clench your fist several times. This should make them very prominent.

Now, with a firm pressure of a finger tip try to press the blood out of the vein in the direction of the wrist. Start from the elbow and move the finger steadily, firmly, and carefully along one of the largest veins toward the wrist. Push slowly and stop when you see that a part of the vein you have covered has lost its distended appearance—that is, you have succeeded in emptying a portion of the length of the vein you have covered.

Now, try to do the same thing but in the opposite direction—moving your pressing finger along the vein *from* wrist *toward* elbow.

What do you see to be the difference in result of these two efforts?

[The pressure from elbow to wrist empties at least a *part* of the vein. The blood is pushed back toward the hand and no more flows in from above to replace it—because of the vein valves. In the other direction the moving finger may push the blood along but it is quickly replaced by flow from the hand region.]

To the student: (b) What could account for the fact that blood pressed out of a segment of the vein toward the wrist is not replaced, although in the opposite direction it is?

[Clearly something prevents the flow of blood in the vein from elbow to hand but not in the opposite direction.]

283

To the student: (c) Repeat the experiment from elbow toward hand. Then release the finger pressure and watch carefully the part of the vein that had been cleared of blood. From which direction does blood move back into the vein?

[Blood moves from the direction of the wrist toward the elbow and quickly fills the undistended segment of the vein. This motion is so rapid that it may be necessary to repeat this operation several times before the class is convinced that this is so.]

To the student: (d) What can you conclude about the direction of flow of blood in the veins from these observations? Show how your conclusions come from the observed evidence.

[In the case of pressing from wrist to elbow the blood could be seen to flow back into the undistended segment from the wrist end. When the blood is pressed back from elbow to wrist, the blood did not flow back. Clearly the blood can flow easily in this vein from the wrist toward the elbow but not in the opposite direction.

[At this point you should tell the class that the result is due to valves in *all* the veins which prevent the flow of venous blood *toward* the extremities. Their own manipulations have demonstrated the existence of such structures.]

To the student: (e) What is the functional significance of the valves of the veins? Of what importance is one-way flow in this portion of the circulatory system?

[Students cannot be expected to answer this question fully on the basis of the information given. The class should be brought to see, however, that the demonstration of the valves suggests the function of the venous system as a whole—the return of blood to the heart. By way of increasing their comprehension of the significance of the valves, it may be well to tell the class that blood pressure is low in the venous system, that the valves assist in the flow of blood through the veins by eliminating back flow to a large extent.]

INVITATION 37

SUBJECT: *Embryonic circulation*

TOPIC: *Persistence as evidence of function*

To the student: (a) The flow of blood through a human embryo is quite different from the blood flow in a human after birth. The blood goes through different organs and goes through them in a different order. Even the pathway through the heart is different in the two cases. In this Invitation we shall work with the differences in pathways through the heart.

Figure 16 shows that *before* birth very little of the blood from the heart goes to the lungs. Most of it goes out to the rest of the body and back again to the heart. In a sense, the lungs are bypassed. The large, *heavy* arrows of the diagram indicate the path taken by most of the blood passing through the heart. The medium-sized arrows show the path taken by a small quantity of blood. The small, white arrows indicate routes along which there is *very little* flow.

Notice the opening between the auricles—the *oval window*. This opening allows most of the blood coming to the heart from the caval vein to pass directly to the *left* auricle. Note also that *some* of the blood that does not take this bypass route, and therefore enters the right ventricle, is also shunted to the left soon after leaving the heart. This occurs by way of a short vessel joining the pulmonary artery and the aorta—the *arterial duct*. These two bypasses prevent most of the blood from flowing through the pathways to and from the lungs. What is the possible significance of this bypassing the pulmonary (lung) circulation in the embryo?

[From the facts at hand, students should suggest that a pulmonary circulation in a fetus is necessary only to nourish the lung tissue and that a full pulmonary circulation is unnecessary since the fetus does not breathe. It is important to point out, however, that such a statement rests on the assumption (common for inferences of this kind) that the organism is efficient—that it has developed through evolution in such a way that most

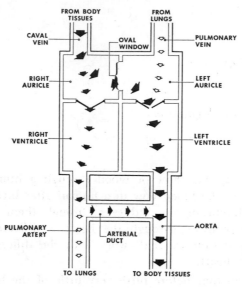

Figure 16

of its organs and actions (even what does *not* happen) are "useful," not wasteful or irrelevant to the rest of what goes on.]

To the student: *(b)* After birth the structure of the heart changes. With birth and the onset of breathing the oval window shuts and the arterial duct closes off. What changes in the path of blood flow occur as a result of these changes in structure? What is the significance of these changes accompanying birth?

[From an examination of Figure 16 students should quickly see that the closure of the oval window and the arterial duct make it necessary for *all* the blood passing through the heart to traverse the pulmonary circulation. Figure 17 shows the path of circulation that is newly established at birth.]

To the student: *(c)* These changes not only route blood through the lungs; they also cut off *all* the blood flow through the placenta. Then the placenta becomes detached from the uterus of the mother. The doctor then cuts the umbilical cord so that the placenta is no longer connected to the newborn child. What do these data suggest concerning the *relation between* the functional significance of the placenta to the embryo and the lung to the newborn child?

[In the ensuing discussion students can be brought to see that if the

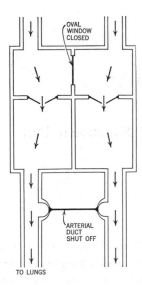

Figure 17

evidence is interpreted in terms of the conception of function, we see the probability that the two organs, the lung and the placenta, may have a common function in relation to the blood. The conspicuous evidence is that as one organ—the placenta—ceases to function, the other—the lung—begins to function. The interpretation would then be that the two organs have a common functional significance—that birth is accompanied by a replacement of placental function, or one aspect of it, by pulmonary function. Arguing from the hypothesis that a replacement of function of placenta by lung occurs at birth, students should quickly see that it is very likely that the placenta serves the fetus as a respiratory organ.

[In conclusion the teacher should retrace the course of this discussion in order to make clear the process of reasoning involved. It is *assumed* throughout that some of the *needs* of the fetus and the newborn baby are the same and that the structure of each performs the function in a different way in the two different circumstances. On the basis of these assumptions we infer that the placenta carries out a function in the fetus that in the newborn will be carried out by the lung.]

INTERIM SUMMARY 4

The Concept of Functional Part

PART AND WHOLE

In Invitations 32 to 37 we have seen the operation of a principle of enquiry which gives special weight to the whole, one which treats parts as subservient to the whole and to be investigated by asking what role they serve in the economy of the whole.

In this conception the "whole" has first place. It is a "going concern" with a certain character or nature. That character or nature is expressed through a number of capacities and activities characteristic of it. Thus the character or nature we call "animal" is expressed through a catalogue of capacities and activities as familiar as it is venerable, that is, ingestion, digestion, distribution and assimilation, excretion, locomotion, integration, reproduction, and so on. The character or nature of a specific animal would be expressed through specific versions of these genetic traits plus certain others which set the species apart from other species.

These capacities and activities, in turn, make certain demands. There are conditions that must be held within bounds and needs that must be supplied if they are to be maintained. It is here that the "parts" play their role. They are the servants of the whole, supplying its needs as well as constituting its visible existence.

In this conception the notion of "parts" is very flexible. For the purposes of one stage of investigation, they may be taken to be such gross parts as the entire circulatory system, the digestive tract, the nervous system, and so on. At another stage, we may focus down, treat each system, for the time being, as a "whole," and investigate its parts, the organs. Organs, in turn, may be treated as wholes while we investigate, as their parts, the tissues, the variety of cells, even the microstructures, which compose and maintain them.

The pattern of enquiry which flows from this conception of part and whole is two-pronged. First, there must be some preliminary general knowledge of the whole: a grasp of its character and nature, and a detailing of this character and nature as we have noted. We have already seen examples of this procedure: detailing of animal character in terms of ingestion, digestion, and so on; specification of human character in terms of a modified animal character plus those further behaviors which set men apart from other animals.

This general character of the whole is discovered through the classical process called induction. In this process numerous instances of each kind of thing are scrutinized in their normal state and condition. From this repeated observation of them comes the state of mind called "experience"; from experience, in turn, comes the inductive leap that discards individual variability and the incidental and takes hold of what is central and characteristic.

This process is not a guaranteed one. Our sample of instances may be biased (atypical). The selective observations of different investigators will select in different ways, depending on their past histories and interests. Hence, disagreement as to the essential character of the "kind" under investigation will arise. (This is the principal source of the shades of opinion and the diversities of organization which are found wherever men develop classificatory schemes in which "kinds" are distinguished and defined.) As we shall see, however, the pattern of enquiry used to investigate parts not only requires this prior, inductive investigation of the whole but progressively corrects it.

THE PATTERN OF ENQUIRY: PART RELATIVE TO WHOLE

When we have our preliminary, inductive grasp of the whole reasonably well established, we are ready to turn to investigation of the parts. The leading question we are to ask in each such investigation is clear enough: What is the role of each part in the whole economy? What does it do *for* the whole?

What is not at all clear, however, is what data we need to answer such a question and how these data are to be interpreted. It is at this point that the conception makes its crucial commitment, sets forth the notion which is at once its greatest strength and its sorest point. That notion is briefly and simply this: The *structure* of every part, the *location* of every part, and the observable *actions* of or in every part are all appropriate to, neatly fitting for, the role it plays in the whole (see also pages 00-00).

This crucial notion brings the functional conception to life, makes it an operative principle of enquiry, by telling us what data to seek and what questions to ask in our laboratory. If the structure, location, and action of each part *is* appropriate to its role, then, from knowledge of structure, location, and action, we should be able to infer that role. These matters, then, are what we should seek to discover as our data: What is the shape, the architecture, the detailed structure of the part we are investigating? Second, we ask, What are its neighbors and how is it connected with them; what does it pass on to them, or do to them? Third, What are its activities? That is, what perceptible motions are there (as in the heart)? What do we see entering them? Leaving them? Happening within them? What chemical changes? Physical changes?

We now also know what to do with these data when we have them. We are to treat them as indicators, evidence of what the part does for the whole. We are to interpret the structure of the part as being what it is because that is the structure which will best (or effectively) enable the part to serve its function. The connections of this part to other parts similarly exist as the connections which best enable it to play its role. So too, the motions and changes which take place in the part.

One fine example of the use of this principle of enquiry is Harvey's investigation of the role of the heart. First, he examined its structure: the chambers into which it was divided, the movable flaps which guard the entry into each chamber, the arrangement of the fibers which compose its walls. By a closer study he identified these fibers as muscular. He then asked himself what such a structure and arrangement would do. He traced the consequences of contraction of the muscle fibers: They were so arranged that their contraction would result in an overall reduction in the volume of the chambers of the heart. He then took note of the consequences of such a constriction on the blood the heart contained. He saw that, in view of the connections of chambers to one another, and in consequence of the arrangements of the flaps at entry and exit, the blood would be impelled through certain pathways from chamber to chamber and thence out.

Having thus inferred what he could from structure, Harvey turned to the question of connection of the heart with its neighbors, saw the emergence of vessels leading to and from the lungs, and others leading to and from the remainder of the body. By further study of the structure of these vessels (whether their walls were muscular; if so, whether as strong as or weaker than those of the heart; whether they too had valves; and if so, permitting flow in what direction) he sought data that would check his inferences from the structure of the heart and lead to independent inferences which would either be consistent with or opposed to the inferences made from the structure of the heart.

Finally, although there is no significance to the order in which structure, action, and location are examined and interpreted, he turned to the visible actions of heart and blood vessels, noted the order and sequence of contractions and expansions, the alternate paling and reddening of heart wall and blood vessels, and so on. Here he had still a third body of data from which to infer function and with which to check his other inferences.

Thus, he came to the conclusion that the role of the heart is to pump blood in a constant circulation from lungs to heart to body, thence back to the lungs.

This, of course, is not the end of the work, for he had not yet connected this movement of the blood to the whole economy, the need or needs of the body which, ultimately, this circulation serves. This matter remains to be firmly established, but there is already a clue—the clear, great emphasis in the body on flow through the lungs. All the blood passes through the lungs at each complete circuit, whereas only a fraction goes to any other single part (note here the further use of data about connections and neighbors). So, the next problem for enquiry is indicated. What happens to the blood in its passage through the lungs?

If and when that question is answered, we shall note other organs which are more richly supplied with blood vessels than the ordinary and seek to discover what special events take place there. Finally, we shall try to trace the connection between what happens to the blood in lungs and other major places and the needs of the body in general.

KNOWLEDGE OF PART AND WHOLE

When enquiry controlled by the functional principle is well done, a kind of knowledge emerges which is quite different from that brought to being through the principle of causal lines. From the latter come items of knowledge such as the following, each of them from a different research, reported in different scientific papers and recorded in different sections of a textbook:

1. Chemicals produced by some bacteria stimulate and direct the movement of white blood cells.

2. In cases of local infection by many bacteria, the physical state of the neighboring blood vessel walls is altered. In consequence, materials flow out which do not do so ordinarily. For example, fibrinogen,

the precursor of fibrin strands, may flow out in considerable quantity.

3. Various bacterial products are known to increase the phagocytic (engulfing) activity of white blood cells.

From enquiry directed by the functional conception, on the other hand, would come something like the following:

When bacteria invade a local area, as by a scratch or puncture wound, a number of processes are set in motion which serve eventually to wall off the infected area, prevent the spread of the invading organisms, and destroy the developing colony of invaders.

The statement would then go on to report the detailed steps by which these ends are achieved. In so doing, it would include all the items which emerged as fruits of causal-line enquiries, but these details would be related to one another and to the whole organism by exhibiting the roles they played in protecting the body against invading bacteria.

Note that the two familiar words "organ" and "function" have their origin in this concept of part and whole. "Organ" means much more than merely a distinguishable part; it means an agent, a subordinate, a something in the service of the whole. "Function" means much more than a mere doing; it means a doing which is seen and understood and phrased in terms of the service it renders to the whole.

It is in this concept of the organism that the central place of anatomy and physiology as sciences of great dignity has its origin. Anatomy is responsible for discovery and description of the structure, topography, and architecture of parts. Physiology is responsible for discovery of motions and actions and for interpretations of all the data on structure, location, action, to a conclusion about function.

STRENGTHS AND WEAKNESSES

There are shortcomings and objections to this classical (going back to Aristotle and Galen) concept of the organism, just as there are weaknesses in the concept of independent causal strands. There are also major advantages. Let us examine some of each.

Its principal virtue consists in the fact that it brings the "organism," the whole, back into the field of enquiry. The whole is lost in the concept of independent causal strands and this loss has consequences for enquiry.

A second and less obvious advantage of the classical conception over that of independent causal lines is this: It permits an integration of many biological sciences, the interconnecting of many different lines of enquiry. We have already seen that it requires an integration of anatomy and physiology, the two being joined to give us the data and the conclusions about functions. It also serves to relate psychology and ecology on the one hand and physics and chemistry on the other to the sciences of anatomy and physiology. It thus brings into existence an organized "whole" of biological knowledge which mirrors the very "whole," the living organism, with which it deals.

Knowledge of the physics and chemistry of living things is joined very simply to our knowledge of their anatomy and physiology. Just as we investigate organs as composing and serving functions in the whole, so we investigate physical components and processes and chemical components and processes as composing and serving functions within the organization of the tissues, cells, and organs which comprise them.

At the other end of the line, ecology and psychology are similarly organized into the structure of biological knowledge. Once we have, for example, studied the nesting, migrating, and nurturing behavior of birds, we can turn back and ask how all these complex behaviors are achieved. What organs serve to initiate the flight northward? What is it that turns the energy of the bird to the building of a nest?

When the involved organs are located, we can ask still further questions which relate the physics and chemistry of the bird's cells, as well as its organs, to its behavior as a bird. How do changes in the proportion of dark to daylight, or seasonal changes in temperature, lead to further chemical and physical changes within the organ which, in turn, lead to still further chemico-physical changes triggering flight, or nesting, or mating, or brooding?

With so much to be said in favor of the classical concept of the organism, it is hard, at first, to see what weaknesses it may have and what objections to it can be raised. But weaknesses there are and objections, indeed violent ones, have been raised.

One of the most serious objections to the concept of function as a principle of enquiry is that it treats organisms as if they were works of fine art. They have finish, completeness. Everything about each organism is as it should be. It could not have been otherwise; it would, presumably, have great difficulty in surviving a transition to something other than it now is.

There are two ways we can take this criticism; one is important to our understanding of it as a principle of enquiry, the other is not. We can take it as a statement of fact: The organism is not finished and perfect; it is not inflexibly what it is. Certainly these critical assertions are true. The organism has had a history of change. There are remnants in its body of past organizations, now pointless and irrelevant, "functionless." Moreover, we know that there are mechanisms built into the living organism by which further change is at least promised if not made inevitable. These changes will often occur piecemeal. Consequently, some organisms, even large populations of them, may possess characteristics (parts) which are merely tolerable in the present working scheme rather than effective, necessary elements of its organization.

Although these assertions are sound, they do not, for this reason, constitute a criticism of the classical concept as a principle of enquiry. Let us grant that many organs of the body are less than perfectly adapted to their roles. Grant further that some organs, whatever their source in past or projected conditions of the organism, are presently without a function. It is still the case that fruitful knowledge will be disclosed by asking the questions and seeking the data prescribed by the principle.

If the part which concerns our research *is* well adapted in structure, action, and location to its role, then our research will disclose that role. If the organ is less than perfect, if it includes remnants of past roles and aspects that presently have no role, our research will have loose ends. Some data collected will find no place in our final formulation of the role played by the part. Moreover, we may need to verify our first conclusions by studying many more neighboring and connected parts than would otherwise be required.

It may even be the case that our interest will have lighted on a structure that has no function, that is not, in the sense of the classical principle, an organ at all. In that case the data we require will not be forthcoming. The thing under investigation will refuse to answer the question we put because, in this case, it has no answer. This very silence, though, is evidence we can use. It points to the possibility that in this case we *are* dealing with a part which does not "belong," and this, too, is proper knowledge of the organism with which we are dealing.

In short, if we admit the desirability of finding answers which are only probable, in which we can have a *degree* of confidence rather than absolute sureness, the classical conception remains a fruitful principle of enquiry despite its limitations.

INVITATIONS TO ENQUIRY, GROUP V

THE SELF-REGULATORY ORGANISM: Homeostasis; Dynamic Equilibriums; Organismic Behavior; Adaptive Change of Equilibriums; Interconnections of Homeostases

Interim Summary 5, The Whole as Determiner of Its Parts

INVITATION 38

SUBJECT: *A thermostatic model*

TOPIC: *The concept of homeostasis*

[The student has already been introduced to instances of physiological regulation. In Invitation 24, for example, he studied the way in which the thyroid and the pituitary interact to keep the rate of secretion of thyroxine constant within close limits. In Invitation 25 he saw a complex instance of physiological regulation in the operation of the pituitary-gonad system. However, the ideas of dynamic equilibrium and of physiological regulation have not yet been explored.

[In this and the next few Invitations we begin the study of regulatory mechanisms as such, confining ourselves at first to the simplest kind—a homeostatic mechanism which maintains one substance in a dynamic equilibrium about a single "normal" point. In this Invitation we introduce a model of regulation, a thermostat. In Invitations 39 through 41 we turn to one important physiological instance. These Invitations deal with the control of blood sugar. In later Invitations we explore some of the complications of the simple homeostatic mechanism.]

To the student: (a) In this Invitation we shall explore something nonbiological—a thermostat. We do so because what a thermostat does will show us in a simple way something which is extremely important to our own bodies.

The name *thermostat* tells us what the device does. *Thermo-* denotes heat; *-stat* is like the word *static*, which means *unchanging*. A thermostat, in short, is a device that helps keep the temperature of something from changing. As we shall see, our bodies are equipped with many similar mechanisms. We, too, have "stats," mechanisms that help keep one thing or another constant or unchanging—or, at least, *almost* unchanging.

The central feature of a thermostat is a mechanical sense organ, one which is sensitive to temperature. Figure 18 shows one such mechanical sensor. It is a coiled-up strip made of two layers of metal. One layer is made of a metal that expands considerably when it is heated and contracts considerably when it is cooled. The other layer

296

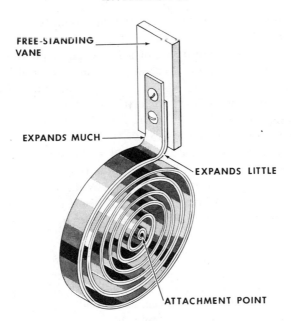

FREE-STANDING VANE

EXPANDS MUCH

EXPANDS LITTLE

ATTACHMENT POINT

Figure 18

expands and contracts very little. Notice the arrangement in Figure 19, in particular that the coiled-up strip is firmly attached at the center while the other end is connected to a vane which is free to move. Now imagine that the coiled-up strip is heated—because, for example, it is in a room which becomes two or three degrees warmer. What will happen to the free-standing vane?

[It will be swung to the *right*. Even if a few students have an immediate insight into this action, it would be wise to trace it through. The student with the insight could be asked, for instance, How do you know? or Can you show us why it swings to the right and not to the left? With assistance from the teacher, the student would then indicate the push which occurs because the outer layer of the coil expands while the inner layer, relatively, does not.]

To the student: (*b*) What would happen to the vane if the coiled-up strip had cooled instead of heated?

[It would swing to the left, of course—a point many students will infer, not from a grasp of the mechanism, but from the order of the questions and the previous answer. It would therefore be wise at this point to persuade a student who had not responded to query (*a*) to attempt a tracing of the events that lead to the left swing.]

To the student: (c) Now let us install our temperature sensor in a set-up where it will achieve something. Figure 19 shows such a situation. Suppose that *A* is the living room of a house and that the furnace and air conditioner in the basement can send warmed or cooled air into the living room as shown in the diagram. Notice, too, that our temperature sensor is connected to a source of electricity. It is wired in such a way that, depending on the position of the free-standing vane, the electricity will be sent to the furnace or to the air conditioner or to neither. Suppose, now, that the sensor is adjusted to be straight up and down when the room is at 70°. Suppose that the room air is at 70°. Now assume that it is a hot summer day. What will begin to happen to the air in the room? What, in turn, will begin to happen to the vane of the heat sensor? What will this action lead to? What will happen then?

[These questions are put together for economy. In class they should be taken one at a time to clarify the following sequence of events: (1) Room air warms, (2) sensor moves to the right, (3) further increase of air temperature, (4) sensor closes switch to air conditioner, (5) air conditioner begins to operate, (6) room air begins to cool, (7) after sufficient cooling, the sensor vane is withdrawn from the switch contacts and shuts off the air conditioner.

[It is important to emphasize steps (1) through (4). Otherwise the students will gain the impression that thermostats (and homeostatic mechanisms) maintain perfect constancy. If they receive this impression, the one

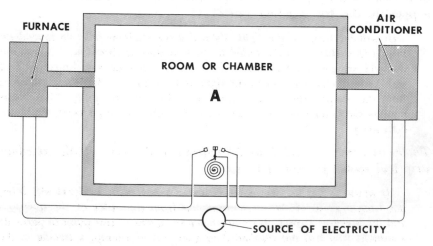

Figure 19

basis for understanding the conception of a *dynamic* equilibrium will be lost.]

To the student: (d) We have now seen in detail one operation of our mechanism. It is not a perfect operation for it did not maintain the room temperature at exactly 70°. Review the steps in its operation. Where is it most in error? That is, does it tend to let the room warm too much or cool too much?

[In this case the room warms considerably before cooling is effective. The cooling process, however, is reasonably well controlled.]

To the student: (e) *Two* factors contribute to this error. Try to find both of them.

[In the first place, the room *must* warm to some degree to actuate the thermostat. Then the thermostat and air conditioner take some time to become effective, during which time there is further temperature rise in the room.]

To the student: (f) One of these sources of error is often said to be "built-in." That is, it cannot be removed in any straightforward way. Which of the two is this "built-in" error?

[The time delay after the sensor is actuated can be corrected for: We simply design the thermostat to close sooner than cooling is wanted. However, the fact that the room must warm to some extent before actuating the sensor is "built in." The sensor responds to a *change,* not to a constant condition, and this is characteristic of most, if not all, biological sensors.]

To the student: (g) Suppose now that it is winter and the thermostat operates in the other direction to turn on the furnace. Suppose further that the furnace delivers heat to the room by sending very hot steam into *heavy* cast-iron radiators. In this operation, what are the two sources of error in the effectiveness of the system? And in which direction does each operate?

[Here, again, there is the "built-in" error, the temperature drop to actuate the sensor. The second source is the temperature difference between air and radiators which persists for some time after the furnace goes off. The point of this exercise is not to trace the simple phenomenon of heating but to emphasize (1) that a necessary error source exists in either case, and (2) that fluctuation in *both* directions from equilibrium is likely to occur.]

To the student: (h) We now see that to call this mechanism a "stat" is something of an exaggeration, since temperature is not kept at an unchanging point. Instead, it changes a little in one direction, swings back toward where we want it, and then may swing a little too far. Still, the temperature is kept very close to where we want it, and certainly it is kept much closer than it would be if we had no such mechanism.

The complete system or mechanism we now have, involving the sensor, the furnace, and the air conditioner, is called, in biology, a *homeostatic* system. This name is a little more cautious than "stat" by itself, for *homeo* means "similar to" or "almost like." So a homeostat or homeostatic mechanism is a system that keeps something *almost* or *nearly* unchanging.

Controls in the living body act in much the same way as thermostat. We tend to say that our body temperature remains at 98.6°C, for example, but in fact there is continuous rising and falling around the 98.6° mark as our bodies' thermostatic mechanisms operate. There is a name for such pendulumlike swings about a normal or middle point. They are called dynamic equilibriums. *Dynamic* means changing or moving. *Equilibrium* means balance. So a dynamic equilibrium is a balance among changes.

Now let us see what major components are necessary to make up a dynamic equilibrium. First of all there is something, some *quantity*, which is held in equilibrium. In our example this quantity can be thought of as the *amount of heat* in the room. Something must happen to this quantity, however, before we can begin to think of it as being held in dynamic equilibrium. For instance, what happens to this quantity on a cold winter day? What happens to this quantity on a hot summer day?

> [It is important here to think about loss or gain of a quantity. Hence, if students tend to say that temperature drops or that temperature rises, call their attention to the statement above that temperature can be thought of as due to the quantity of heat in the room. The useful formulation of answers is to say that there is a loss of heat in winter, and a gain of heat in summer.]

To the student: (i) What happens *after* the heat gain on a summer day if our thermostatic system is working? After the heat loss on a winter day?

> [Again, it is important to deal with the idea of quantity rather than degree. Thus, our air conditioner *removes* some quantity of heat from the room. Our radiators *add* some quantity of heat to the room.]

To the student: (j) Now let us make a summary list of the factors we have identified as belonging to the idea of a dynamic equilibrium.

[The list is as follows:
1. Some quantity to be held constant
2a. Undesirable loss of the quantity
2b. Undesirable addition to the quantity
3a. Compensatory addition to the quantity
3b. Compensatory subtraction from the quantity]

To the student: (k) Let us make a similar summary list of the major components of a homeostatic mechanism which can maintain a dynamic equilibrium.

1. A sensing mechanism
2. A source of compensatory replenishment
3. A means for compensatory subtraction

[In later Invitations these general ideas will be spelled out in biological terms: A source of compensatory replenishment, for example, will be spelled out as ingestion or addition from storage; a means for compensatory subtraction will be spelled out as excretion or storage.]

INVITATION 39

SUBJECT: *Control of blood sugar*

TOPIC: *Maintenance of dynamic equilibrium*

[In this Invitation the student encounters a physiological dynamic equilibrium and some parts of the homeostatic mechanism that maintains it. Storage, conversion, and excretion are introduced as means for removing an excess of the substance maintained in equilibrium. Reconversion, withdrawal from storage, cessation of excretion, and "sparing" are introduced as means of maintaining an equilibrium in the face of reduced intake.

[This Invitation is a long one, with many successive questions raised. It is *not* intended for use in a single class session. It may be broken into a number of pieces at almost any point after query *(c)*.]

To the student: (*a*) You probably already know three important biological facts about sugar. First, simple sugar (glucose) is the main fuel for living cells, tissues, and organs. It is used in large amounts when our large muscles are active. Second, the normal diet of an animal such as man or the dog contains a large amount of sugars and starches. Third, these sugars and starches are converted to simple sugar in the intestine and from there pass into the blood.

Suppose that a number of normal animals are fed one balanced meal a day, in the early morning, and then spend the following 12 hours in active exercise. During this period of time a blood sample is taken from each animal at 1-hour intervals and the percentage of glucose in it is measured. Just before the meal it was found that the concentration of blood sugar was 0.09%. Draw a graph to show what would happen to the blood sugar concentration in the course of the day if only the facts we have stated affected the sugar concentration.

[Any *descending* line, whether exponential or straight, would be an acceptable crude answer. However, such an answer would overlook the initial (sugar-depleted) condition and the time necessary for digestion to occur. Hence the following query may be desirable.]

To the student: (b) Did you think of the fact that the animals had been without food for 24 hours? And that digestion takes time? If not, what changes in our graph should we make? If so, what points on the graph represent (1) the blood sugar level before the meal? (2) the blood sugar level when absorption of sugar from the intestine is taking place?

[Taking account of these factors the graph should show an initial low, an increasingly rapid rise to a peak, and then a steady drop through the remainder of the time points.]

To the student: (c) A biochemist made accurate estimates of the amount of absorbable sugar in the meal and of the rate at which the animals burned sugar during the day. On the basis of these data he plotted a curve to show what the blood sugar level should be at each hour of the day. He assumed that only the ingested sugar was available to the animal and that sugar would disappear from the blood only as the animal's muscles and other living tissues utilized it for fuel. His graph looked like Figure 20.

The biochemist then made hourly measurements of the *actual* blood sugar, taking the blood sample from a leg. His results were quite different from those his assumptions had led him to expect. The "expected" and "actual" curves are compared in Figure 21.

Look first at the parts of the two curves between *B* and *C*. Assume that the sugar in the meal is being digested and absorbed from the intestine into the blood at the rapid rate the biochemist expected. Assume, too, that the rate at which the animals are burning the sugar is also as he expected. How, then, can we account for the difference

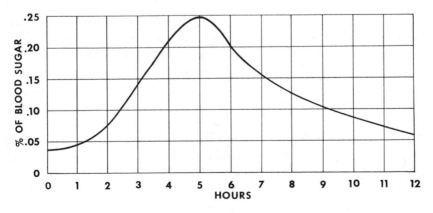

Figure 20

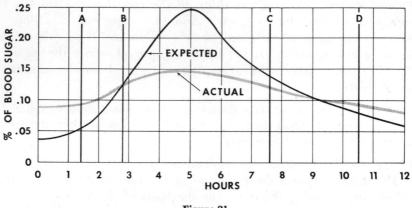

Figure 21

between "expected" and "actual" in the sugar concentrations of blood taken from the animal's leg?

> [Clearly something more than utilization as fuel is removing sugar from the blood. This much can be inferred from the two graphs. If this problem is posed to students who know some broad details of physiology, the next query will probably be unnecessary.]

To the student: (d) Clearly, something more than utilization is taking place. Otherwise the blood sugar level would have risen much higher. What might it be, considering that the experimental animals are animals much like yourself?

> [Three major possibilities are mechanical storage; chemical storage (conversion of sugar to something else) ; and excretion.]

To the student: (e) The experimenter knew that the circulatory pathway of blood which passes through the intestinal wall has a special feature. The vein which carries the blood away from the intestine (the portal vein) does not carry the blood directly to the heart. Instead, it passes first into the liver. Only after this passage into the liver is the blood collected into another vein (the hepatic vein) and carried to the heart. This is shown in Figure 22.

In view of this special arrangement of the circulation from the intestine, what likely possibility would account for the difference between expected and actual measurements of blood sugar in the period between *B* and *C*?

> [Removal of excess sugar in the liver.]

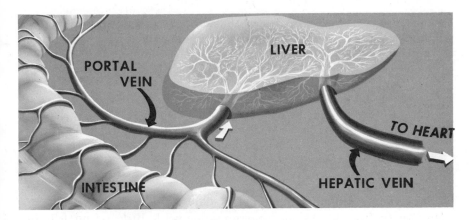

Figure 22

To the student: (f) How can we test this hypothesis? What data would we need and how could we get them?

[The possibility could be tested by comparing blood sugar concentrations in the portal vein with those in the hepatic—during the period of active digestion. The point is spelled out in separate questions in order to emphasize, without overt language, the order of experimental reasoning, from the problem, to the data wanted, to means for obtaining the data.]

To the student: (g) If our hypothesis is a sound one, what should our data show? If our hypothesis is mistaken?

[If the hypothesis is sound, the hepatic samples should show a sharp reduction in concentration over the portal samples. If not, the hepatic samples should be very little lower, or immeasurably lower, than the portal samples.]

To the student: (h) Our experimenter found that hepatic samples were sharply lower in sugar than portal samples. He concluded that there probably was storage of blood sugar in the liver. His interpretation was challenged by another biologist who said that he could see no justification for supposing that *storage* took place in the liver. It was just as likely, he said, that the liver *destroyed* the excess sugar. The first experimenter pointed out that his original curves supplied some basis for preferring a *storage* interpretation to a *destruction* interpretation of the experimental data. Look at these curves (Figure 21), especially the part from the origin of the curves to *A,* and the portion after *D.* What evidence supports the storage hypothesis?

> [The areas referred to represent the hours *after* a long period of exercise plus the period of rest and *before* replenishment of blood sugar by ingestion. In these periods the blood sugar level had not dropped to the level it should have reached if sugar utilization had not been compensated for in some way.
>
> [Of course, storage is not thereby proved. Nor is there any evidence, as yet, to distinguish between mechanical storage and storage by conversion. The important point, as far as scientific enquiry is concerned, is to emphasize once more that between absolute certainty and ignorance there is a wide range of more and less probable interpretations of data with which science deals.]

To the student: (i) What new kind of data would enable our experimenter to support his storage interpretation with little or no fear of contradiction?

> [The phrasing of this question in terms of "support his interpretation" instead of "prove" continues the point made before. The relatively obvious new kind of data would show that sugar *in liver tissue* increased as it decreased between portal and hepatic vein—and vice versa.]

To the student: (j) The experimenter sought just such data. He did not find an increase of glucose (blood sugar) in the liver. He did find, however, that the concentration of another substance (glycogen) rose in the liver as glucose level between portal and hepatic vein dropped. He also found that as bodily activity continued in the *absence of a meal,* glycogen concentration of the liver dropped and glucose appeared in the hepatic vein. When it was also shown that the glycogen molecule is like a long chain of glucose molecules he decided that his case was well supported and announced the conclusion that the liver served as part of a homeostatic mechanism which tended to keep blood sugar in a dynamic equilibrium.

To the student: (k) Some time later another investigator showed that the largest normal amount of glycogen stored in the liver was enough to maintain the blood sugar level for only about 5 hours. Yet, as the original curves show, the blood sugar level is maintained for a much longer time, despite continued muscular activity. Clearly there is more to the homeostatic mechanism for blood sugar than liver storage. From the following facts consider what else may be at work.

1. The large users of glucose are the skeletal muscles.

2*a*. On a diet containing more sugar than is utilized but containing little fat, mammals become fat.

2*b*. During long periods without food, the blood sugar level may be maintained at about 0.06%. In such a period, fat disappears from the body. In addition, the size of muscles and other organs may grow smaller as their proteins disappear.

> [The fact that skeletal muscles are the largest users of fuel suggests that glucose—as glycogen—may well be stored in muscle as well as in liver. This does, of course, occur.
>
> [2*a* and 2*b* suggest that sugar is convertible into fats and incorporated into proteins, as well as capable of storage in the sugar-like form of glycogen. The facts suggest, conversely, that fats and proteins may be broken down to supply blood sugar when the level drops far enough for long enough. This is the interpretation most students tend to give. However, it is worth trying for a less commonly noted possibility, which is raised in the following query.]

To the student: (l) The facts we have noted led a number of physiologists to suspect that fats and proteins could be broken down in the body to yield glucose, which could then be used as a fuel. It has been shown that these breakdowns do, in fact, occur. However, another and quite different possibility arises if we shift our attention. So far, we have been thinking about sugar homeostatically. That is, we have concentrated on the steady level at which it is maintained in the blood. From this point of view, we tend to think mainly about what can store or otherwise remove sugar from the blood when it is in oversupply and how it can be replenished when the rate of use exceeds the rate of intake. Now let us concentrate on another aspect of the situation. Let us concentrate not on the blood but on the muscles and their need for a continued supply of fuel. Our lives depend on our ability to move; hence we can suppose that evolution has tended to ensure an adequate supply of fuel to our muscles under most circumstances.

Suppose, then, that the supply of glucose dwindles to the point where it cannot supply the fuel required by our muscles. If muscles are to maintain their activity, what must happen then?

> [The answer is, of course, that under these conditions another metabolic pathway—one which used some other fuel—might open up. Thus fatty acids might be metabolized rather than the remaining blood sugar.]

To the student: (m) Turn back to the idea of a dynamic equilibruim of blood sugar. How would the opening of a way to use some other fuel help to maintain this equilibrium?

[This would be an example of "sparing": The remaining sugar is spared, thanks to the intervention of a process which can use a substitute.]

To the student: (n) So far, we have seen nothing but one or another form of storage as the way in which the equilibrium of blood sugar could be maintained when intake of sugar was high. What other, quite different, possibility is there for taking care of an excess?

[The other possibility is, of course, excretion.]

To the student: (o) When the possibility of excretion was investigated, it was found that the blood sugar level must become surprisingly high above normal before sugar appears in the urine. In short, the body "prefers" the storage to the excretion pathway for dealing with an excess of sugar. What is the obvious adaptive advantage of this in the case of a fuel such as sugar?

[The excess may well be useful at a later date.]

To the student: (p) Now let us sum up. First, list the ways in which the body maintains a dynamic equilibrium in the face of an *excess* intake.

[1. Relatively simple storage (glycogen)
  (a) In a central storage (liver)
  (b) At the sites of major use (for example, muscle)
 2. Conversion (for example, into fats, also destructive conversion, as in query *(h)*)
 3. Excretion
 4. (Not distinguished in this Invitation) Increased utilization]

To the student. (q) What are some ways of maintaining an equilibrium in the face of intake which is *less* than normal?

[1. Removal from simple storage
 2. Reconversion or conversion
 3. "Sparing"
 4. Reduced utilization
 5. Reduced excretion (if excretion normally occurs)]

INVITATION 40

SUBJECT: *Blood sugar and the internal environment*

TOPIC: *Fitness of models*

[A good, well-taught student who has done Invitations 38 and 39 may note two distinct incoherencies between our treatment of the principle of homeostasis in Invitation 38 and the treatment of blood sugar in Invitation 39. First, the introduction to the idea of homeostasis suggests, quite correctly, that a substance held in dynamic equilibrium is so held because an excess or defect would be, in itself, harmful. This concept is suggested in Invitation 38 through the idea of uncomfortably or unhealthfully high or low temperature. By contrast, we have shown no evidence that an excess or defect of blood sugar is, in and of itself, similarly undesirable. Rather, we have put the emphasis on the need for a continuing supply of sugar for metabolism.

[In fact, the significance currently attached by physiologists to the equilibrium of blood sugar has this emphasis. Blood sugar is primarily treated as glucose-in-transit rather than as a condition of the "interior environment" as Claude Bernard originally conceived it.

[However, in one direction at least, blood sugar level behaves like an important factor of an internal environment: A persisting drop of blood sugar to a point of 0.05% or below produces serious symptoms—weakness, sweating, blushing and blanching, loss of sphincter control, convulsions, coma, and death. However, a similar deviation from normal in an upward direction produces no such serious symptoms.

[We omitted this point originally because Invitation 39 was already very long. However, its omission can be used to advantage with exceptionally good students. It provides the occasion for a different kind of Invitation— an invitation to be constructively *critical* rather than docilely submissive to texts and other official sources of information. Such a critical-constructive invitation has two components: *(a)* the detection of inconsistency, incoherence, or omission; *(b)* the correction of them.

[We have suggested that these activities are only for very good students. This restriction holds, at least, for students normally schooled to docility— as, unfortunately, is often the case. Critical-constructive work may not re-

309

quire extra intelligence but it does require extra motivation, for it is of great value only if it is self-directed work.

[In the present case, such self-directed work would be initiated as follows:]

To the student: (a) Apparently, the author of our Invitations has made some serious slips. In Invitation 38, where we worked on a thermostat, furnace, and air conditioner system, he claims to have given us a model of homeostasis and dynamic equilibrium as they work in living organisms. Yet, it seems to me that there are important differences between that model and the picture we saw in Invitation 39 of blood sugar control. Would you like to check me on this impression by reading and thinking over those two Invitations tonight and letting me know what you find?

[The student would, we hope, find two matters of importance. First, the point already made—no evidence that the blood-sugar level is itself of importance. Second, no mention is made of a sensing device such as the thermostat in the model of Invitation 38. (If he finds others, we would appreciate being told about them.)

[Whether the student succeeds or fails in this first effort to be constructively critical is less important to his intellectual development than the fact that he is made aware of the existence and legitimacy of such work. Hence, if he does fail, it is not only proper but desirable that these incoherencies be pointed out to him. The second step would then proceed as indicated in the next student assignment.

[Because the idea of critical-constructive work flies in the face of the usual long conditioning of children to docility, two emphases have been built into the assignment as phrased in (a). There is, first the clear point that the author of a piece of printed matter may be in error. There is, second, the suggestion that the "live" teacher (the teacher present in the classroom as against the remote teacher who wrote the book) may also be in error or need help: ". . . it seems to me . . ."; ". . . would you check my impression. . . ."

[When this first stage in the critical-constructive Invitation is completed (either by the student's own successful discovery or by being shown), the second self-directed step can be taken, as follows.]

To the student: (b) I have two books in the laboratory which might help us find out what the case is here. Would you like to take them home and see if they have anything to tell us about our problem?

[We advise that this second (constructive) phase be limited to the first of the two problems stated—that of whether the blood sugar level is itself of importance. The other incoherence—the absence of discussion of the sensing mechanisms—raises a problem so very complex as to be inaccessible except to a highly exceptional student. At least three hormone mechanisms are involved. In addition, there are nervous factors, emotional factors, and the general idea of mass action and the reversibility of chemical reactions.

[Notice that the *degree* of self-direction and self-motivation can be controlled. At one extreme, the student may be given one or two books and no further help. At the other extreme, he may be shown exactly what pages to read. Between these extremes he may be given apparently casual help by remarks about index and table of contents, together with the suggestion of key words to be used in entering the index.

[As we have remarked, motivation is of prime importance in this case. Hence, special reward of the student who undertakes the work is important. The special reward should be intrinsic rather than extrinsic. That is, it should grow out of and be related to his effectiveness at the task. In the present instance, two possibilities for such intrinsic, special rewards are built into the situation. First, as the assignment is phrased, the student has the potential reward of *teaching the teacher*. He has been asked to help the teacher by looking up something of which the teacher is uncertain. Having done so, he must, of course, have a chance to bring the task to its climax—to instruct the teacher on the material he has discovered.

[The second built-in intrinsic reward is the opportunity to *emulate* the teacher, that is, to teach his fellow students. This can be brought to realization by asking the student, once he has done his first work of self-directed search and reading, to spend one more evening working out notes from which he would make a report of his "researches" to the class.]

INVITATION 41

SUBJECT: *Blood sugar and insulin*

TOPIC: *Sensing mechanisms of homeostasis*

[For general use, this Invitation directly follows Invitation 39.]

To the student: (a) In Invitation 38 we examined a model of a homeostatic mechanism—one involved in the maintenance of temperature in a house. It had three parts: a thermostat, a furnace, and an air conditioner. What was the function of each of these parts?

[1. Furnace: can add (heat)
2. Air conditioner: can remove (heat)
3. Thermostat: "senses" departure from equilibrium and controls 1 and 2]

To the student: (b) In Invitation 39 we examined a homeostatic mechanism for maintaining blood sugar in dynamic equilibrium. For each of the functions you have named (removal, addition, and sensing), name at least one mechanism for performing the analogous function in the case of blood sugar.

[Removal: Storage in liver
Storage in muscle
Conversion to other substances (for example, fats)
Excretion (only when rise in blood sugar is considerable)
Addition: Release from muscle and liver glycogen
Conversion of fats and proteins
Possible utilization of other fuels ("sparing")
Sensing: ?]

Obviously, one problem remains. What and where is the sensor which controls these processes? Is there an equivalent of a thermostat in the case of blood sugar?

Doctors have long been familiar with a disease called diabetes in which the level of blood sugar rises far above normal. Because of certain pancreatic changes in animals that died of diabetes, it was

312

suspected that this organ had something to do with the control of blood sugar in normal animals and that diabetes was due to its failure to function normally. What simple plan of experiment could test this possibility? What results would be expected if the hypotheses were sound? Unsound?

[Removal of the pancreas should lead to acute symptoms of diabetes if the hypothesis is sound. Such symptoms should not follow if the hypothesis is unsound.]

To the student: (c) This experiment was performed by Mering and Minkowski, two physiologists working in Alsace. They found that diabetes symptoms appeared in all cases in which the difficult surgery had been successful. Some part of the pancreas, then, seemed to be the "sugar-stat" for the whole body. Or, at least, some physicians and physiologists so supposed. Much evidence was found after the work of Mering and Minkowski to suggest that a rise in blood sugar stimulated the pancreas to secrete additional amounts of a hormone into the blood, which then, somehow, led to removal of blood sugar by the liver and other organs. Conversely, it was suspected that a drop in blood sugar would lead to secretion of less hormone by the pancreas, which in turn would stop stimulating the liver to store sugar—even, perhaps, lead to a reversal of action by the liver, so that it would return stored glucose to the blood.

Some physiologists were suspicious of the simplicity of this interpretation. One such physiologist, Samuel Soskin, working in Chicago, was impressed by the many factors involved in the supply and withdrawal of blood sugar: liver storage and liver release of blood sugar; similar storage and release by muscle; conversion of fats and proteins to glucose and the reverse; changes in the rate at which sugar was used as fuel by muscle and other tissue. He suspected that the control of these many operations might be much more complex than the simple "pancreas theory" suggested. He was especially impressed by the speed with which storage by the liver followed injection of glucose.

What complication of the overall control is suggested by the speed of reaction mentioned? Why?

[The rapidity with which the liver responds suggests that there is not time enough for reaction by the pancreas, transport of the pancreatic secretion to the liver, reaction by the liver. Hence there is the possibility that the liver itself responds to the concentration of sugar in the blood passing through it.]

To the student: (d) Soskin and his coworkers planned and executed an ingenious experiment to test the possibility that the liver might be its own "stat." They removed the pancreas from a number of experimental animals and replaced the organ with an "insulin pump." (Insulin was the name given to the hormone secreted by the pancreas.) This insulin pump was adjusted to pour a *constant* amount of insulin into the blood—just enough to maintain a normal sugar level. These animals were then fed a large meal rich in sugar.

What should happen to the blood sugar level after such a meal if Soskin were right about the ability of the liver to "control itself"? What should be the results if the view were correct that pancreatic secretion of insulin was the "stat"?

[Clearly, if the liver can "control itself," then with a nearly constant supply of insulin the curve of blood sugar over the 2 or 3 hours after the meal should closely resemble that of normal animals. If the "insulin theory" is correct, the blood sugar should show a marked and sustained rise after the sugar-rich meal—a typically diabetic response.]

To the student: (e) Under the conditions of this experiment, the blood sugar level rose very little more than it does in normal animals with pancreas intact. What conclusion shall we draw? Why can we *not* draw the conclusion that the liver is necessarily its own "stat"?

[We can draw the conclusion that blood sugar control can occur without a fluctuating supply of insulin. We cannot conclude that the liver does the sensing job itself, for the obvious reason that the entire remainder of the body may be involved.]

To the student: (f) Soskin and his collaborators later found very good evidence that the liver was its own "stat." Thus, there appeared to be at least two "stats" involved. Some aspects of blood sugar withdrawal and replenishment are controlled by the insulin from the pancreas. Some aspects seem to be under the direct control of the liver. But there is more to the story.

The disease diabetes is controlled in humans by regular doses of insulin. If too much insulin is taken by the patient, his blood sugar falls below normal. If a diabetic patient in such a condition sees or hears something that makes him angry, his blood sugar will often rise to a normal level or even higher! This can occur even though the patient does nothing about his anger. What additional pathway for control of blood sugar is suggested by this?

[Since seeing, hearing, and the experience of anger are *nervous* reactions this suggests that there is a nervous pathway which can stimulate release of sugar from storage.]

To the student: (g) Blood sugar control is known to be even more complex than these several experiments suggest. Furthermore, much about it is as yet unexplained.

INVITATION 42

SUBJECT: *Blood sugar and hunger*

TOPIC: *Organismic behavior as homeostatic*

[In this Invitation we take advantage of the fact that blood sugar equilibrium involves eating to introduce the idea of behavior of the whole organism as part of homeostatic control of equilibrium. We begin with hunger as a stimulus to appropriate eating, then suggest other instances.]

To the student: (a) We have seen that depleted blood sugar can be replenished by (1) withdrawal from storage; (2) conversion of fats and proteins to sugars. What obvious third route for replenishing our supply of sugar is there?

To the student: (b) Increased ingestion, simple eating of sugar, is an obvious way to replenish a low supply. If, however, eating of carbohydrates is to be an economical and appropriate addition to the homeostasis of blood sugar, under what conditions should we feel impelled to eat? When would eating be wasteful or unhealthful? How can this wastefulness be partly corrected?

[We should feel impelled when sugar level drops. Eating when the body's supply is adequate is hardly adaptive. Its wastefulness can be partly corrected by storage, which we have considered in earlier Invitations.]

To the student: (c) Humans and other animals are impelled toward eating by an experience we call hunger pangs—pain-like feelings in the stomach. These feelings seem to arise from contractions of the stomach when they occur with a certain rhythm. What data might enable us to determine whether hunger pangs are, in fact, part of the homeostatic mechanism for maintaining our blood sugar level?

[There are two possible lines of discussion here. The students may suggest that we take blood samples of animals (for example, humans) when they do and when they do not feel hungry, and determine whether the onset of hunger coincides with a drop in blood sugar. On the other hand, previous training may impel the students to go directly to experimental manipulation of one of the factors—blood sugar being the manipulable member of the pair. The following queries follow from each of these alternatives.]

To the student: (d) Suppose we take blood samples as you suggested and find that hunger pangs occur only when the blood sugar drops. This would be good evidence that the sensation of hunger is closely tied to blood sugar homeostasis. But show that there is a good reason for *not* concluding from such evidence as this that a drop in blood sugar is the stimulus to hunger pangs.

[This touches the point that coincidence (correlation) between two variables is suggestive but hardly conclusive evidence for a causal or stimulus-response relation between the two. Students may not interpret the phrasing of query (d) as asking for a general rule. If not, invite them to formulate the general rule. A direct, simple, and frequently used phrasing (put in terms of blood sugar) runs as follows: Many things we have not measured may happen at the same time as the drop in blood sugar. We could be reasonably sure only if we tested all of them and found that only a drop in blood sugar level always occurred with the onset of hunger pangs.]

To the student: (e) Suppose we found that, in fact, a drop in blood sugar was always or almost always followed by hunger pangs. What further experimental procedure might help us to be more certain that the fall in blood sugar is a stimulus to hunger pangs?

[Experimentally "remove" sugar from the blood, trying to keep other relevant factors constant.]

To the student: (f) One way to do this would be to withdraw large quantities of blood from an animal, remove some of its sugar, and then return the blood to the animal. This would be difficult, complicated, and risky, however.

One investigator pointed out that it might, in fact, leave us with exactly the same problem which led us to plan such an experiment. What could he have had in mind?

[Any such radical effort is likely to change many factors other than the one intended. We would have, again, many possible changes but knowledge of only one, hence the temptation to ascribe results to the known change.]

To the student: (g) If you review what we found out about blood sugar level in Invitation 41, you will discover what might be a much more reliable and a much simpler way to reduce blood sugar level experimentally. Try to find it.

[Like Invitation 40, but directed in this case to a whole class and in a much more restricted and structured situation, this query invites the student to engage in close reading for a purpose other than that which the document to be read was written to serve. This kind of search and reading is obviously a highly desirable skill and habit. What the student will find if he reads alertly is the possibility that injection of insulin can be used as a device for obtaining rapid and sharp drop in the blood sugar level.]

To the student: (h) When insulin is injected into the blood there is a sharp drop in blood sugar, and this drop is quickly followed by hunger pangs. Moreover, the injection of glucose promptly stops the pangs. What are you tempted to conclude from these data together with the data from naturally occurring drops in blood sugar?

[Certainly we have here very persuasive evidence for supposing that a drop in blood sugar level is at least one stimulus to hunger pangs. However, ...]

To the student: (i) There is still room for doubt—indeed, for the same doubt we had when our data consisted only of the coincidence of natural drop in blood sugar with the onset of hunger pangs. Show that this is the case.

[In the absence of highly detailed knowledge of the role of insulin in the entire homeostatic mechanism of blood sugar, we are still left with the following situation: The role of insulin may involve an alteration of many factors involved in carbohydrate metabolism. One of these, rather than the drop in blood sugar, may be the stimulus to hunger contraction. However, we have left little doubt that hunger pangs are related to carbohydrate metabolism generally, and perhaps directly to the drop in blood sugar.

[At this point we leave the discussion as developed thus far and return to the possibility that the student response to query (c) was to suggest immediate recourse to experimental control of sugar level rather than first obtaining data on "natural" drops and the occurrence of hunger pangs. This suggestion would first be pursued through analogs of queries (e) through (h) in order to clarify the experimental procedure, and so on. Then:]

To the student: (j) Even if we are sure that insulin injection leads to a drop in blood sugar level, what else do we know about the effect of this substance—what else may it be doing in the body? In view of such possibilities how sure can we be that a drop in blood sugar level under normal conditions leads to hunger pangs?

[Thus we show that, just as in queries (d) to (i), experimental procedures are needed to complement "natural" observations; so here, "natural" observations are needed to complement the experimental procedure. It is important to note, in this regard, that the popular faith in the "experimental" or manipulative as invariably superior to the "natural" must be taken with more than one grain of detailed reasonableness. The experimental procedure may give us the potentialities for control of unknown factors left untouched by the "natural" observation. Only the potentialities are given, however. To make them real requires a large and detailed body of relevant knowledge which, in the case of the insulin technique described, was not available. Moreover, the experimental manipulation, however well controlled, showed only a possible route by which hunger pangs could be roused. It does not show us that this route is necessarily the one used by the body under normal or all conditions. As far as present knowledge is concerned, the phenomenon of hunger remains complex and obscure.]

To the student: (k) Let us assume for the time being that a drop in blood sugar does give rise to hunger pangs, hence to ingestion and replenishment of our carbohydrate supply. This means that the behavior of the whole organism with relation to its environment may be just as much a part of a homeostatic mechanism as detailed chemical changes within organs and tissues.

Show that the southward migration of birds in autumn and their return north in spring may also be a homeostatic mechanism.

What about the change in the coats of many fur-bearing animals with change in the seasons?

INVITATION 43

SUBJECT: *Basal metabolic rate*

TOPIC: *Adaptive change of equilibriums*

[Thus far, we have treated the regulative organism as if all it did was to maintain important constancies—for example, keep blood sugar from shifting far from its "normal" concentration. The flexibility of the organism goes much beyond this. It not only resists environmental changes by maintaining a constant in the face of them; it also will change a constant adaptively in response to environmental pressures. We deal with one such adjustment now.]

To the student: (a) In mammals the metabolic release of energy goes on at all times—even when we are at rest or asleep. Thus our body temperature is maintained, and heartbeat, nervous activity, and many other bodily processes continue to operate. The rate at which this quiet or resting metabolism goes on is homeostatically maintained. In most mammals, including man, it is so constant under ordinary conditions that deviation from the constant is used by physicians as a symptom of illness. The main controller of this *basal* or resting metabolic rate is the thyroid gland, through is secretion of a basal metabolism-controlling hormone called thyroxine. (We worked with this substance and organ in Invitation 24.) The greater the amount of thyroxine produced by this gland, the higher the basal metabolic rate.

The activity of the thyroid gland is controlled, in turn, by a neat interaction with the anterior lobe of the pituitary gland. A hormone from this gland stimulates the thyroid. Thyroxine from the stimulated thyroid, in turn, tends to inhibit secretion of the anterior lobe hormone. Hence, between these two factors, thyroxine secretion undergoes very short pendulum-like swings around its point of equilibrium.

Thus far, this sounds like just another simple case of homeostasis. Note this, however: If humans (and other mammals with a constant basal metabolism) move to a distinctly colder climate and stay there for awhile, their basal metabolism rises to a new constant level. If humans who have lived in a cold climate move to a warmer one, then their basal rate goes down! What must have happened to the homeostatic mechanism in such a change? Go back to the analogy of the thermostat controlling a furnace and an air conditioner. What second, additional control do we see operating in the case of change in basal metabolic rate?

> [In terms of the analogy, we not only have a thermostat that controls means of addition and withdrawal of the equilibrium factor, we also have something that controls the thermostat, that changes its setting upward or downward as adaptive need requires.]

To the student: (b) So in addition to thyroid-pituitary control of one another and thyroid control of basal metabolism, there may be an additional mechanism behind all this which controls the thyroid-pituitary control of one another. If you set out to investigate the way this mechanism had its effect, what *two* major possibilities would you try to test?

> [The additional or second-order control mechanism could act on the pituitary-thyroid system in either of two ways. It could change the sensitivity of the thyroid to pituitary stimulation. That is, a rise in metabolic rate could be obtained if the thyroid were stimulated to produce more thyroine by a given concentration of the pituitary substance. Conversely, we could obtain more thyroid substance if the inhibitory effect of thyroxine on pituitary were delayed or decreased.]

To the student: (c) What is the adaptive value of this control of a control—this mechanism for resetting the "thermostat"?

> [The obvious adaptive value concerns the metabolic requirements for maintaining constant temperature under different prevailing conditions. However, there may be much more to it than this.]

To the student: (d) Until now, we could think of the body as consisting of many different, distinct, and independent homeostatic mechanisms, each one tending to its own business. One such mechanism controlled blood sugar. Another controlled basal metabolic rate. And there are many others; for example, one controls the acidity of the blood, another the proportion of sodium and potassium ions in the blood, still another our body temperature. Such a view of the organism is very much like the view we obtain from the idea of organ and function, where we thought of each organ as having its specific part to play in the body as a whole. Thus it was just like a simple automobile engine or other machine, with each part doing just one thing and incapable of doing something different.

This is a very neat and simple way to think about the body, but let us see if it is good enough. Note carefully that we have seen a blood sugar homeostasis in which insulin and other factors controlled blood sugar. We have seen a basal metabolism homeostasis in which pituitary-thyroid controlled basal metabolism. Now consider this: Blood sugar is the principal fuel for body metabolism. If that is the case, what can we expect might well follow from a change in the thyroid control of basal metabolism?

> [If basal metabolism depends mainly on the utilization of sugar and the control of basal metabolism is changed upward or downward, then sugar utilization will also change. This, in turn, will probably call for a shift in the input-output rates of blood sugar.]

To the student: (e) Are, then, the homeostatic mechanisms for sugar, for basal metabolism, for thyroid secretion, insulin secretion, and for pituitary secretion separate or connected? If we carried this idea to its extreme, what would we have to say about every organ and every action in the body?

> [The homeostatic mechanisms named are clearly interrelated. If we carry this idea of interrelation to its limit, we would face the probability that every organ and every action of the body affects every other organ and every other action.]

To the student: (f) What new pattern of research should we carry out to test this possibility and learn more about the body's control of itself? If, for instance, we can change, experimentally, the equilibrium point of blood sugar, what, in general, should we look for?

> [This concept of the completely interrelated organism leads to the following pattern of enquiry. We remove an organ or change the level of some equilibrium and then look for any and every other possible change in other organs and actions, in order to discover the compensatory and adjustive changes of which the body is capable. Thus, if we could experimentally alter the blood sugar level, we would conceive of this alteration as changing a variable of the body to which every other variable of the body was related. To put this in a mathematical form: We would have to consider the body as a single vast equation in which every organ and action was a factor. In such an equation, a change in any variable must lead to change in at least one other and could lead to a change in many or in all. Furthermore, the "constants" we have called dynamic equilibriums may well be constant only under some conditions. Under other or more extreme conditions (changes in other variables or "constants") they, too, may turn out to be variables.]

INVITATION 44

SUBJECT: *The stress reaction: adrenaline*

TOPIC: *The self-regulating organism*

To the student: (a) Let us look at another example of the idea of the self-regulating organism. The inner core of a gland called the adrenal secretes a hormone called adrenaline. When we are quiet and relaxed very little adrenaline can be found in the blood—something like 0.000001%, a millionth of a per cent. If, suddenly, we are startled, angered, or see something dangerous, the amount of adrenaline in the body suddenly rises until there is a hundred times as much as when we are quiet and peaceful.

One rapid effect of this great increase in blood adrenaline is to stimulate the liver to convert large quantities of its stored glycogen into glucose—which passes into the blood. What is the adaptive value of this action?

> [It amounts to a quick and greatly increased mobilization of fuel resources for muscular (and other tissue) metabolism. The student's phrasing can be, quite legitimately: It prepares us to run or fight better.]

To the student: (b) The sudden increase of adrenaline with fright, anger, or anticipated danger has many other effects. It increases the rate of heartbeat and of breathing. It constricts (narrows) the blood vessels of the skin and digestive tract while enlarging the diameter of blood vessels of the heart and skeletal muscles. It even leads to chemical changes that shorten the time necessary for blood clotting. Show how each of these changes fits with the increased blood sugar as an adaptive change preparing us for fight or flight.

> [Increased heart and respiratory rate contribute to oxidative metabolism in the muscles. The vasomotor changes shunt a much greater proportion of blood to the heart and skeletal muscle. The shortened blood-clotting time makes an obvious potential contribution.]

To the student: (c) Look now at the ways in which this "stress reaction" of adrenaline occurs. What does it tell us about the idea of a "normal" or stable blood sugar? Heart rate? Blood pressure? What does it say about the independence or interaction of homeostases in general?

[It shows that the equilibrium points for blood sugar, heart rate, blood pressure, and so on, are not fixed but changing—though here, as against the rise of basal metabolic rate with colder prevailing climate, we have a short-term rather than a long-term change. In general, it shows us once again how far goes the interconnection of the various "stabilities," actions, and functions of the body; in short, how far the body as a whole can change most or all of the actions of its parts so as to maintain the whole organism in an effective relation with the environment.]

INTERIM SUMMARY 5

The Whole as Determiner of Its Parts

EXAMPLES OF SELF-REGULATION

When blood sugar drops to an "uncomfortable" level, events are triggered which tend to remedy the situation. One mechanism leads to release of stored sugar by the liver. We are impelled by another to stop what we are doing and seek out food. (There is evidence, indeed, that some animals seek out the particular kind of food they need.) Still a third mechanism stops down the rate at which our muscles consume the sugar that is available.

Conversely, when too much candy has raised the blood sugar level, similar but contrary processes are evoked. The kidney extracts sugar for excretion. The liver stores as much as it can; so do muscles. Hunger pangs cease. Our appetite is "spoiled."

This homeostatic regulation, however, is not the whole story. A more profound change can occur. The body can *"reset its thermostat."* If whim or circumstance puts us on short rations for a time, the thermostat behaves, at first, as it did before. It irks us by way of hunger pangs and lassitude and calls for stored sugar from the liver. If we stay on the short rations long enough, however, the impulsion called hunger ceases; the muscles readjust the ways in which they take energy from available materials; the liver releases its store only when blood sugar drops to a level measurably lower than the level which formerly evoked release. And this new state of affairs persists even if we return to a "fuller life"—until we have stayed on it for awhile.

Enquiry has disclosed many instances of this kind of reset of a thermostat or second-order control which legislates a new "normality." When a muscle is called into activity, many capillary beds open which normally are closed. When exercise ceases, the beds close down again.

325

If the muscle is used repeatedly over a sufficient length of time, however, the number of beds which supply it, even in its resting state, is generally increased, and an exercise call for extra supply is answered by a proportionate increase.

A similar shift of norm occurs for the basal rate at which our body releases energy for maintenance. It rises if we stay long enough in a cooler climate and drops if we spend sufficient time in a warmer one. It occurs, too, in the case of water utilization by the kidneys and, again, in the case of body temperature. In brief, the body can shift its norms as well as correct departures from them.

These instances seem commonplace. In the first place, they are only second-order instances of a host of ephemeral adjustments to need and circumstances which are of the essence of being alive. In the second place, they are changes in rate, in degree, which we tend to take for granted. But let us look further.

At first glance, mountain sickness appears to be but another instance of second-order change in degree. When we move to a high altitude, distress—shortness of breath, fatigue, nausea, and so on—usually ensues. The amount of oxygen in the volume of air we breathe is so much less than at sea level that our demand, even under the quietest circumstances, exceeds the available supply. In two weeks or three or four, however, this is changed. The thin air of the heights is now adequate: We extract from it as much oxygen as we require.

Many of the changes which bring about this more satisfactory state of things are only changes in degree: increase in the amount of oxygen-carrying pigment in the blood, increased blood pressure and circulatory rate. But at least one extensive study provides evidence of change of a different kind, a change in the very architecture and chemistry of an organ.

The organ is the lung. Under sea-level conditions its cells are only epithelial: They constitute a membrane. The oxygen of inspired air and the carbon dioxide brought by the blood are thought to move across it by ubiquitous physical means: solution in water, then diffusion from regions of higher to regions of lower concentration. The inspired air has a higher oxygen "tension" than has the blood which flows through the membranes of the lung spaces; the blood has much more carbon dioxide than does the inspired air. In consequence, exchange occurs simply as a result of the relative frequencies of molecular collisions.

After adjustment to high altitude, all this is changed—if we accept the research of J. S. Haldane. He undertook measurement of oxygen tension—in the ambient air, in the lung spaces, in the blood flowing to and away from the lungs—among members of an expedition to Pikes Peak. When members of the expedition reached adjustment to altitude —no longer showed symptoms of mountain sickness—Haldane's data showed oxygen tensions *lower* in the alveolar air of their lungs than in the passing blood. In short, oxygen was moving *against* the concentration gradient. The indicated conclusion was that under the condition of oxygen want, sufficiently long maintained, the very structure and action of lung membrane was transformed. It was no longer an epithelium across which oxygen "moved itself" but a secretory organ which moved oxygen "forcibly," by the expenditure of energy (active transport).

Though Haldane's data have been challenged (because of the doubtful accuracy of the assays of alveolar air) his conclusion remains a tenable possibility.

There are similar cases which have not been challenged. There is compensatory over-growth of one kidney if the other is damaged or removed. Similar compensation occurs in testes. There are well-documented instances of increase in auditory acuity after loss of vision. (In this case it is not clear, however, whether we are dealing with the structural change of ear or nerve, or with "learning.")

Karl Lashley reports similar flexibility in local functioning of the brain. Animals are taught a certain behavior. Then, by trial, the brain area which is the locus of this learning is located and destroyed. When the animals recover from the microsurgery they are retrained on the same problem. They learn it—in another area of the brain.

Now, three more cases—from experimental embryology.

1. Before the outgrowth of nerves, the forelimbs of amphibian embryos are transplanted to a site forward or to the rear of their normal position. Development proceeds. Each transplanted limb receives a full complement of nerves from the spinal cord. But they are *not the nerves they would ordinarily have received.* Instead, they are nerves from a different segment of the spinal cord, nerves which would, in "normal" circumstances, have gone elsewhere to perform another service.

2. Remove one developing eye from an amphibian; transplant it to another, in line with one of its eyes and immediately adjacent to it. The two rudiments grow, make contact, and *reassort their constituent cells.* From the two rudiments come not two eyes, but one, its parts in harmonous relation with one another.

3. The middle kidney is removed from a chick embryo. By treatment

with alkali and a material which digests protein it is disintegrated into a mere suspension of separated cells. The suspension of separated cells is then placed in a culture medium. The cells grow, multiply, migrate, cling to one another, and in three days or thereabouts the culture dish contains identifiable units of a kidney. "The discrete cells," say the authors of this work, "are thus capable of re-establishing the structural pattern of their tissue of origin." (A. and H. Moscona, *J. Anat.*, **86,** 1952, 287–301.) Says another scientist of a similar capability in the cells of an *Amblystoma* embryo: There was directed cell migration, selective cellular adhesions, mutual assimilative inductions. (Johannes Holtfreter, from "Growth in Relation to Differentiation and Morphogenesis," *Symposia Soc. Exptl. Biol.*, No. 2, 1948.)

Here, then, are a number of instances of the flexibility of the organism. There is not one concentration of sugar which is "right" for the organism but several, depending on the condition and activity of many (or all?) other chemicals, physical factors, and structures of the body. There may be not one anatomy of the lung, but two—perhaps three or more—depending, again, on the condition of the company it keeps, the state of other organs, the condition of the outward world with which the organism interacts. There is not one normal size for a kidney, not even a normal average, but several sizes, depending on the demand its brother parts make. The same holds for the blood supply to muscles, the size of muscles, even the size of that principal muscle, the heart. The organization of the brain is flexible, its parts multivalent, pluripotent. When a part is lost or damaged, its role may be taken over by another part.

In the embryo, the "normal" fate of a given cell or group of cells is subject to change. Instead of becoming what it usually becomes, it conforms to the organization in which it may be put by change of circumstance—even radically altered circumstance—and plays the part that fits that organization. Paul Weiss says, with moving imagery, ". . . newt belly skin grafted to an axolotl head has . . . complied with the locality (of the host) but has done so in a manner characteristic of the donor. . . . The cells have reacted to the lateral head field of the *axolotl* to the best of their *newt* knowledge." (*Principles of development*, Henry Holt, New York, 1939.)

THE NEED FOR A THIRD PRINCIPLE OF ENQUIRY

Such events as these point to the usefulness of a principle of enquiry in biology, one that will open doors to which the classical conception has no key. The classical conception in its pristine form depends on a fixed structure, a fixed function, and a fixed relation between the two. A fixed function is the knowledge object at which the enquiry aims. A definite structure (and action) is required in order to yield the data which this pattern of enquiry must use. A fixed relation between the two is the ground on which the data of structure, locus, and action are interpreted to yield knowledge of function.

By contrast, such flexibilities and varipotencies as we have just described point to a new knowledge-object which requires a new kind of data and a new pattern for its interpretation. Instead of knowledge about *one* collection of regular, recurrent parts, we will want knowledge about changeable parts within the set; even, possibly, of changeable sets. For example, we would no longer ask, "What is *the* cellular structure of the lung?" Rather, we would ask (by analogy to Haldane's report), "What is the *repertory* possessed by this organ, the *several* cellular arrangements and architectures of which it is capable?"

This new question would bring in its wake two related ones: (1) What conditions evoke this, that, or the other item in the repertory? (2) What are the complex, subtle processes by which one item of the repertory, one cellular architecture, is replaced by another?

By questions such as these we could seek out the flexibilities of our "parts." We also want to seek out the flexibilities of our "whole." This can be done by an analogous shift of the problem we pose. We would no longer ask, "What is the function or role of this organ?" Instead, we would ask, "When parts X, Y, and Z do thus and so, what role does part A play?" We would then ask, "When X, Y, or Z is changed in such-and-such a way, what correlative change takes place in A—what new roles does it play to make of A, X, Y, and Z an integrated whole?"

Such a pattern of research reflects the conception of the self-regulatory organism.

SECTION THREE

The Teaching of Biology

CHAPTER

5

Teaching Strategies and Styles

It is essential for the teacher of BSCS biology to develop teaching strategies and a teaching style conducive to developing thinking and enquiry processes in students. It is essential because of the changing nature of biological science and the consequent need to change educational goals.

Biology is changing from an empirical science to a science in which the invention of concepts and conceptual structures is taking a progressively more important role. Because of this change, the significance of the *products* of biological research—the meaning of information that has accrued from biological investigation—becomes more and more dependent on a grasp of the *processes* of biological science, the ways (both intellectual and technological) in which the body of biological information has been attained. The word "enquiry" has been used by the BSCS to designate these processes.

In addition, the very structure of biology as a field of knowledge, or discipline, has changed and is changing. The major conceptions by which we organize the body of biological information are becoming more refined, and the interrelations between these conceptions are being discerned and developed. These major organizing concepts and theories have been identified by the BSCS writers in the "themes" that are woven through the versions. One reason for the changes in and development of these conceptions are changes in biological investigation. Hence an understanding of enquiry is necessary in order to understand the changes in the major conceptions by which we organize biological information.

A general reorientation in educational aims and means is required as a consequence of the developments in modern biology. This reorientation entails helping both the teacher and the high school student develop the competencies and habits related to enquiry processes, since these processes are the key to understanding the nature of science. Enquiry in science is also closely related to the development of the ability to carry on self-directed learning and to the development of thinking. Both of the latter are emphasized as essential aims of education in a contemporary, democratic society.

Because of the foregoing, it is necessary to examine teaching strategies and styles in order to identify and analyze those that are most conducive to achieving the aim of developing the competencies and habits related to enquiry processes in biological science. Teaching strategies, as used in this book, refer to the approaches used by teachers to involve students with subject matter in order to accomplish certain ends or learning objectives. On the surface, "strategy" seems to be merely a different name for teaching "methods" or "techniques." However, the term "strategy" entails a new meaning, which rather precisely relates means and ends. Within each teaching approach there are strategies that are more conducive to certain specified ends than to others. For example, questioning is a general approach that is used by almost all teachers. However, certain types of questions encourage mere recall of facts, whereas other types of questions encourage students to analyze, interpret, relate, apply, and evaluate; that is, different types of questions encourage a variety of types of thinking processes. If one uses the appropriate types of questions, more complex thinking will be encouraged. Thus the meaning that distinguishes "strategies" from "methods" is the understanding of specific forms of teaching approaches *in relation to* specified desired outcomes.

Teaching styles refer to particular teachers' use of teaching strategies. It is much more difficult to identify and describe teaching styles than teaching strategies, since the former are probably closely related to teachers' personalities. Given a continuum of authoritarian-democratic or closed-open personalities, probably most teachers can develop an effective style which utilizes strategies conducive to the development of thinking and enquiry in science. It may be that a relatively few persons at the extreme "authoritarian" end of the continuum cannot; it may also be that a relatively few persons at the extreme "democratic" end of the continuum quite "naturally" use teaching strategies conducive to thinking. However, for the majority of teachers, one of the major tasks of their professional preparation is to help each individual develop his own teaching style by *practicing* the various teaching strategies that tend to facilitate the development of thinking. More specifically, the major task of professional preparation for biology teachers is to develop a teaching style based on practice of strategies which help students develop competencies related to enquiry in biology.

The purposes of this chapter, then, are twofold: (1) to describe various teaching strategies consistent with the aim of developing an understanding of enquiry in biology; and (2) to suggest activities that can be used in teacher education that will help teachers develop a teaching style which includes competence in the use of these strategies.

STRATEGIES FOR LABORATORY ACTIVITIES

Below is a description of an activity that has been used successfully in BSCS teacher preparation programs. It includes two procedures for dealing with the same biological phenomenon. It is desirable to actually carry out Procedure A, but if this is not possible read it and Procedure B with questions such as the following ones in mind. What are the differences in the amount and kind of student participation? What enquiry processes are the students engaged in? How do these relate to probable resultant learnings of each? Which do you think is the preferable approach and for what reasons?

Procedure A

Present the "students" with Petri dishes containing a circle of starch-covered file folder and several split grains of corn which have been placed outside down on the starch-covered cards. Explain that the corn grains were soaked in water for 24 hours, then sliced in half, placed on the starch-covered cards, and left for several days.

State again that the card on bottom of the dish is covered with starch and describe how to carry out the iodine test for starch. Let the students carry out the starch test on fresh pieces of card. (*Expected results:* entire surface of card will turn color.)

Now have students remove the corn grains from the dishes and apply the starch test to these cards. (*Expected results:* card will turn color *except* for areas slightly larger than the perimeter of each corn grain.)

Discussion (divide the class into groups of 3 or 4 students). What are some possible explanations for the appearance of noncolored (that is, nonstarch) areas of the cards? Within the small groups, have the students suggest and discuss possible explanations. Then have each group present all their possible explanations to the class as a whole. Write each possible explanation (hypothesis) on the board. Group the hypotheses. For example, several hypotheses may pertain to the corn grain "doing" something to the starch; several may pertain to the starch being absorbed by the corn grains.

(*Variation:* Some students may already know the "correct" hypothesis and hence this is not a real problem for them. When this occurs, place slices of cork similar in size to the corn grains on starch-covered cards and treat in the same way as the corn grains and other starch-covered cards.)

Now ask for possible ways of investigating each type of hypothesis. (Don't spend too much time on this, since it is not the main point of this activity which is to examine different teaching strategies.) The results of this discussion should provide students with a rough outline of possible experimental designs.

Procedure B

Read the following to the "students":

"Enzyme Activity in Corn Grains

"You may have noticed that certain vegetables and fruits vary in taste. Some carrots are rather tasteless, while others are very sweet; with increasing age a banana changes from a rather starchy-flavored to a sweet-flavored fruit. In each case it is obvious that in the sweet-tasting fruits and vegetables additional sugar is formed from insoluble carbohydrates. The following demonstration will show that enzymes capable of digesting starch to sugar are present in living tissues.

"Take a soaked (germinating) corn grain, cut it lengthwise with the razor blade, and test the cut surfaces with iodine solution. What food is present in abundance? What other foods known to be present in corn are not demonstrated by the test?

"Your teacher will take one of the Petri dishes containing starch agar and one with plain agar and flood them with iodine solution. What difference do you observe?

"On another Petri dish of starch agar there are two or three corn grains which have started to germinate. Each has been cut lengthwise and the cut surface placed on the agar. They have been on the agar for about 2 days. On a third Petri dish containing starch agar are cut corn grains which had started germinating but which were killed before they were put on the agar. The cut corn grain will be removed from the surface of the starch agar by your teacher, who will flood the surface of each dish with iodine solution for 2 or 3 minutes and then pour off the excess.

"How do you account for the difference in the areas in the agar on which germinating corn grains are located? What kind of food would you expect to find in the clear areas? What do you observe in the Petri dish which contained the germinated but killed corn grains? Would dry corn grains work as well as moist, germinating ones?"

Discussion. What is the difference between the procedure we used (Procedure A) and what has just been read? How could you modify such an exercise so that the students are more actively participating in the enquiry processes?

Again have students form small groups to discuss these questions. Follow this by reports from each group and general discussion of the proposed plans.

Now, consider Procedures A and B in terms of the suggested questions and others that no doubt have occurred to you. Discuss these with other persons if possible. Through your consideration and discussion you should develop a rather clear idea of the differences between "illustrative" and "investigative" laboratory activities.

The essence of investigative laboratory work consists in the fact that students are given biological problems, scaled to their current abilities, which they pursue in a way similar to the way a biologist would pursue the problems. An experience of this type differs from traditional laboratory exercises in that the latter serve mainly to illustrate biological facts and principles. The illustrative type of laboratory exercise is, of course, still important and useful. But much greater use should be made of investigative activities.

Since investigative laboratory activities provide students with real (for them) biological problems, they are characteristically more open than the illustrative type of exercise. Four degrees of openness can be identified with respect to laboratory exercises.[1]

The degrees of openness, in increasing order, are as follows. The least open exercise poses a problem, suggests means of investigating this problem, and presents data resulting from that investigation. (This is an open, investigative exercise only if the students do not already know the "answer" from other activities such as reading the text.) The task of the student is to interpret the data in light of the way the problem was formulated and the way the investigation was carried out. The

---

[1] Adapted from Schwab, Joseph J. 1962. "The Teaching of Science as Enquiry" in J. J. Schwab and P. Brandwein, *The teaching of science*, Harvard University Press, Cambridge, Mass.

next most open exercise poses a problem and suggests ways and means of dealing with it. The student carries out the investigation and interprets his results. Third, the exercise can pose a problem, but the methods of investigation and the interpretations are left open. In the most open exercises, the problem also is left open—the students are merely presented with some biological material as a subject of investigation and are allowed to proceed on their own.

The last type of exercise can be a frustrating experience if used in the very first part of the course. However, it need not be frustrating if it is used later in the course, if it has been preceded by the other types of exercises, and if a beginning understanding of enquiry has developed.

With the more open types of exercise, the artificial distinction between mind and hand is eliminated. This is especially true of the most open type in which only a subject for investigation is presented. Too often, both teachers and high school students think of the laboratory as a place in which they *do* something and the classroom as where they read and think—if "remembering" can be properly considered "thinking." But in the most open investigative laboratory, the very openness of the problem leads to discussion. Several problems are posed, different principles of enquiry come to light as these problems are discussed, ways of attacking each of these problems are considered, and so on. In short, the discussion closely parallels what a team of research biologists might actually do.

If the class as a whole is engaged in such a discussion, the result may be the organization of several teams of four or five dealing with separate problems. When the problems have been investigated—experiments designed relative to formulated hypotheses, data gathered, and interpretations made—the teams can report to the entire group. On the basis of these reports, discussion can ensue which brings out the factors in investigation that must be understood if an understanding of biological enquiry is to be developed. These factors are as follows.

1. Differences in the fruitfulness of the ways of approaching and investigating the problems.
2. Differences in kinds of data sought and interpretations made relative to formulated hypotheses.
3. Assumptions made in formulation of hypotheses and interpretation of data.
4. The course of subsequent research that would follow from the kinds of approaches used.

There are ways of examining teaching strategies for laboratory activities other than in explicit terms of illustrative or investigative types. One thought-provoking way has been developed by Abraham Shumsky.[2] Although his book pertains primarily to elementary school teaching, the ideas developed are applicable to secondary teaching. It is a book that all teachers should read.

Below are three models for a high school biology laboratory activity based on Shumsky's distinctions among types of teaching strategies. Notice that while the model laboratory activities are quite different, they use the same biological material *(Elodea)* and essentially the same phenomenon (cell membrane permeability) for investigation.

As you read these models, ask yourself questions: What are the differences in the ways the activities are structured by the teacher? In what ways are students participants in each activity? What types of learnings are probably the results of each? In what respects are these models similar to or different from the illustrative-investigative models? Are there differences in degrees of openness? What are your reactions to each model?

Model A

Materials

> *Elodea* leaf
> Slides and cover slips
> Absorbent paper
> Microscope
> Distilled water
> Salt solution, 2%, in dropper bottle
> Salt solution, 5%, in dropper bottle
> Salt solution, 10%, in dropper bottle

Objective. To determine the effect of different salt-water concentrations on the cells of an *Elodea leaf.* (Teacher announces this.)

Introduction. Teacher makes some brief statements about the cell membrane as the gateway to the cell.

Procedure. Teacher lists needed materials on board; distributes instruction sheet that tells students, step by step, what procedures to follow. Teacher reviews with students the procedures for using the microscope.

As students begin to carry out the exercise according to the instruction sheet, teacher moves about the room assisting students. Teacher

---

[2] Shumsky, Abraham, *In search of teaching style*, Appleton-Century-Crofts, 1968.

checks on whether a student is following previously learned procedures for using the microscope. As students begin examination of *Elodea* leaf, teacher asks such questions as: "How many cells do you see?" "What is the shape of each cell?" "What is happening inside of the cell?" "What do you see moving?" "What other changes do you see?"

Students write answers to these questions and other factual questions listed on the instruction sheet and draw diagrammatic sketches of an *Elodea* cell.

Discussion. After students have completed their observations and sketches, all are asked to direct their attention to the blackboard where teacher is ready to begin discussion. "John, what did you see?" John reads the answers he has written in his notebook. "That's right; did the rest of you see the same things?" (One student who begins to say he did not is ignored.) Teacher then explains what the observations indicate about movement of water into and out of cells.

Evaluation. On the following day, students are given a quiz based on the answers to questions about what was observed in the previous day's exercise and on the teacher's explanation of these observations.

Model B

Materials. Same as for Model A.

Objective (unannounced by teacher). To determine the effects of different salt-water concentrations on the cells of an *Elodea* leaf.

Introduction. Discussion of characteristics of cells, such as yeast and cheek cells, which previously have been observed by using the microscope. "Have you ever wondered about the function of the membranes of cells? What might be some functions?" (Teacher lists suggestions on the board as students offer them.) "Let's see if we can find out whether this is one of the functions of the cell membrane." (Teacher selects the function that pertains to the maintenance of fluid in cells.)

Procedure: Teacher distributes instruction sheet that lists needed materials and provides general suggestions for carrying out observations, such as: How will you change concentrations of water around the cells? What kinds of observations will provide evidence about maintenance of fluid, that is, what will you look for? How will you record your observations? As students begin to make observations, teacher moves about the room helping students to decide on their experimental design by raising questions that assist students in focusing on essential steps in the procedure that they apparently have overlooked. Not all students need this assistance, but the ones who do are not made to feel inadequate, since the teacher conveys helpfulness by the way he approaches students and by the way he phrases questions.

Discussion. This begins by first asking different students to list their procedures and observations on the board. Then the observations are compared. Discrepancies in the observations are related to the different procedures to determine whether the procedures account for the discrepancies. After the discrepancies are resolved, students are asked to explain the meaning(s) of the observations. A number of students respond to this, and each response is listed on the board. Students are then asked if there is one generalization that includes all of the explanations on the board. The students agree on one generalization and the discussion is concluded.

Evaluation. The next day students are given a quiz that asks them to state the generalization arrived at the previous day and the evidence (observations) that supports this generalization. They are also asked to indicate any refinements in the experimental design that they have thought of since yesterday.

Model C

Objective (unannounced by teacher). To assist students in becoming involved with a biological problem and to develop ways of investigating that problem. (*Note:* this class has not previously studied *Elodea.*)

Introduction. Glass containers with fresh (pond) water and one or two *Elodea* plants are prepared in advance and are distributed to students—one container for each group of three or four students.

Procedure. Teacher begins discussion by pointing out that these plants are commonly found in fresh water ponds. "Have you ever wondered whether such a plant could survive if transplanted to the ocean?" "What do you think would happen?" Students are encouraged to make some guesses and predictions. As needed, the teacher makes additional statements and/or raises related questions designed to stimulate the students' interest in the general problem. When a number of students have made suggestions, all of which the teacher accepts and lists on the board, each group is asked to see what they can find out and to be prepared to report to the class when their investigation is completed.

Students are allowed to work freely in small groups. Teacher assists, when requested by students, in obtaining needed materials and equipment for their study. If students request help in deciding on a question to be investigated or in designing the investigation, teacher provides it by questions that focus students' attention on specific components of the situation that are relevant to the general problem or on factors that need to be considered in setting up a well-designed experiment relative to their stated problem. Teacher is careful, however, to offer no more help than is actually needed by the group; instead, teacher encourages them to "mess about," if necessary, trying out various things, before they decide on a more systematic investigation.

Discussion. After the investigations are completed, each group reports on the specific questions that it investigated, what was done, and what was found out. Other students are encouraged to ask questions and to make comments and suggestions.

When all groups have reported, discussion turns to a comparison of the different approaches. Points brought out in the discussion are differences in assumptions, differences in the way the problem was viewed by different groups, and how these differences relate to different experimental designs and interpretations of results.

Evaluation. Students are asked to state a different, but related, problem and to design and carry out an investigation that would extend their understanding of the relationship between *Elodea* and its fluid environment.

The three models described should be discussed in terms of the questions suggested previously. It is particularly helpful to relate the strategies used in the models to the kinds of learning outcomes that are likely to result from each.

Another activity that is useful in teacher preparation is to present an open-ended laboratory exercise that is aimed at developing a more thorough understanding of teaching strategies. In this type of exercise the teachers are presented with some biological material, preferably living—a collection from a pond, for example. They are then asked, as individuals or as teams, to develop ways to use this material to illustrate or investigate different problems, principles, and concepts in biology. In addition, they are asked to develop these for specified student groups. The most important specifications are the previous experience of students with biological phenomena and principles and their level of development in terms of enquiry processes.

As is true with investigative laboratory exercises, there are differences in the openness of these exercises. In the first few exercises of this type the instructor should perhaps suggest the problem, principle, or concept to investigate; in later ones, this should be left to the choice of the participant-teachers. Such exercises increase the teacher's scope of experience with biological materials and also point up the fact that the same material can be used in a variety of ways or that different materials may be suitable for similar purposes. More importantly, this type of exercise can provide experience for the teacher in developing his own investigative laboratory exercises. This ability can contribute to his teaching of biology in three ways: (1) he is free from word-for-word dependence on exercises in laboratory manuals; (2) he puts into practice his understandings of biological investigation and discovers ways of conveying this to high school students; and (3) he gains a better understanding of the suitability of different strategies for different purposes and for use with different students. These are probable outcomes if, after development of plans by individuals or small groups, these plans are discussed by the total group. The discussion should focus on questions such as the following ones.

How appropriate is the exercise to the problem, principle, or concept chosen?

What is its feasibility with large and small groups and with a variety of laboratory facilities?

Is it appropriate for use with different student groups or must it be modified depending on the previous experience of the students?

QUESTIONING STRATEGIES

It has been said, "To question well is to teach well." Although this may be a generally accepted truism, it is only in the last few years that serious attention has been given to the art of questioning in relation to the development of thinking. In the past, questions have been used mainly to determine *what has been learned*. Now it is recognized that questions can be the basis of teaching strategies which promote the *development of various intellectual functions*. In this section we shall review some of the recent work on questioning which suggests strategies for an enquiry approach to teaching biology.

Perhaps the most thorough work on questioning in relation to the development of intellectual functions has been done by Hilda Taba.[3,4] The importance of her work is twofold: (1) it is a contribution to the theoretical consideration of thinking, and (2) it provides applications and models for teaching. Although the practical aspects of her work were carried out in connection with elementary school social science, the principles are applicable to other educational levels.

[3] Taba, Hilda, S. Levine, and F. F. Elzey, *Thinking in elementary school children*, San Francisco State College, 1964 (Cooperative Research Project No. 1574).

[4] Taba, Hilda, "The Teaching of Thinking" in *Critical reading*, edited by M. L. King, B. D. Ellinger, and W. Wolf, J. B. Lippincott, 1967.

On the basis of her theoretical construct of thinking, Taba distinguishes three categories of cognitive tasks that represent a sequence from simpler and more concrete to more complex and abstract.[5] The categories of cognitive tasks are concept formation, interpretation of data and inference, and application of principles. Each of these cognitive tasks is further delineated as follows.

Concept formation. Differentiation of properties or characteristics of objects or events; grouping by abstracting certain common characteristics and grouping them on the basis of the similar properties; categorizing and labeling.

Interpretation of data and inference. Identifying specific points in the data; explaining specific items or events by relating points of information to enlarge the meaning and/or establish relationships; forming inferences that go beyond that which is directly given.

Application of principles. Predicting and hypothesizing by analysis of a problem or conditions to determine which facts or principles are relevant; developing informational and logical parameters that constitute the causal links between the conditions and the prediction.

An implication of Taba's formulation of thinking and cognitive tasks is that there are two major types of teaching functions which seem to affect, either positively or negatively, the development of cognitive skills. The first type is questions or statements that have a *managerial* or psychological function and that indicate approval, disapproval, agreement, and the like. The second type is questions or statements that *guide* the discussion and that are related to the logic of the content and to the cognitive operations sought. (Giving direction to or guiding discussion is distinguished from controlling thought. The latter applies to situations in which the teacher provides what the students should do for themselves. Which of the models of laboratory exercises in the preceding portion of this chapter control thought?) Let us examine some specific examples of these two types of teaching functions.[6]

There are four types of managerial questions or statements.

1. They may be supportive: "Yes"; "That's on the right track"; "Let's see if I understand you correctly . . ." (restates, summarizes, or clarifies the student's statement).

2. They may invite more thinking: "I'm not sure I understand your point"; "Could you clarify for us how your point relates to the

[5] *Ibid.,* pp. 143-146.

[6] The examples given are adaptations of unpublished material prepared by Ann O'Neil of San Francisco State College.

topic?"; "Can you give an example?"; "Can you explain to us how you arrived at that interpretation?".

3. Managerial comments also give information which refocuses the content or the limits of discussion: "Remember that we only have data on one type of plant"; "We should remember that Mendel's studies were done before chromosomes had been described."

4. Managerial comments may also be negative: "No, that's wrong"; "You're not thinking"; "I don't agree with you"; "That doesn't answer my question"; "That's not what you were told to do."

Obviously, the last type inhibits both thinking and verbalization, whereas the first three types invite more thinking and discussion.

The teaching function of guiding cognitive operations has been subdivided by Taba [7] in terms of three purposes that can be achieved by different questioning strategies. The first is *focusing*, which establishes the content or topic and the cognitive operation to be performed. The second is *extending* thought on the same cognitive level. The third is *lifting* the level of thought to a more complex cognitive task. Below are examples of each of these purposes.

Focusing. "In that filmed sequence of the frog's behavior, what factors might have influenced what it did?" (Compare with, "What might influence a frog's behavior?")

"What do you observe in that drop of pond water?"

"How does the data obtained by group A compare with that of group B?"

"What are some things to remember in using the microscope?"

"What are some possible explanations (hypotheses) of why the frog did not swallow the robber fly?"

Extending. "Can anyone think of another way we might obtain the needed data?" (Compare with, "What does the lab exercise say about how to obtain the data.")

"What are some other questions we might ask about the changes that occurred in the pond?"

"That's an important generalization. Would anyone state it in a different way?" (Notice that this question allows for assimilation of an important idea and also encourages more students to participate by thinking about and stating the idea in their own terms.)

Lifting. "Now that we've identified some of the characteristics of these animals, can we group them in one or more ways?" (Concept formation, from differentiation to grouping.)

[7] *Op. cit.,* pp. 147-151.

"What part of the data leads you to that interpretation?" (Which cognitive tasks does this shift involve?)

"Assuming that our generalizations about the growth of a yeast population apply to other populations, what can you predict about changes in the human population?"

If one is to incorporate these strategies of questioning into their teaching style, it is essential to practice using them. Several ways of doing this are (1) analyze each of the example questions for its content and cognitive task; (2) provide examples for each of the cognitive tasks described on page 00 and each of the functions described above; and (3) analyze a taped discussion for the different types of questions; are they the best questions for the purpose to be accomplished?

Another view of questioning has been developed by Norris M. Sanders.[8] Sanders states that the basic ideas underlying his study of questions come from *Taxonomy of educational objectives*.[9] However, for reasons stated, he slightly reorganizes and redefines the categories, as follows.

Memory. The student recalls or recognizes information.

Translation. The student changes information into a different symbolic form.

Interpretation. The student discovers relationships among facts, generalizations, definitions, values, and skills.

Application. The student solves a lifelike problem that requires the identification of the issue and the selection and use of appropriate generalizations and skills.

Analysis. The student solves a problem in the light of conscious knowledge of the parts and forms of thinking.

Synthesis. The student solves a problem that requires original, creative thinking.

Evaluation. The student makes a judgment of good or bad, right or wrong, according to standards he designates.

Sanders also reminds the reader "that these preliminary definitions are by no means adequate for distinguishing the categories." They are submitted only as a necessary background to a discussion of the way in which ideas from the *Taxonomy of educational objectives* can be used by classroom teachers.[10]

[8] Sanders, Norris M., *Classroom questions: what kinds?* Harper and Row, 1966.
[9] Bloom, Benjamin, et al, *Taxonomy of educational objectives,* Vol. 1, Longmans, Green and Co., 1956.
[10] *Op. cit.,* p. 5.

The focus of Sanders' book is to describe a practical plan of using questions so that there are varied intellectual activities required of students. It is a valuable source of activities that aid teachers in improving their questioning strategies, since much of the material is based on workshops with teachers.

Another way of viewing questions has been developed by the Great Books Foundation in connection with their Junior Great Books Program.[11] Their categorization of questions is perhaps particularly useful in critical reading of text materials even though the specific examples that they provide pertain to other types of reading materials. This scheme distinguishes three types of questions: questions of fact, questions of interpretation, and questions of evaluation. Questions of fact pertain to statements made by the author, whether or not they agree with the "facts" in the world. These can be answered by referring to statements in the text. Questions of interpretation explore the author's meaning. Responses to such questions can usually be supported by statements from the text even though they cannot be directly answered by the text. Questions of evaluation ask the student to agree or disagree with the author, to state in what respects he agrees or disagrees, and often to state the criteria that are the basis of his evaluation.

The foregoing approaches to types and uses of questions can be very helpful to prospective and in-service teachers in examining and refining their teaching strategies. Using one or more of these sets of categories of questions, the teacher can analyze his own strategies. He can also plan portions of lessons that primarily use one or another type of question and can check whether he does actually use these.

However, mention should also be made of schemes for analysis of classroom interaction which have been developed through various research efforts. One of the best known of these systems is that developed by Flanders.[12] The two major categories in his analytical system are "teacher talk" and "student talk." The former is subdivided into statements and questions that are classified as direct or indirect.

Another scheme for analysis of classroom interaction has been developed by James J. Gallagher and others of the Institute for Research on Exceptional Children at the University of Illinois. This scheme is of particular interest, since it has been used in a study of teachers who were

[11] *A manual for co-leaders,* 1965, Great Books Foundation, Chicago, pp. 14-27.
[12] Flanders, N. A. "Intention, Action and Feedback: A Preparation for Teaching," *Journal of Teacher Education,* 14, 251-60 (summer, 1963).

using BSCS materials.[13] The classification scheme is three-dimensional; the dimension most closely related to our topic in this section is the dimension of style of thinking. Five categories of this dimension are defined: "*description,* or the defining or describing of aspects of a concept or event; *expansion,* which would lead the group off to other lines of thinking or encourage new association; *explanation,* which would focus on reasoned argument through sequential deductive steps of thinking; *evaluation-justification,* which reveals an attempt to make a decision and then explain the reasons for the judgment; *evaluation-matching,* which depends on the presence of previously established criteria for judgment and attempts to match events or circumstances to those criteria." [14]

Because questioning techniques as a basis of teaching strategy are so important to an enquiry-approach to high school biology, BSCS is developing two models that can be used in teacher preparation. Each model consists of a set of the Inquiry Slides along with a taped discussion of a teacher using these slides with a group of students. (These models for teaching were made available during the summer of 1969 and can be obtained on loan from the BSCS.)

STRATEGIES FOR DISCUSSIONS

Even though we have dealt separately with questioning as a basis for teaching strategies, it is obvious that questioning and discussion are closely allied topics, since they provide different perspectives of the same class activity. One way of viewing discussion is presented in Chapter 6 of this book. However, other comments [15] on discussion may also be helpful.

In the biology programs developed by BSCS, the use of the laboratory is an integral part of the student's learning activities. But laboratory work for the sake of laboratory work may not contribute greatly to the learning even though students enjoy manipulating laboratory equipment. The usefulness of a laboratory activity depends on what has gone before to prepare for the investigation and on what questions can grow out of the activity to encourage further investigation. Stated another way, a laboratory activity becomes significant to the degree to which students are involved intellectually, not just me-

[13] Gallagher, J. J. "Teacher Variation in Concept Presentation in BSCS Curriculum Program," *BSCS Newsletter,* No. 30 (January 1967), pp. 8-18.

[14] *Ibid.,* p. 11.

[15] Comments on pages 00-00 are based on material in the Introduction to the Teacher's Handbook of *Biological science: patterns and processes.*

chanically, in the activity. Without discussion, then, the laboratory activities would be a pleasant but unrelated set of activities. The discussion develops greater understanding of the processes of enquiry and relates major concepts arrived at through the laboratory and other activities.

Discussion should be an interplay between the teacher and the student involving a person-to-person relationship. The formation of this relationship requires time, for it involves the recognition and respect of the student as an individual. Listen carefully to what he has to say; ask questions to understand the student's point of view. If your students do not respond right away, be patient. Some of your students may think slowly and carefully, some may not have understood the question, some may be reticent to speak up in class because of past unpleasant experiences. Once a productive interaction between teacher and students has been established, the teacher can begin to encourage similar interaction among students.

Below are some comments received from teachers regarding discussion activities in their classrooms:

"The classes are made up mostly of nonverbal students who are very difficult to work with in discussion periods. These students believe they should not open their mouths unless they are absolutely sure—hence the quiet discussion."

"Comments like 'why,' 'how come,' 'look at mine,' etc. have become common expressions in my classroom."

"But how do you get the students to realize the chains are off?"

"I find my discussion periods go much better if I don't stand in front of the class throughout the discussion; I often sit down, especially when I realize the students are exchanging ideas with one another."

"Discussions should not go beyond the point of student interest."

"They aren't as nonverbal as we assume. They're not verbal in the same way we are, but they have a great deal of verbal strength."

"It is difficult to know when to step aside and when not to."

"Discussions sometimes degenerated into what a few of the more vocal students wanted to say. There should be more small group discussion activities."

"It was very gratifying just to step back and let them express themselves after about six weeks of class."

"My class seemed to take a long time to answer the questions concerning the material. In fact, on some days it seemed like eternity. In the beginning, no one would say anything."

"The idea that the students are to think, and most importantly that they themselves come up with the main idea, is a wonderful experience both for the teacher and the students. This hasn't been easy. But I feel I have attained this by encouraging the students to have ideas and to discuss."

And here are some comments from students regarding discussion:

"But honestly, I think this class teaches you quite a lot. It lets *you* do something. You don't just sit and listen to some teacher talk."

"I think the thing I liked best was when we were in class and we answered a question. The way the teacher replied to it always made me think I was not entirely wrong even though I might have been."

"It was discussion and we could ask about anything we didn't understand."

Besides the necessity for teachers to develop skill in handling discussion, it is valuable to assist students to become discussion leaders also. This not only facilitates small group discussions and thus encourages greater student participation but also tends to encourage students to develop their own questioning skills. This aids in refining their thinking. There follow some suggestions for activities that provide opportunities for students to develop their discussion-leading skills.[16]

1. Give the students opportunities to compose their own discussion questions.

2. When there is a reading or problem-solving homework assignment, ask for a volunteer to lead a discussion on it the next day. Offer to give any assistance the student might need in preparing for his role.

3. Group discussions involve more students and provide good practice in leading discussions. Change groups from time to time in order to create new group interactions.

4. Students who become particularly good discussion leaders might be asked to lead an Invitation to Enquiry. Again, the teacher can offer the student assistance in preparing for this.

5. Some teachers have found tape recorders helpful in improving discussion techniques. The discussion can be recorded and then evaluated by all the participants. Such questions as the following aid in evaluation: Did everyone speak loudly enough for all to hear? Are

---

[16] An approach to developing student discussion and enquiry competencies is being developed by personnel at the Mid-continent Regional Educational Laboratory. Material pertaining to the "Inquiry Role Approach" should be forthcoming in late 1969 or 1970.

arguments supported by facts or by opinion? Is the discussion monopolized by a few? Was known information used to solve the problem? Did the participants recognize and use significant contributions of others?

The Invitations to Enquiry, presented in Section 2, are models for discussion that involves students in enquiry. Their use with students has been discussed in Chapter 4, but comment about their use in teacher preparation are in order here.

Each Invitation contains suggestions to the teacher for a teaching strategy. Despite the "program" format, the teacher is faced with divergent student thinking and he must deal with it in terms of its pertinence to the problem at hand. He himself is forced to think; he must give up his role as the final authority on matters of science. In some instances, student comments will put the teacher in jeopardy unless he understands the approach. The teacher not only must be a judge of the relevancy of expected student responses but also must honor, in an adequate way, the thinking of students who are not "in tune." That is, some students may have thoughts that are beyond the scope of the lesson or in addition to what the writers of the Invitations anticipated. The teacher must cope with these ideas in a way that both promotes the objectives of the lesson and allows students to do critical and logical thinking. The teacher must realize that student contributions *are* the substance of teaching science as enquiry and that unless student thinking and involvement are at a maximum, the goal of the enquiry is negated. For these reasons, prospective or in-service teachers must have the opportunity to practice the Invitations until they feel comfortable with them.

The amount of practice needed to effectively implement the Invitations will vary. Some teachers, accustomed to an enquiry approach, can use the Invitations with little or no practice. Most teachers, however, do require practice in their use. The Invitations should be considered as models of teaching strategies which promote classroom interactions, generate student offered information, reduce teacher exposition, and organize and guide discussions without teacher domination.

Many biology teachers have described new insights into their teaching based on the use of the Invitations. These models have provided teachers with a format for devising new approaches to presenting prelaboratory discussions, for extending laboratory exercises into open-ended experiences, and for keeping the idea of suspended judgment in making decisions before the student. Teachers have reported putting new "twists" on laboratory exercises and of reconsidering the way they present discussions and textbook material because of their experiences with the Invitations.

After one becomes comfortable with the Invitations presented in Section 2, it is desirable to devise one's own Invitations. One way to begin to develop an Invitation is to analyze a research report, such as an article in *Scientific American,* and develop an Invitation based on this analysis. This approach has been used successfully in several in-service institutes during the past few years. Two examples of Invitations developed by teachers in an in-service program are presented at the end of this section.

New Invitations have also been developed by the authors of the Australian adaptation of the BSCS materials. These authors found that since the original Invitations were designed for the first three American versions, it was difficult to integrate them closely into the Australian course. Thus they developed a new series of Invitations, a total of 22, which were designed for use at specific points in their course. Two examples of the Australian Invitations follow the examples of Invitations developed by teachers.

A final comment about class discussion: make sure that each discussion is *resolved* in some way, even if only by a decision to seek further evidence or to withhold judgment.

EXAMPLE: INVITATION TO ENQUIRY

USE OF PESTICIDES; FORMULATION OF PROBLEM

[Based on "Inimical Effects on Wildlife of Periodic DDD Applications to Clear Lake," *California Fish and Game,* January, 1960.]

To the student: (a) In December, 1954, one hundred western grebes were reported dead on the shores of Clear Lake in the coastal mountains of northern California. (Grebes are a species of diving water bird which feed on fish.) No other animals or plants seemed to be involved in this unexplained die-off. The question is, "Why are the grebes dying?" If you were a research employee in the Department of Fish and Game, what hypotheses could you offer that might lead to an explanation?

[Most likely hypotheses are: (1) Grebes have a contagious disease; (2) grebes have been poisoned without the poison affecting other organisms in the area; (3) overpopulation, overcrowding, or some other mechanism related to a population explosion; and (4) cyclic die-off due to aging or completion of breeding cycle.]

To the student: (b) Some of the grebes were sent to the Department of Fish and Game Disease Laboratory. Infectious disease was not detected. In March, 1955, and again in December, 1957, more dead grebes were reported at Clear Lake and autopsies on specimens submitted to the laboratory were also negative for infectious disease. A few weeks after the December, 1957, die-off two sick grebes were sent to the disease laboratory. Again, no infectious disease was found.

Two sections of fat from the birds were submitted for toxicological examination. Results indicated that DDD, a pesticide very similar to DDT, was present in the fat at the unusually high concentration of 1600 parts per million (ppm). From this evidence, which of our previously stated hypotheses seem most probable? How could this hypothesis be tested further? What might be some difficulties in obtaining data relevant to this hypothesis?

[The poisoning hypothesis seems most probable. Some of the difficulties in obtaining required data are: undesirability of killing off a portion of a population already severely depleted; possibility of study of effects of DDD on live birds is difficult because stresses associated with penning, etc., might affect tolerances.]

To the student: (c) Since securing further poisoning data on the grebes poses so many difficulties, let us assume that DDD poisoning is the cause of death of the grebes. Have we really solved anything? What questions can be asked now?

[What is the source of DDD? Why does it seem to cause only the death of grebes?]

To the student: (d) Let us consider the question of source of DDD. What are some ways that grebes might be getting DDD? Which of these sources seems most probable?

[In the water, in their food, being sprayed from the air, etc. Remind students that no other organisms at Clear Lake have shown observable effects. Point out that our information thus far does not give us much basis for determining which of the possible sources is most likely; therefore, we need additional information.]

To the student: (e) Here are some additional data. For many years previous to 1949, there had been a serious gnat problem at Clear Lake. This had an adverse effect on the resort business. Studies were made of possible control methods and DDD was selected as the best answer. In 1949, a carefully calculated application was made that gave a final dilution of approximately 0.014 parts DDD per million parts of water. This controlled the gnat problem until 1954 when a second application was made. The treatment was repeated in September, 1957; the same dilutions were used as in the previous applications. How could these facts relate to the death of the grebes? Restate the problem as you see it now.

[Since the DDD was applied to gnats and turns up in grebes, this suggests that it is being transmitted through a food chain of gnats-fish-grebes. The problem now is, are the fish in Clear Lake the DDD poisoning source for grebes?]

To the student: (f) What data do we need to test this possibility?

[Students should indicate that fish in the food chain need to be examined for DDD.]

To the student: (g) In 1958 the Department of Fish and Game began a study that involved three separate collections of animals (primarily fish) from Clear Lake. These samplings were then subjected to chemical analysis. Partial results of their work are shown in Table 1.

TABLE 1. *Range of DDD Contamination of Specimens Collected at Clear Lake*

Species	Number Analyzed	Parts per Million DDD	
		Visceral Fat	Edible Flesh
White catfish	82	1700-2375	22-221
Largemouth bass	19	1550-1700	5-138
Brown bullhead	62	342-2500	12-80
Black crippie	15	1600-2690	60-115
Bluegill	100	175-254	5-10
Hitch	54	—	11-28
Sunfish	1	—	5
Sacto blackfish	32	700-983	7-20
Carp	50	40 [a]	51-62
Total fish	415		
Frogs	9	5 [a]	—
Grebes	7	723 [a]	—

[a] Composite samples
Source. Adapted from Table 4 of the original report.

What do these data suggest?

[Numerous statements that translate the data can be expected, such as: many fish show unusually high ppm DDD content; there is considerable variation in amounts of DDD in the different types of animals; there is a much greater concentration of DDD in the visceral fat than in the edible flesh of most of the fish; carp show about same concentrations in fat and flesh.]

To the student: (h) What new problems are posed by these results?

[What is the source of the DDD in the fish? Why is the concentration in the visceral fat so much higher than in the flesh of most of the fish? Do grebes eat all the types of fish examined or do they eat only certain ones?]

To the student: (i) Even though we have formulated more questions than we have data to answer adequately, given the information we do have, what seems to be the most plausible explanation of the death of the grebes? Support your position with the information provided and also indicate what additional data would be needed to further substantiate your position.

[At this point it is probably best to have students form small groups for discussion; follow this by each group reporting its interpretation to the class.]

EXAMPLE: INVITATION TO ENQUIRY

MIGRATION OF NEWTS; FORMING HYPOTHESES, DESIGNING EXPERIMENTS, INTERPRETING DATA

[Based on O. V. Twitty, "Migration and Speciation in Newts," *Science,* **130**, 3391, December 25, 1959, pp. 1735–43]

To the student: (a) Let's suppose you live on a ranch or a farm with a stream cutting through the property. You notice that during the spring of the year the stream is inhabited by a large population of salamanders. Examining the situation more closely, you observe that for awhile eggs are present, and shortly after this, small, immature salamanders can be seen swimming and feeding all along the stream. Several weeks later you return to the same area and the salamanders are gone. Over a period of several years you observe this same phenomenon each spring, that is, migration of the animals to the stream, breeding, and emigration from the stream. Interested in this pattern of behavior, you consult the literature and find out that not only do many animals carry on this activity each year, but some animals possess what are called "homing instincts," that is, the ability to locate *specific* areas where breeding or other activities may occur. You begin to wonder, "Do salamanders possess homing instincts, that is, do these animals return to the same stream year after year, or are their movements into the water of a random nature?" You decide to investigate this problem, but before going any further, what hypothesis might you formulate?

[If salmanders do possess "homing instincts", then the same individuals should be found in the stream each year.]

To the student: (b) How would you go about setting up an experiment to test this?

[The following possible suggestions may be made by the students:
1. Watch the salamanders very closely; someone else may foresee the difficulty involved in doing this.
2. Collect, mark, and release the salamanders—at this point you may wish to ask how they should be marked. Some may suggest pen, nail polish,

356

and paint—remind them it must be somewhat permanent; tagging may be another idea, or if some have done some reading in areas of this sort, they may mention toe clipping or tail notching.]

To the student: (c) Going a step further, since the stream is quite extensive, perhaps you want to determine what particular part or parts of the stream certain individuals may inhabit. How could you set this up?

[Possible suggestions:

1. Rope off the area or areas.
2. Place markers on the bank to indicate certain segments.
3. Screen off the zones; you may wish to encourage other students to consider this. Guide them to the idea that this would restrict their movements and interfere with their random movement—possibly another variable.]

To the student: (d) What other factors would you have to consider or maintain to make your data as valid as possible?

[1. Method of collecting, marking, time of day, etc., would have to be accomplished the same way each time so as not to create new variable factors.

2. Adequate numbers, that is, marking as many animals in each given area as possible since some may die, etc.

3. Making sure that none of the procedures make conditions unfavorable, thereby influencing the results.]

To the student: (e) This same situation was encountered by a well known embryologist, C. V. Twitty, who in his later years became a naturalist. His experimental area was a ranch in Sonoma County, California, which was traversed by a stream called Pepperwood Creek. Twitty found four species of salamanders belonging to the genus, *Taricha,* migrating into the stream each spring to breed. He set up 58 stations, each 50 yards apart, covering approximately a 1½ mile strip of the creek. The first year he collected and, using the toe-clipping method, marked the salamanders he found at each station, releasing them in the areas where they had been captured. Each spring for 5 years he collected as many animals as possible and kept a set of data on the recaptures picked up at each station. The date collected at station No. 9 are presented below. What interpretations do these data suggest?

STATION No. 9—originally 262 males marked

DATA:		
1955	61% recaptured	
1956	60% recaptured	Total of 85% of
1957	35% recaptured	original animals
1958	50% recaptured	recaptured
1959	41% recaptured	

[The data leave many questions unanswered, for example, "Would the same results have been obtained using females?" "What happened in 1957?" "Where were the original animals during the years that they were not recaptured?" These could lead to further individual investigations. On the other hand, a couple of positive interpretations can be made from the data:

1. The large percentage of total recaptures seems to indicate that male salamanders do possess "homing instincts."

2. The life span of salamanders is or can be as long as 5 years. It also would indicate that they have some mechanism for "remembering" their home area over long periods of time.]

To the student: (f) To fill in some of the gaps in the previous data, Twitty carried out some other experiments where females, as well as males, were marked; some were released up stream, some down stream, and some over varying distances on land. In each case, the substantial number of recaptures in the home segments of the stream tended to provide more positive evidence for homing instincts. Twitty then hypothesized that if these animals did possess homing instincts, they should be able to properly align themselves with the stream when placed on land. This type of alignment is called orientation. He found that at distances up to 700 yards, there was almost immediate orientation in 80-95% of the cases. Beyond 700 yards, the initial movements were random, then eventually these too became oriented in the proper direction. The final phase of Twitty's work was to try to determine what mechanism might be responsible for these "homing instincts." What possible suggestions can you make?

[Sight, hearing, smell and, if some of your students have done some reading on animal navigation, they may suggest kinesthetic sensation, that is, "memorizing" the topographical features over which the animal travels to reach his home area.]

To the student: (g) What hypotheses could you make regarding each of these?

[1. If sight alone is involved, then removing this sense would make orientation impossible.

2. If smell alone is involved, then removing this sense would make orientation impossible.

3. If hearing alone is involved, then removing this sense would make orientation impossible.

4. If the kinesthetic sense is involved, then changing the topographical features would make it impossible for orientation to occur properly.]

To the student: (h) What types of experiments could you suggest to test these hypotheses?

>   [1. Sight—covering eyes, blinding, surgical removal.
>   2. Smell—plugging the nostrils, inactivating the olfactory nerves.
>   3. Hearing—plugging the ear canals, sever the auditory nerves.
>   4. Kinesthetic—change physical features of the environment.]

To the student: (i) These are some of the same things that Twitty experimented with on separate groups of salamanders. First, he discounted hearing because tests indicated that salamanders have a very poor sense of hearing, and certainly orientation and homing of the type exhibited by these animals would depend on a very keen sense. After surgically removing the eyes and allowing complete recovery, these salamanders were displaced $\frac{1}{4}$ to $\frac{1}{2}$ mile from their home site in the stream. To test the kinesthetic mechanism, he built a platform on top of the bottom substratum of the creek, and tilted it in the opposite direction of the natural contour. Collecting animals from both up stream and down stream, he placed them in the center of this 6 foot platform and released them to observe their orientation. For the test of smell, he plugged the nostrils with vaseline. The results are presented below. How would you interpret these?

Surgical removal of eyes—1 year later, 10% recaptured in home segment. Kinesthetic test on platform—almost complete orientation in right direction. Nostrils filled with vaseline—little or no orientation.

>   [Although the data itself is rather inclusive, the recapture of sightless salamanders and orientation of the "platform group" seems to indicate that these two senses are not solely responsible for homing instincts. The complete lack of orientation by those animals with the sense of smell blocked off appears to present more positive evidence of what's involved in homing abilities. It might be wise at this point to carefully evaluate the technique used here. Twitty felt that perhaps it may have been such a traumatic experience, blocking the nasal passage, that it was impossible for them to achieve orientation. In other words, it may have been the method used, rather than the lack of smell, that brought about these results. Now the question is, "Is orientation and homing ability dependent on the sense of smell?" This is yet to be determined.]

Although these are well-thought-out invitations, they might have been developed in other ways. What modifications can you suggest that would emphasize other aspects of enquiry? When you have thought of as many modifications as possible, compare these with the Invitations in Chapter 4, Group I. Such a comparison might suggest additional ways of handling the two examples.

INVITATION A14 [17]

(This invitation should make the discussion of water transport more real to the students. It also makes some points about the use of physical systems as an aid to understanding processes in organisms. The invitation should be done *after* students have read the text section on movement of water in plants. It can be followed (for students who know some physics) by problem No. 4. See the notes in Part B of this guide for setting up demonstrations.)

TOPIC: *Use of physical systems as analogues.*
SUBJECT: *Capillarity and movement of water in plants.*

To the student: (a) In the previous invitation, our aim was to see if we could use a physical system to help understand the loss of water from plants. It is clear from the test (pp. 377–384) that biologists have tried to understand the movement of water up the plant in a similar way. Look at demonstration A. These are capillary tubes standing in water. What do you notice about the water level in them?

   [The water level is higher in the narrower tubes.]

To the student: (b) The water is held at a level in the tubes above the basic water level by surface tension. It has been suggested that surface tension is the force that pulls water up a plant. What structures in a plant might correspond to these capillaries?

   [Depending on how well they have understood the text, students may suggest xylem vessels only, or they may suggest the spaces in the leaf cell walls, or they may suggest both. If they do not suggest xylem vessels, leave out the first part of "To the student" Part C—continue from "can you see any problem. . . ."]

To the student: (c) Although the xylem vessels are finer than any of the capillaries we have set up, measurements show that they are still not fine enough for water to rise in them to the height of a tall tree.

---

[17] Invitations A14 and A15 are from the Teacher's Guide to *Biological science: the web of life*, Australian Academy of Science, Canberra, A. C. T., 1967, pp. 218-225.

360

However, the spaces between cellulose fibers in the leaf cell walls are very fine indeed. These can be thought of as tiny capillaries, and measurements show that they are quite fine enough to hold water at the height of a tall tree. But can you see any problems in supposing that the capillaries that raise the water are at the end of the branches of the tree, with xylem vessels below? How does this differ from the simple capillaries in demonstration A?

[The capillaries in the demonstration are continuous. In the suggested system we have the fine capillaries at the top drawing water through wider vessels below.]

To the student: (d) Look at demonstration B. The fine capillary on top of the larger vessel is the same diameter as the straight fine capillary. Lift the larger vessel slowly and watch the water level in the capillary. What does this suggest about the leaf capillaries and xylem vessels of the plant?

[As the larger vessel is lifted, the water is held in the apparatus and the water level in the fine capillary remains at the same height above the water level in the dish. Hence we can suggest that very fine capillaries in the leaf *could* hold water in the xylem.]

To the student: (e) If we left the apparatus in position with the capillary supporting a column of water in the larger vessel, water would evaporate from the capillary. Would you expect the water level in the capillary to drop as water evaporated? (We would have to keep the water level in the *large* dish the same by adding water as it evaporated.)

[No—the level depends on capillary action, so more water should be drawn up as water evaporates. Students may be unsure of this, and as this idea is crucial to the understanding of the system in plants, it may be advisable to mark the level in the capillary, leave the apparatus set up, and observe it for several days. (Keep up the water level in the dish—it is the height of water in the capillary above the water level in the dish that is maintained.)]

To the student: (f) Now empty the apparatus, and dip the large vessel back in the water to the level it reached full of water before you emptied it. What happens? What does this suggest about the xylem in the plant?

[The vessel does not fill. This suggests that capillary action in the leaves will not fill empty xylem vessels, but will only hold water in the xylem, or draw it through, if the xylem is already full of water.]

To the student: (g) Does this raise any problems in understanding the movement of water in the plant? Would there be any time in the life of a tree when vessels were empty and had to be filled.

> [No. Tall trees grow from seedlings (or from cuttings); the plant does not grow first and then fill with water. Hence the xylem vessels are normally full of water. (Pulling up the larger water vessel in the demonstration might be compared with growth of the tree.)]

To the student: (h) It has been shown experimentally that if a length of the stem of a plant is frozen then allowed to thaw, the shoot above the frozen area wilts and dies. As water in the xylem freezes, air comes out of solution and remains as bubbles in the xylem when water thaws. Can you relate this observation to our ideas of how water might move in the plant?

> [The experiment indicates that water does not continue to move in the plant unless there is a continuous column of water in the xylem. This is what we would expect on the basis of the model, so increases the likelihood that water does move in the way we are suggesting.]

To the student: (i) Demonstration C shows a physical system that was set up by the biologist, H. H. Dixon, as a model to show the way water might move in a plant. The pot at the top is porous (explain this term if necessary) and water evaporates continually from the pores. There are three parts to the apparatus: the pot, the glass tubing, and the reservoir of water at the bottom. To what do these correspond in the plant?

> [Porous pot—cellulose cell walls
> Tubing—xylem vessels
> Reservoir—water in soil]

To the student: (j) If we tried to make the tubing in our apparatus as high as a tall tree, the water column would break. Does this mean that water does not move in a tree in the way we have suggested?

> [No. Xylem vessels are not glass tubing—nor are leaves porous pots.]

To the student: (k) In what way do, or might, the xylem vessels in a tree differ from the piece of tubing we have used in the demonstration? Could any of these differences account for the greater efficiency of a tree?

> [There are many xylem vessels (this will not explain the different heights to which water can be raised). The xylem vessels are much finer than the capillary we have used, so we would expect the tensile strength of water in these columns to be much greater. The vessels are made of lignin not

glass—this might also affect the tensile strength of water in the vessels. The capillaries in the pot are of a different substance (that is, not cellulose) and may not be of the same size as the capillaries in the leaf cell walls—the forces in the leaf may therefore be greater than the forces in the pot.]

To the student: (1) Comparing the plant with this physical system has led us to ask a number of questions: How great would the tensile strength of water be in lignin tubes the diameter of xylem vessels? How small are the spaces between cellulose fibers in the leaf cell walls and how great would surface tension forces be in them? This is one of the values of using physical systems to investigate processes that occur in living things: by comparing the system with the organism we are often made to ask questions that take us back to look more carefully at the organism.

INVITATION A15

(This invitation is tied to Exercise 22.1 and should be followed immediately by Exercise 22.2. The invitation has two aims: to allow students to see how the action of parts of the body can be inferred from structural and experimental evidence, and to increase their familiarity with the idea of the circulation of the blood. Any hypothesis about the action of structures in living organisms is almost always based on indirect evidence; Harvey's work is a classic example of the way in which such evidence can be collected and interpreted to give a consistent picture.)

TOPIC: *Interpreting indirect evidence: inference of the action of organs from structural and experimental evidence.*

SUBJECT: *The circulation of the blood.*

REQUIREMENTS: *Student's Manual, Part 2*

To the student: (a) You know that vertebrates and some invertebrates have hearts to which blood vessels are connected. You also know that our hearts beat continually so long as we are alive. Insofar as you know, what are the functions of the heart and blood vessels?

[Students will almost certainly "know" that the blood vessels carry blood around the body, and that the heart pumps or pushes. Note that we are not asking here for the functions of the blood, but only for those of the heart and blood vessels.]

To the student: (b) What evidence do you know of for these ideas?

[Students may find it very hard to give evidence: they know because they have been told. They may think of some evidence—for example, that blood escapes from a cut blood vessel (blood escaping from any wound is not acceptable; how do they know it comes from the blood vessels?). Any evidence suggested should be examined to see whether it *is* evidence for the proposition. Some possible suggestions—for example, that death results from heart failure—may be found to contain so many other assumptions that they are not particularly useful. (In this case, the assumption is that the blood transports substances necessary for life.) Others—for example, that the blood stops circulating if the heart stops beating—may be found to be inferences from the "known" functions of the heart and blood vessels, and not observations. Genuine evidence, if any, should be listed and referred to when relevant later.]

To the student: (c) It seems that it is not easy to think of evidence for this apparently simple idea—that the blood circulates in the blood vessels, impelled by the heart. This idea was first put forward by an English doctor, William Harvey, in 1628. In this invitation we shall examine how Harvey established this idea, paying special attention to the kinds of evidence on which he based his reasoning.

Harvey's theory was by no means the first theory of the action of the heart and blood vessels. The ancient Greeks had dissected various animals (including human bodies) and knew that there were vessels containing blood connected to the heart. In the second century A.D., the anatomist Galen did many very thorough dissections, and his ideas became the basis for ideas about the heart and blood vessels up to the time of Harvey. He distinguished carefully between two kinds of blood vessels; arteries, which had thick walls, and veins, which had thin walls. At the beginning of Exercise 22.1 there is a diagram summarizing the most generally accepted idea of the action of the heart and blood vessels at the time of Harvey. When you look at it, and throughout this invitation, you must remember that microscopes were not used at this time. Hence in dissections the blood vessels appeared to dead-end in the tissues.

In Exercise 22.1 you will examine some of the evidence Harvey used to develop his theory of the circulation of the blood—the structure of the heart. You should try to decide what inferences he could draw from it.

[Students now do Exercise 22.1. It would be useful if they could look at the "Galenic scheme" before the class period when the exercise is to be done. Note that if some or all are to do the "further investigation" on valves in veins this is better completed before the invitation is resumed.]

(Continuation of invitation. Students should have with them their Student's Manual, Part 2, and their answers to the questions in Exercise 22.1. It may be useful to have a dissected heart available in case there are points of disagreement or points that have been overlooked.)

To the student: (d) What conclusions did you reach about the possible directions of flow of blood into and through the heart?

[It is suggested that students build up a diagram as they discuss this. Comparison of this diagram with Figure 22.1-1 may help to raise points that may have been overlooked—did they look for holes in the system, for example? Remember to ask whether their observations of the beating heart had any relevance.]

To the student: (e) It seems that we can infer the direction of blood flow from the structure of the heart. Do you think this is sufficient evidence that the blood *does* flow in the directions we have inferred?

> [It is good evidence—but it would be better to have some observations of moving blood to back it up. Students may or may not be satisfied with the evidence as it stands. If they are satisfied, they could be asked how they can be sure that the beating of the heart does not move the valves in some other way.]

To the student: (f) Harvey looked for further evidence that the blood does move into the heart by the veins and out through the arteries. He cut open a live snake, and observed the beating heart. Here is his account of the experiment he did (see Student's Manual, Part 2, Appendix). Do you think Harvey now had sufficient evidence to conclude that the blood enters the heart by the veins and leaves it through the arteries?

> [The structural and experimental evidence together leave little room for doubt.]

To the student: (g) In the experiment just quoted, Harvey mentioned that the pulse in the artery returned when the ligature was removed. Other questions that Harvey considered were the action of the heart itself and the origin of the pulse. Many people of his time believed that the heart filled like a bellows—it worked to distend itself, thus drawing in blood. Then when it relaxed, blood was able to flow out and was propelled along the arteries by movements of the arteries felt as the pulse. Did your observations on the frog's heart support the hypothesis that it works to distend itself and that the blood flows out when it is relaxed?

> [The students will probably have concluded the opposite (a less remarkable piece of observation on their part than Harvey's because they have not been brought up on the bellows theory).]

To the student: (h) Harvey's interpretation of his observations was that the blood flowed into the heart when it was relaxed, then was propelled out when it contracted. He also concluded that the *pulse* in the arteries was due to their distension by blood forced into them by the beating heart. What evidence might he have for this conclusion?

[There is the evidence already mentioned: that the pulse stops when the blood supply from the heart is cut off by a ligature and returns when the ligature is removed. Students may suggest other observations that could be made—they will do well if they can think of any Harvey didn't think of. List any suggestions to compare with Harvey's observations.]

To the student: (i) As well as the observations on the snake's heart, Harvey made the following observations. Consider them and see whether you think his conclusion was justified.

1. The pulse in the arteries leaving the heart coincides with the contraction of the ventricles.

2. When the ventricles stop contracting, the pulse stops. If the ventricles contract weakly, the pulse is feeble.

3. If an artery is cut or punctured, the blood spurts from it when the left ventricle contracts (or when the right ventricle contracts if the artery to the lung is cut).

Can you think of any other interpretation of these observations than that the pulse is due to the distention of the arteries by blood forced into them by the heart beat?

[It may be possible to think of other interpretations, but they are much less simple. We could perhaps suppose a very rapid transmission of some message from the ventricles to the arteries—but then they would presumably have to squeeze to spurt blood out, and cut arteries are observed to be distended as the blood spurts. There is no need to press students for alternative interpretations; the object of the question is to get them to examine the evidence carefully.]

To the student: (j) Harvey now had a clear idea of the action of the heart: that blood flowed into it while it was relaxed and was propelled out when it contracted. He also had a clear idea of the direction of blood flow: that it entered the heart by the veins and was expelled through the arteries. However, this raised another problem. If you look again at the diagram we have drawn, you will see that blood enters the left auricle through veins from the lungs. Where could this blood have come from?

[The most obvious answer is from the arteries that carry blood to the lungs. Another possibility is that it is made in the lungs; if the students suggest this, they could be asked "from what?".]

To the student: (k) If, as Harvey supposed, the blood in the veins came from the arteries, how could it get from arteries to veins?

[Students may reply "through capillaries." If so, remind them that no one had seen capillaries when Harvey was working. Alternatively, they may suggest that the blood soaked through somehow.]

To the student: (*l*) Harvey was not able to make any more definite suggestions about how the blood got from the arteries to the veins in the lungs except to suggest that it soaked through in some way. So he turned his attention to the blood entering the right auricle through the veins. At this time it was believed that this blood came from the digested food via the liver. Can you see any possible objections to this hypothesis?

[Students may not be able to think of any—or they may repeat the quantitative objections of Harvey—for example, what do you make blood out of between meals? Some store in the liver?]

To the student: (*m*) According to Harvey's scheme, all the blood that entered the heart through the veins left again through the arteries. It could not get back into the veins from the heart. Yet arteries and veins are continuously filled with blood. Now can you see any objections to the idea that the blood entering the heart through the vena cava is made from digested food?

[They should now be able to see that blood would have to be made continually in rather large quantities.]

To the student: (*n*) Harvey made some calculations. He had observed more than three ounces of blood in the heart of a dead man. So he assumed that the heart might hold two or three ounces—or even an ounce and a half. Each time the heart contracts, some blood is forced out: he supposed that only an eighth of the blood it contains is forced out (he believed that more than this was forced out, but was arguing from the minimum amount so that he could not be thought to be exaggerating). He observed that the heart beats, at the very least, a thousand times in half an hour (in someone with a pulse rate of 70 beats per minute, which is about average, it beats more than 2000 times in half an hour). Using these figures—the heart holds at least 2 ounces, loses at least $1/8$ of its contents at each beat, and beats at least 1000 times in half an hour—what is the least amount of blood that could be lost from the heart in half an hour?

[250 ounces.]

To the student: (*o*) Remembering that this was a very conservative estimate—the real figure probably being many times this—and that this figure is for only half an hour, Harvey would certainly seem to have been justified in rejecting the idea that all the blood entering the right side of the heart came from the digested food. So where could it have come from?

[Again, from the arteries.]

To the student: (p) Harvey was convinced that the blood must get somehow from the arteries to the veins, so that it was the same blood flowing through the heart over and over again. He described this as "a motion as it were in a circle." We call it the circulation of the blood. But because this was a very new idea, and because he could not give an explanation of how the blood moved from the arteries to the veins, he looked for still more evidence. This theory required that the blood should always flow away from the heart in the arteries and toward it in the veins. Although he had shown that the blood entered the heart through the veins, he had not yet shown that it did not reverse directions in the limbs, for example, and flow into the tissue via the veins as was supposed in the Galenic scheme. Do you know of any arguments why this should not be true?

[If any students have done the "Further Investigation" to Exercise 22.1— or even read it—this is their cue. Otherwise it is necessary to describe the experiment in detail.]

To the student: (q) Harvey demonstrated that the valves in the veins (which had been discovered by others who had not recognized their significance) would only allow the blood to flow toward the heart. He then did a further experiment which he described as follows. As you read (or listen) try to work out what further evidence this gave for this theory (see Student's Manual, Part 2, Appendix). How do you think Harvey interprets these results?

[The observation that the veins become gorged with blood only when the ligature is loose enough to allow blood into the arteries, and tight enough to prevent it flowing back to the heart in the veins, is very good evidence that the blood moves from the arteries to the veins then back to the heart.]

To the student: (r) The weight of his supporting evidence was such that Harvey's ideas became accepted almost at once by most scientists. It is interesting to notice that all his evidence was indirect. Can you think of any way in which we might today demonstrate the circulation of the blood directly?

[Well, can you?]

To the student: (s) It is inevitable that most if not all the evidence for the action of parts of *living* organisms must be indirect evidence. We simply cannot get inside a living organism and watch it work; it is extremely difficult to put instruments inside it and be certain that we have not altered its working. Hence a physiologist who studies the workings of living things must be able to find convincing indirect evidence.

Another important thing is that although Harvey worked out a convincing theory of the circulation of the blood, there were several observations he could not explain (as well as being unable to explain how the blood moved from arteries to veins). He was not able to account for the observation that arteries had much thicker walls than veins, nor could he account for the observation that had been made by others that the blood in the arteries (and in the pulmonary veins) was a more vivid red than the blood in the veins (and pulmonary artery).

(It is desirable that Exercise 22.2 on capillary circulation be done as soon as possible after completing this invitation.)

PLANNING THE COURSE

The Teacher's Guides of the various BSCS versions provide many suggestions for planning the course, from a given lesson to the entire year's work. For this reason, only one major point with regard to course planning is made here: the BSCS course should be considered a dynamic interaction between the students and subject matter which is guided by the teacher. A corollary of this is that each version, or any combination of the BSCS materials which the teacher might select for use, should be considered as aids to the teacher in structuring the course so that the interaction between student and subject matter is likely to occur in the most effective way. Another corollary is that the subject matter is *not* the textbook; the subject matter consists of biological entities and events that have been investigated in various ways resulting in concepts and generalizations about those biological entities and events.

Many teachers have found that the most successful way to structure their course so that there is maximum interaction between students and subject matter is to use the textbook primarily as a resource which is read after the students have engaged in a topic through laboratory and discussion. As stated in the Australian version Teacher's Guide,

"Students should be encouraged to regard the text as a supplement to their student's manuals and practical records, as a record of the data and interpretations of other biologists, and is designed to be read by the student in his own time." [18]

Below is an example from the Australian Teacher's Guide that illustrates a way in which the course can be structured by integrating a variety of materials and activities. It also illustrates a way of structuring a course in relation to the major ideas which the student will arrive at as a result of the course work. (Another illustration of structuring the course in terms of major ideas is provided in the Teacher's Handbook of *Biological science: patterns and processes*.)

"*Major Ideas:* There is strong evidence that the stated objectives of the course can best be achieved by a course based on laboratory work which involves direct, personal observation and experimentation by the students. It is only by becoming active participants in the processes of science that students can learn what it is all about.

"During the course students will also be introduced to some of the major generalizations of biological science. These generalizations have been set out in the form of 'major ideas' which were arrived at after much discussion between teachers and research biologists. The major ideas therefore form the core of the course; they are the generalizations which it is hoped *students will form for themselves* as a result of their work. The activities set out for the course are designed to give students the experience which will allow them to form these ideas." [19]

[18] Teacher's Guide, *Biological science: web of life*. Australian Academy of Science, Canberra, A. C. T., 1967, p. 11.

[19] *Ibid.*, p. 6.

Example [20]

Ideas	Suggested Student Activities	Notes
The Basis of Science Is Enquiry	*The Laboratory* (Introduction to Student's Manual)	Homework Reading
Careful observation is basic to scientific investigation.	*Exercise 1.1 — Observing Living Things.*	2 periods observation, plan Ex. 1.2 first part of 3rd period; discuss Ex. 1.1 remainder
Although familiar organisms can be divided into the broad categories "plants" and "animals," such classifications are artificial and some organisms are hard to place.		
The scientist interprets his data. Experiments are designed to yield data capable of interpretation.	*Exercise 1.2 — An Experiment: The Germination of Seeds.* *Invitations A1 and A2 —Soil Conditions and Plant*	Students read before class. Half a period planning, short periods on succeeding days until data ready to discuss. Read notes on use of invitations.
It is sometimes useful to consider organisms as members of the functionally defined categories of producers, consumers, and decomposers.	*Text Reading—Producers, Consumers and Decomposers* (pp. 1,2.). *Guide Questions: Nos. 1, 2.* *Exercise 1.3 — Interrelationships of Producers and Consumers.* *Exercise 1.4—Action of Decomposers.*	Homework, Students to record answers to Guide Questions. Set up Ex. 1.3 and 1.4 now, discuss later. Leave 1.4 for many weeks.

[20] *Ibid.,* pp. 16-18.

Ideas	Suggested Student Activities	Notes
Matter is cycled from the environment through organisms and back again.	*Invitation A3 — Materials Used by Growing Plants.* *Text Reading—The Foundation of Life* (pp. 2–5). *Guide Questions:* Nos. 3 and 4. *Problem:* No. 3.	Schwab Inv. 3 may be substituted. Check answers to guide questions before discussing Problem 3.
Energy from the sun is incorporated into organisms, and transferred from one to another. Energy is eventually dissipated.	*Text Reading—Energy and Living Things* (pp. 5–8). *Problems:* Nos. 1, 2.	Do at least one. Ex. 1.2 and 1.3 will be ready for discussion about now. Discuss preliminary data from 1.4 soon, but leave set up.
	Text Reading—The Web of Life (pp. 8–12). *Guide Question:* No. 5. *Problem:* No. 4.	Homework.
The scientist uses instruments as an aid to making observations.	*Exercise 1.5 — Use of the Microscope: Introduction.* *Exercise 1.6 — Use of the Microscope: Biological Materials.*	These two exercises can be done earlier or later than this. Before Ex. 3.5 if this is to be done, otherwise any time before Ch. 6. Up to 4 periods.
	Problems: Nos. 5, 6.	See notes. Ex. 3.5 may be set up now, or Ex. 1.4 may be used later to illustrate succession.

STRATEGIES OF EVALUATION

Since Chapter 8 deals with evaluation in detail, only two points are made here.

1. Recent evaluative studies of the teaching of BSCS biology [21] suggest that two things are critical in determining what occurs in class and hence what students are likely to learn. These are the teacher's conception of his role and the teacher's understanding of the rationale of BSCS biology. This chapter and subsequent ones in Section 3 are designed to help the teacher formulate a conception of his role as a guide to learning and to suggest activities that will help the teacher to develop competence as a guide. The rationale of BSCS biology is dealt with explicitly in Chapter 2 and is implied in much of the material in these chapters on the teaching of biology.

2. Evaluation of students can have two major purposes: to *judge* their achievement and to *aid* them in assessing their learning progress. The latter should be given much more emphasis than has heretofore been the case. Much of Chapter 8 is devoted to describing and illustrating strategies for assisting students to assess their progress.

SELECTED READINGS

Many recent publications deal with teaching strategies. Although most of those in the following list do not deal with science teaching, they contain many fruitful ideas which could be applied to the teaching of biology.

Association for Supervision and Curriculum Development. *Evaluation as feedback and guide.* Washington, D. C., 1967.

Fearnside, W. W., and W. B. Holther, *Fallacy; the counterfeit of argument,* Prentice-Hall, 1959.

King, Martha L., and others, (editors), *Critical reading,* J. B. Lippincott, 1967.

Mager, R. F., *Developing attitude toward learning,* Fearon Publishers, Palo Alto, Calif., 1968.

Raths, Louis E., and others, *Teaching for thinking: theory and application,* Charles E. Merrill, 1967.

Torrance, P., and R. Strom, *Mental health and achievement: increasing potential and reducing school dropout,* Wiley, 1965.

Webster, Staten W., *Discipline in the classroom,* Chandler Publishing Co., San Francisco, 1968.

[21] *BSCS Newsletter* No. 30 (January, 1967). This issue is devoted entirely to evaluation studies.

CHAPTER

6

Discussion in the Teaching of BSCS Biology

Because many teachers use discussion as a basic classroom activity, it is worthy of examination as an activity. Are there characteristics of discussion which are conducive to certain kinds of learning? Are there different forms or qualities of discussion which produce certain kinds of learning?

One of the main emphases of BSCS biology has been the expressed goal of conveying an understanding of the processes of biological science as well as of its products. The phrase "science as enquiry" is used to designate those processes. Nevertheless, these two—process and products —are not to be considered as two separated matters. The processes make sense only in connection with concrete investigations aimed at solutions to real and concrete scientific problems. Conversely, a body of *modern* scientific knowledge is never thoroughly and correctly understood until something of the enquiry which produced it is also understood. This follows from the extent to which modern science, including modern biology, has moved to a theoretical level—that is, the extent to which facts are sought under the guidance of conceptual frames of reference and are couched in terms of these concepts.

For example, to understand the current knowledge of carbohydrate metabolism requires much more than the details of a number of chemical exchanges. We need to understand the conceptions of structure-and-function, of homeostasis, of the regulative organism—and we need to understand them, not abstractly, but as the working tools of biological investigations. Hence it becomes important, if not indeed necessary, to work through some of the key rescarch (in one form or another)

SOURCE. Adapted from *Eros and education* by J. J. Schwab and Evelyn Klinckmann. Copyright 1958 by the authors. Permission granted to the BSCS for publication in this book.

which has led to this knowledge of carbohydrate metabolism. It needs to be worked through, moreover, in such a way that the major problems dealt with by the researchers are dealt with by the students, until they, the students, come to understand what went into solving these problems.

For a thorough understanding of bodies of knowledge discussion is an important tool. A greater depth and degree of understanding characterizes knowledge gained in this way. Further, discussion of a certain sort is indispensable for the development of the intellectual arts and skills required to traverse a pathway of understanding. This is true because discussion itself can be an engagement in and practice of the activities of thought and communication involved in those intellectual arts and skills.

There is another reason why discussion of a certain kind can be a valuable means to developing an understanding of the processes of biological science. It is that discussion can utilize the "energy of wanting" in the pursuit of educational goals.

Appetite, or wanting, leads us to action. This energy of wanting is as much the energy source in the pursuit of truth as in the pursuit of other contemporary values—pleasure, power, fame, friendship, or security. Therefore, a means of education that utilizes this energy source is an effective means—although, of course, other criteria also must be used in selecting the best educational means.

Discussion is especially appropriate as a means of utilizing the energy of wanting because it can tap motivating forces. It provides something in its initial phases to which the energy of the student can attach. This is true because the students' desire to like and be liked can be directed toward the teacher if the initial phase of discussion is conducted properly. In later phases of discussion this energy, originally attached to the teacher as a person, can be transferred, by appropriate discussion techniques, to the desire to want and to practice the qualities which the teacher as an educated person exemplifies.

Discussion of the type that is a means of understanding processes of biological science and can tap motivating forces in the student is a situation composed of four factors. The central factor in this situation is a certain kind of face-to-face relation between teacher and student. The other three factors are enabling or supporting factors. These are the administrative organization of the school, the physical conditions of the classroom, and the curriculum materials. These supporting factors are important as they aid in establishing the kind of face-to-face relation which is central as a beginning to constructive discussion; they also are important because they can facilitate the channeling of the effects of this relation toward desired outcomes of the curriculum.

The first and central factor in discussion is a certain kind of face-to-face relation between teacher and student. This must be established from the first day, but its characteristics should continue to be apparent throughout the school year. The desired relation is based on a direct interpersonal relation between teacher and student. By this is meant that teacher and student meet as individual person to individual person. The key to this kind of interpersonal relation can be called "reciprocity of evocation and response." Let us try to illustrate this by describing what might occur on the first day of a class—what, in fact, should occur on the first day.

Some students will be curious about, have an interest in, the teacher as a person; that is, the teacher "evokes" interest in some students. As the teacher presents his introduction to the course work, he will glance from student to student and note responses in certain students which indicate this interest. As he notices it, the teacher can, in turn, respond by momentarily halting the movement of his eyes and for a brief period directing his remarks to each student who shows an interested response. By doing this the teacher shows that the student has evoked his interest and he is responding by recognition of this student as an individual person. We should notice that reciprocity of evocation and response requires recognition of and liking for students as *individuals;* it also requires recognition of and liking of indivdual *qualities* of persons.

This kind of interpersonal relation cannot be established the first day with all the students. In fact it is not desirable to do so, since it is only over a period of time that different students will show an interested response. In addition, over a period of time different qualities of students will become apparent to the teacher, and he can respond to them in different ways. Students are quite aware of their differences; if a teacher treats all alike, a lack of respect for the teacher can ensue. Thus by recognition of only a few students the first day, and by recognizing others because of different qualities on subsequent days, respect as well as liking for the teacher can begin to be established.

Let us describe one concrete example of recognition of different qualities of students. A student has shown little interest in either the teacher or what has been going on in class. The topics of classwork have been primarily related to biochemistry. The teacher suspects that the disinterested student is mainly interested in outdoor activities—in camping, hunting, and fishing. One day the classwork deals with the problems of field research in ecology. The teacher directs more questions than in previous class periods to the disinterested student and finds by his response that the hunch about the student's interest is correct: He shows an interest in this topic he had never shown in biochemistry. One of the culminating points of the classwork on the problems of field research is the fact that certain biological problems cannot be dealt with in the laboratory because this changes factors which are key variables in the field situation. The student shows even greater interest in this because it is the first time he has realized that science is not always conducted by white-coated men in laboratories. A whole new aspect of biology has been opened for him, and from then on he shows greater interest in all the classwork.[1]

Let us return to the question of the teacher showing a liking for students and responding to being liked by students. This has sometimes been criticized as indicative of a weakness on the part of the teacher. It may be—if the teacher *needs* these indications of liking for his self-assurance. But the mature teacher who does not need them can still welcome them. Moreover, the teacher is not manifesting indiscriminate liking, which would include what is infantile in the student. He manifests a liking for what is a central characteristic of growing up—the student's potential maturity of intelligence and of *wanting to know.* This is a prime and eternal characteristic of the young which teachers *should* like—but which we too often ignore or even destroy, so that often the youth of high school age has the last remnant with which teachers can work.

It is difficult to describe accurately the quality of the initial establishment of a genuine interpersonal relation involving reciprocal evocation and response, but to describe the extremes which are the two pitfalls of such an attempt will help to define the limits of the desirable relation. One pitfall is an over-anxiety on the part of the teacher to be liked. This is perhaps most often a fault of the beginning teacher. An over-anxious teacher will overstate his wish to respond to or to like the student. Then respect is lost. The other pitfall is coldness on the part of the teacher. This may be a simple coldness or may take the form of impatience or of being preoccupied with other matters. In each case, the teacher is likely to reject the curiosity and interest of the student, which may lead to the loss of some liking, since the student once rejected is less likely to try again.

We can review the characteristics of this first phase of discussion by making a check list of these characteristics clarified by examples of

---

[1] This, like the other illustrations presented, has been slightly fictionalized, but it is based on the actual experience of a teacher and student.

behavior typical of them. Notice that some of the items listed are positive descriptions of our discussion characteristics, some are negative.

TEACHER-STUDENT INTERACTION IN EARLY PHASES OF DISCUSSION

A. Reciprocity of evocation and response

1.(+) Teacher shows signs of recognizing students as individual persons. (Looks directly at individuals to whom speaking; lets gaze move to one student after another while speaking, momentarily stopping gaze at each one. This indication of genuine awareness should not be confused with the "gimmicky" version typical of some television masters of ceremony.)

2.(—) Teacher shows signs of relating to class as an amorphous mass. (Few or no direct glances at individuals; eyes do not move from one student to another; looks at back of room, out of window, or apparently at students but with glassy stare indicating no real recognition.)

3.(+) Teacher shows signs of recognizing individual qualities of students. (Speaks and relates to different individuals in different ways. On subsequent days, this might be shown by the teacher's remembering comments of individual students from one day to the next and referring to them where pertinent. This might also be shown by directing questions to students with certain interests when they relate to the topic and when these students have not previously participated actively.)

4.(—) Teacher shows coldness toward students. (Lack of spontaneity, or naturalness as a person; impatience; preoccupation.)

5.(—) Teacher shows over-anxiety to be liked. (Is gratified by obvious and insincere flattery; lets students "walk all over him"; does not involve his authority when necessary for fear of being disliked.)

As described, the establishment of a genuine interpersonal relation leads to a mutual liking between teacher and students. But a second stage in teacher-student interaction usually occurs and is essential if the necessary respect for the teacher is to be established. This is the stage of provocative testing of the teacher's judgment. This phase is essential since the adult-youth difference is not a sufficient basis for respect as it may be in the parent-child relation. In the teacher-student relation, the student justly demands the right to admit the adult to the role of teacher. [This is a just demand because the adolescent student does not admit to an inevitable and necessary childishness. Since a child (as against an adolescent) does admit this, the adult-youth difference is sufficient in a parent-child relation: with this admission, the adult status is a solace to the child.]

The provocative testing of the teacher's judgment can occur in a multitude of ways, but all have essentially the same characteristic of being an inappropriate contribution to the business at hand. This may take the form of pretended eagerness; contributing an esoteric application or source; presenting irrelevant opinions or facts, manufactured questions, irrelevant skills, or irrelevant bits of knowledge. An example of presentation of an esoteric application or source is the exceptionally brilliant student who brings up for discussion a matter which may be related to the topic in hand but which is far beyond the current understanding of the other members of the class. If the teacher succumbs to the temptation to discuss what to him may be more challenging and interesting, even though the class as a whole will not benefit from it, he has failed in his judgment. Very occasionally such a discussion, in brief, may be beneficial to the class as a whole in that it may provide glimmerings of possibilities which the average students have not yet suspected. Such a discussion extended or indulged in too often is, however, a failure on the part of the teacher.

An example of presenting irrelevant skills is the student with unusual ability in painting who discovers his biology teacher is interested in this form of art. The student then uses every opportunity to interject his special skill into the discussion to attract the attention of the teacher. If the teacher responds by providing the attention, without appropriate indication to the student that his contribution is irrelevant, the teacher has failed in his judgment.

The purpose of these various offerings is to test the teacher's judgment—even though the students, as a rule, are not consciously aware of this. Thus the proper response to these provocations is an immediate, clear communication to the student of the teacher's awareness of the questionable nature of the proffered contribution. Along with this, however, the teacher must be careful not to reject the student along with the offer. For example, if the contribution is esoteric, the teacher may suggest that the student ask after class; he may wait patiently while a contribution is presented, then merely continue with the business at hand, but without revealing by facial expression, posture, or gesture either contempt, impatience, or a patronizing patience. Another possible response is to indicate the connection of the contribution with another aspect of the material without blaming the student for not seeing this connection. Or the teacher may even postpone consideration of the contribution, with the permission of the student, on the grounds of the prevailing interest of the group or of personal privilege.

A somewhat different way of handling provocative testing is to compliment a student occasionally when he makes a particularly relevant contribution or expresses himself especially well. In this way students become aware of good examples of relevance and use of language and may be encouraged to improve their own contributions accordingly. Such occasional compliments ·also indicate that the teacher can distinguish between the relevant and the irrelevant, the poorly expressed and the well expressed, and thus may eliminate to some degree the students' need to test the teacher's judgment.

If such provocative testing is successfully handled by the teacher, such success will be evident by the subsequent behavior of the students. This behavior will entail a continued participation, but with fitting contributions; it will also entail the students' ignoring opportunities of reopening tangential points or questions. With partial success, the students will continue to participate with fitting contributions but will immediately remind the teacher of his promise to deal with the earlier, inappropriate contribution. ("Last week you said we could talk about . . .") Lack of success will be indicated by the continuing provocative behavior of the students. Occasionally the latter may continue to occur with particular students. This may not be as much a failure on the part of the teacher as an inability on the part of the student to establish a working relation with any person who to him is an authority-figure. Continued patient handling, as suggested, may lead the student to discover the profit and pleasure in a relation with an adult. In other cases, only professional aid which is outside the teacher's task and for which, as a teacher, he is not responsible, may be the answer.

A third phase in teacher-student interaction involves another kind of testing on the part of students. This is the testing of the psychological space—the degree and kind of tolerance the student may expect from the teacher—of the class. This very often takes the form of students' showing eagerness to talk and attract attention, to contribute to the early class discussion by offering at least superficially appropriate contributions. These contributions may reveal awkwardness or error or may be a misapplication of intelligence or judgment. In this case the student needs to be reassured that his attempts will be accepted as attempts, not as definitive measures of his powers or limitations. Such acceptance makes possible future growth.

The appropriate response of the teacher to this form of testing is to be measuredly and discriminatingly kind while providing correction and assistance. To withhold correction and assistance is to reveal an indiscriminate permissiveness inconsistent with the role of teacher; the student wants to be taught, he wants to be accepted as a *learner*—not just accepted. A lack of kindness or actual destructiveness is revealed by demanding inappropriately adult standards, by expecting the student to show adult responsibility toward words, choices, and decisions. The students are not adult; one of the things they must be helped to learn is adult standards.

Let us present an example. The students are asked for a common language word that can be used for the statements technically known an *hypotheses*. One student says *idea* and is rejected out of hand by the teacher. This is demanding inappropriately adult standards pertaining to meanings of words. The way the teacher should work with the proffered contribution is to assist the student to see that "idea" is too broad in its meaning, but that by properly qualifying "idea" he would have a term that meets the requirements of the original question—for example, "a *possible* idea."

These two ways of handling contributions—demand of inappropriately adult standards and indiscriminate permissiveness—are the two pitfalls that set the bounds of appropriate response by the teacher to the contributions which test the psychological space of the class.

We can review the characteristics of teacher-student interaction in the later phases of discussion, in which the two types of "testing" behavior occur, by making a check list of these characteristics.

TEACHER-STUDENT INTERACTION IN LATER PHASES OF DISCUSSION

B. Provocative testing of teacher's judgment

1. Students test judgment by inappropriate contributions. (Pretended eagerness, esoteric application or source, irrelevant opinion or fact, fake interest, manufactured questions, display of irrelevant skills, irrelevant bits of knowledge.)

2.(+) Teacher responds to such tests by indicating awareness of questionable usefulness of proffered contribution. (Suggests student ask after class if contribution is esoteric; waits patiently, then continues; indicates connection of contribution with other aspect of problem; postpones contribution on ground of prevailing interests of group or of personal privilege.)

3.(+) Teacher does not reject student with offer. (When responding as illustrated under 2, the teacher does not reveal contempt, impatience, or patronizing patience by expression or posture; does not blame student for not seeing lack of relevance to matter under discussion; does not peremptorily postpone consideration without permission of student.)

4.(+) Students participate with fitting (appropriate) contributions.

C. *Testing of psychological space of class (psychological space: degree and kind of tolerance student may expect from teacher)*

1.(+) Students show eagerness to talk and attract attention by at least superficially appropriate contributions.

2.(+) Teacher responds by being measuredly and discriminatingly kind. (Does not reject student; does not "make fun of him" or "ride" him because contribution is not precisely appropriate.)

3.(+) Teacher responds by correction and assistance. (Continues to work with the student who has made the contribution until the student has moved at least slightly toward a better understanding than he has; does not shift question immediately to another student if contribution is not exactly what teacher had in mind.)

4.(—) Teacher shows inappropriately adult standards. (Expects students to show adult responsibility toward meanings of words, choices, decisions.)

5.(—) Teacher shows indiscriminate permissiveness. (Does not correct and assist students.)

6.(+) Students show responsibility toward words, choices, decisions.

If the three preceding stages—the establishment of genuine interpersonal relation, the successful handling of provocative behavior and of behavior testing tolerance and acceptance—have been successfully traversed with a class, the teacher has in his hands a potent instrument. It is at this point that a liking and respect for the teacher as a person must be transferred from the person of the teacher to the qualities and capacities of the teacher as an educated person. There must be a shift from liking a person with certain qualities and capacities to a liking for the qualities and capacities in their own right. The student who is not permitted, encouraged, and helped to make this shift has been enslaved or used rather than taught. If the shift is made, students will first recognize, then show *pleasure in practicing* the qualities and capacities which the educated teacher possesses. That is, the student begins to *emulate* the teacher's characteristics as an educated person rather than only like and respect him for those characteristics. To evoke such emulation is the ultimate art of teaching.

A check list of characteristics pertinent to this aspect of discussion follows.

D. *Movement from mere liking or respect to discrimination of certain qualities and capacities in the teacher*

1. Students show liking and respect for the teacher as a *person*.

2. Students show recognition of *qualities* and *capacities* exhibited by the teacher when he is a "model" of an educated person. (Working with others—the students—to gain clearer understandings; confident in his competence yet showing humility with regard to the extent and certainty of his knowledge.)

3. Students show signs of enjoyment in *practicing* the qualities and capacities of the teacher as an educated person. (Conjoint work with other students as well as with the teacher, as in helping another student better understand something, in suggesting that another student might have something to contribute on a certain point; earnest pursuit of progressively clearer and more extensive understanding; humility with regard to what he does know, especially if he is a superior student.)

This brings us to a consideration of the necessary attributes of the teacher in his role as teacher. These must be considered because many of these attributes are identical with the qualities and capacities of the teacher as an educated person, referred to before. In addition, the teacher has other, administrative, roles, which may or may not contribute to the establishment and maintenance of that kind of interpersonal relation which is the central factor in discussion.

First the teacher must be competent. He must possess powers that are potentially accessible to the student but that the student does not yet possess. That is, the teacher must be a model. The powers he must possess are those characteristic of an educated person—mastery of the content and structure of his subject matter field; awareness of the broad significance, if not the details, of interconnections between different subject matter fields; an enthusiasm for learning which shows as a continued desire for an active seeking after further knowledge; a humility and open-mindedness about what he does and does *not* know so that he is neither afraid of a challenge to his competence nor ashamed to present materials in class which tax his abilities as well as the abilities of the students; the possession of fundamental verbal skills—the ability to "read" the structure as well as the content of a book or article and thus be aware of different levels and subtleties of meaning; the ability to speak and write cogently, concisely, and meaningfully in relation to one or more types of audience.

The second attribute of the teacher in the role of teacher is that he presents tasks which are "real" to the student. One quality of a "real" task is that it be difficult enough to challenge the student but not so difficult that he becomes lost before he starts; it involves a well-judged increment of movement from possessed abilities to those about-to-be-possessed. Too, some of these tasks should be difficult enough to require the teacher's participation in solving them, and some of these, in turn, should be so difficult that the teacher, too, must visibly move from abilities possessed to abilities about-to-be-possessed. The teacher's work on such problems is one of the most important ways in which he can discharge the responsibility of being an "ally," a responsibility which will be described in a moment.

The second quality of a "real" task is that it should have a desirable outcome which is visible to other students as well as to the student who successfully completes the task. For example, if a student is set the task of interpreting a body of evidence for a scientific conclusion, he should have a chance to present to others his reasoned argument from evidence to conclusion. This sort of presentation has two important effects. In the first place, it changes a mere "assignment," a chore, into something others can both approve and use. Thus the student becomes a valued contributor to others, not one who seeks merely the teacher's approval. Second, his success affords believable evidence to other students that they, too, can develop effective and useful competences.

The third attribute of the teacher as teacher is that he must be his students' ally, that is, one who helps the student become capable of solving the sorts of problems which comprise the subject taught. This is achieved, first, by being the sort of capable model described. That is, the teacher comports himself in such a way that he is visible to the student, not merely as one who has a stock of "answers," but as one who knows how to obtain answers or arrive at solutions to the problems he poses in the work of the class. Then, he shows students how he works, what his methods or other resources are, and how they can be used. Finally, he gives the student *support,* he gives him a hand over the rougher spots, by helping him see where he has gone wrong, by showing him better alternatives; and he affords usable encouragement by showing confidence (based on good judgment) in the student's ability to master the problems he faces.

The teacher can fail to be an ally in two different ways. On the one hand, he may merely "take over" for the student, do his work for him. This may supply the student with "answers" but leaves him, in the end, no more competent than he was at the beginning. On the other hand, the teacher will fail as an ally if he does too little and demands too much—if he merely assigns tasks but does not help the student see how to accomplish them; if he deals coldly with the student and withholds encouragement; if he tries to force students into his mold rather than leave them scope for individual ways of doing things.

The attributes of the teacher as teacher have been defined because the teacher has other roles as well—administrative roles which are important to consider here since they may adulterate or completely negate the capacity of conjoint action to a desirable end, which is, in effect, the teacher-student relation we have described. These administrative roles are primarily vested in the functions of calling roll and of maintaining discipline (the policeman), of assignment planning (the taskmaster), and of examining (the inquisitor).

There are various ways in which these functions can be handled to minimize these administrative roles. Numerous methods of taking roll deemphasize the notion that the student is merely a subject to authority, a mere subordinate to discipline. There are also ways of maintaining discipline other than by invoking authority. These are ways which appeal to "necessary rules of the game"—the standards of behavior necessary if constructive classwork is to be carried on.

One means for minimizing the roles of taskmaster and inquisitor in the ordinary school situation is by forming a staff or committee which is responsible for course planning and examining. The strength of such a device is that it assures the student that what he is asked to do has been devised by a meeting of minds, by discussion, and conjoint decision on what is appropriate to the course. This removes the structure of the course from the possible onus of being a series of tasks imposed by the whim of one individual teacher. Similarly with the function of examining; a committee-established examination permits classwork to be willing and independent since its purpose is not primarily to meet a test arbitrarily set by the person who teaches. The student recognizes the examination as a product not of possible whim but of community judgment. His teacher then appears as a source of aid toward the student's achievement of the aims of the program.

Such committee-established examinations, however, if too far removed from the classroom, may promote aims inconsistent with or contrary to the aims of the classroom teacher. This has occurred with many state-wide examinations and college-entrance examinations. Thus *the teacher must be a part of the community* which is active in devising both course structure and examinations. Without this membership, aims common to teacher, course planner, and examiner are difficult if not impossible to achieve. The classroom teacher is a key participant, since very often his direct familiarity with students and with particular situations allows him to discern aims that are not seen clearly by the subject matter specialist and to see limitations of aims that may not be apparent to a professional examiner.

To review attributes of the teacher we can again make a check list.

ATTRIBUTES OF TEACHER

E. In role of teacher

1.(+) Is competent. (Has powers that are characteristic of an educated person and that are seen as attainable by students; is a model.)

2.(+) Presents "real" tasks. [(*a*) Tasks are difficult enough to challenge students and to employ and challenge the competence of the teacher. (*b*) Tasks have visible outcomes that are usable by other students and that elicit their approval.]

3.(+) Is an ally. [(*a*) Teacher shows that he has problem-solving competences. (*b*) Shows students his methods and resources. (c) Helps and encourages students.]

F. In role of "policeman" (one administrative role)

1.(+) Some method used for tasks such as roll taking which de-emphasizes notion that student is a subject, subordinate to discipline.

2.(+) Obtains discipline by appeal to necessary "'rules of the game" rather than by invoking authority.

Note should be taken of the fact that the kind of teacher and learning we are describing can be seriously impeded or much facilitated by the physical characteristics of the classroom. The three most important characteristics are as follows. First, seating should be such that students can speak to each other, as well as to the teacher, face-to-face. Second, the seating should afford some degree of physical shielding of the students. (For example, a desk or table which leaves the lower part of the body invisible and which enables the student to hide his hands occasionally.) Third, the physical distance between students should not be so great as to require shouting or leave faces indistinct.

Discussion, as we have outlined, is

". . . in one sense only a systematization, a conscious and controlled development, of what every good teacher tends to do and what he would want to do, had he only the appropriate climate in which to function and a curriculum which provided the necessary raw material of discussion—problems to be solved, dignified occasions for deliberation on policy and action, and unsentimental occasions for apprehension of works of art. He wants something more for his students than the capacity to give back to him a report of what he himself has said. He wants his students to possess a knowledge or skill in the same way that he possesses it, as a part of his best-loved self. He does not want his teaching to be mere phonography, although the administrative structure of education often prevents his being more. He wants to convey not merely what he knows but how he knows it and how he values it." [2]

The aim of a good education in biology includes not only knowledge attained, both of the products and the processes in biology, but the desire for knowledge and the ability to seek it. Hence, the energy of wanting, initially manifest in liking and respect for the teacher, must be shifted first from the teacher to the qualities the teacher possesses as an educated person. This energy must be shifted finally to the objects or materials of biological science. That is, the student must not only develop certain qualities and capacities in himself, but he must develop an interest in the subject matters of the major fields of knowledge that will cause him to continue to study them, pursue them, for the intrinsic pleasure of learning. Certain qualities of discussion are appropriate to this aim.

Does the discussion arrive at a specific understanding of a specific object of knowledge, such as the Watson-Crick model of the DNA molecule? Have the students been guided bit by bit from a general understanding of the three-dimensional structures of molecules, through more specific understandings of why certain chemical radicals join with certain other chemical radicals, through a knowledge of precursor units, to an understanding of the chemical "logic" of DNA given the known evidence regarding molecules present in cell nuclei and the chemical properties of these molecules? (This, of course, is a condensed description of one possible way of developing an understanding of the model of the DNA molecule.) Additionally, have the students gained some understanding of the model-character (see Invitation 10) of DNA? Do they understand that this model is a *construct* which accounts for and explains known phenomena and which suggests further lines of research to confirm or clarify our understanding of DNA? If such questions regarding the understanding of the model of the DNA molecule can be answered in the affirmative, the discussion that has led to such understanding has one essential quality. This can be called the *substantive* quality of discussion; the discussion has efficiently led to a specific understanding of a specific object of knowledge.

[2] J. J. Schwab. Eros and education: a discussion of one aspect of discussion. *J. General Education*, 8, (1), 65 (October, 1954).

A second quality of discussion can be described by the following questions. Does the discussion bring out the problems encountered by researchers who have, for example, studied DNA? Does it illustrate the processes by which these problems were solved? In short, is the discussion an example of the way in which research scientists have moved toward an understanding of DNA? If the discussion has done these things, it has a quality we shall call *exemplary* since the discussion exemplifies scientific enquiry.

Finally, does the discussion stimulate two or three students to try the activity of approaching specific understandings by means of the processes of thought and communication which have been exemplified in the discussion? Do one or more students begin to attack other problems in their course in ways similar to those in which research scientists have solved problems? For example, do they begin to ask questions about the appropriateness of different hypotheses to a stated problem? Do they begin to seek and discern assumptions implicit in the experimental design and in interpretation of data? Do the students begin to exchange and discuss their answer to such questions? Do they *constructively* challenge each other's interpretations and analyses of the problems under consideration? If students begin to respond in such ways following a discussion, the discussion has had a *stimulative* quality.

No one or simple combination of these qualities is sufficient. Rather, a dynamic balance of the three (the substantive, exemplary, and stimulative functions of discussion) is necessary. The balance is a dynamic mean which will vary from group to group, from teacher to teacher, and from time to time in the course of discussion. If the group is a low-energy group, more of the stimulative will be required; if it is a high-energy group, much substantive and exemplary material is necessary. If the teacher has insights into his subject matter and/or the course of discussion—if he readily "sees" important concepts and generalizations without clearly understanding how and why these insights are valid, if the discussion proceeds progressively toward its goal without the teacher clearly understanding how and why it so proceeds—he will need consciously to give closer attention to the exemplary. If the teacher is remote from his students, if his enthusiasm is not contagious, he needs to attend more consciously to the stimulative.

As in attempting to describe the quality of interaction between teacher and student, we cannot describe the dynamic balance of the functions of discussion accurately in its fullest expression. Its boundaries can be outlined, however, by describing the extremes that delimit the desirable balance. One extreme is over-emphasis on the substantive. This is illustrated by a teacher who recites a dramatic historical narrative of the sequence of development of a certain concept (such as the cell principle) which does not permit the student to grapple with the intellectual problems involved, but merely tells him so-and-so found this and came to this notion, and so on. Such an historical narrative then culminates, with a "Book of Revelations" quality, in the grand statement of the modern cell principle. The student is left with no notion of the pathways of understanding that led to this principle, partly because he has not been asked to cope with the problems faced by the actual researchers and partly because the end point, the modern cell principle, is so over-dramatized that any glimmerings of the processes leading to it that the student may have gained are forced into the background.

A second extreme is over-emphasis on the stimulative function. In this over-emphasis the mere activity of communications is substituted for organization and development; the powers of communication are confused with the powers involved in the objectives of sound education. The best "talker" holds the floor the most, partly because the others as well as himself are fascinated with his ability to talk. Knowledge trends to be drawn only from the limited experience of the students; the content of the discussion is restricted to what the students already know. Another variation is exchange of personal experiences with no movement toward better understanding or solution of a commonly recognized problem.

The third extreme is the discussion that is exclusively exemplary. This can take two forms: the lecture, and the presentation by the teacher of his "self." In the "lecture" the teacher begins by requesting questions regarding the assignment. After a few of these, he then proceeds to elaborate the background of the problem, exhibits the principles of attack, analyzes the evidence presented, and shows the cogency of the conclusions drawn. In short, the teacher traverses a pathway of understanding by telling the students all the understandings that the students should be working with themselves. In the "presentation of self" the teacher treats himself as the object to be taught. He portrays himself as an example of the objective scientist, or of the observer of facts. This is the "good" (that is, the "fascinating") teacher who is often so popular with students for a brief period of their immaturity.

To complete our analysis of discussion, we can summarize the qualities of discussion in a check list, which appears at the end of the chapter.

QUALITY OF DISCUSSION

G. Qualities apparent in discussion

1.(+) The discussion is substantive. (It arrives at a specific, intended understanding.)

2.(+) The discussion is exemplary. (It is a defensible instance of movement toward the intended understanding.)

3.(+) The discussion is stimulative. (It stimulates at least a few students to try the activities involved.)

4.(+) There is a dynamic balance of the three qualities.

> *(a)* For low-energy groups: more stimulative.
> *(b)* For high-energy groups: more substantive and exemplary.
> *(c)* For teacher with insights: more exemplary.
> *(d)* For remote teacher: more stimulative.
> *(e)* For average group with well-balanced teacher: early in the course, the stimulative should be the most, the substantive next, the exemplary the least; later in course, the exemplary should exceed the substantive; still later, the exemplary should exceed the stimulative.

H. Extremes apparent in discussion

1.(—) Over-emphasis on substantive. (The dramatic dialogue; exemplary functions are buried under the function of arriving at a specified end; therefore no cognitive map of the route persists. Emphasis on the learning of concepts in isolation to the exclusion of *how they have been developed* and *how they are used* in biological investigation is one example.)

2.(—) Over-emphasis on stimulative. (The bull session; the activity of communication mistaken for its organization and development; powers of communication confused with the powers involved in the objectives of sound education; the amount of knowledge drawn from the limited experience of the students is mistaken for enlightenment. This may also occur when the teacher is merely a showman, fascinating the students with his wizardry.)

3.(—) Over-emphasis on exemplary.

    (a) The lecture. (After a few questions from students about the assignment, the teacher elaborates the background of the problem, exhibits the principles of attack, analyzes the evidence presented, and shows the cogency of the conclusions drawn.)

    (b) Presentation of self. (The teacher dramatizes himself as a sufficient model of the scientist.)

Check List

A. 1.(+) Teacher shows signs of recognizing students as individual persons.

2.(−) Teacher shows signs of relating to class as an amorphous mass.

3.(+) Teacher shows signs of recoginzing individual qualities of students.

4.(−) Teacher shows coldness toward students.

5.(−) Teacher shows over-anxiety to be liked.

B. 1. Students test judgment by inappropriate contributions.

2.(+) Teacher responds by indicating awareness of "test."

3.(+) Teacher does not reject students but only their inappropriate offer.

4.(+) Students participate with fitting (appropriate) contributions.

C. 1.(+) Students show eagerness to talk and attract attention by at least superficially appropriate contributions.

2.(+) Teacher responds by being discriminatingly kind.

F. 1.(+) Some method used for roll taking which de-emphasizes notion that student is a subject.

2.(+) Obtains discipline by appeal to "necessary rules of the game."

QUALITIES OF DISCUSSION

G. 1.(+) The discussion is substantive.

2.(+) The discussion is exemplary.

3.(+) The discussion is stimulative.

4.(+) There is a dynamic balance of the three qualities:

(a) For low-energy group: more stimulating.

(b) For high-energy group: more substantive and exemplary.

(c) For teacher with insights: more exemplary.

(d) For remote teacher: more stimulative.

(e) For average group with well-balanced teacher early in course: stimulative most; substantive next; exemplary least.

(f) For average group later in course: exemplary exceeds stimulative.

3. (+) Teacher responds by correction and assistance.

4. (−) Teacher shows inappropriately adult standards.

5. (−) Teacher shows indiscriminate permissiveness.

6. (+) Students show responsibility toward words, choices, decisions.

D. 1. (+) Students show liking and respect *for teacher as a person.*

2. (+) Students show recognition of *qualities* and *capacities* exhibited by teacher when he is a "model" of an educated person.

3. (+) Students show signs of pleasure in *practicing* the qualities and capacities of the teacher as an educated person.

ROLES OF TEACHER

E. 1. (+) Is competent.
2. (+) Is an ally.
3. (+) Presents "real" tasks.

H. 1. (−) Over-emphasis on substantive. (The dramatic dialogue.)

2. (−) Over-emphasis on stimulative. (The bull session)

3. (−) Over-emphasis on exemplary.
(a) The lecture.
(b) The presentation of self.

CLASSROOM STRUCTURE

I. 1. (+) Students can speak to each other and to teacher face-to-face.

2. (+) Tables or desks are arranged so students have some physical shielding.

3. (+) Distance between individuals in group is not excessive.

Key:

3 = High degree or frequently
2 = Lesser degree or about half the time
1 = Slightly or occasionally
0 = Opportunity for, but lacking
___ = No indication since no opportunity for observation

7

Reading the Text

There are different ways of reading and analyzing a textbook. Of initial importance, of course, is understanding the literal meaning of the written material. Sometimes students need assistance with this because they have not learned to deal with variations in density and style of writing. Beyond the literal meaning of the written material, however, intelligent reading—true literacy—requires that the student be able to read, analyze, and question the text for different kinds of understandings.

First, let us consider the problem of understanding the literal meaning of the written material. Textbooks are generally written in a style different from that of other types of books; in addition, different textbooks have different styles, and within a text there may be differences in density of writing. Like other texts, the BSCS versions differ somewhat in style, and within each version the density of writing varies from one section to another. Some sections are written in a *succinct* style that wastes few words and sentences. There is little of the common textbook style in which something is said, then illustrated, then said again. Rather, what has to be said is said once and for all. This means that the student who has been habituated to the more "spread-out" meanings of usual texts will very likely find himself lost before he has read more than two or three sentences in some parts of some chapters— simply because he does not realize that each word of each sentence performs a function not performed by any other word there. He has not encountered such *dense* writing very often. Hence, he has not learned to vary the tempo of his reading and his concentration to suit it.

Many educators believe that the tendency of textbooks to be "spread out" has had undesirable educational consequences—precisely those we have mentioned—stunting of the students' ability to read materials of varying style and density. It has also tended to stunt students' capacity for sunstained and concentrated attention to meanings.

A net educational gain could come, therefore, from an adequate introduction to the student of ways of reading dense exposition. Let us present two examples of writing which have different densities and suggest procedures for comparing them so that students will begin to understand and be able to read different densities of writing.

Begin by telling the students about the fact that there can be different densities of information per unit of prose in different kinds of writing. Tell them about the differences in reading tempo and attention required by different densities.

Then demonstrate this. First present the following paragraph or one from your BSCS version which is similar to it in density.

"It is common experience that living things grow by taking materials from their surroundings and converting them into living matter. For example, in a seed that falls from a tree to the ground there is a tiny embryo plant weighing a small fraction of an ounce together with less than an ounce of stored food. As the tiny plant absorbs the food in the seed it begins to grow. Roots emerge and reach down into soil, while stem and leaves rise into the air. After all the food that was stored in the seed has been used, the plant still continues to grow and to increase in weight."

(Write the paragraph you use on the board.) Ask the students to read the given paragraph silently and decide how many ideas it contains. Essentially, there is only one: Living things convert materials from their surroundings into living matter. If they think there are more, make the following analysis.

Read the first sentence and ask them to state, even more briefly, its central idea. Encourage them to make it as tight as possible (for example, living things convert surrounding material into living matter). Then ask them how many words were used to convey the idea in the original sentence. There are twenty, whereas our example uses only eight. Here, then, we already have an example of a difference in meaning density of better than two to one.

Now point out to the students that the denser, shorter sentence is just as clear as the longer one—perhaps even clearer—but that it requires much more attention. Each word must be noted. Not one can be skipped without losing the meaning of the sentence.

Now ask the students, What does the next sentence do? On first hearing the question, they will probably mishear you as asking, What does the next sentence *say?* That is, they are likely to give you a summary instead of noting and saying that what the next sentence *does* is to begin to give an illustration of the idea contained in the first sentence.

You can use their mishearing as a vivid new example of the need for attention to short, dense statements. Then you can go on to show that the remainder of the paragraph is but a development of an illustration of the one idea contained in the first sentence. You can end by pointing out that the whole paragraph (some hundred words) has its meat in the one eight-word sentence about conversion by living things.

Now, present a second paragraph which is written very densely, such as the following:

"The smallest unit of matter which possesses all the chemical properties of an element is called an *atom.* Since there are 92 elements in nature, there are also 92 different kinds of atoms. All atoms are built of varying combinations of three fundamental types of particles which are called *protons, neutrons,* and *electrons.* It is convenient to regard the structure of an atom as a miniature version of the solar system. In each atom there is a central dense mass, analogous to the sun, and around this central mass, in definite orbits or shells, there are smaller bodies analogous to the planets. This is a highly simplified and diagrammatic concept of atomic structure, but will suffice for our purposes."

Ask the students to try to shorten (make more dense) the first sentence, the second, and the third. It is very unlikely that they will succeed in saving more than a word of two. Then have them try to estimate the number of ideas conveyed in these three sentences. They will almost inevitably underestimate, for there are many. A quick survey shows the following in the first sentence alone:

1. Some units of matter are called atoms.
2. There are other units of matter.
3. Some of these other units are smaller than atoms.
4. Every element has a number of characteristic properties.
5. Some of these properties are considered chemical properties.
6. There are other properties which are not "chemical."
7. The atom is the smallest unit of matter that retains all the chemical properties.

The density of meanings indicated by this list can be brought home to the students by telling them that the following questions can be answered from the first sentence, even though a quick reading of the sentence would not reveal this fact.

1. Are there units of matter smaller than atoms?
2. Larger?
3. How do we define an element? (By a list of chemical properties)
4. Does matter have properties which are not chemical?

Voice some of these questions, one at a time, and let the students try to find an answer in the sentence. If they do not, show them the phrase or clause that provides the answer.

By such an exercise you may teach your students something about meanings and about reading which all their previous schooling may have omitted.

In addition to grasping the immediate meaning of the written material, we should read texts and analyze them at different levels for different kinds of understandings. After the students have learned to read text materials of different densities, and as they read to ask questions which help them to do this, then other aspects of reading the text can be considered. There are different kinds of questions which can be asked of text materials. Questions can be used for at least two purposes other than clarifying surface meaning.

Questions that extend and elaborate the meaning of text materials need to be asked. One type of question appropriate to this purpose is the question which applies text content to other situations or areas. Another type of question asks the relation between material currently read and material in previous chapters or other course work. A third type of question appropriate to this purpose asks the student to speculate on a problem or question suggested but not answered by the text material or other previous course work. By such questions the students can bit by bit begin to compare and contrast, relate and apply, the material from one area of biology to another.

Questions at the ends of chapters in each version are designed for this purpose—to extend and elaborate the meaning of text materials. Nevertheless, students very often are unfamiliar with such questions. They are used to being asked only questions which can be directly answered by a word or sentence stated in the text. Hence they may need some help in answering questions which go further. One way of providing such help is to develop a *sequence* of questions beginning with a question that can be answered directly by the text material (this also emphasizes the importance of knowing specific information as a basis for further understanding) and culminating with a question (or questions) that concerns (or concern) past learning or future problems. By working out such sequences of questions during the early part of the course, the students should begin to be able to ask their own questions as they read later chapters.

The third level of reading and analyzing the text involves understanding the broader relations among areas of biology (and hence chapter content). This level also requires development of an understanding of concepts which help us to organize much of the specific content of biology and which interrelate the major areas of biology. The BSCS themes—genetic continuity, complementary of organism and environment, and so on—are such organizing concepts. If students are to begin to understand biology as a field of knowledge, they must begin to understand these major conceptual schemes, which help us organize biological knowledge. This is the reason for identifying the themes and for attempting to weave them into the text and laboratory materials.

Many of these organizing concepts are (intentionally), however, only implicit in the written materials, and students therefore will need help in discerning them and understanding them. For example, if the current text materials deal with genetics, and populations have already been studied (or vice versa), the idea of genetic continuity can be used to interrelate and give better understanding of the facts and principles of these two biological topics. By appropriate questioning, students may even be led to formulate the notion of genetic continuity if they are not already familiar with it. Such questioning might begin with the significance of genetic phenomena to successive generations of offspring—similarity and diversity, stability and variability. Then this significance to generations in a family line might be compared with the significance to larger related groups and finally to an entire population. If similarities and stabilities are stressed, the notion of genetic continuity will probably occur to some students. If diversity and variability are stressed, the concept of evolution can be brought into the discussion and related to the concept of genetic continuity.

Many other examples of ways in which students can be guided to an understanding of broad organizing concepts could be given. The possibilities are infinite, even though the concepts and areas of biology are relatively few, since the ways this can be done will depend on the total sequential development of the version and the things stressed by the teacher throughout the course.

The important thing to keep in mind is that unless the teacher makes some attempt to encourage students to think explicitly about these broad concepts and the interrelations between the areas and phenomena of biology, the students are not likely to begin to understand them. There is much evidence that unless the broader and perhaps vaguer learnings are specifically taught for they are unlikely to be learned by students. This means that such aims as (1) the understanding of the meaning of conceptual schemes and their functional use in the organization of knowledge and (2) the understanding of scientific enquiry must constantly be kept in mind by the teacher. Much classwork must be spent in guiding the students from the literal meaning of explicit text material to these broader understandings.

One final point should be made. To talk to the students *about* these broad understandings is not as effective as stimulating the students to think about relations among phenomena and principles so that they begin to formulate these broad understandings on their own. Guided discussion is a good means for accomplishing this. When students have begun to attain these broad learnings, a fuller, more detailed presentation by the teacher or from some other source (such as a monograph) will be more valuable.

It is apparent that not all text material can or should be dealt with at all of the three levels of analysis described. Aiding the student to learn to read different densities of text material is of primary importance. Without this he does not have the substantial foundation of facts and principles necessary to a great breadth of understanding of biology. Reading the text should not, however, stop with an understanding of the literal meanings of specific facts and principles. Numerous portions of the text can be read and analyzed at the second level—extension and elaboration of the explicit text material. And some portions should be read and analyzed at the third level—understanding of the broader relations among areas of biology and of the meaning and use of conceptual schemes to organize biological knowledge.

If, in the beginning of the school year, the first level of reading is stressed and the other two gradually introduced, by the end of the year it should be possible to spend proportionally more time on the second and third levels of reading and analysis. If this is done, it should be possible for the students to have begun to develop a *synthesis of understanding* of biology as a science which includes many details as well as broad generalizations.

8

Evaluation in BSCS Biology

STUDENT PERCEPTIONS OF "TESTS"

Techniques of "testing" have an important bearing on the way students approach the learning process, irrespective of the subject matter or the particular curriculum involved. "Tests," and the "grades" resulting from them, are more often than not regarded by students as rewards for their efforts. Furthermore, the student is quick to perceive the basis on which he is rewarded. Thus, no matter how lofty or desirable the objectives of a course of study may be, the key to success, insofar as the student is concerned, is what is tested for. If memorized details and repetition of what text and teacher have said are the main demands of tests, they are what the student will concentrate on. Thereby his attainment of more important educational objectives, no matter how much desired by the teacher, is thwarted or wholly circumvented.

Take, for example, the following two test items.

Sample 1. All of the following are classes of the Phylum Mollusca, EXCEPT

        A. Pelecypoda.   *B*. Cephalopoda.
    C. Arthropoda.   *D*. Gastropoda.
        E. Amphineura.

Sample 2. Which one of the following is a hormone secreted by the pancreas?

        A. Trypsin.   *B*. Erepsin.
    C. Insulin.   *D*. Amylase.
        E. Pepsin.

401

Such questions measure rote memory, and rote memory merely of terminology.

If, however, the *tasks* required of the student in the test situation are compatible with, and truly reflect, the more important goals, the student will direct his efforts accordingly. Consequently it is crucial that the tasks given to the student genuinely reflect the kinds of learning expected of him.

Another aspect of the student's perception of evaluation techniques lies in the purposes for which they were used. If the student views tests as primarily used to *judge* him and considers that such judgments, when accumulated, can influence his entire future success in our society, then any evaluation technique becomes a serious threat which can inhibit rather than contribute to his learning. A very strong position on this negative influence of tests has been stated by Paul R. Lohnes.[1]

". . . My contention is that the measurement record attached to students in most of our secondary schools today is inadequate and harmful; that it involves errors of commission by sponsoring invidious comparisons, burdening teachers, and erecting a barrier between the teacher and the student; that it involves error of omission by ignoring important traits of individual differences in adolescents and providing inadequate interpretations of the traits it does report."

If, on the other hand, the student perceives evaluation techniques as a means of checking his progress, as an aid in finding out what he has learned well over a period of time, then "tests" can be a valuable technique for maximizing learning. The importance of these differences in students' perceptions of tests is so great that further discussion of the purposes and use of tests is warranted.

PURPOSES OF "TESTS"

There are two general purposes for which "tests" are used in high school classrooms. By far the most common of these is to provide a "grade" for each student which purportedly reflects his achievement in the subject. The second, less common, purpose is to use tests in ways that will assist *all,* or almost all, of the students to attain mastery in terms of the objectives of the course. Underlying these two purposes are different concepts and assumptions about learning and student abilities.

[1] Lohnes, Paul R., "Reformation through Measurement in Secondary Education," *Proceedings of the 1967 Invitational Conference on Testing Problems*, Princeton, N. J.: Educational Testing Service, 1968.

Underlying the more common use of tests is a set of concepts developed during the last few decades. These concepts were outcomes of approaches to educational measurement which incorporated assumptions about learning that are now being questioned. One of these is the meaning of IQ scores. For too long we have assumed that the concept of IQ reflects inherited and fixed capacity to learn. More recently the meaning of IQ has been challenged. In much of the recent literature one finds statements such as the following.[2]

"Teachers' perceptions of intelligence tests are in the process of change and are becoming more realistic. It is recognized that *we do not know how to measure native potential intelligence.* What is measured is rather learned behavior, that is, the product of the interaction between life experience and native intelligence. Teachers recognize that the present intelligence tests are of limited range. For instance, they fail to tap the abilities which enter into creative functioning. . . .

"In contrast to previously implied assumptions, intelligence does not grow and unfold on its own. Its growth is not a predetermined curve. It is the educational impact of the environment on the child's inherited potential which results in the abilities called intelligence. Maternal, sensory, and school, deprivation may cause slowing of the growth of intelligence. A rich environmental stimulation may facilitate growth. . . ." (Italics added.)

Another concept underlying the common use of tests for judgment is that of the normal curve. The normal curve is a mathematical construct based on chance and random activity. It has been found that when certain characteristics of biological and physical phenomena are measured, these measurements, when graphed, sometimes approximate a normal curve. The meaning of this is at least twofold. The factor or characteristic measured is "affected by a great many causal determiners, each of which acts to increase or decrease the magnitude of the factor in each instance by a small amount, and independently of the other determiners" (Ch. 19, p. 2.) In short, the factor or characteristic is the result of a great many determiners acting as though at random. Also, there is no reason to expect that *all* sets of measurements will approximate a normal curve.

These meanings have critical implications for use of the normal curve in testing and grading. First, because many sets of grades do approximate a normal curve, we tend to interpret this as indicating

[2] Shumsky, A. *In search of teaching style.* New York: Appleton-Century-Crofts, 1968. pp. 185-186.

that students' *abilities* in that subject area are distributed according to a normal pattern. However, such a distribution probably results from (a) the way in which the test was constructed, and (b) the range of aptitudes of those students for that subject matter. We will return to aptitudes shortly. Second, because of our assumption that a normal distribution on test scores reflects a range of students' abilities, we tend to *expect* only a small portion of students to achieve well in a given course.

But consider a newer concept of the meaning of the normal curve and the resultant changes in assumptions about students and their potential for learning.[3]

"We have for so long used the normal curve in grading students that we have come to believe in it. Our achievement measures are designed to detect differences among our learners, even if the differences are trivial in terms of the subject matter. We then distribute grades in a normal fashion. . . .

"Having become 'conditioned' to the normal distribution, we set grade policies in these terms and are horrified when some teacher attempts to recommend a very different distribution of grades. . . . Finally, we proceed in our teaching as though only the minority of our students should be able to learn what we have to teach.

"There is nothing sacred about the normal curve. It is the distribution most appropriate to chance and random activity. Education is a purposeful activity and we seek to have the students learn what we have to teach. If we are effective in our instruction, the distribution of achievement should be very different from the normal curve. In fact, we may even insist that our educational efforts have been *unsuccessful* to the extent to which our distribution of achievement approximates the normal distribution."

What Bloom argues for in this article is the concept of "learning for mastery" as a basis for our educational practice. "Most students (perhaps over 90 percent) can master what we have to teach them, and it is the task of instruction to find the means which will enable our students to master the subject matter under consideration."[4]

There are two concepts that are critical for this approach. One is

[3] Bloom, Benjamin, "Learning for Mastery," *Evaluation Comment*, University of California at Los Angeles, Center for the Study of Evaluation of Instructional Programs, Vol. 1, No. 2 (May, 1968). This paper will be published as a chapter in Bloom, Hastings, Madaus, *Formative and summative evaluation of student learning*, McGraw-Hill.
[4] *Ibid.*

the effect of *teacher expectation* on student learning and the other is the notion of *aptitude as the amount of time required* by the learner to attain mastery of the learning task.

Many persons who have been close observers of teaching in the classroom have suspected that what the teacher expected of each student often made a critical difference in what that student learned. Teacher *expectation* is no doubt a factor in learning, which for too long has been overlooked. Recently, several studies provide experimental evidence that what a teacher expects of a student is closely related to the achievement of that student.[5-9] Results of these and other studies provide a basis for adopting, as an operating principle of teaching, the notion that students can learn more than previously has been expected.

Quite different, but supportive of the notion of teacher expectation, is the concept of aptitude formulated by John Carroll.[10] Previously it has been found that students do have a range of aptitudes with respect to a given subject area and that this range of aptitudes is distributed normally. Hence when students in that subject area are given the same instruction in terms of amount, quality, and time available for learning the end result on a measure of achievement is a normal distribution of scores. This has been interpreted to mean that aptitude scores provide a measure of the maximum amount of learning a student can be expected to achive in a given subject matter area. However, using Carroll's concept of aptitude as the *amount of time* required to attain mastery, we must consider the possibility that all students could attain mastery in a given subject area if each is given sufficient time and assistance. Those with the greater aptitude will need less time and assistance; those with lesser aptitude will need more, but all potentially can achieve a similar level of mastery.

The implications of these ideas are (1) that the biology teacher must begin to examine the basic assumptions about students and their

[5] Merton, R. K. "The Self-Fulfilling Prophecy," *Social theory and social structure,* Free Press, 1957, pp. 421-436.

[6] Rosenthal, R., and Lenore Jacobson. "Teachers' Expectancies: Determinants of Pupils' IQ Gains," *Psychological Reports,* 1966, 19, 115-118.

[7] Rosenthal, R., and Lenore Jacobson. *Pygmalion in the classroom.* Holt, Rinehart and Winston, 1968.

[8] Sampson, E. E., and Linda B. Sibley. "A Further Examination of the Confirmation or Nonconfirmation of Expectancies and Desires," *Journal of Personality and Social Psychology,* 1965, 2, 133-317.

[9] Schwab, W. B., "Looking Backward: An Appraisal of Two Field Trips," *Human Organization,* 1965, 24, 373-380.

[10] Carroll, John, "A Model for School Learning," *Teachers College Record,* 1963, 64, 723-733.

learning which he uses as a basis for teaching, (2) that all high school students can achieve mastery of BSCS biology, or any other subject area; and (3) that "tests," if used in new ways, can contribute to the process of learning rather than merely judge the outcome of that process.

In the remainder of this chapter, techniques of evaluation are discussed as means of assisting students to attain mastery of BSCS biology.

TECHNIQUES OF EVALUATION

The point of view of the preceding discussion is that all students can attain a high level of achievement in the desired outcomes of BSCS biology and that evaluative techniques can assist them in this achievement if evaluation is viewed by the student as a useful means of checking his progress in attaining the desired outcomes. What, then, are the desired outcomes?

The BSCS curriculum has direct implications as to the kinds of learning expected of students. Two broad aims are built into the course materials themselves. Each of these is crucial in the learning process if the nature of biology as an investigative science is to be communicated to the student at all effectively.

These two broad dimensions are interwoven with one another throughout all the course materials and together define the goal of student achievement in the BSCS context. These dimensions can be summarized as follows.

1. *Substantive course context:* The body of biological knowledge specific to each version.

2. *Scientific process:* The process of enquiry, requiring development on the part of the student of certain skills and abilities requisite to genuine understanding of the nature of scientific enquiry.

The biological themes common to the three versions of the course serve as integrating principles contributing to the unity of biology as a science. These themes, in themselves, represent an interweaving of the two dimensions of the BSCS program.

Any test, no matter what kind, is a *sampling* of tasks. Therefore, we can never hope—in a single test or even a battery of tests—to encompass the entire domain to be evaluated. Rather, we try to see to it that the sampling is *representative* of this domain. Hence a valid achievement test in the BSCS context must be *representative* in the sense of measuring both dimensions of the course we have described. Thus, an achievement test that takes account only of subject matter content and excludes the process dimension will be inadequate, unrepresentative.

Similarly, when it comes to testing for substantive content, mere ability to repeat this information is not the goal of instruction. The student must be able to *use* his knowledge in a variety of ways. Hence, a good achievement test should go beyond mere recall to test for ability to apply knowledge.

In preparing test items, problems of achieving clarity and a proper level of difficulty constantly present themselves. Hence, it is very easy to overlook the question of what outcomes or objectives are being tested. Therefore, it is important to identify the different outcomes for which we should test. These can then be kept in mind as guides in constructing test items.

Four kinds of outcome (some with subdivisions) are relevant to BSCS biology.

The Ability to RECALL INFORMATION *and to* MAKE MINOR REORGANIZATIONS *of Materials Learned.* This ability is familiar and self-explanatory. As we have suggested, items of this kind should constitute only a very small part of tests.

Ability to SHOW RELATIONS *between Different Bodies of Knowledge Learned at Different Times or in Connection with Different Topics.* One of the outstanding characteristics of organisms is that they are organized—their various parts, levels, and activities are closely related to one another. Hence, in addition to knowing something about the glucose molecule and something about the responses of plants to light and moisture conditions, we strive for a further understanding on the part of students: We want them to understand these matters in relation to one another.

Understanding of Materials Learned as Demonstrated by ABILITY TO APPLY *Knowledge in New Situations.* In a sense, all outcomes beyond the first one (recall) are tests of understanding. However, a test which shows that a student can *apply* what he knows is of special importance. In a sense, it is the minimum test of knowing, since even a machine such as a tape recorder can repeat what has been said to it —and we would hardly say that a tape recorder "knows" what it has recorded.

This objective has two major subdivisions:

A. Ability to apply nonquantitative bodies of knowledge.

B. Ability to apply and manipulate quantitative materials.

Ability to Use Cognitive Skills Involved in an Understanding of Scientific PROBLEMS. This ability is, in a sense, the opposite pole of the outcomes previously listed. These previous outcomes all concern themselves with the remembrance, organization, and application of knowledge. The fourth objective, on the other hand, has to do with *arriving* at this knowledge, and understanding it as the outcome of scientific research. (See also Chapter 2.)

In the following pages these general objectives are elaborated and examples in the form of multiple-choice items are provided.

ABILITY TO RECALL AND REORGANIZE MATERIALS LEARNED

A. SIMPLE RECALL

Examples

1. List the names of the stages in mitosis.

2. Why are bats classified as mammals and not as birds?

3. How many different kinds of atoms are there in the amino acid alanine? How many of each kind are there?

4. How is the approximate age of fossils determined?

5. Which of the following statements could apply to both an open-sea biome and a grassland biome?

    (a) The amount of carbon dioxide available for photosynthesis varies greatly with temperature.

    (b) Light penetrates to depts of 500 feet through levels inhabited by photosynthesizers.

    (c) First- or second-order consumers grow to large size.

    (d) Most producers are near the microscopic level in size.

    (e) Pressure may be as high as 1000 pounds per square inch.

B. MINOR REORGANIZATIONS OF MATERIALS LEARNED

Examples

1. What are some of the characteristics that are typical of populations and not of individuals?

2. In what respect are the rickettsiae similar to the viruses?

3. What are some differences between artificial and natural selection? In what ways are artificial and natural selection similar?

4. Which of the following pairs of organisms is best adapted for living in a dry, warm, sunny, and sandy environment?

(a) Clubmosses and millipeds.

(b) Fungi and insects.

(c) Herbaceous plants and reptiles.

(d) Psilopsids and cephalopods.

(e) *Welwitschia* and reptiles.

5. Explain why the element neon is not found in chemical compounds.

6. Why is it necessary to reserve judgment in deciding whether viruses are alive? Name nonliving characteristics of viruses.

7. What are some of the difficulties in classifying all living things as either plants or animals?

ABILITY TO SHOW RELATIONS BETWEEN BODIES OF KNOWLEDGE

Two of the following examples use the word "significance"; this is a quick and useful way to "structure" or delimit a question about relations. A very common misuse of the word "significance" often leads to vague and unsatisfactory test items. This misuse takes the form, "What is the significance of sex?" Obviously this is a subject on which whole books have been and could be written. On the other hand, to ask, "What is the significance of sex in the production of raw materials for natural selection?" specifies a distinct and limited field of attention for the student.

Examples

1. What is the significance of the processes of mitosis and cell division to the continued life of the individual cell?

2. What is the significance of mitosis and cell division to the growth of a multicellular organism?

3. Name some ways in which man has altered his own environmental conditions in constructing cities.

4. What are some of the ways in which fungi are economically important to man?

ABILITY TO APPLY KNOWLEDGE IN NEW SITUATIONS

This is one of the most important of our four kinds of outcomes. For that reason more numerous and varied examples are given.

A. ABILITY TO APPLY NONQUANTITATIVE BODIES OF KNOWLEDGE

Examples

1. It is believed that complex molecules were formed in the "hot thin soup" of the early seas. These reacted to form large "aggregates" which eventually showed the attributes of life. How would it be possible to determine when the transition from the nonliving to living clumps of molecules occurred?

2. The questions following relate to the making of a "key" which can be used to classify the animals pictured in Figure 1.

The animals pictured can be classified into two groups in several different ways by using different features of the animals as the basis for classification in each case.

CLASSIFI-CATION	GROUP A	GROUP B
1	I, II, III, V	IV, VI
2	I, II, III	IV, V, VI
3	I	II, III, IV, V, VI
4	II, III	I, IV, V, VI

(a) The five features that can be used for classification are:

(i) Gills versus lungs
(ii) Paired appendages versus no paired appendages
(iii) Scale covering versus nonscale covering
(iv) Egg-laying versus live-bearing reproduction
(v) Warm-blooded versus cold-blooded

(1) What feature is the basis for classification 1?
(2) What feature is the basis for classification 2?
(3) What feature is the basis for classification 3?
(4) What feature is the basis for classification 4?

(b) The biologist could put the animals in the order I, III, II, V, VI, IV, on the basis of increasing

(1) completeness of scale covering.
(2) number of appendages.
(3) rank in a food chain.
(4) speed of locomotion in natural habitat.
(5) usefulness of appendages in land locomotion.

3. The fossil record shows that both body size and fang length of sabre-toothed cats increased up to the time the cats became extinct.

Figure 1

In the light of this information, it would be most reasonable to believe that

    (a) a large body and long fangs had adaptive value.

    (b) large, long-fanged cats ate all the smaller cats.

    (c) only large cats with long fangs were fossilized.

   (d) the mutation rate in sabre-toothed cats steadily increased.

   (e) the climate was gradually warming during this period.

4. The following questions relate to a respiratory disease which causes a thickening of the lung lining in several species of animals.

   (a) The first species to suffer ill effects from this disease would be those which

      (1) ferment maltose.

      (2) synthesize fat.

      (3) have open circulatory systems.

      (4) have low metabolic rates.

      (5) have high metabolic rates.

   (b) The ill effects of the thickening of the lung lining could be reduced by an increase in the number of

      (1) blood platelets.   (2) red blood cells.

      (3) white blood cells.   (4) enzymes.

      (5) Rh factors.

   (c) In order to help a human male victim of this disease, a blood transfusion is planned. However, the victim has type A blood and no type A blood is available. Which of the following blood types could be substituted without harm to the victim?

      (1) AB only.   (2) B only.

      (3) O only.   (4) AB or B only.

      (5) AB, B, or O.

   (d) Unfortunately, the transfusion produced a serious unfavorable reaction. The most reasonable explanation for this is that

      (1) the blood donor was a female.

      (2) the blood donor was not of the same race as the victim.

      (3) the donor's blood contained too many white blood cells.

      (4) the victim's blood type changed between the time it was tested and the time the transfusion was made.

      (5) antagonistic blood factors were present, even though the types were compatible.

   (e) If the victim lived, the harmful waste products resulting from the transfusion would eventually be eliminated through the victim's kidneys by a process of

    (1) assimilation.   (2) filtration.

    (3) osmosis   (4) reabsorption.

    (5) respiration.

5. The following questions relate to the study of evolution.

  (a) After studying a variety of authoritative scientific writings, the biologist would be correct in deciding that all the following serve as evidence in support of the theory of evolution EXCEPT:

    (1) Mammal embryos at one stage in their development have gill pouches.

    (2) Organisms preserved as fossils differ from presently existing organisms.

    (3) Some animals possess organs which have no useful purpose.

    (4) The structure of all chordates is similar.

    (5) There is an abundance of protozoa in the world at present.

  (b) On a geological expedition, the biologist finds, at a depth of 500 feet on an exposed rock face, the fossil skeleton of a bird that had a wing span of 20 inches. At a depth of 200 feet on *another* exposed rock face, the biologist finds a fossil skeleton similar to the first one in all respects except that the wing span is only 10 inches. The biologist hypothesizes that the bird with the larger wing span lived much earlier in geologic time than did the second bird. Of the following, the best way to test this hypothesis would be to

    (1) search the rock layers in which the bird fossils were found for the fossils of other organisms whose relative ages are known.

    (2) examine the mutated chromosomes of the two birds with an electron microscope.

    (3) expose the two fossils to radioactivity and then measure their ages with a Geiger counter.

    (4) measure the thicknesses of the two rock layers in which the fossils were found.

    (5) X-ray rock samples from the two locations where the fossils were found.

6. In addition to testing ability to apply knowledge, the following questions also illustrate a test of ability to interpret data, an objective listed as part of our fourth outcome. However, the illustration is used

here rather than under that outcome because the interpretation called for here leads not to new scientific knowledge but only to knowledge of specifics to which Mendel's principles apply.

    (a) Assume that Mendel raised several hundred red-flowered plants for 5 generations, crossing some and allowing others to self-pollinate. If only red-flowered plants appeared, it can be correctly inferred that the

        (1) plants were homozygous for flower color.
        (2) plants were heterozygous for flower color.
        (3) gene for red flowers was dominant.
        (4) gene for red flowers was recessive.
        (5) genes for red and white flowers were attached to each other.

    (b) If a single white-flowered plant appeared in the sixth generation, it is most reasonable to conclude that

        (1) meiosis was abnormal.
        (2) mitosis was abnormal.
        (3) a mutation occurred.
        (4) crossing-over occurred.
        (5) random segregation did not occur.

    (c) Suppose that Mendel crossed purebred red-flowered plants with purebred white-flowered plants and obtained all red-flowered offspring (F_1 generation). The gene for red flowers must have been

        (1) assorted.
        (2) sex-linked.
        (3) dominant.
        (4) probable.
        (5) carried on the same chromosome as the gene for white flower.

    (d) Assume that Mendel then crossed 2 of the red-flowered offspring (F_1 generation) noted in *(c)* and obtained 8 plants (6 with red flowers and 2 with white flowers) in the F_2 generation. It can be correctly concluded from these data that

        (1) a mutation occurred in the F_2 generation.
        (2) meiosis was somewhat abnormal in this case.
        (3) the F_1 plants were heterozygous.

(4) the ratio of red-flowered plants to white-flowered plants in F_2 generations from an identical cross will be 3:1.

(5) the probability of white-flowered plants in the F_3 generation is 1/4.

(e) Assume that Mendel then crossed more of the red-flowered offspring (F_1 generation) noted in (c) and obtained 4000 plants (2980 with red flowers and 1020 with white flowers). It is reasonable to conclude from these data that

(1) a mutation occurred in the F_2 generation.

(2) the laws of chance were not operating.

(3) the genes for white flower and red flower were carried in the same chromosome.

(4) the gene for white flower is incompletely dominant.

(5) the probability of white-flowered offspring from the F_1 cross is 1/4.

(f) Suppose that Mendel grew pea plants from 32 pea seeds he had collected from a red-flowered F_1 plant. If the first 18 plants to flower had red flowers, the probability that the nineteenth plant would have red flowers was

(1) 0/4.   (2) 1/4.   (3) 9/16.   (4) 3/4.   (5) 1/1.

Assume that Mendel discovered 1 purple-flowered pea plant in his garden among a strain of white-flowered plants which had been breeding true for several generations. When this purple-flowered plant was crossed with white-flowered plants half the offspring had purple flowers and half had white flowers.

(g) Of the following, it is most probable that crossing the purple-flowered plants thus obtained would have yielded offspring in the ratio of

(1) 1 purple:0 white.

(2) 3/4 purple:1/4 white.

(3) 9/16 purple:7/16 white.

(4) 1/2 purple:1/2 white.

(5) 0 purple:1 white.

Assume that Mendel actually obtained 796 purple-flowered plants and 405 white-flowered plants from the crossing of the purple-flowered plants.

(h) The most reasonable explanation for the phenotype ratio observed is that

    (1) random segregation of chromosomes was occurring.
    (2) seeds homozygous for the purple gene died.
    (3) the gene for purple flower is dominant.
    (4) the genes for flower color occur in pairs in most seeds.
    (5) the gene for white flower leads to mutant plants.

B. ABILITY TO APPLY AND MANIPULATE QUANTITATIVE MATERIALS

Examples

1. The following questions apply to the so-called typical curve of population growth.

    (a) What would be the shape of this curve if the population were one of deer and all the coyotes and wolves in its neighborhood were destroyed about midway in the development of the population?

    (b) What would happen if the available water were materially reduced early in the growth of the population?

2. A scientist analyzes a corn plant and finds that it contains 360 g of the element carbon (C). The scientist wishes to determine what weight of air supplied the carbon in the corn. [Assume that at "standard conditions" of 0°C and 1 atmosphere the percentage by weight of carbon dioxide (CO_2) in air is 0.05%.]

    (a) Which one of the following items of additional information will the scientist need in order to determine the weight of air which supplied the carbon in the corn plant?

        (1) Density of CO_2 at standard conditions (2.0 g/l).
        (2) Density of air at standard conditions (1.3 g/l).
        (3) Percentage by volume of CO_2 in air (0.03%).
        (4) Percentage by weight of C in CO_2 (27%).
        (5) Molecular weight of CO_2 (44).

    (b) In order to determine from the data given the weight of air which supplied the C in the corn plant, the scientist must make the assumption that

        (1) the soil in which the plant was grown contained no C .
        (2) CO_2 is soluble in water.
        (3) CO_2 in the air is the source of C for the plant.
        (4) the corn plant was grown in a sealed container.
        (5) the corn plant was supplied with an adequate amount of water.

(c) The weight of CO_2 which supplied the amount of C found in the corn plant is most nearly

(1)   97 g.   (2) 360 g.

(3)   470 g.   (4) 720 g.

(5) 1300 g.

(d) The weight of air which supplied the amount of C found in the corn plant is most nearly

(1) 7×10^1 g.   (2) 5×10^3 g.

(3) 3×10^6 g.   (4) 5×10^7 g.

(5) 4×10^{10} g.

(The items in this set are interdependent to the extent that answering one item incorrectly will jeopardize one's chances of answering succeeding items correctly. This situation is encountered in all complex scientific problems, where errors at any particular point will render the ultimate solution incorrect).

3. Using echo-sounders (instruments which transmit sounds and detect the return of accompanying echoes), scientists have discovered in the ocean a "phantom bottom" which reflects sound. (Assume that the speed of sound in sea water at 50°F is 1600 m/sec.)

(a) If the temperature of water did not vary with depth, which one of the graphs [see Figure 2] would correctly relate the depth of the "phantom bottom" to the time required for a sound to travel from the surface to the "phantom bottom" and back again?

At a certain location in the ocean where the temperature of the water at the surface is 50°F, soundings were taken at noon and at 9:00 P.M. on several days. The following data were recorded: The time for sound to travel from the surface to the "phantom bottom" and back at noon and at 9:00 P.M is given in the right-hand columns.

DAY	NOON (seconds)	9:00 P.M. (seconds)
1	1.01	0.51
2	1.00	0.50
3	1.03	0.52
4	1.02	0.49
5	1.00	0.50

(b) If the temperature of the water was 50°F at all depths, the depth of the "phantom bottom" at noon was most nearly

(1)   400 m.   (2)   800 m.
(3) 1600 m.   (4) 2400 m.
(5) 3200 m.

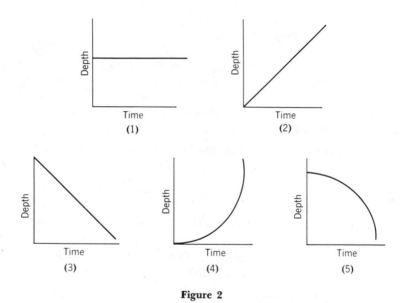

Figure 2

4. Graph A in Figure 3 shows the relation between the speed of sound in sea water and the temperature of the water. Graph *B* shows the actual relation between temperature and depth in sea water at the location where the soundings were made.

(a) On the basis of all the information given, it can be correctly concluded that the

(1) temperature of sea water increases with increasing depth.
(2) speed of sound in sea water decreases as the temperature of the water increases.
(3) speed of sound in sea water increases with increasing depth.

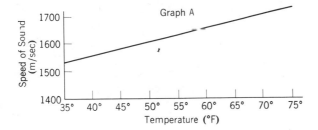

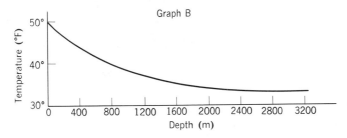

Figure 3

    (4) "phantom bottom" was actually deeper at noon than was calculated in 3(b).

    (5) "phantom bottom" was actually *not* as deep at noon as was calculated in 3(b).

(b) Which, if any, of the following could account for the difference between the time required at noon and at 9:00 P.M. for sound to travel to the "phantom bottom" and back?

    (1) The temperature of the water decreases in the evening.

    (2) The speed of sound in water decreases as the temperature of the water decreases.

    (3) The "phantom bottom" rises in the evening.

    (4) The "phantom bottom" descends in the evening.

    (5) None of the above.

    5. Questions (a) to (f) and graphs A, B, C, and D shown in Figure 4 concern factors that affect photosynthetic activity.

(a) Photosynthetic activity is expressed in terms of the amount of oxygen released per hour by the plant, mainly because oxygen

    (1) combines readily with most substances.

    (2) is needed by all living things.

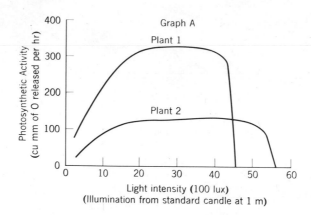

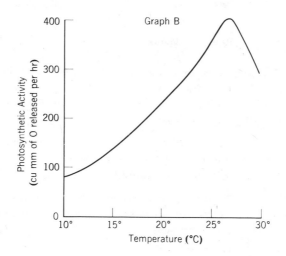

Figure 4

(3) is the principal product of photosynthesis.
(4) is a readily measurable byproduct of photosynthesis.
(5) is constantly released by all plants.

(b) Plant 1 would be better suited than would plant 2 for living in

(1) artificial light.
(2) light having an intensity of 5000 lux.
(3) light of all intensities.

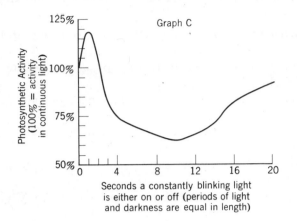

Seconds a constantly blinking light
is either on or off (periods of light
and darkness are equal in length)

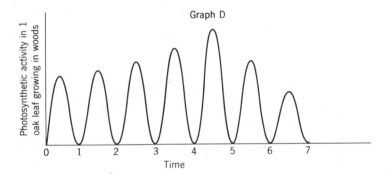

Figure 4 (Continued)

   (4) the dark.

   (5) the shade.

(c) Photosynthetic activity is greatest at a temperature of

   (1) 10°C.   (2) 20°C.   (3) 25°C.   (4) 27°C.   (5) 30°C.

(d) Photosynthetic activity is greater in

   (1) continuous light than in blinking light.

   (2) blinking light than in continuous light.

   (3) light which blinks on or off every second.

   (4) light which is on for 20 sec and off for 20 sec than in continuous light.

   (5) light which blinks on or off every 10 sec than light which blinks on or off every second.

(e) The photosynthetic activity of a plant in light blinking on or off every 10 sec can be

(1) increased by either increasing or decreasing the length of light and dark periods.

(2) decreased by either increasing or decreasing the length of light and dark periods.

(3) increased only by increasing the length of light and dark periods.

(4) increased only by decreasing the length of light and dark periods.

(5) decreased only by decreasing the length of light and dark periods.

(f) What probably accounts for the variations shown in Graph D in the photosynthetic activity of the oak leaf?

(1) The variation in CO_2 available.

(2) The rotation of the earth.

(3) Changes in the amount of H_2O available.

(4) Changes in temperature.

(5) None of the above.

6. Questions (a) to (c) pertain to the human reproductive cycle. The graph in Figure 5 shows the relation between time and uterine development in adult human females.

(a) Fertilization would be most likely to occur during the phase illustrated by the portion of the graph numbered

(1) I.   (2) II.   (3) IV.   (4) V.   (5) VI.

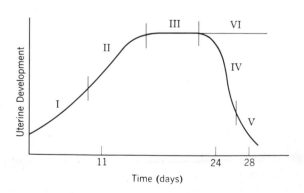

Figure 5

(*b*) What portion of the graph correctly illustrates uterine development after the embryo becomes attached to the uterine wall?

(1) I.   (2) II.   (3) IV.   (4) V.   (5) VI.

(*c*) Degeneration of the corpus luteum would begin during the phase illustrated by the portion of the graph numbered

(1) I.   (2) II.   (3) III.   (4) V.   (5) VI.

(Questions 3–6 pertaining to graphical interpretation illustrate a progression from a set in which the information needed in answering the questions is supplied to a set in which fairly complete background knowledge is required in dealing with the questions. It is interesting to note that difficulty is not a direct function of the amount of background knowledge the examinee must bring to the question.)

ABILITY TO USE SKILLS INVOLVED IN UNDERSTANDING SCIENTIFIC PROBLEMS

There are six major subdivisions of these outcomes. In the logical order of scientific enquiry, they are as follows:

Ability to discern problems
Ability to formulate and screen useful hypotheses
Ability to infer what data to seek in testing hypotheses ("If . . . , then . . ." logic)
Ability to plan experiments appropriate to a problem
Ability to interpret data (draw appropriate conclusions)
Ability to identify assumptions and principles

There are two sides to each of these subdivisions. On the one hand, there is the ability on the part of the student to *contribute* to these scientific activities, actually discerning a problem, or conceiving a hypothesis, or coming to a conclusion from data. We can call this the creative or constructive mode of those objectives.

On the other hand, there is the ability to follow, understand, and judge the soundness of these scientific operations when they are reported in a summary of research or in a scientific paper. Thus, for example, we might ask the student to examine a body of data and the conclusion drawn from them; we would then ask the student why certain other plausible conclusions were less defensible or desirable than the one in fact drawn. We can call this the analytical mode of these objectives.

These subdivisions in their two modes are listed and numbered in the following table in what we suspect is the order of their increasing difficulty for most students.

CONSTRUCTIVE MODE	ANALYTICAL MODE
1. Ability to interpret data	Ability to follow and judge interpretations of data
2. Ability to infer what data to seek ("If . . . , then . . ." logic)	Ability to understand the relevance of data sought to problem posed
3. Ability to design experiments to test hypotheses	Ability to understand, analyze, and judge designs of experiments
4. Ability to formulate and screen useful hypotheses	Ability to discern the appropriateness of hypotheses chosen for test
5. Ability to discern problems in a situation	Ability to identify the problem under attack in a piece of research
6. Ability to identify assumptions and principles	

Examples of some of these follow. It should be noted that it may be useful to test one or another of these skills separately. However, such a separation is exceedingly artificial from the point of view of the way in which research is actually carried out and reported. Hence, some of the examples involve a number of these skills taken together.

A. (THE ANALYTICAL MODE; MOST OF THE SUBDIVISIONS, TAKEN TOGETHER)

Questions 1–8 refer to van Helmont's account of an experiment he performed in the seventeenth century:

"That all vegetable matter immediately and materially arises from the element water alone I learned from this experiment. I took an earthenware pot, and planted in it a willow shoot weighing 5 lb. After 5 years had passed, the tree weighed 169 lb. The earthenware pot was constantly wet only with rain or (when necessary) distilled water; and it was ample in size and imbedded in the ground; and, to prevent dust flying around from mixing with the earth, the rim of the pot was kept covered with an iron plate coated with tin and pierced with many holes. I did not compute the weight of the leaves of the four autumns. Finally, I again dried the earth of the pot, and it was found to be the same 200 lb minus about 2 oz. Therefore, 164 lb of wood, bark, and root has arisen from the water alone."

In van Helmont's day it was thought that there were only four "elements," earth, water, air, and fire. It was further thought that air and fire had no weight.

1. Van Helmont thought the 164 lb of matter in the plant came

   (a) entirely from the earth in the pot.
   (b) entirely from water provided to the plant.
   (c) from both the earth in the pot and the water provided.
   (d) from matter in the air which entered the plant.
   (e) from matter dissolved in the water provided to the plant.

2. Van Helmont probably assumed all the following EXCEPT:

   (a) Air in the atmosphere contains a considerable amount of dust.
   (b) Matter is neither created nor destroyed.
   (c) Gases in the atmosphere are a source of matter for the plant.
   (d) Rain water contains no dissolved matter.
   (e) The leaves produced had appreciable weight.

3. Van Helmont would probably have realized that his conclusion was wrong, if he had

   (a) measured the weight of water given off by the plant.
   (b) determined the weight of the pot used in the experiment.
   (c) more accurately weighed the tree at the beginning and end of the experiment.
   (d) not provided water to the tree in addition to rain water.
   (e) weighed the leaves produced in the four autumns.

4. If van Helmont had found that the earth in the pot weighed considerably less at the end of the experiment than at the beginning, he would probably have concluded that

   (a) the matter in the tree came entirely from the "element earth."
   (b) the matter in the tree came entirely from the "element water."
   (c) earth is soluble in water.
   (d) the matter in the tree came from both water and earth.
   (e) water can enter a plant only through its roots imbedded in earth.

5. On the basis of the assumptions of his day regarding the four "elements," van Helmont

   (a) should have grown the tree in an air-tight container.
   (b) reached an incorrect conclusion because he failed to weigh the leaves produced.
   (c) made an error in failing to weigh the water used by the tree in the experiment.

(d) should have burned the tree which grew and found the weight of the ashes produced.

(e) was correct in the plan and performance of the experiment.

6. Van Helmont distilled the water he provided for the tree because

(a) he probably thought that undistilled water would kill the tree.

(b) distilled water is less dense than undistilled water.

(c) he did not wish to introduce an unknown weight of the "element earth" to the pot.

(d) this was the easiest procedure to follow under the circumstances.

(e) he did not realize that the pot already contained water-soluble minerals.

It is now known that there are over a hundred chemical elements, all of which possess weight; that earth, water, and air are each composed of two or more of these elements.

7. It could be hypothesized that elements originally present in the air became part of the weight of the willow tree. This hypothesis

(a) is not in agreement with the information.

(b) has already been ruled out by the results of van Helmont's experiment.

(c) was tested by van Helmont in his experiment with inconclusive results.

(d) is in agreement with van Helmont's original idea about the source of matter in the plant.

(e) could be tested by growing plants in a sealed container with provision for weighing all matter that enters and leaves the container.

8. Van Helmont's incorrect conclusion resulted mainly from

(a) wrong ideas about the composition of air.

(b) inaccurate weighing of the "element earth."

(c) faulty reasoning from correct assumptions.

(d) failure to weigh the leaves produced.

(e) using a tree rather than a flowering plant in the experiment.

B. (1, ANALYTICAL MODE)

What conclusion was Redi justified in reaching on the basis of the results of his experiments? Why are conclusions ( ) and ( ) not as well justified?

C. (ANALYTICAL MODE, IN GENERAL)

In an early chapter of your text Einstein is quoted. He talks about "the face of a watch and the hidden works" to suggest one aspect of the method of science. In the case of Darwin's contribution to biology, what are some of the matters that correspond to the "face of the watch"? To the "hidden works"?

D. (2, CONSTRUCTIVE MODE)

A biologist hypothesizes that the pituitary and ovary influence each other in producing the uterine cycle in human females. Which of the following observations would support this hypothesis? Which one would *best* support this hypothesis?

(a) Removal of the ovary is followed by degeneration of the uterus.
(b) Removal of the pituitary is followed by death.
(c) The ovary produces hormones.
(d) The pituitary evidently controls a large number of body functions.
(e) Uterine development takes place only when both pituitary and ovary are present.

E. (2, 3, 4, CONSTRUCTIVE MODE)

The following questions relate to a biologist's study of human reproduction.

1. The biologist hypothesizes that the pituitary and ovary influence each other in producing the uterine cycle in females. Which of the following observations would best support this hypothesis?

(a) Removal of the ovary is followed by degeneration of the uterus.
(b) Removal of the pituitary is followed by death.
(c) The ovary produces hormones.
(d) The pituitary evidently controls a large number of body functions.
(e) Uterine development takes place only when both pituitary and ovary are present.

2. Which one of the following procedures would provide the best test of the biologist's hypothesis?

(a) Check all the endocrine glands of the body.

(b) Determine the relative acidity of hormones from the pituitary and from the ovary.

(c) Compare the amount of hormone secreted by the pituitary with the amount secreted by the ovary under identical conditions.

(d) Note what happens to the uterus when varying amounts of hormone from the pituitary and from the ovary are injected into the body.

(e) Keep a careful day-to-day record of uterine changes following fertilization.

3. The biologist finds that the secretions of the pituitary evidently influence the ovary in a number of different ways. Two reasonable hypotheses to explain this finding are that the

(a) ovary reacts in several different ways to the same hormone.

(b) ovary is more directly influenced by external conditions than is the pituitary.

(c) pituitary is producing enzymes.

(d) pituitary produces several hormones.

(e) pituitary secretions contains hemoglobin.

F. (1, FIRST TWO ITEMS, THEN 4; BOTH CONSTRUCTIVE MODE)

The biologist obtains 40 test tubes of thyroxine solutions. He adds enzymes removed from cancer cells to 20 of the tubes. All the tubes are tested every 5 min for the amount of thyroxine present and the results are plotted in the graph shown in Figure 6.

1. How long does it take for half the thyroxine to disappear in the tubes to which the enzymes were added?

(a) 12 min.   (b) 15 min.
(c) 18 min.   (d) 22 min.
(e) 25 min.

2. The biologist is now justified in concluding

(a) that some cancer cell enzymes destroy thyroxine.

(b) that some cancer cells synthesize thyroxine.

(c) that high temperatures destroy thyroxine.

(d) that thyroxine is highly unstable.

(e) none of the above.

3. Of the following, the biologist's next step should be to determine whether or not

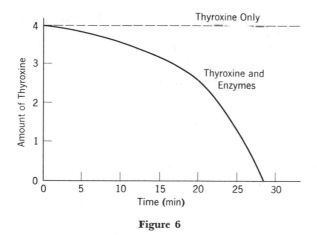

Figure 6

(a) enzymes from normal cells will destroy thyroxine.
(b) thyroxine can be artificially synthesized.
(c) thyroxine breaks down naturally even when no enzyme is present.
(d) the cancer cell enzymes will destroy bacteria.
(e) the cancer cell enzymes can be destroyed by heat.

G. (3, 2, 6, CONSTRUCTIVE MODE)

We suspect that light stimulates plant growth, that is, that the greater the amount of light received by a plant, the taller it will grow.

1. What experiment should we perform to test this hypothesis? Be careful about controls.

2. What data would we expect if our hypothesis were true?

3. Suppose we worked entirely with corn plant. Regardless of the results we obtained, would we test the hypothesis further by using other kinds of plants? Why or why not?

H. (4, ANALYTICAL; 3, CONSTRUCTIVE)

The following questions relate to a competent biologist's cancer research. He reasons, in part, as follows:

I. Since nearly all cells contain nuclei, and

II. since the nuclei contain the genes, then perhaps

III. cancer cells contain mutant genes that produce enzymes that destroy growth-regulating hormones;

IV. thyroxine is a growth-regulating hormone, and

V. thyroxine is secreted by the thyroid gland, which is a part of the endocrine system.

1. Which of these statements is a general hypothesis that the biologist is likely to investigate experimentally?

(a) I. (b) II. (c) III. (d) IV. (e) V.

2. One test of the biologist's general hypothesis would be to determine

(a) whether or not enzymes from cancer cells destroy thyroxine.
(b) whether or not enzymes affect the metabolic rate in cells.
(c) the effect of cell enzymes on the thyroid gland.
(d) the reactivity of the cell enzymes.
(e) the secretion rate of the thyroid gland.

I. (1, CONSTRUCTIVE MODE)

The following questions relate to an investigation of the cause of a disease in cattle.

1. A biologist injects some blood from a diseased cow into a healthy cow. If the healthy cow develops the same disease, which, if any, of the following conclusions is justified?

(a) A microorganism causes the disease.
(b) Something in the blood causes the disease.
(c) The blood from a diseased cow will cause the disease in a second healthy cow into which it is injected.
(d) The disease is caused by some food both cows have eaten.
(e) There is insufficient evidence to reach any of the above conclusions.

2. The biologist isolates a pure strain of microorganisms from the blood of a diseased cow. Following injection of this microorganism into a group of fifty cows, ten of them contract the disease. It is, of the following, most reasonable to conclude that the

(a) microorganism did not get into the blood of the 10 cows that remained healthy.
(b) microorganism is not the cause of the disease.
(c) strain of microorganisms was too weak to cause the disease.
(d) 10 cows which remained healthy were naturally immune to the disease.
(e) 40 cows which got the disease ate something poisonous.

J. ANALYTICAL AND CONSTRUCTIVE MODES TAKEN TOGETHER

When natural fats are introduced into the body of an organism we quickly lose track of them since they mix with similar substances already present in the organism. If, however, radioactive tritium isotopes are incorporated into the molecules of fat in the diets, we can trace the ultimate fate of these fats. The amount of radioactive fat present in a sample of fat can be found by burning the fat and determining the amount of heavy water (water containing tritium) in the water which results from the combustion. Assume all stored fat in the organism is contained in fat storage spots known as fat depots.

A group of mice are fed a carefully weighed diet which contains 20% radioactive fat. The mice are decidedly underfed, and it is expected that all the fat fed will be immediately burned for energy. At the end of 10 days on the radioactive diet the mice are killed. The experimenter determines the total amount of fat as well as the percentage of radioactive fat from the diet present in the carcasses. The amount of heavy water present in the body fluid is measured and the amount of radioactive fat which on combustion would give rise to this amount of heavy water is computed.

The accompanying table summarizes the data obtained for the group of mice as a whole.

Weight of mice at beginning of experiment   250 g
Weight of radioactive fat consumed   35 g
Total fat in the carcasses at end of experiment   12 g
Percentage of carcass fat which is radioactive diet fat   50%
Weight of heavy water in the body fluid   0.01 g
Weight of radioactive fat which when burned will give rise to heavy water in the
    body fluid   7 g

1. What evidently was the experimenter's original hypothesis?

   (a) Animals on a starvation diet will burn fat in their diets immediately.
   (b) Fat storage is a continual process in all animals.
   (c) Tritium is an essential part of the diet of an animal.
   (d) Radioactive material will kill mice.
   (e) The bodily processes of an animal on a diet containing tritium will be greatly changed.

2. Which of the following is a necessary assumption in this experiment?

   (a) Fats containing tritium are identical in all respects with natural fats.
   (b) Mice are better able than other animals to withstand doses of tritium.

    (c) Substances containing tritium will remain separated from similar natural substances.

    (d) The organism will treat fats containing tritium exactly as it treats natural fats.

    (e) Plants will react to tritium exactly as animals do.

3. How much radioactive fat from the diet was burned for energy by the mice?

    (a) Less than 7 g.   *(b)*   7 g.

    (c) Between 7 and 30 g.   *(d)* 30 g.

    (e) More than 30 g.

4. What assumption is necessary in order to determine, from the data given, the amount of radioactive fat from the diet burned by the mice for energy during the experiment?

    (a) Fluid excreted by the mice is 20% radioactive.

    (b) All fat not stored is burned for energy.

    (c) The mice weighed less at the conclusion of the experiment than at the beginning.

    (d) A specific amount of energy results from the burning of 1 g of fat.

    (e) A greater amount of radioactive fat is burned for energy than is stored in fat depots.

5. An experiment is performed which is identical in all respects to the one described, except for a lengthened feeding period. Under these conditions all the following may occur EXCEPT:

    (a) Diet fat may be stored in fat depots.

    (b) The fat deposits in the carcasses of the mice may contain radioactive fat.

    (c) Heavy water may be found in the body fluid of the mice at the conclusion of the experiment.

    (d) The weight of radioactive fat eaten may be greater than the weight of the mice at the beginning of the experiment.

    (e) The weight of heavy water in the body fluid at the end of the experiment may be greater than the weight of fat eaten.

6. It can be correctly inferred from the information given that

    (a) most of the fat in the fat deposits at the end of the experiment came from fat consumed during the experiment.

    (b) some of the radioactive diet fat was burned for energy.

(c) nonradioactive fat is denser than radioactive fat.

(d) no combustion (burning) of fat occurred in the mice during the 10-day feeding period.

(e) most of the fat in the fat deposits at the end of the experiment was originally present in the mice.

7. The most precise conclusion that can be drawn from the experiment is that mice on starvation diets

(a) burn all fat in their diets immediately for energy.

(b) burn half the fat in their diets and store the other half.

(c) starve to death.

(d) can be kept healthy by giving them proper dosages of tritium.

8. The experiment suggests which of the following hypotheses as the most reasonable one to test in further experimentation?

(a) The proportion of stored fat to burned fat decreases as the starvation feeding period is lengthened.

(b) Animals under starvation conditions store almost all the fat in their diets.

(c) Radioactive materials are not absorbed into living organisms.

(d) Tritium is a fatal poison which is immediately effective in all amounts.

(e) Tritium atoms lose their radioactive properties when they are involved in chemical reactions.

As we remarked earlier, tests have two functions. On the one hand, they measure the effectiveness of learning and teaching. On the other hand, they become one of the most effective ways through which students come to understand what is expected of them and are motivated to move in these directions.

Because tests are so important, we recommend that some means be used to ensure that tests used in the course of a year are adequately representative of all the subject matter and all the objectives which define the course. One useful method is to construct a check sheet which can serve as both a reminder of what to test for and a record of what has been tested for in previous examinations. One such check sheet is given on page 000.

Finally, a few words about the actual construction of tests. In general, two types are useful for short examinations: the so-called "objective" test item which can be scored by a clerk or a machine, and the short essay.

Test Plan (Content and Ability Coverage)

TEST _____

NO OF ITEMS _____

TEXT CHAPTERS, CONTENT CATEGORIES, THEMES, OR LEVELS OF BIOLOGICAL ORGANIZATION; FOR EXAMPLE:	RECALL	ABILITIES			INTERPRETATION OF DATA	"IF ..., THEN ..." LOGIC	DESIGN OF EXPERIMENTS	HYPOTHESIS MAKING	DISCERNING PROBLEMS	IDENTIFYING ASSUMPTIONS
		APPLICATION								
		SHOWING OF RELATIONS	NON-QUANTITATIVE	QUANTITATIVE						
Chapter 1										
Chapter 4										
Genetics										
Populations										
Homeostasis										
Molecular Level										
TOTALS										

Among forms of objective items are true-false items, multiple-choice items, and matching items. Of these we recommend the multiple-choice form as the most flexible and useful.

In constructing the set of four or five alternative answers for a multiple-choice item, one or two rules are usefully kept in mind. First, it is of course necessary that the individual item be clear and definite in its meaning. Second, it is important that wrong alternatives have differing degrees of obviousness in their "wrongness." No more than one member of the set of alternatives should be transparently irrelevant to the question put. The others should be possible or plausible. Otherwise, the test will not "discriminate" adequately. It may separate the utterly unprepared from all others, but fail to distinguish between the fair, the good, and the excellent student.

When the alternatives of a multiple-choice item are so graded in their degree of "rightness" or completeness, it becomes clear that such items can be used and graded in two different ways. The commonest way, but not the best for all purposes, is to consider *one* response as the only "right" one, all others being treated as wrong. This simplifies scoring and is the common practice where machine scoring is used.

Much more is accomplished, however, in the way of rewarding, motivating, and guiding students through tests if we reward students for less-than-perfect responses as well as for the best response. This can be done, ideally, by constructing the four or five alternatives of an item in such a way that one of them is wholly irrelevant or erroneous, another contains a degree of or approximation to a sound response, a third is sound in all but one respect, and a fourth is best of all. Then, when such an item is scored, a student may receive no credit, 1 point, 2 points, or 3 points, depending on his response.

Such an ideal broad-range item is too difficult to construct to be practicable for all cases. It is not at all difficult, however, to construct sets of alternatives in which there is a second-best as well as a best response. Moreover, infrequently, we will construct items with two second-best choices, or containing three items out of the four or five which can be graded in quality. Such items as these should be striven for. Then, a test consisting of six multiple-choice items can be used to provide appropriate rewards over a considerable range of scores for a considerable number of students.

One of the most important uses of such a test is in a classroom "post mortem." The items are taken up one at a time, and students are asked to explain why one response is good but not the best, why another is better, and so on. If the test is to be used for purposes of "grading," then the students' reasons for answers should be as important, if not more so, in assigning the grade as his initial score on the test. If the test is used by students' to assess self-progress, then let each rate himself on the basis of his choices and his reasons for them.

Another technique for emphasizing the use of tests as a means of assessing self-progress is the following procedure: After students have responded to the test items, number the papers and remove the names, keeping a record so that papers can later be returned to the students. Give each paper to another student. Each student then corrects the paper before him through use of his text, notes, memory or whatever he requires. If "grades" are to be assigned, the student is graded on his ability to correct another's paper, not on his own initial score.

There are many techniques other than the multiple-choice test for assisting the student in assessing his progress in attaining the desired outcomes of BSCS biology. Some of these are the following.

1. Evaluation of laboratory skills. Plan an interesting laboratory activity that can be done in a number of short steps and that is related to the work the class is doing at the time. You may want to put out all the necessary equipment or you may want the students to gather the appropriate equipment as part of the test. If the class is large, sectioning it into groups will save time and also tend to maintain student interest. As the class watches, ask for a volunteer (volunteers if there are groups) to perform the first step. How efficiently and/or accurately does the student accomplish the task? (Let him make the judgment or, if he is sufficiently confident, let him ask others to do so.) Then ask for another volunteer to perform the second task, and so on until the laboratory activity has been completed. In this way, each student has a chance to volunteer for some task he knows he can perform. During the course of the year, structure such lab evaluations so that students can measure their improvement in laboratory techniques as well as their ability to perform new techniques. Also relate the techniques to adequacy of experimental design and to obtaining of wanted data.

2. Evaluation of discussion. Divide the class into groups. Give each group the same discussion problem. Allow each group time to read, discuss, and reach some conclusions. One member of the group should take notes on the remarks and conclusions of the group. The teacher can circulate from group to group, assisting when asked. Such assistance is best provided by raising questions that help students to focus on important aspects of the discussion problem. When sufficient time has been allowed for group discussion, bring the class together. Have one member from each group report on his group's remarks. The class as a whole can then point out strengths and weaknesses in terms of the factors considered relative to the problem, how these factors were related, whether any conclusions reached were warranted, etc. Each group can judge its own performance in light of the comments by the class as a whole.

3. Reading and writing. (a) Select a short reading and ask the students to pick out and state in a sentence the main idea of the selection. (b) Read a paragraph aloud and ask students to write or tell what the main idea is. (c) Have students select sections following procedures similar to (a) and (b). (d) Ask students to construct an essay-type question concerning a lab activity, a reading, or a discussion and then write a short essay answering their question.

4. Short-essay items. One of the most important cautions is this: See to it that the question is so phrased that it adequately delimits, "structures," the student's attention, so that he is helped to focus on the point or points with which he is expected to deal. One example of a delimited essay item follows.[11]

"John prepared an aquarium as follows: He carefully cleaned a ten-gallon tank with salt solution and put in a few inches of fine washed sand. He rooted several stalks of weed *(Elodea)* taken from a pool and then filled the aquarium with tap water. After waiting a week he stocked the aquarium with ten one-inch goldfish and three snails. The aquarium was then left in a corner of the room. After a month the water had not become foul and the plants and animals were in good condition. Without moving the aquarium he sealed a glass top on it.

"What prediction, if any, can be made concerning the condition of the aquarium after a period of several months? If you believe a definite prediction can be made, make it and then give your reasons. If you are unable to make a prediction for *any* reason, indicate why you are unable to make a prediction (give your reasons)."

Notice that this essay item presents a limited situation but one involving a complex of biological factors. The essay question based on this situation asks the student to do a particular thing—it does not merely ask him to discuss what might happen to the aquarium in several months. Specifically, the essay question requires the student to select appropriate principles and extrapolate beyond the situation given.

[11] B. S. Bloom (editor), 1956, *Taxonomy of educational objectives,* Longmans, Green and Co., London, page 131. Adapted from Test 1.3B, "Application of Principles in Science," Progressive Education Association, evaluation in the Eight-Year Study, Univ. of Chicago, 1940.

5. When the student feels a measure of confidence in an evaluation situation, he can be exposed to many different types of evaluation techniques. Tape recorders, films, overhead projectors, flannel boards, and graphing materials can be used effectively as well as group competitions, puzzles and similar activities. Interest, enthusiasm, and hard work are all indications of a genuine involvement that should be recognized and sometimes rewarded. Be on the lookout for the following:

How do students react to suggested readings?
Do they notice related articles in newspapers and magazines?
Do they mention related television programs?
Who comes at odd times to talk about science?
Who brings in living materials to use in the classroom?
Are students raising questions as they use reading materials?
Do the students enter more readily into class discussion?

6. An important consideration in the construction of any test which is to be used for "grading" is the range of difficulty of the items. It has been common practice in the past to construct tests so that a "satisfactory" student performance would consist of successful responses to a fairly high percentage of items—75%, for example, or 65%.

The weakness of a test constructed in this way consists in the fact that it may not adequately challenge the highest achieving students (since both the best and the less effective students may get scores in the range of 90 to 100). Similarly, but considered from the point of measuring and certifying student achievement, such a test will not discriminate among students in the upper range of achievement. The implication of this is that if a test is to be used for certifying students, it is best to construct it so that a "satisfactory" performance is approximately 55% of the items. This is the practice used in the development of the BSCS tests.

7. If tests can be constructed and used cooperatively by several teachers, many advantages accrue. First, the relations between teacher and student are much improved, since part of the stigma of being an inquisitor is removed. (This point is discussed at greater length in the essay on discussion.) Second, a larger group of students' scores can be pooled to obtain data about the quality of the test, and in order to discover and set fair and just standards. Third (and of exceeding importance), far better tests can usually be constructed in this way.

This third, major advantage accrues because the work is divided among several persons and the time thus saved can be used for critical sifting of the test materials. For example, the group of teachers might first meet briefly to decide what objectives and content sections should be tested for. Then the work is portioned out—on the basis, say, of objectives or subject areas. When each teacher has prepared his items they are put together, and copies of all of them are distributed to each teacher. Each teacher then "takes" the test. Following the "take," responses are compared and items discussed. In this way ambiguities are discovered and repaired, and group judgment is brought to bear on determining what objectives are actually being tested for and whether the test is a good sampling or representation of what was intended.

8. Whenever a teacher uses new materials and techniques he is interested in evaluating his progress as a teacher. If you attempted to construct a review (quiz or test) that will help you determine what you accomplished, it might look something like the following:

What material did I cover?

How many hours did my class spend in the laboratory?

Do the above three questions really describe what I have accomplished? Are they merely facts that I might find in my notes?

What have the students learned?

    A. Much   B. Little

    C. Some little, some much

In what activities am I most successful?   D. None of the above

    A. Discussion   B. Laboratory

    C. Evaluations   D. Sometimes one, sometimes another

Do I feel that multiple choice questions are a good way to evaluate accomplishment?

    A. With certain material   B. Never

    C. If it is well constructed   D. If I am accustomed to this
                                          type of test

What would I learn about my progress if I tried to write a description of what goes on in my class?

If I taped some class sessions, would I be able to determine the contributions I make to discussions, the type of questions I ask, and the model I provide?

Would I gain an idea of my progress if I simply examined my personal reactions to the course? Do I like it? Do I feel successful? Do I enjoy the students?

Should the opinions and attitudes of the students be a part of my evaluation?

Do I want to set my standards so high and make the test so difficult that I may fail and consequently feel unsuccessful?

Do I want to evaluate myself in the same way I evaluate students?

Would all of the foregoing types of evaluation cover the different
aspects of the course?

Do I feel that someone other than myself would be better able to
evaluate my progress?

ACTIVITIES FOR TEACHERS

Since assessment of student progress toward mastery of the desired
outcomes of high school biology can be critical in assisting the student
to attain that mastery, it is essential that the teacher develop as much
competence as possible in this teaching strategy. Below are some suggestions for activities that will assist the teacher in mastering evaluation
strategies.

Try out with a small group of students the evaluation techniques
listed on pages 00 to 00 (Nos. 1–5).

Examine each of these approaches in terms of the two purposes
of evaluation and the different assumptions about students which
are discussed in the first two sections of this chapter.

For other suggestions and information pertaining to evaluation see
the following:

Stothart, Jimmy R. "Test Building as an Institute or Workshop Activity," *American
Biology Teacher*, April, 1967, pp. 277-281.

Lee, Addison E., et al., *Resource book on testing for the laboratory blocks.*
BSCS Special Publication No. 6.

BSCS Newsletters No. 10, No. 19, and No. 30.

Lee, Addison, E. (editor) . *Research and curriculum development in science education,*
Vol. 1, *The new programs in high school biology.* Science Education Center, The
University of Texas at Austin, 1968.

9

Visits to Museums, Zoos, and Similar Institutions

Most of America's high schools are within a reasonable distance of a natural history museum, aquarium, zoological garden, or botanical garden. These institutions vary markedly in character; some are large institutions with expert staffs and an active education department while others are modest organizations under the guidance of a director with little formal training or background in the biological sciences.

A resourceful teacher should be able to utilize these institutions, when available, to enrich and extend the education experiences of his students in a variety of ways. The few suggestions made here are examples of the many useful things that could be done. Obviously the teacher will have to depend on his own resources to a greater extent if the nearby institution does not have a well-developed educational program.

Although an unplanned visit to a museum, zoo, or botanic garden can be a valuable educational experience for students, the value is greatly enhanced through advance planning. As a very early step in such planning, the teacher should visit the institution and should become acquainted with its displays and facilities. He should estimate the strengths of the institution in terms of his own teaching program. For example, the Climatron of the Shaw Botanical Gardens in St. Louis is an excellent resource for the teaching of ecology, and the Cleveland Museum of Health is a comparable resource for instruction in human anatomy and physiology; neither, however, could serve as an appropriate alternative to the other.

After estimating how the facilities of the nearby institution could contribute to his teaching program, the teacher should consult with the staff member of the institution who is responsible for the coordination of school visits. He would discover, for example, that at the American

Museum of Natural History in New York City a very competent staff, including a number of former high school biology teachers, has organized a series of specific programs for students at various educational levels and is prepared to make available a variety of special services to the teacher and his students.

Even if the local institution does not have staff resources available to the teacher, it is still important to discuss a proposed visit with the officials of the institution because of potential scheduling problems. For example, in most natural history museums the overwhelming number of school visitations is concentrated in the spring of the year; in some small museums, space is at such a premium that the unannounced arrival of a class during the peak season constitutes a serious disruption.

If the contact person at the institution is not familiar with the educational program of the teacher, it might be helpful for the teacher to describe what he hopes to accomplish through use of the institution's facilities. The teacher might leave a copy of the textbook he is using at the institution for study. If the institutional staff seems receptive, he might suggest how the institution could assist him (and other teachers) in providing improved educational services for local students.

There are three basic ways in which institutions such as museums, zoos, aquaria, and botanic gardens can contribute to the education of students enrolled in school programs:

1. The resources of the institutions are delivered to and utilized in the school.

2. The students visit the institution under organized school auspices, usually in classroom units.

3. The students visit the institution individually, that is, in a non-assembled arrangement.

With reference to the first scheme, three examples are offered. For many years there has been available to the schools of the St. Louis area a series of study materials and exhibits delivered, on request, from a central audiovisual center. The Florida State Museum (Gainesville) has circulated displays illustrating biological principles which are placed in corridors of local schools. The Prince of Wales Museum (Bombay, India) circulates, among schools, specimens of local birds and mammals in plastic tubes for students to handle and study.

The classroom visit to the museum, aquarium, zoo, or botanic garden requires the greatest amount of planning. Before leaving the school the students should, at least, be briefed about the purpose of the visit. In addition, it is usually valuable to provide some written materials. Suggestions for these are given below.

Upon arriving at the institution, local specific briefing may not only be desirable but essential. However, this should be kept to a minimum. A long lecture conducted on the grounds of an exciting and stimulating institution will not only deprive the students of an opportunity to explore the place but will distract them from the lecture being presented.

Many teachers find it highly desirable to prepare in advance (with or without the assistance of the institution being visited) an open-ended guide sheet or questionnaire for students to complete during the visit. Properly constructed questionnaires can be enquiry oriented and will be attractive and fascinating to students as a kind of intellectual treasure hunt. Review of these papers can provide for stimulating classroom discussions for several subsequent periods after the visit. If several alternative questions are given to different groups of students, the educational experience through the subsequent discussions is considerably enhanced.

Obviously something similar can be achieved by using open-ended guides or questionnaires and by assigning them to students to pursue on an individual basis, which is the third basic way to use the institution. Here there would not be an assembled classroom visit to the institution but, instead, students would visit on an individual (or team) basis to complete the assigned work.

In planning for either an organized class visit or for independent student visits, it is usually desirable to involve the students in planning their observations. The teacher might begin by naming certain exhibits that are in the museum. For example, one museum [1] has habitat exhibits of raccoon, coyote, wolf, bear, beaver, moose, elk, caribou, Dall sheep, and white-tailed deer. This would be followed by asking students to pose questions related to topics currently, or recently, under study which might be answered by observations of these exhibits. If, for example, ecology is the current area of study, students might suggest questions such as: What plants are characteristic of the habitats of each of these animals? What information is provided regarding the latitude and/or elevation of the habitat? What, if anything, can be observed that provides information on the food web of which the animal is a part? Using these general questions as a starting point,

[1] The James Ford Bell Museum of Natural History, Minneapolis, Minn.

students can then suggest more specific questions and the types of observations they should make to help answer the questions.

A variation of this approach is to ask several interested students to plan the visit. The students, with the aid of the teacher if needed, would develop a guide sheet for the other students. In addition to including questions that provide a focus for observations during the visit, space on the guide sheet should be provided for questions that occur to the students during the visit. These can then be used as a starting point for subsequent independent visits and/or for follow-up in the course schedule.

The discussion that follows the visit should not only attend to the observations made but also to the inferences and reasoning the students use in relating their observations to the questions. For example, there may not be sufficient evidence to determine latitude or elevation directly, but there may be enough to provide a probable answer if students reason from other information they have as well as what was observed. If they become aware of the inferential process used in answering some questions in contrast to questions that can be answered by information gained from direct observation, there is an added educational outcome that has considerable value.

The most frequent targets for natural history museum visits seem to be of two kinds. Many museums have excellent habitat groups, that is, displays illustrating animals and plants in their natural ecological environments. Some examples are: American Museum of Natural History, New York City; Rochester (N. Y.) Museum of Science; New York State Museum, Albany Smithsonian Institute, Washington, D. C.; Charleston (S. C.) Museum; Florida State Museum, Gainesville; Louisiana State Museum, Baton Rouge; Chicago Natural History Museum; Nashville Children's Museum; University of Michigan Museum, Ann Arbor; Minnesota Museum of Natural History, Minneapolis; Milwaukee Public Museum; University of Kansas Museum, Lawrence; University of Nebraska Museum, Lincoln; Denver Museum of Natural History; University of Colorado Museum, Boulder; and California Academy of Sciences, San Francisco. There are also many others.

Habitat groups lend themselves to a variety of educational functions most of which are not self-explanatory and will need to be brought out by the teacher. Such habitat displays are valuable in acquainting students with the kinds and varieties of animals and plants and their environments. They can be instructive in the areas of adaptation, evolution, biogeography, natural selection, food pyramids, and many other topics. To exploit these internally hidden educational opportunities will require, in most instances, planning and resourcefulness on the part of the teacher.

The other common type of exhibit of many natural history museums is the display of series of mounted mammals, birds, and reptiles. Other than the identification of kinds of animals, the educational value of such exhibits is relatively limited, especially if the specimens are not grouped into a pattern of biological interest. However, thoughtfully organized guide sheets can elicit intellectual challenges not immediately apparent in the long rows of specimens. Fairly common approaches are the discovery by students of the differential arrangement of sense organs (eyes and ears) in carnivores and herbivores and the relationship between the structure of beaks in birds and their feeding habits.

Museums, zoos, acquaria, and botanic gardens certainly are places in which students can learn about the variety of animal and plant life and can gain an appreciation of the distribution of biota over the face of the earth. For many students—and probably for most—the simple inspection of animals and plants in three dimensions is an extremely important educational experience even though its impact is difficult to measure.

There seem to be two major educational values to be gained from museums, zoos, aquaria, and botanic gardens. The first (obvious value) is to acquaint students with the subjects—plants and animals in three dimensions—of biology. No picture of a polar bear is as impressive as the polar bear habitat group in the Denver Museum of Natural History, and that display is superceded in instructiveness by the living polar bears in the New York Zoological Gardens.

The other important value is the potential use of these institutions in enquiry instruction. For the most part, these institutions have not yet realized their tremendous potential in this kind of instruction, although a few have developed "junior curator" or "treasure hunt" activities. For the foreseeable future, however, exploitation of these institutions for enquiry instructions will depend on the inventiveness of the teacher. One example of such use follows.

In most zoological gardens the animals are arranged systematically: for example, a reptile house, hoofed animals, pachyderm house, and a primate house and so on. Many zoos also have animals grouped geographically: African Plains, Australian mammals, South American mammals, and the like. But these displays with obvious intent can also be used for additional educational purposes. For example, in the Brookfield Zoo of Chicago there is an enclosure containing a group of North American Timber Wolves. At first glance the educational impact is limited to that indicated above. However, these wolves have lived together in this particular enclosure for a number of years and have developed a distinct social hierarchy. There is a dominant male, obvious by his behavior and the behavior of the other wolves in the enclosure, and the other animals have their clearly defined places in the hierarchy. The connection between this exhibit and a classroom discussion of social hierarchy in animals is clear; the wolves could serve to make a classroom lesson real and exciting.

It is not only the public displays that can be used for educational purposes. Sympathetic curators or keepers are sometimes willing to show a class the behind-the-scenes operations. Thus the manner in which animals or plants are preserved, classified, and studied in museum collections or the way that illnesses of animals from all over the world are treated by the zoo's veterinarian are topics of great fascination to many students and, for some, may serve to introduce them to lifetime careers that they otherwise would never have discovered.

CHAPTER

10

Use of Visual Aids in Teaching
the Biological Sciences

Why bother to use films or, for that matter, why bother with any visual aid at all? Visual aids are difficult to locate and they take up valuable class time. If they are ordered, they usually do not show up at a proper time, and if they do, the projection equipment is usually broken. The classroom is too light or the classroom is too hot, so why bother? Most teachers have probably felt some of these frustrations at one time or another, and yet, many (if not most) continue to make some use of films, film strips, slides, and overhead transparencies. Visual aids can certainly supply enrichment experience for students in almost any classroom and yet, only too often, they are accepted as a form of passive entertainment.

Probably no other subject area lends itself as well to visual presentation as does biology. Biological phenomena are interesting to watch, and living things are natural camera subjects. If used properly, visual aids become valuable tools for teaching, since it is often easier to describe something with pictures than to describe it verbally. With motion pictures the dimension of movement is added—movement through time and space as well as movement of organisms themselves. Events that take days and even months to occur can be shown on the screen in a few moments. Conversely, events that occur too rapidly for the human eye to perceive can be captured on film and can be observed carefully over a longer period of time. A class of students can be transported to the far corners of the world or enter into the remoteness of a single cell. Phenomena and experiments well outside the capabilities of the ordinary high school laboratory can be presented through the use of motion pictures, slides, film strips, overhead transparencies, and the like.

447

Although visual aids may add to the classroom experience, their use also presents the teacher with a variety of problems. Perhaps of paramount importance is the fact that time is valuable. In the United States, most teachers have approximately 150 hours of contact with each class of students. With the expansion of knowledge, the broader subject areas, and the greater emphasis on laboratory work, the teacher and the class have little time for materials that do not contribute effectively to the learning experience. Situations differ widely, but many teachers simply do not use visual aids because of the problems of scheduling. It is difficult for a teacher to schedule a film in May for use with a class in December. Often teachers have trouble locating a particular film that will illustrate exactly what is to be taught and, in many cases, the teacher has no opportunity to preview the film before it is scheduled. There are still many schools where it is almost as difficult for a teacher to locate and schedule projection equipment as it is to arrange for appropriate films or other visuals. There are other problems as well associated with the use of visuals, and each teacher could list a variety of things not discussed here.

The question, then, that each teacher must face is whether to use or not to use visual aids. If they are to be used, what will be used and when will it be used? Can the visual aid better contribute to a given objective than another activity? Will it actually contribute to the students' understanding of biology? Is the use of a film a meaningful learning activity for the student, or is it simply unrelated entertainment? Used intelligently and judiciously, films, film strips, and slides can expand the horizon of biology teaching for both students and teachers. The pitfall to avoid is the use of visual materials as time-consuming frosting in an already crowded course schedule.

How, then, might visual aids be used most effectively? If used with careful planning, they can become very versatile. There is no reason why a single film must be shown in its entirety, nor is there any reason why a teacher cannot turn off a sound track when it does not add to the lesson. There is no reason that an entire film strip need be shown, nor that an entire set of slides be presented. By judiciously selecting visual and audio portions to be presented to the students, teachers can adapt many visual aids to a given purpose. If this is done, the most didactic film presentation can often be converted into a means of stimulating student discussion and interaction. A teacher can do much to encourage this by posing questions, by stopping the projector at selected points, and by inviting student discussion. An adequate presentation depends on being able to spend time in preparation just as one would prepare a laboratory or other class activity.

Of obvious importance is the need to determine the best time to introduce a particular film or other visual aid. For example, a film on Mendelian genetics, used before the class has had an opportunity to discuss inherited characteristics, probably would be lost as a meaningful experience for the students. This is not to say that films should never be used to introduce a topic or subject area. Some can, just as others may be used to summarize the end of a unit. There are many ways in which films can be used, and this is best determined by the teacher. Most teacher's guides for a given course of study indicate films that may be useful with a particular unit. This does not mean that the teacher should feel compelled to use the visual aids where they are indicated or that he should not feel free to substitute when better materials are recognized.

Perhaps it is most important to identify areas for which a visual supplement is best for a specific teaching purpose such as to illustrate a biological phenomenon that is difficult to observe in the laboratory or to provide a situation in which students can deal with a particular aspect of enquiry. Once these areas are identified, the teacher must determine what materials are available and how they can best be brought into the classroom.

We have already suggested that many problems exist with the scheduling of materials. The ideal situation would be that each teacher have his own visual aids or, at least, that each department within a school maintain its own library of films, film strips, slides, and overhead transparencies. This may not be as impossible as it seems. Although 16 mm sound films are extremely expensive as well as time consuming, 8 mm loop films are quite inexpensive. Often twenty 8 mm loop films can be purchased for the price of one 16 mm sound film. Projectors also are less expensive for the smaller format. With the advent of Super 8 mm loop films, screen images are quite acceptable and, in many cases, approach those of the 16 mm film. Likewise, 35 mm film strips are quite inexpensive. Slides and overhead visuals cost slightly more, but slides and overheads can often be made by teachers themselves to perform a specific task. If resources are limited, it is a good idea to build a library of one type of visual aid to the exclusion of others.

By considering what visuals are important and what can be most useful in a limited amount of time, individual teachers or departments can determine exactly what would be advantageous to have. Appropriate visuals may be acquired over a period of years with a few titles being added in each new budgetary period. Most departments will probably need no more than 40 or 50 individual packages of visual aids. In order to give teachers a choice of what is available, these must be purchased judiciously.

KINDS OF VISUAL AIDS

There are many different kinds of visual aids, and they can be put to many different uses. The original creators of the films or film strips often do not envision the use to which the teacher ultimately puts these materials.

Descriptive Films

A vast majority of films fall into this category. In the area of life sciences, they illustrate biological phenomena, often with a narrative track that describes the events depicted on the screen. There are many of these films that can be useful in the classroom. Ecology films, for example, place the students in ecosystems other than those found close to home. Films on behavior can provide a glimpse into the complex ways in which organisms respond to their environment and to each other. Many biological phenomena, which are difficult to describe in words, can be depicted in a short time on the screen. Mitosis or the development of an embryo are good examples. Through the use of time-lapse and stop-motion photography, descriptive films can be very useful. Good examples can be found in the catalogs of the major film distributors.[1]

Techniques Films

Techniques films are useful in introducing students to new techniques that may be difficult to describe in words. They can also be useful to the teacher in learning these techniques. Many film distributors carry films relating to biological techniques. In this area, 8 mm films are both practical and inexpensive.[1]

Films that Emphasize Enquiry

A few films have a narrative of enquiry. These are films that utilize the processes of science to describe a phenomenon. Some examples are: "Controlled Experiments," produced by Indiana University; "The Scientific Method," produced by Encyclopaedia Britannica Films

---

[1] See Appendix 5 for lists of film distributors and film libraries.

"Using the Scientific Method," produced by Coronet Films; and "Scientific Method in Action," produced by the International Film Bureau.

Single Concept Films

Single concept films are usually found in the 8 mm format. They are short films that describe a given phenomenon or structure and are generally descriptive in nature. Many distributors carry this kind of films.

Single Topic Inquiry Films

These films are designed to stimulate student participation. This is a relatively new idea in film making and only a limited number of distributors carry such films. Foremost in this area are the BSCS Single Topic Inquiry Films and the SRA Inquiry Films. Unfortunately, the SRA films deal primarily with physical rather than biological phenomena.

Film Strips

Film strips have been available for a long time and evidently are very popular. For the most part, they can be placed in the same general category as motion pictures. They have an advantage over motion pictures in that they cost considerably less. They have the disadvantage of not having the added dimension of motion.

Slides

There has been wide use of 35 mm slides, $2\frac{1}{4}$ slides, lantern slides, and other formats as lecture aids for many years. Occasionally, sets of slides are put together which resemble a film strip presentation, but more often than not, individual slides are assembled by the teacher for use with a particular lesson. Slides have a distinct advantage of being very flexible in that they can be placed in any order and can be added to or subtracted from at the discretion of the person using them.

Overhead Transparencies

Overhead transparencies are rapidly gaining acceptance in many classrooms throughout the United States. The main disadvantage to overhead transparencies is that they are relatively expensive when purchased in their prepared format. However, overhead transparencies are quite often easily prepared by the teacher, utilizing copying devices already on hand. A number of distributors furnish biological material that can be copied for overhead use. Overhead transparencies have one great advantage that no other medium has: the teacher can draw directly on the transparency, either before class time or during class discussion.

VISUAL AIDS AND THE BSCS

When the BSCS set out to develop films for use in biology classrooms, teachers, educators, and research scientists were consulted on what they felt would be the ideal characteristics that visual materials might possess. These characteristics were identified as follows.

1. Visual aids should limit their subject matter to a single phenomenon or a few related phenomena.

2. They should involve subject matter and techniques that cannot be handled in the normal high school laboratory. These might include experiments that are too difficult to undertake or procedures that take too long to develop. They may involve concepts that involve movement either in time or space.

3. Film slides, film strips, transparencies, and projectors should be easily accessible. Ideally they should be available to the teacher in his own room or, second best, within his own department. These materials must be as inexpensive as possible in order to be widely available.

4. Motion pictures should be four to five minutes long, although by selective stopping of the film the presentation might take much longer.

5. Sound tracks should be limited or eliminated. Natural synchronous sound might be beneficial when sounds are important to the understanding of a concept.

6. The materials must actively involve students in a meaningful way to promote their knowledge of biological concepts and their understanding of the processes of science.

In order to be an important adjunct to the teaching of biology, visual aids must incorporate a teaching strategy. Primarily, films, film strips, and the like are illustrative in nature. Both visual and audio portions of the films describe to the students a given phenomenon or related phenomena. Occasionally, producers of films will include teacher's notes describing ways in which teachers might use a film along with a list of questions that might be asked on the content of the subject matter covered in the film. We all recognize that one way in which visual aids may be used is to illustrate a lecture. Teachers may go further in using prepared materials by developing their own teaching strategy. Different ways in which films may be used include: as the basis for a discussion; for individualized instruction; as testing vehicles; as an introduction to the laboratory; and as data sources for a lab itself. The advent of the 8 mm loop cartridge projector and the many loop films which are now available have expanded considerably the way that visual aids can be utilized in the classroom. Today there are few biological phenomena that cannot be illustrated with a simple 8 mm loop film. Also there are many film strips available on almost any biological subject. Slides and overhead transparencies are harder to locate in many subject areas. How these materials are used is largely up to the teacher; yet, few have the time to develop the teaching strategies that would make the most effective use of the visuals in question. To meet this need the BSCS is beginning to develop materials for specific teaching purposes.

The Biological Sciences Curriculum Study has released a series of Super 8 mm loop films which provide students with more than a passive, observational experience. The BSCS Film Committee first conceived of the Single Topic Inquiry Film Program as a series of short 16 mm films which would illustrate significant biological concepts and phenomena which could not be duplicated for the students in the laboratory. With the advent of the 8mm loop cartridge, the BSCS directed its attention toward the use of this format. As the films developed, questions were added that could be answered by the students and, later, the films were arranged in an enquiry format. They pose questions, raise problems, and present experimental data that challenge student participation. The films provide data concerning biological problems in a sequential format so that the students themselves develop the concepts in question. Teachers are asked to study the films and teacher's guides before using them. The film is stopped on selected photographs, statements, charts, and questions. Classroom discussion is solicited. Students are asked to pose further problems and are invited and encouraged to provide their own ideas for experiments. The films are available only in the Super 8 mm cartridge format. Each film is approximately 4 minutes in length; however, adequate and proper presentation of a Single Topic Inquiry Film may require one or more class periods. While each film contains biological information, it also provides students and teachers with an experience in the processes of science, stressing science as enquiry. Forty films have been developed covering the nine basic themes of BSCS Biology. There are films on animals, plants and protists and films dealing with levels of organization from the molecular to the world biome.

SECTION 4

The Background of Biology: Physics, Chemistry, Statistics

11

Energy

Some of the most effective and best known current research in biology is approaching the secrets of life through the study of chemistry. For example, photosynthesis is understood in terms of the capture, within a leaf, of light energy during the chemical combination of carbon dioxide and water to form glucose. During respiration, as you well know, the glucose is recombined with oxygen, producing again carbon dioxide and water, and releasing the energy which was stored in the glucose for other life processes. The adenosine diphosphate—adenosine triphosphate (ADP-ATP) cycle is recognized as a means of converting and transferring energy thus obtained. There are many steps, but the net result is that the energy goes into the formation of "high-energy" phosphate bonds (in Chapter 16 we shall discuss bonding) as ADP is converted into ATP. In the chemical combination represented by ATP the energy is made available for use.

Chemical operations such as these show us the importance of the physical sciences for biology teachers. Chemical combination and re-combination, oxidation and reduction, and other chemical changes involved in the absorption, transfer, and release of energy are aspects of biology. Because one of the significant biological roles of chemical change is the capture, storage, change, and utilization of energy, we need to understand something about energy.

There is still another reason for the biologist to be concerned with energy. Chemical changes require, store, transfer and release energy. Energy is an indispensable part of the chemical changes through which the cell develops, grows, repairs, and reproduces itself—the processes of biosynthesis. In these processes energy becomes the key to understanding how atoms bond into molecules, what elements will combine with one another, and what three-dimensional arrangement the atoms will form in a given molecule; it also determines how electrons are arranged within the atom itself.

Yet another reason for studying energy is the fact that the biologist encounters it at a macro-level as well as at a micro-level. Energy as heat that can be sensed and energy as visible light are parts of the environment of all organisms. Hence, such manifestations of energy, energy differences, and energy changes as temperature, light intensity, and color are significant for biology.

ENERGY AND THE THEORETICAL ASPECT OF PHYSICS

Energy, in one way the physicist treats it, is a part of the science called mechanics. Mechanics is a science, dealing with the motions and interactions of material bodies, which has reached a highly *theoretical* stage of development. In the earliest stages of the growth of mechanics, scientists simply described the phenomena which they studied. They weighed bodies, rolled them, dropped them, let them bump into one another. Then they described the changes that the motions of these bodies underwent in the varying circumstances: what pathways the bodies followed in their motions, whether they slowed down or sped up and, if so, how much.

Since Isaac Newton, mechanics has gone a long step further. Instead of being content to describe at great length all the many ways in which bodies behave, scientists have tried to capture these numerous and varied behaviors in a small network of *concepts*. This is a theoretical development of the science, and it is sought because of two major advantages that it confers. First, it enables a science to be much more precise, because the concepts can be defined as exactly as we please. We can borrow an example from elementary plain geometry to say that the sum of the interior angles of a triangle is one straight angle, or 180°. This is a theoretical statement and it applies to what might be called *perfect* plain triangles. This statement makes no concessions to our inability to draw perfectly straight lines or to measure angles with perfect precision. With concepts thus defined, it becomes possible to specify the conditions and operations for experimental measurements (and their significance) with greatly increased precision.

The second advantage of such a theoretical development is that it enables us to disover many interrelations among phenomena which we might not otherwise find. Before Newton's system of concepts was developed, the motion of a projectile near the earth's surface and the motions of the planets were only dimly and imprecisely recognized as similar to one another. With Newton's definition of the concept of force, however, these remote and apparently different kinds of motions could be seen as being of the same kind, differing only in the conditions under which they occurred.

The point of this introduction is that to discuss energy as physicists understand it, we must write and read about it in the kind of language typical of theoretical science. We must use terms such as *force, mass, velocity, speed,* and *momentum.* To read it otherwise will almost inevitably lead to misunderstanding and puzzlement. This injunction is not a heavy burden, however, since only three characteristics of theoretical language need be kept in mind. Let us observe what they are.

The first important characteristic is this: most of the nouns of a theoretical language (*energy* or *force,* for instance) are conceptual. That is, these nouns do *not* refer to things, qualities of things, or events that are immediately discriminated and experienced through our senses. They refer, instead, to ideas—ideas that are precisely and strictly defined. These ideas are applicable to observable things, qualities, or events, but they are not the same as these events. For example, forces are involved when we push, pull, or lift, and pushing, pulling, and lifting are ways of applying force, but a push or a pull is not purely and simply identical with force. The force associated with a push is only that carefully restricted aspect of the push which we measure when we follow the instructions given in the definition of force.

The second notable characteristic of a theoretical language is this: most of its concepts are defined *in terms of one another.* Recall our earlier reference to "a small network of concepts." This characteristic makes it difficult to talk about one concept, such as energy, without discussing others. When we try to consider one concept by itself, we become imprecise, incomplete, or both. Specifically, in order to talk about energy, we must talk about work. To define work, we must talk about force, and to discuss force, we must define momentum.

The third characteristic of the theoretical language used in physics is that most of its *concepts are quantitative.* That is, the most precise or useful definition of one of its concepts is achieved not by stating what the concept represents but by stating the changes it undergoes in relation to changes in other concepts.

FORCE AND MOMENTUM

It is easy to describe the sort of event in which the concept of momentum is involved. Suppose a big ball, weighing about 100 lb, is rolling slowly toward us. (Here we are really interested in *mass* rather than *weight;* but weight *at a given place* is a measure of mass. We shall discuss concept of mass more carefully later on.) If this big ball is rolling slowly toward us, we may try to resist its motion by stopping it, or swerving it from its path. We find that a certain effort is required. Now suppose the same ball rolls toward us twice as fast as before. Again we try to resist its motion by stopping it or swerving it from its path. It requires more effort than in the first case. If a third ball, weighing twice as much as the first, rolls toward us at the same speed as the first ball, we shall find that, again, more effort than in the first case is required to stop or swerve it.

Something differed (speed or velocity) in the first case and the second; there was another difference (weights of the objects) between the first and third cases. The differences were *experienced* as differences in the effort required to stop each ball—the second and third required more effort than the first. These experienced differences are close to what the physicist means by momentum.

The two sources of difference in expended effort—weight and velocity —are enough to permit us to give the physicist's definition of momentum. The momentum of a moving body is that which changes when the velocity of the body or the mass of the body changes. More tersely: momentum is proportional to mass and to velocity taken together (for the time being, we still leave mass undefined; but remember that weight is a measure of mass.) Algebraically, if we use P for momentum, we write $P = mv$.

With this definition of momentum in hand, it will now be easy to describe the sort of experience which will lead us to the physicist's definition of force.

Let us reexamine the three moving balls. Suppose we have managed to slow each of them by a certain amount. It took effort to do so in each case, and the required effort differed between the first and second cases, and between the first and third cases. This experience of differing effort is close to what the physicist describes with the aid of the concept of force. To put it crudely, force was what we exerted in order to slow the moving ball (or stop or swerve it).

Now remember how momentum, P, is defined: $P = mv$. Hence, to stop, slow, or swerve the ball is to change *one* of the quantities (velocity) that defines the quantity called momentum. Hence, a force, again crudely put, is that which changes a momentum. (Note that a swerve is also a change in velocity, because velocity is defined as some particular speed in *a certain direction;* a swerve changes the direction, hence it changes the velocity.)

Now let us broaden our experience to permit a more precise definition of force. As one of the balls (any of the three; it doesn't matter in this case) comes toward us, we exert a steady effort for as long as it takes to slow the ball a certain amount—say 10%. In a second trial we try to double the effect (slow the ball by 20%), yet keep the sensed effort constant. We find that we can increase the effect by increasing the time through which the effort is exerted. For a third trial, let us exert the same effort, but apply it just half as long as in the first case. The result is just half the effect as in the first case—the ball slows by 5%.

Amid the differences among these three trials, a constant factor can be found. When the effort is constant, the change of velocity per *unit* time also remains the same. When we change the sensed magnitude of our effort, the change in velocity per unit time also changes, as we discovered in our trials. The change (or increment) per unit time increases as the effort increases; it decreases as the effort does.

Thus we find that there is something, an effort, which by experimental measurement is roughly proportional to the *rate* of change it causes in momentum. We can express this algebraically (using the symbol Δ to mean a change) as

$$\text{Effort} \approx \frac{\Delta mv}{\Delta t} = \frac{\Delta P}{\Delta t}.$$

The physicist's definition of force is a precise parallel of this approximate relation. It reads:

$$F = \frac{\Delta mv}{\Delta t} = \frac{\Delta P}{\Delta t}.$$

With two small changes, we can write the definition of force as it appears in many textbooks. First, we must remember that only the velocity factor of momentum is changed in a particular trial. During each trial, the ball does not change; hence its mass m remains the same throughout the time it takes to complete a trial t. So we may write the former equation

$$F = m\frac{\Delta v}{\Delta t}.$$

Then we need only note that $\Delta v/\Delta t$ (the change of velocity with respect to time) is the definition of *acceleration*. Thus our final definition of force takes the form

$$F = ma.$$

WORK AND ENERGY

With force and momentum defined, we can proceed to the physicist's concept of *work,* which leads us immediately to a conception of energy.

Imagine a brick at rest on a small elevator in a basement (below the earth's surface). Assume that the elevator can raise the brick to exactly the height where the elevator platform will be even with the surface of the ground. When the platform is level with the ground's surface, the elevator will stop. Observation of this changing situation will now show us something about the nature of work. Bear in mind that the physicist's concept of work is somewhat different from the ordinary notion.

When the elevator and the brick are *at rest* at the bottom of the basement, the elevator will exert a force on the brick exactly equal and opposite in direction to the force of gravity (which we represent by the weight of the brick—we shall soon see why). The forces must be equal and opposite, so that the *net force* of the brick is zero; otherwise the brick would not be at rest, but would accelerate in one direction or another, as is required by the last equation above. If the force exerted by the elevator were greater than the downward force of gravity, the brick would accelerate upward. Let us start the elevator. It begins to exert a somwhat greater force upward than the downward gravitational force; the net force is no longer zero; and the brick "feels" a force in the upward direction. It begins to move upward—that is, it is being accelerated.

As soon as we have applied a small force in excess of the gravitational force (so that the brick is moving upward at low velocity), we can reduce the upward force until it again *exactly equals* the gravitational force. The net force is again zero, and no further *acceleration* will be imparted to the brick. Again refer to the last equation above. However, the brick will continue to rise with uniform *velocity* (assuming no friction, which would be another force, of course) because, being mass, the brick has the property of *inertia;* that is, if at rest, it will continue its uniform motion in the absence of application of a force.

Right at the surface of the ground, we slip a trowel under the brick, which is still rising with uniform velocity. We raise the trowel, keeping pace with the brick but imparting no additional force to it beyond an upward force equal to that of gravity. At 6 ft above the earth's surface we stop the brick's upward motion, holding the trowel halted under it. (The fact that this "thought experiment" would be virtually impossible to perform need not concern us, because the precisely defined *conception of work* is obtained by considering ideal conditions, conditions relevant to the conception we are considering. Later we introduce modifying factors to fit circumstances of practical application, but these do not affect the conceptual definition.)

In the act of raising the brick at constant velocity from the earth's surface to a height of 6 ft, *work* as conceived by the physicist was done. The act of raising the brick fulfilled the physical definition of *work* in two ways: First, the brick was displaced a certain distance. Second, it was displaced by an applied force. In brief, the physicist's definition of work is "That quantity which is measured by the magnitude of an applied force times the displacement produced by the force." Or

$$W = F_{app}h.$$

Thus we see, first, that work done is proportional to displacement. Raising the brick half as far means that we have done half as much work (if the applied force remains the same). Raising it twice as far means we have done twice the work. Second, note that once the brick had begun to move upward, the force we had to apply to keep it moving upward at a uniform speed was just equal and opposite in direction to the gravitational force on, or weight of, the brick. Hence, if the brick had twice the mass (and twice the weight) the force would have been doubled too. Then the work done would have been doubled. So work done is proportional to the force applied.

There is a corollary to these two points. When we have reached the top of the lift, continuing to hold the brick aloft but stationary, the definition tells us that *no* further work is being done—at least on the brick —since there is no further displacement of it. The fact that we may grow tired by holding the brick aloft, but at rest, makes no difference in the physical work on the brick. (Work is being done through oscillatory displacements within the muscle fibers and their molecules. But this is work done on molecules or parts of the fibers, not on the brick.)

Let us consider another point. The elaborate conditions under which the brick was lifted were designed to ensure that no acceleration was applied by us to the brick, once it was in motion. As we exerted ourselves to displace the brick, we added nothing to its momentum. Yet common sense cries out that the work we did in applying the force must have imparted something to the brick. Biological considerations lead us to note that we expended energy in lifting the brick. Surely this energy did not simply disappear.

POTENTIAL ENERGY

The physicist agrees with our intuitive observation that in doing work we must have "added something" to the brick. In his system of conceptions, he can show that *potential energy* was added. Let us see what this means.

Recall that the force which we applied to lift the brick was only one of the two forces involved. There was also the gravitational force which we call the weight of the brick. (Throughout this discussion, we shall use "weight" as meaning *force, not* the mass of the object in question.) Once the brick has been lifted above the earth's surface, the weight of the brick can be used to do work. All we need to do is to release the brick, and it will be displaced downward by its weight as far as we have lifted it—to the surface of the earth. Thus, again, work is done— a force acts through a displacement—this time a gravitational force. In a sense then, by raising the brick we "charged" or "loaded" the earth-plus-brick system, much as we might load a gun. We gave the earth-plus-brick system a potential (a possibility) for doing work. We could show that *all* the work that we did in raising the brick could be recovered (under ideal conditions) when the brick is allowed to fall. When the raised brick is released, the stored work reappears. The brick accelerates, taking on downward momentum. This moving brick can do work—think of a pile-driver.

When a body or system of bodies—a stretched rubber band, a wound-up spring, an elevated brick—gains the capacity to do work by having work done on it, we say that the body or system has *potential energy*. Potential energy is always spoken of and measured in terms of the work that could be done if the energy were released, or in terms of the work that was done in storing the energy in the first place, which is quantitatively the same.

In our brick example, the potential energy is obviously mechanical. Potential energy is also found in many forms which are not so obviously mechanical. The electrical energy stored in a charged condenser or the chemical energy locked in a lump of coal or a teaspoon of sugar are examples. In physical theory, however, electrical, chemical, and all other forms of energy are at least implicitly equated to mechanical work, work that could be done if the energy were released by displacement of some weight through some distance.

KINETIC ENERGY

Any body in motion can do work. At the bottom of its fall the impact of the brick might drive a nail, crack a walnut, or shatter itself. The brick in motion therefore possesses energy. The energy a body has by virtue of its motion is known as *kinetic energy*. If the speed of a moving brick were doubled, its kinetic energy would be four times as great. If two bricks are stuck together and fall at the same speed as a single brick, the two will have twice the kinetic energy of one brick by itself. Many experiments with different objects have established that the kinetic energy of a body is proportional to its mass times the square of its velocity. (Mass is not the same as weight, although we often determine the mass of a body by weighing it. *Mass is a quantity of matter* whereas *weight is a force,* the gravitational attraction—of the earth or any other bit of matter—for the matter of the object we are considering. The mass of a golf ball is greater than that of a ping-pong ball. The mass of two bricks is approximately twice that of one brick. However, the weight of any of these objects will depend on where it is located in the universe, although their masses wil be independent of location. Thus the ping-pong ball in the gravitational field of a dense star could weigh more than the golf ball in near-empty space.)

The definition of kinetic energy is

$$KE = \frac{mv^2}{2}.$$

Now, what about the kinetic energy of the brick when it reaches the ground after having fallen from 6 ft above it? When all the necessary units have been properly defined, we find that kinetic energy may be expressed in the same terms as potential energy. Moreover, as the brick reaches the ground, it has just as much kinetic energy as it had potential energy at the beginning of its fall, if we can neglect the work which the brick had to do in stirring the air during its fall—in other words, we neglect air resistance. Notice that at the beginning of the fall, the brick had no *kinetic* energy, while at the instant it strikes the ground, it has no *potential* energy. In fact, if we were to calculate or measure the kinetic energy at any point of the fall, and compare it with the potential energy which had been used up, we would find that they are equal. This is a simple illustration of the *principle of conservation of energy.*

Perhaps we can clarify and summarize our discussion of energy with three graphs in Figure 1. Note that in these graphs the independent variable—the height—is plotted along the vertical axis, and the dependent variable, work or potential energy—along the horizontal axis. In Figure 1a, the ordinate OT represents the height at which our original brick was held still above the ground. The arrow lenght TI represents the work done in lifting the brick from the ground to that height, and also the potential energy that the brick has when in that position. At any other height, say OH, the arrow length HF represents the work done, or the potential energy stored, during the process of lifting the brick that far. The curve OI (a straight line when work is being represented this way) plots the variation of potential energy as height varies.

Figure 1b repeats Figure 1a, with two additional curves (lines) TJ and IJ. What do these curves mean when the brick is falling and is at height OH above the ground? Again, the length of arrow HF (never mind the arrowhead at E for the moment) represents the potential energy the brick still has. The length of arrow FD corresponds to potential energy which has been converted to kinetic energy ("used up") during this much of the fall, and the length HE represents the kinetic energy the brick has at the moment in question. Notice that FD = HE; therefore, the kinetic energy that has been generated during the fall from T to H equals the potential energy that has vanished. Note also that HF + FD = HF + HE = OJ = TI; this means that at any height OH the sum of the potential energy HF remaining and the kinetic energy HE obtained equals the potential energy TI the brick had at its highest point, and the kinetic energy OJ it has on arrival at the

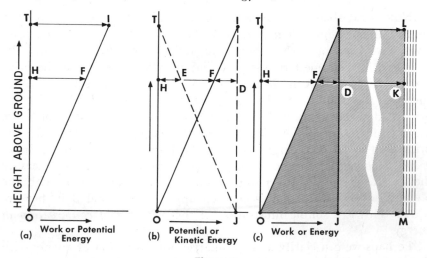

(a) Work or Potential Energy   (b) Potential or Kinetic Energy   (c) Work or Energy

HEIGHT ABOVE GROUND

Figure 1

surface of the ground. The curve *TJ* shows the variation of kinetic energy with height during the fall, and the curve *IJ* shows that the sum of the potential and kinetic energies is constant during the fall. Figure 1*b* thus indicates something of the meaning of *conservation of energy*.

CONSERVATION OF ENERGY

We have not yet answered the question of where the potential energy came from. True, the concept of potential energy was introduced in conjunction with the notion of doing work, lifting the brick. If the principle of conservation of energy is valid, however, the energy must have existed somewhere, in some form, before we did the work. The biochemist tells us where. Our bodies broke down glucose, releasing the chemical energy stored there; we can suppose that before being stored in glucose the energy was stored in our breakfast oatmeal; the oatmeal got its stored energy from sunlight.

Figure 1*c* is a crude picture of the total energy involved in the lifting process. When the brick has been lifted through height *OH*, the potential energy stored then in the system is represented by arrow *HF*. Arrow *FD* indicates the energy which must still be delivered to the brick to elevate it to height *T;* but arrow *FK* represents the real amount of energy that must be processed biologically during the rest of the lift, since the body is not 100% efficient. Arrow *DK* represents energy involved in respiration, maintaining body temperature, circulating the blood—in short, remaining alive. The total energy which must be processed for lifting and living is indicated by *OM*. The energy utilized by the organism in its life processes (not counting that which is used in doing the work of lifting) is represented by *IL*. The vertical band of dashed lines at the right edge of Figure 1*c* illustrates the fact that the location of this edge may depend strongly on factors in the organism, such as fatigue or emotional state, or in the environment, such as temperature.

Before leaving this discussion of energy, let us note that the principle of conservation of energy tells us in part what to expect just after the brick shatters against the ground. The kinetic energy of the fragments, the potential energy the fragments have during their bounces, the increased heat energy in the fragments and the floor, the energy that moves air particles, giving rise to sound, and which moves the ground in slight vibrations, must all add up to equal the kinetic energy the brick had on arrival at the ground.

We cannot test the principle of energy conservation in such a simple, uncontrolled situation as the shattering of a brick. However, many carefully designed experiments have been devised and performed. The result of the evidence they provide is that our confidence in the principle is great—so great that if anything appeared to violate the principle, we would probably prefer to conceive of some new form of energy to reconcile the experiment with the principle, rather than give it up.

Actually, for energy to be conserved in all cases, we must treat matter itself as a form of energy. Albert Einstein showed that if m grams of matter disappeared, E ergs of energy are produced, in accordance with the equation $E = mc^2$, where c is the velocity of light in centimeters per second. (This equation is also correct if E is expressed in Joules of energy, m in kilograms of mass, and c in meters per second.) The reverse of this process also occurs, with energy becoming matter. The matter-energy relation is significant and widely used for studies of nuclear energy (loosely called atomic energy). The matter-to-energy or energy-to-matter conversions are usually so quanitatively unnoticeable that the early science of chemistry could state a law of *conservation of mass*. In fact, most chemical calculations even today do not take into account the relatively small energy-mass interconversions. For this reason mass-energy conversions are almost irrelevant to an elementary discussion of energy in biological systems, although they might be more important in specialized treatments.

SUMMARY

$P = mv$ Momentum of an object of mass m, moving at velocity v.

$F = ma$ The force on an object of mass m equals its mass times the acceleration imparted by the force.

$a = \dfrac{\Delta v}{\Delta t}$ The acceleration of a body is the change in its velocity with respect to time.

$W = F_{\text{app}}h$ The work done on a body is the applied force that moves it against another force times the displacement h in the direction opposite to that of the resultant force whose effects F_{app} works against. W is a measure of the potential energy of a body released at h.

$KE = \dfrac{mv^2}{2}$ The kinetic energy of a moving body equals its mass times the square of its velocity divided by 2. Kinetic energy at the end of a fall equals energy of the body at the start of the fall minus potential energy at the end of the fall.

$E = mc^2$ A formula permitting us to convert matter into energy; it is necessary for the conservation of energy principle to be valid in every case.

Conservation of Energy Principle. Energy cannot be created or destroyed, only converted from one form to another.

CHAPTER 12

Energy in Light

In the preceding chapter we discussed several forms of energy, and we noted that they are convertible to one another. We noted also that for this reason the physicist can express all forms of energy in the terms he uses to describe mechanical energy. The importance of energy in biological studies was indicated briefly. Now it is time to consider in some detail a kind of energy which is of paramount importance for biology—the energy of light, and sunlight in particular.

Of course, we can sense visible light and we can see it arrayed into a spectrum in a rainbow or by means of a prism or a grating in a spectroscope. The boundaries of the visible spectrum are red at one end and violet at the other. We now know that visible light is simply a limited portion of a very broad *electromagnetic spectrum* of radiant energy. Beyond the boundaries of the visible spectrum, laboratory techniques reveal the existence of *ultraviolet* light at the one end and *infrared* light at the other. There is no qualitative difference—only a quantitative one—between violet and ultraviolet light, or red and infrared, or for that matter between red and violet visible light. Usually we call "light" only the portion of the electromagnetic spectrum that is sensible to us through the eye, but if we can detect nonvisible extensions of this portion by optical means, we may call the extensions light, too, as we do infrared and ultraviolet.

We shall see shortly that we can always associate a wave length and a frequency with electromagnetic energy. In fact, it is only the measurable differences in wavelength and frequency that make us say that the various kinds of light have only a quantitative difference and not a qualitative one. For the present, our discussions will emphasize wavelength, since that is easy to picture in interpretations of laboratory investigations of light phenomena. We shall see in this chapter that the notions of wavelength and frequency are always interrelated and, in some contexts, frequency will be more useful.

Ultraviolet light energy is characterized by shorter wavelengths than those of violet light, while infrared light energy is associated with longer wavelengths than those of red light; red light wavelengths are, of course, longer than violet wavelengths. There is a continuous increase in wavelength as we move from the ultraviolet toward the infrared through the spectrum. It appears that there is no wavelength that cannot, in theory, exist; the spectrum not only has no "gaps" in it but also may be extended without limit at either end, so that it contains wavelengths any shortness or length. However, we should not assume that this means that we are able to produce or detect all possible waves, although we can now study vast portions of the spectrum, and new instruments have been developed to extend the range that we can study.

Two portions of the spectrum which we cannot sense but which can be detected indirectly include radio waves and x-ray. Radio waves are much longer than light waves; x-ray are much shorter. For these portions of the spectrum, as with all other parts we have learned to study, improvement of techniques and instruments enables us to move the wavelength boundaries farther apart. The names we give to portions of the spectrum and their extensions depend partly on the ways in which we can detect them; for example, we call whatever we can detect with radio apparatus radio waves.

These remarks about the electromagnetic spectrum are illustrated by Figure 1. The spectrum shown there extends from wavelengths of less than 10^{-11} m to those more than 10^6 m long. To get some feeling for these lengths, note that the diameter of a red blood corpuscle is about one million times greater than 10^{-11} m, while 10^6 m is about the straight line distance from Boston to Detroit.

What characteristics of the electromagnetic spectrum and what part of its range are of the most interest to biologists? The eye detects

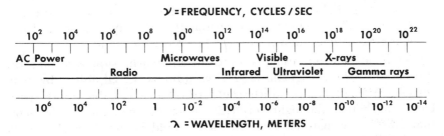

Figure 1

only what we call the visible spectrum. X-rays readily penetrate the flesh. Sunlight contains all the colors of the visible spectrum. White paper reflects all these colors almost equally well; a green leaf, on the other hand, reflects primarily in the green, absorbing much of the other light whose wavelengths correspond to colors. Man and other organisms can live in the presence of visible and infrared light—in fact, they must—but radio waves except at very high intensities do not interact perceptibly with their substance, or affect their biological functions. On the other hand, man's exposure to ultraviolet light, x-ray radiation, or gamma radiation must be restricted, or it may be lethal.

Our interest will be primarily in sunlight and will be primarily directed to how it is reflected or absorbed and used by living tissues, and secondarily to how it is generated in the atoms of the sun, propagated to the earth, transmitted, and absorbed, or reflected by the earth's atmosphere. We discuss light propagation first.

PROPAGATION OF LIGHT; THE WAVE MODEL

How does light travel through a vacuum? Energy is transmitted by a source (for example, the sun) and does work at points far removed from this source. There is no connecting matter between the source, which could be regarded as the cause, and the distant phenomenon, which could be regarded as the effect. What we seek is some conceptual model to account for action apparently "at a distance," without a sensible transmitting medium. We would like to "connect" the cause with the effect, to explain the transfer of energy from source to receiver.

During the 17th century, Christian Huygens commenced the development of a *wave theory of light;* Huygens assumed that a subtle medium filled all space, including vacua, and that light traveled as waves in this subtle medium. By "subtle," Huygens simply meant that the medium could not be weighed, bottled, or detected by any experiments not involving light waves which the medium was hypothesized to account for. This medium was known as the *ether* and the wave theory of light is sometimes known as the ether theory of light.

Sir Isaac Newton, using a different approach, supposed that the transfer of light energy through a vacuum is accomplished by light corpuscles, small bodies which could travel through the empty space. He thought that the light corpuscles were not material in the sense that objects are material; that is, there would be no detectable transfer of mass when light is absorbed by a body. By supposing variations in the size or shape of his hypothetical corpuscles, Newton could account for many properties of light, such as color and partial reflection This ability of the theory endorsed by Newton gave the alternative wave theory of light—Huygens' theory—little chance to develop until the experiments of Thomas Young in the late 18th and early 19th centuries. A variation of Newton's theory, called the *corpuscular theory of light* is still maintained. It has many features that Newton's theory lacked, however, and lacks many that his theory had.

Neither the wave theory of light nor the corpuscular theory of light by itself is entirely satisfactory. First, we may note that we feel uneasy about subtle media such as the ether. We note that this is an *ad hoc* provision established just for the purpose of carrying light waves. Moreover, well toward the end of the 19th century, A. A. Michelson and E. W. Morley conducted a very sensitive experiment designed to show whether the earth in its orbital path around the sun changed its velocity with respect to the ether which was supposed to pervade all of space. Their negative result had several far-reaching effects among which was the discarding of the notion of ether. The corpuscular theory of light, on the other hand, failed in its attempts to explain interference and diffraction phenomena which, as we shall see below, are very nicely accounted for by the wave theory. Present notions, as we shall see, involve *photons* that may be thought of as short, discrete trains of waves which possess both particlelike and wavelike properties.

Despite our disbelief in the ether, however, the wave theory of light is such a nice small network of concepts adequately explaining or interrelating a large number of optical phenomena that we continue to use it where it is useful. We simply disregard the question of what medium might be carrying the waves through space and speak of the light in metaphorical terms; we regard it *as if* it were being carried through space as a wave motion in a subtle medium. In other connections, as will be discussed in later chapters, we shall use the photon theory of light wherever the particlelike properties seem to dominate.

Thomas Young was a physician who did scientific experiments as a hobby. He made several important contributions to the early development of physics, among them his experiments with potential and kinetic energy. He first proposed the term "energy" to designate the group of concepts to which that term now applies. In this chapter we shall consider Young's contributions to the study of light.

Young discovered that some phenomena of light are better explained by a wave model of light propagation than by the corpuscular model. Figure 2 diagrams one of his experiments. *F* is a source of light, *P*

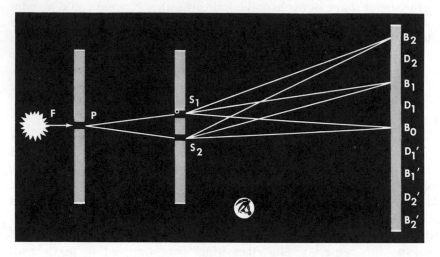

Figure 2

is a slit (viewed end on in the diagram) which provides a narrow source of light for the slits S_1 and S_2. W is a wall or screen some distance from the slits. The slits are narrow parallels, quite close together. With his eye in the position shown, Young observed on the screen a series of bright bands, B_0, B_1, B_2 . . . , B_1' B_2', . . . fanning out in a plane perpendicular to the direction of the length of the slits. This is illustrated in Figure 3, a photograph of the bands on a screen.

We can duplicate Young's experiment crudely with very simple equipment. A flashlight or automobile tail light, some distance away, can be used in place of F and P, and the retina of the observer's eye will take the place of the screen. The slits can be made by pinching two safety razor blades together and, holding them tightly so that they

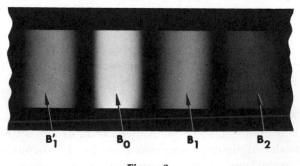

Figure 3

remain parallel, draw them along a straight edge to cut through the paint and silver on the back of a piece of old mirror. Such closely spaced parallel slits can also be drawn in a black paint coating which is dried on a piece of clean glass. If the mirror with the slits is held as in Figure 4, close to the eye (black side to the eye) while we look at the light, we should easily perceive a row of bright images of the light source spread out in the same direction as the separation of the slits—that is, on a perpendicular to the mark made in the paint on the glass. The appearance should be somewhat like the drawing in Figure 5. In general, the brightest dot will be the one exactly in the direction of the source, usually the center dot in the row. The other images will be arranged symmetrically on either side of the bright central dot, fading in brightness in both directions as they lie farther from the central dot. The distance of the slits from each other will determine how far the images are separated from one another. If the slits are fairly far apart because the razor blades were made of thick metal, the dots may not appear distinct at all, but may blend together to form a band of light. Also if the distance source is not a single color, such as a red tail light, but white light, as with a flashlight, you may see some spreading of each image into a spectrum. In general, the dots then would appear bluish on the side next to the central dot and reddish on the side away.

If instead of using a nearby flashlight as a light source you use a distant street light, you may not perceive separate images at all, but only a streak of light spread out in the direction of the separation of the slits. However, careful examination of the streak will show that parts of the image appear faintly colored. The center of the streak, directly toward the light source, will be white, but the end of the streak will appear reddish. Both the repetition of images and the appearance of colored sections are different manifestations of the phenomenon which has come to be known as *two-slit interference*.

Young's explanation of two-slit interference can be most easily

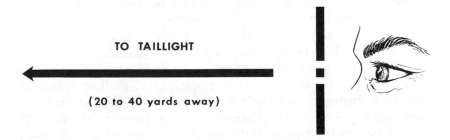

TO TAILLIGHT

(20 to 40 yards away)

Figure 4

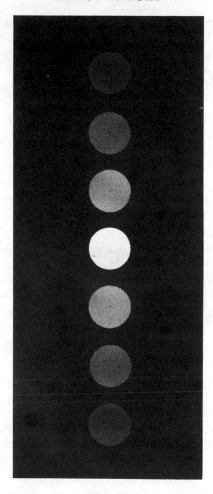

Figure 5

understood by examining an analogous phenomenon in a ripple tank.
Ripple tanks are extensively used in the Physical Science Study Com-
mittee course in physics, and the text of that course (*Physics,* D. C.
Heath, Boston, 1960) may be consulted for more details.

A ripple tank is a device enabling us to study wave phenomena by
examining what happens to water ripples in a shallow tank. The
ripple tank analogue of two-slit interference is illustrated in Figures
6 and 7. They show two overlapping sets of concentric wavelets. The
two wave sources correspond to the two slits through which light passes.
Where wave crests or troughs coincide, there is *constructive* interference,

Figure 6
From *Physics,* D. C. Heath, Boston, 1960; ESI.

and the disturbance travels out to the edge of the picture with augmented crests and deeper troughs than if only one wave source had been involved. Where a wave crest coincides with a trough, there is *destructive* interference, which essentially cancels both crest and trough. It produces regions of no disturbance, the nodal lines which run more or less radially from the region where the sources are.

We can consider the upper edge of the photographs analogous to the screen in the double-slit experiment. Where the nodal lines intersect the upper edge, little or no energy arrives, since the nodal lines mark regions of no disturbance. In the optical case, analogues of these lines would appear as dark bands or spots on the screen. Between the nodal lines, wave trains move outward from the sources, carrying energy from the sources to the screen. In the optical case the regions where these wave trains strike the screen will be bright.

In Figure 7 it is apparent that if we connect the two sources by a short line segment and draw a perpendicular bisector to this segment the perpendicular will run through the center of symmetry of one of the wave systems where there is constructive interference. The point where this bisector strikes the screen corresponds to B_0 in Figures 2 and 3.

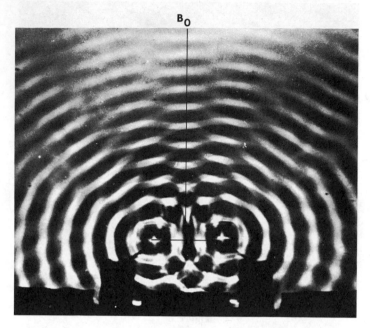

Figure 7

From *Physics,* D. C. Heath, Boston, 1960; ESI.

The different appearance of Figures 6 and 7 is produced by the difference in wavelengths of the ripples. The ripples in Figure 7 are of longer wavelength; one result is that the angles between corresponding pairs of nodal lines are larger in Figure 7 than in Figure 6. Thus the separations between B_0, B_1, B_2, and so on, are generally larger than in Figure 6; that is, longer wavelengths produce images that are more separated.

We can see how constructive interference in a wave model for light would produce the bright regions by referring back to Figure 2. The wave systems from S_1 and S_2 spread in all directions, but we are particularly interested in how they look along the lines S_1B_2 and S_2B_2, from the slits to one of the bright images. If the distances along these two lines are such that the waves always meet crest to crest and trough to trough at B_2, then there is constructive interference, and B_2 will be bright. This is illustrated in Figure 8 by the waves drawn from S_1 and S_2 to B_2.

Obviously, the requirement for brightness at B_2 is that the difference in the distances S_1B_2 and S_2B_2 must be an *integral number of wavelengths.*

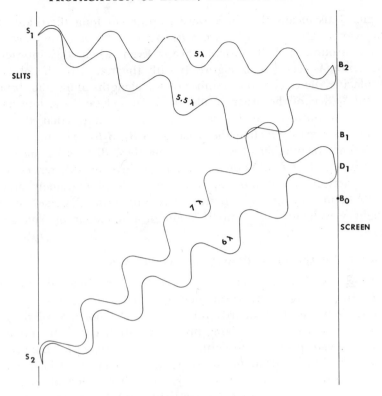

Figure 8

In Figure 8, S_1B_2 is 5 λ (λ is the symbol indicating wavelength), while S_2B_2 is 7 λ, a difference of 2. Constructive interference would also occur at B_1, where the difference would be 1 λ, and so on.

At D_1, however, the distances S_1D_1 and S_2D_1 are such that the waves always meet at D_1 crest to trough, or trough to crest; that is, the waves always cancel each other—there is destructive interference. This requires that S_1D_1 and S_2D_1 differ by 0.5 λ. In Figure 8, S_1D_1 is 5.5 λ long, and S_2D_1 is 6 λ long. The requirement for destructive interference, then, is that the path differences be odd multiples of half-wavelengths— ½ λ, ³⁄₂ λ, ⁵⁄₂ λ, and so on.

Figure 8 illustrates constructive and destructive interference of waves in the wave model of light quite well; however, it is misleading in one respect. The wavelengths of visible light are short when compared with our ordinary perceptions of distance. For example, when we produce yellow (sodium) light by spilling a little salt water into a Bunsen burner flame, the wavelengths of this light are about 5.9 ×

10^{-5} cm. This means that in a wave train 1 cm long there are about 17,000 waves. Or, if the distance S_1B_1 is about 30 cm, the wave train contains around a half-million waves. Even when the interference phenomenon is seen (as in Figure 4) with the retina of the observer's eye replacing the screen, the number of wavelengths along the distances S_1B_1 and S_2B_1 is of the order of 50,000. To understand the interference phenomenon however, it is not important to know that there are many or few waves traveling the distances S_1B_1, S_2B_1 and the rest; only the difference in number of waves traveling these distances is important.

Despite the fact that Young's experiment involves differences of one or two parts in a half-million, it is the basis of extremely accurate methods of measuring wavelengths. Most apparatus designed to measure light wavelengths is a modification or refinement of Young's experiment.

Parameters for the Light Propagation Wave Model

What we say in this section about visible light is applicable to all wavelengths of radiation throughout the electromagnetic spectrum; however, since we are primarily interested in the portions of this spectrum that supply energy to living matter, we shall frame our discussion in terms relevant primarily to light. An adequate model for the propagation of light energy from its source to its receiver must have enough variable elements to account for all known phenomena of light. The phenomena of most interest to us are light's speed, intensity, and color. Let us see how a wave model for the propagation of light accounts for these phenomena. It is important to remember that even though this model may account in some measure for all the phenomena it may not be the only model that will do so; in fact, in Chapter 15 we shall consider another model.

Figure 9 is a diagram of a train of waves. The points marked A are all said to be in the *same phase*. The points designated B are also all in the same phase. The pair of points indicated by C and C' and the pair indicated by D and D' are in *opposite* phase. Note that to be in the same

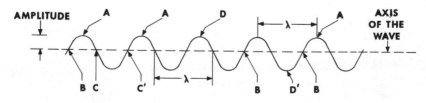

Figure 9

phase the points must be the same distance from the axis of the wave and must be on the same side of the axis, and the curve representing the wave through the points in phase must have the same slope at such points. The two points marked C and C' have different (opposite) slopes. Those indicated by D and D' are on opposite sides of the axis; therefore the points in both pairs are not in the same phase.

The distance along the wave from any point on the wave to the next point in the same phase defines the *wavelength* of the wave. Points A are all separated by integral multiples of λ, as are points B. Points C and C', however, are only $\frac{1}{2}$ λ apart, whereas D and D' are separated by a distance of $\frac{3}{2}$ λ. Points in opposite phase along a wave train are always odd multiples of half-wavelengths apart.

As indicated in the diagram, the height of the crests above the axis (or the depth of the troughs below the axis) defines the *amplitude* of the wave.

We should emphasize that neither the wavelength nor the amplitude of a wave need be fixed from wave to wave. Generally, interference phenomena are seen when waves are the same wavelength and together in a certain region. If the interference is constructive, the amplitude of the resulting wave will be enhanced; destructive interference, on the other hand, will result in smaller amplitudes than that of either of the interfering waves. In what follows, however, we shall consider only waves where the wavelength and amplitude remain constant for one particular wave train.

Figure 10 indicates the meaning of the speed of wave. We are not here concerned with speed as such, but the speed of a wave furnishes an easy way to define *frequency,* a characteristic of light waves which will concern us. The figure represents a series of "snapshots" of a wave, taken every 0.01 sec. In the original drawing the wavelengths were 1 in. long and, for convenience in visualization and calculations, we shall simply regard the vertical lines in the figure as being 1 in. apart. Of course, this does not represent the actual size of light waves which, as we have seen, are many thousands of times shorter. The heavy line marks the successive positions of one crest C as it progresses toward the right. In 0.04 sec, it will move 1 in. to the right, or 1 λ from its $T = 0$ position. In 0.20 sec, C will have moved 5 in. (5 λ) to the right. In 1 sec, C will have traveled 25 in., or 25λ to the right. In short, the velocity of C, and hence of the wave, is 25 in./sec.

Consider the excursions of the black dot, labeled O, which we shall call the oscillator, during this time. In a water wave, O could be a cork which is bobbing on the surface of the water. At $t = 0$, the oscillator is

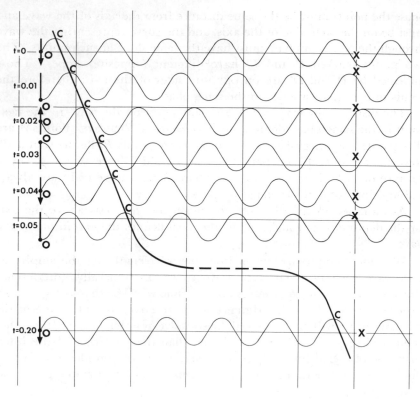

Figure 10

in mid-position, moving downward. At $t = 0.01$ sec, it is momentarily stopped at the bottom of its motion. At $t = 0.02$ sec, O is again at its mid-position, but moving upward, while at $t = 0.03$ sec, it is momentarily stopped at its high position. At $t = 0.04$ sec, O will again be at its mid-position, this time moving downward, having completed one entire circuit of its motion. In 1 sec, O will have repeated a complete circuit 25 times. We thus say that the oscillator has a frequency of 25 cycles/sec, 1 cycle for each wave produced in the wave train during that second. The number of vibrations per second is known as the *frequency* of an oscillator. It is also called the frequency of the wave itself, since it represents the number of crests per second which pass through any fixed point, such as X in the diagram.[1] In our example, if ν is the frequency in cycles per

---

[1] The broadcasting industry, which is much concerned with frequencies, of course, is introducing the term *Hertz*, abbreviation *Hz*, instead of the more cumbersome phrase "cycle per second." One cycle per second equals one Hertz.

second, λ is the wavelength in inches, and v is the velocity of the wave in inches per second, then we have the relation

$$\nu\lambda = v.$$

(ν, the Greek letter nu, is the symbol of frequency; quite often frequency is represented by n, the number of cycles per second; however, n has many other uses, so we shall use the Greek letter symbol.) The equation does not change, of course, if λ is expressed in centimeters and v in centimeters per second. In the situation represented by Figure 10, $\nu\lambda = 25$ oscillations/sec, $\lambda = 1$ in. and $v = 25$ in./sec. Using the centimeter-gram-second (cgs) system, we obtain $\nu = 25$ oscillations/sec, $\lambda = 2.54$ cm, and $v = 63.5$ cm/sec. Cgs units are usually used for defining parameters.

We have said earlier that some element of the wave model must correspond to each important characteristic of light: velocity, color, and intensity. Obviously, the velocity of the wave can correspond to the velocity of light—just about 3×10^{10} cm/sec in a vacuum or in air. The velocity of light in a particular medium is always a constant; therefore, if we know either v or λ we can immediately find the other.

The color of light can be associated with either its wavelength or its frequency. For instance, yellow (sodium) light, seen when salt water is dripped into a Bunsen burner's flame, has a wavelength in a vacuum or air of about 5.9×10^{-5} cm. From our equation, using cgs units, we obtain:

$$\nu\lambda = \nu(5.9 \times 10^{-5}) = v = 3 \times 10^{10}$$

or

$$\nu = \frac{3 \times 10^{10}}{5.9 \times 10^{-5}} = 0.5 \times 10^{15} = 5 \times 10^{14} \text{ cycles/sec.}$$

Thus, the "sodium oscillator" vibrates 5×10^{14} times per second. 5×10^{14} may also be interpreted as a number of wavelengths 5.9×10^{-5} cm long that can be fitted into the distance 3×10^{10} cm, which light travels in a vacuum in 1 sec. Either the frequency or the wavelength interpretation could be used to distinguish this type of light in a vacuum. Frequency is really better for this purpose than wavelength, because frequency, unlike wavelength, is the same for each color, whatever the medium in which we examine the light. To see the importance of this, let us apply the equation $\nu\lambda = v$ to sodium light in glass. In glass the velocity of light is approximately 2×10^{10} cm/sec. (Actually there is some dependence of velocity on color in different media; otherwise the colors in white light could not be separated into a spectrum by a glass prism nor into a rainbow by water droplets. This variation is slight and need not concern us here.) For sodium light in glass, we have $\nu\lambda = 5 \times 10^{14}\lambda = v = 2 \times 10^{10}$, or

$$\lambda = \frac{2 \times 10^{10}}{5 \times 10^{14}} = 4 \times 10^{-5} \text{ cm.}$$

This calculation tells us that the wavelength of sodium light in glass is only about two-thirds as long as the wavelength of sodium light in air or vacuum. But the frequency remains the same.

Table 1 shows the approximate frequency, wavelength, and velocity values for three colors of light in vacuum (or air), glass, and water.

TABLE 1

	Vacuum	Glass	Water
Far red:			
λ (cm)	7.6×10^{-5}	5×10^{-5}	5.8×10^{-5}
v (cm/sec)	3×10^{10}	2×10^{10}	2.3×10^{10}
ν (cps) *	4×10^{14}	4×10^{14}	4×10^{14}
Sodium yellow:			
λ (cm)	5.9×10^{-5}	4×10^{-5}	4.6×10^{-5}
v (cm/sec)	3×10^{10}	2×10^{10}	2.3×10^{10}
ν (cps)	5×10^{14}	5×10^{14}	5×10^{14}
Far violet:			
λ (cm)	3.9×10^{-5}	2.6×10^{-5}	3×10^{-5}
v (cm/sec)	3×10^{10}	2×10^{10}	2.3×10^{10}
ν (cps)	7.7×10^{14}	7.7×10^{14}	7.7×10^{14}

* Cps is the abbreviated form of cycles per second.

Using a wave model to explain the propagation of light, we have tied the speed of light to the speed of the wave, and the color of light to the frequency or wavelength in a vacuum. We must still account for the *intensity* of light, the energy carried per second by the waves onto an imaginary surface placed at right angles to the direction of the light beam. We have not yet used the amplitude parameter, so we can consider that energy per second (energy per second is called power) transported by our wave model is proportional to the amplitude of the wave.

It should be noted that we have really defined what we mean by the amplitude of a light wave, a nice problem, since light does not form material waves. When rigorously treated, amplitude may be defined in terms of various electromagnetic quantities, and the exact expression for power depends on which of these quantities are used. For our purpose, we can adopt the view that the power P is proportional to *the square of the amplitude times the square of the frequency, or*

$$P \propto \nu^2 A^2.$$

We shall not here justify the statement that ν and A are both squared in this relation. However, we can obtain an intuitive grasp of the plausibility of the assertion that P is proportionate to some combination of amplitude and frequency. Note that A is a measure of the size of the wave disturbance as each wave arrives; ν, on the other hand, specifies how often such disturbances arrive. Disturbances of a certain size will convey more energy, as they arrive the more frequently.

(In Chapter 15, where we shall associate light energy with light emission and absorption by matter, we shall find it convenient to modify the wave model. The new model will not use the amplitude parameter in accounting for light intensity.)

The biologist's interest in light or radiant energy stems from our great dependence on energy from the sun. To have some idea of the amount of solar energy that we receive, we can note that at the distance of the earth from the sun, a square meter of surface placed normal (at right angles) to the sun's rays receives about 20 kilocalories per minute—that is, at our distance from the sun, the energy streaming through space is 20 kcal/min/m². Kilocalories are the "large" calories that we worry about in diets. A kilocalorie is the amount of heat required to raise the temperature of a kilogram of water from 15 to 16° C. The amount of solar energy that falls on the entire illuminated hemisphere of the earth can be calculated to be about 2.5×10^{15} kcal/min or 4×10^{13} kcal/sec. One ton of TNT releases about 10^6 kcal of energy and this has become a unit by which we express the energy of atomic bombs. A one-megaton atomic (hydrogen) bomb has the energy equivalent of one million tons of TNT, or 10^{12} kcal. A 40-megaton bomb could release forty times as much energy or 4×10^{13} kcal. We see, then, that the sun pours onto the earth the energy equivalent of a 40-megaton bomb every second. (This energy is distributed over the surface of a hemisphere nearly 8000 miles in diameter with an area of 100 million square miles, rather than over the area of 1000 to 1500 square miles as would be the case with bombs.)

Using the wave model for the propagation of light energy, we saw that frequency and wavelength account for color and can be used to distinguish among regions of the electromagnetic spectrum the portions we label visible light, ultraviolet and infrared light, radio waves, and the rest. We now ask, How is energy distributed in the solar spectrum? Does red sunlight carry more energy than blue? Does ultraviolet sunlight deliver more energy to us than visible sunlight? Do all kinds of sunlight penetrate the earth's atmosphere equally well?

These questions and some others are best answered by considering some graphs. Figure 11 shows some spectral energy curves that are related to incoming radiation from the sun. Two curves are of interest in the figure—the curve for solar irradiation measured outside the earth's atmosphere, and the one for solar irradiation measured at sea level.

The notion of a "black body" is used by physicists to provide a standard energy radiator. A black body may be constructed as a hollow sphere or cube of roughly 1-foot dimensions. A small hole is made in the wall of the otherwise completely enclosed vessel; then the sphere is placed in a furnace and raised to a given temperature. The energy curve of the radiation coming through the hole is said to be the energy curve of a black body at that temperature. We need only note that outside the earth's atmosphere the energy curve for the sun looks very much the same as that of a black body at a temperature of 6000° K. ("K" denotes Kelvin or absolute temperatures. Kelvin temperature is centigrade temperature plus 273° so that, for example, 100° C = 373° K.)

These irradiation curves plot wavelengths along the abscissa. Wavelengths are expressed in units of 10^{-5} cm for convenience. The ordinates use a complex unit which requires a word of explanation.

Its numerator is a simple one, one-tenth of a kcal. The denominator, on the other hand, contains three terms, each of which calls for explanation: "sec," "m²," and "10^{-7} cm." The "sec" takes care of the fact

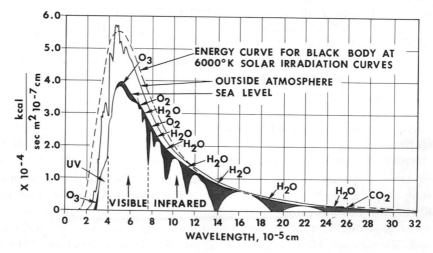

Figure 11

that we are interested in the energy arriving from the sun *per unit time;* hence, kcals per *second.* The "m²" takes care of the fact that we are interested in the energy per unit of time *per unit of surface;* that is, per *square meter.* The "per 10^{-7} cm" factor is an expression of what statisticians call the "graininess" of our sampling procedure. (Physicists often call it band-width, but there are other meanings for that term, so we shall use graininess.) Graininess indicators tell us how much error we may expect in our curve as a result of assigning a value from one measurement (or an average value) to a whole band or interval of wavelengths. In general, the "finer" the grain, the smaller the interval and the less the error.

For example, we might let each energy measurement represent the value of the energy carried by all the wavelengths that lie in a 10^{-5} cm wavelength band, such as all those that lie between 4×10^{-5} cm and 5×10^{-5} λ. Or we could make the "grain" finer, and let one measurement give us the value of energy carried by all wavelengths ranging from, say, 4.0×10^{-5} cm to 4.1×10^{-5} cm. The next measurement would give us a value for the energy carried by wavelengths of 4.1×10^{-5} cm to 4.2×10^{-5} cm, then from 4.2 to 4.3, and so on. For the curves in Figure 11, the wavelength ranges were only 10^{-7} cm.—4.00×10^{-5} cm to 4.01×10^{-5} cm, 4.01 to 4.02 and so on. The wavelength range 10^{-7} cm is a "millimicron." A micron is a unit of length equal to 1-millionth of a meter. A millimicron is 1-thousandth of a micron, or 1-millionth of a millimeter.

Figure 11 helps us to compare solar irradiation outside the atmosphere with that at sea level. Note that at wavelengths just a bit shorter than visible light, for instance, 3.5×10^{-5} cm, less than half the energy striking the top of the atmosphere penetrates to sea level. This points up the fact that our atmosphere is an effective shield against ultraviolet radiation, and explains why we sunburn more easily at high altitudes than at low altitudes. In addition, we note that both curves peak in the *visible* region—that is, they indicate the most energy per millimicron (millimicron is a unit indicating one-thousandths of a micron, or one millionth of a millimeter). Undoubtedly evolutionary processes have developed vision to utilize this most plentiful supply of radiant energy. Note also that both curves tend to peak toward the short wavelengths, or blue end of the visible spectrum. This helps to explain why daylight looks whitish or bluish white.

The deep notches, marked off by regions of shading, also indicate sea level reductions in incoming radiation. Each notch is labeled with the symbol for a gas which is present in the atmosphere—O_2 (oxygen),

O_3 (ozone), H_2O (water vapor), or CO_2 (carbon dioxide). These gases reduce the "transparency" of the atmosphere to certain wavelengths of energy. Thus the water vapor in the atmosphere makes it almost completely opaque for energy carried by wavelengths of about 14×10^{-5} cm. Ozone slightly reduces the transparency of the atmosphere for most of the visible part of the spectrum.

The cause of this reduction in transparency is that the gases selectively absorb some of the wavelengths. Much the same effect is demonstrated by the glass in the taillight of an automobile. The red glass is transparent only for the red wavelengths; it absorbs practically all the other visible wavelengths. If we were to put red glass over our sampling area, and put a notch in the curve of irradiation to correspond to radiation that did not pass through the glass, the notch would cover all the visible spectrum except the red. In Chapter 15 we shall discuss how the molecules of gases can absorb light, and what happens to the radiant energy thus trapped.

Figure 12 shows how the atmosphere transmits light of wavelengths lying roughly between 1 and 7 μ (microns; a micron is 10^{-6} m or 10^{-4} cm). Note that each dip in this curve can be correlated with one of the notches in the sea-level solar irradiation curve.

REFLECTION AND ABSORPTION OF LIGHT BY LEAVES

What part of the energy carried by sunlight is used most effectively by green leaves in the process of photosynthesis? We can suspect immediately that the energy carried by wavelengths corresponding to green light plays very little part in photosynthesis, for since the leaves appear green, they must reflect green light. If leaves reflect all or most of the green light falling on them, this energy cannot be used by the leaf in glucose production.

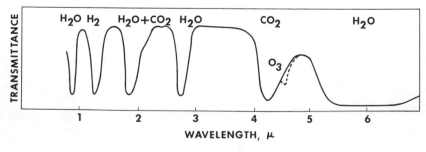

Figure 12

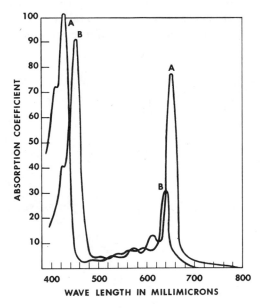

Figure 13

Figure 13 shows absorption spectra of chlorophylls *A* and *B* in ether solution. Note that both chlorophylls absorb very strongly in the *blue-violet* regions, and quite strongly in the shorter *red* (toward the orange) wavelengths. The absorption in the green and yellow regions —from 500 to 600 mμ—is very low; that is, the light energy of these colors is reflected.

SUMMARY

$S_2B - S_1B = n\lambda$

        Where S_2B and S_1B are distances from slits to screen in Young's double-slit interference experiment. When n is an integral number, there will be constructive interference, and B will be bright. When n is an odd multiple of $\frac{1}{2}$ λ, there will be destructive interference, and B will be dark.

$\nu\lambda = v$ The formula relating the velocity of a wave to λ, its wavelength, and ν its frequency. We find ν, the measure of the frequency of the wave, or the oscillator which generates it, most useful, since it does not change as the wave travels in different media, as do λ and v.

$P \propto \nu^2 A^2$ The power (energy per second) carried by a particular wavelength of energy is proportional to the square of the amplitude times the square of the frequency.

CHAPTER
13

Molecules and Atomic Weights

The philosophers Leucippus, Democritus, Epicurus, and Lucretius suggested the notion of atomism five hundred years before the Christian era, but modern atomic theory really began at the start of the nineteenth century. Modern atomic theory is based on the work of investigators who showed that the chemical combination of elements usually involves definite proportions by weight of each element that forms the combination. In terms of what we now know, this means, for example, that in combining to form water, hydrogen and oxygen are always combined in a weight ratio of eight parts oxygen to one part hydrogen. Carbon and oxygen form the compound CO_2 in a weight ratio of about three parts carbon to eight parts oxygen. In ammonia, three parts by weight of hydrogen combine with fourteen parts of nitrogen. The general principle that elements combine in certain definite weight ratios is known as the *law of definite proportions*.

Dalton believed that the law of definite proportions was valid, but he carried the concept on which it is based a bit further in the *law of multiple proportions*. To see what this law means, let us consider data on the weights of combining elements for carbon with oxygen, and for sulfur with oxygen, as given in Table 1*a* and Table 1*b*. Dalton did not use the modern names we use for elements and compounds, but this need not concern us.

In the first table, note that when the same weight of carbon forms both carbon monoxide and carbon dioxide, the required weights of oxygen are reciprocal multiples of each other. Alternatively, if the same weight of oxygen is used in making the two compounds, then the required weights of carbon are reciprocal multiples of each other. The table also illustrates the law of definite proportions, for we note also that when we vary the carbon in CO_2 (taking half as much in the last row of Table 1*a* as in the second row) the oxygen required is also halved.

491

TABLE 1A

C	O	Compound
(1) 12	16	Carbon monoxide (carbonous acid gas)
(2) 12	32	Carbon dioxide (carbonic acid gas)
(3) 12	16	Carbon monoxide
(4) 6	16	Carbon dioxide

TABLE 1B

S	O	Compound
(1) 32	16	Sulfur monoxide
(2) 32	32	Sulfur dioxide
(3) 32	16	Sulfur monoxide
(4) 16	16	Sulfur dioxide

In Table 1*b* we note that when the same weight of sulfur is used in making both sulfur monoxide and sulfur dioxide, the combining weights of oxygen for the two cases are reciprocal multiples of each other. As in Table 1*a,* we note that the law of definite proportions is illustrated in rows 2 and 4: When we vary the sulfur in sulfur dioxide—this time taking half as much in row 4 as in row 2—the oxygen required is halved also. Dalton was able to present such evidence as is indicated in Tables 1*a* and 1*b* for many different compounds.

Dalton viewed the experimental data that led to the laws of definite proportions and multiple proportions as indicating that matter was composed of definite minimum quantities—atoms—and that the difference between two or more compounds of the same two elements was determined by the numbers of atoms of the two elements which went into the creation of a single particle of the compound. We now call these compound particles *molecules;* Dalton called them compound atoms, as distinguished from the atoms of elements, which he called simple atoms. We can illustrate Dalton's atomic theory by looking again at the combining weight data of Tables 1*a* and 1*b*. We repeat these data on the left side of Table 2, where, for convenience, the numbers represent quantities in grams. On the left side of Table 2

TABLE 2

Gram Amounts			Atomic Amounts	
C	O	Compound	C	O
12	16	Carbon monoxide	1	1
12	32	Carbon dioxide	1	2
12	16	Carbon monoxide	2	1
6	16	Carbon dioxide	1	1

in the first and second rows we present combining weight data using the same amount of carbon for each compound. The law of multiple proportions is then illustrated by the fact that *twice as much oxygen* is needed to produce carbon dioxide as to make carbon monoxide. In the third and fourth rows, the combining weight data show that the same amount of oxygen is used for each compound. Again, the law of multiple proportions is illustrated, this time indicating that only *half as much carbon* is needed for carbon dioxide as for carbon monoxide when the amount of oxygen used for each is constant. The data in the second and fourth rows of the table are entirely consistent, except, of course, that the fourth row represents the production of only half as much carbon dioxide—22 g as compared to 44—out of half as much carbon and half as much oxygen. (Which law does this illustrate?)

In the two columns at the right of Table 2, the first and second rows show the data as if we were producing the smallest possible amounts—that is, only single molecules—of the compounds. Thus, in a molecule of carbon monoxide, one atom of carbon is combined with one atom of oxygen, while in a molecule of carbon dioxide one atom of carbon is combined with two atoms of oxygen. We note in passing that 16 g is four-thirds of 12 g; thus row 1 implies that one atom of oxygen weighs four-thirds of 12 g (the weight of one atom of carbon). Or, on an atomic weight scale, if carbon weighs 12 atomic weight units, oxygen weighs 16 atomic weight units.

We can now illustrate one of the most serious difficulties of Dalton's atomic hypothesis by considering the last two columns of the third and fourth rows of Table 2. Can we not base our conclusions about the atomic weights on the assumption that we used the same amount of oxygen in a molecule of each compound, rather than on the assumption that the amount of carbon is constant? The answer is, of course, that we can, as far as these data indicate. We know already that the third and fourth rows exhibit the law of multiple proportions just as well as do the first and second rows. We reach different conclusions about the atomic weight of carbon, however, depending on which assumption we make. We have already noted that the first and second rows give the atomic weights of carbon as 12 units, if the atomic weight of oxygen is taken as 16 units. Rows 3 and 4, on the other hand, give the atomic weight of carbon as 6 units, if oxygen is taken as weighing 16 atomic weight units. How can we decide which value to use?

Dalton tried to resolve this difficulty by studying a great many compounds of oxygen and carbon, as well as other elements, and by assuming that nature would favor simple molecules over complex ones. This simplicity principle is assumed, for example, in Dalton's notion that a molecule of water should consist of one atom of hydrogen combined with one atom of oxygen; he knew of only one compound of hydrogen and oxygen, and he believed it should have the simplest possible combination of atoms. The simplicity principle was inadequate, however, and data other than combining weights were required to remove the difficulty. We shall examine some of these data in the rest of this chapter; more complete treatment will be given in Chapter 16.

At this point, we again emphasize the distinction between weight and mass. In discussing chemical combinations we normally speak of weights—combining weights, atomic weights, molecular weights, etc. But refer again to our discussion on force and momentum; weight *at a given place* is a convenient measure of mass. The chemist is not really interested in how strongly the earth attracts the masses of oxygen and hydrogen which he is causing to combine into water. However, since the weighing process is his most convenient method of determining mass, he speaks of combining weights rather than combining masses.

GAY-LUSSAC'S LAW OF COMBINING VOLUMES

At about the same time as Dalton's work, Joseph Louis Gay-Lussac was considering data which Dalton tended to ignore. Ultimately these data, together with the combining weight data, were to help answer our questions about a choice of basis for determining atomic weights. Gay-Lussac noticed that when two or more of the substances involved in chemical reactions are gases, the volumes of the gases that combine or are produced are always integral multiples of each other. (It is assumed here, of course, that the gases whose volumes are being compared must be at the same temperature and pressure.) Not all the details of Table 3a are drawn directly from Gay-Lussac's work, but the data presented are typical of his observations.

TABLE 3A

---

(1) 2 vol. hydrogen + 1 vol. oxygen → 2 vol. water vapor

(2) 1 vol. nitrogen + 3 vol. hydrogen → 2 vol. ammonia

(3) 2 vol. carbon monoxide + 1 vol. oxygen → 2 vol. carbon dioxide

(4) carbon (solid) + 1 vol. oxygen → 1 vol. carbon dioxide

Gay-Lussac was able to show that compounds of nitrogen and oxygen, sulfur and oxygen, and several others, obeyed this law, although in many cases the combination of the elements into a particular compound might be only indirectly achievable.

AVOGADRO'S HYPOTHESIS AND AVOGADRO'S NUMBER

In 1811, Amedeo Avogadro published his most important contribution, which rested squarely on the works of Dalton and Gay-Lussac. Avogardo made two daring hypotheses: (1) Equal volumes of gases at the same temperature and pressure contain equal numbers of molecules; and (2) the atoms of some *elements,* when in the gaseous state, are combined into diatomic *molecules.*

Let us consider what these hypotheses mean when they are applied to the data of Table 3a. For convenience, let us assume arbitrarily that the "unit" whose integral multiples comprise each "volume" contains one million molecules; we shall see in the next chapter that this would be quite a small volume. The first hypothesis, when applied to line 1 of Table 3a, indicates that two million hydrogen molecules combine with one million oxygen molecules to produce two million molecules of water vapor.

Here Avogadro's first hypothesis taken by itself encounters trouble. We are supposed to find two million particles of water vapor as a product from one million particles of oxygen as an ingredient. If all the molecules of water vapor are alike, if Dalton's atomic notions are correct, and if water as a compound contains the element oxygen, then each water vapor molecule must contain at least one oxygen atom. In other words, we must get two million oxygen atoms out of one million oxygen molecules. This, then, is the reason for Avogadro's second hypothesis, which turns out to be necessary for several of the ordinary gaseous elements—oxygen, nitrogen, hydrogen, chlorine, and fluorine.

We can easily adapt this discussion to lines 2, 3, and 4 of Table 3a. For example, in the production of ammonia, each molecule of nitrogen used goes into making two ammonia molecules; therefore, nitrogen molecules must be decomposable into at least two atoms—we know that they are diatomic. Moreover, for every three molecules of hydrogen, we get two molecules of ammonia; therefore, hydrogen molecules, too, must be capable of being decomposed into atoms.

The shorthand for writing chemical equations incorporates everything we have just said. Table 3b repeats the information of Table 3a, but with Avogadro's interpretations added. In line 1, for example, the subscript 2 in H_2 indicates that two atoms of H—hydrogen—are com-

TABLE 3B

(1)	$2H_2 + O_2 \rightarrow 2H_2O$
(2)	$N_2 + 3H_2 \rightarrow 2NH_3$
(3)	$2CO + O_2 \rightarrow 2CO_2$
(4)	$C + O_2 \rightarrow CO_2$

bined into a molecule of hydrogen, and the subscript of O_2 shows that the oxygen molecule also contains two atoms. Moreover, the coefficients 2 of H_2 and H_2O in line 1, and the implied coefficient 1 of O_2 can be read in either of two consistent ways: Two molecules of hydrogen combine with one molecule of oxygen to produce two molecules of water vapor, or, since the substances may all be gaseous, two volumes of hydrogen combine with one volume of oxygen to produce two volumes of water vapor. We can read the rest of the table in the same general manner, except that it is not customary to interpret line 4 as involving gaseous carbon, since under all ordinary laboratory conditions the element carbon is a solid.

We note in passing that the equations in Table 3b are balanced—that is, there is the same total number of atoms of a given element (no matter what compounds it may be found in) on both the left and the right sides of the equations. Thus, there are two diatomic molecules of hydrogen (four hydrogen atoms) on the left side of the equation in line 1, and there are two atoms of hydrogen in each of the two water molecules (again four hydrogen atoms) on the right side. Similarly, there are two atoms of nitrogen and six atoms of hydrogen on each side of the equation in line 2, and so on.

One great strength of Avogadro's hypotheses is that they combine both the volume and the weight data. Another achievement is that they readily provide a way to determine molecular weights when substances are in gaseous form, thus solving the difficulty Dalton's hypothesis faced. For example, if oxygen *atoms* weigh 16 atomic weight units, then oxygen *molecules* weigh 32 units. If carbon atoms weigh 12 units, then carbon monoxide (CO) molecules weigh 28 units. Or, the ratio of the weight of an oxygen molecule to that of a carbon monoxide molecule is

$$\frac{O_2}{CO} = \frac{32}{28} = 1.14.$$

According to Avogadro's first hypothesis, equal volumes of these gases at the same temperature and pressure should contain equal numbers of molecules; therefore, the ratio of the *density* of oxygen to the density of carbon monoxide should also be 1.14. From the *Handbook of Chemistry and Physics* we find that the density of oxygen is 1.429 g/l, while that of carbon monoxide is 1.25 g/l; thus $1.429 \div 1.250 = 1.14$. We generalize this numerical example according to Avogardo's first hypothesis into the equation

$$\frac{\text{density of gas sample}}{\text{density of known gas}} = \frac{\text{molecular weight of sample gas}}{\text{molecular weight of known gas}}$$

This equation enabled Stanislao Cannizzaro during the 1850's to determine many atomic and molecular weights by comparing densities of gases of unknown molecular weights with those whose weights had been measured.

Much other evidence supports the general validity of Avogadro's hypotheses. To see one type of such evidence, we define *gram molecular weight:*

The gram molecular weight is that quantity of a compound or an element which has a weight in grams equal numerically to the number expressing the molecular weight of the substance; thus 32 g of oxygen is 1 g mol wt of oxygen, and 28 g of carbon monoxide is 1 g mol wt of carbon monoxide. (This quantity is also called a *mole*.)

It follows from Avogadro's first hypothesis that gram molecular weights of various substances should all contain the same number of molecules; we call this number N_A. N_A is known as *Avogadro's number*, and it has been possible to determine this number in several different ways. These determinations all show that 6.0249×10^{23} is the number of molecules in a gram molecular weight of any substance.

CHAPTER 14

Gases and Kinetic Energy

In Chapter 11 we discussed momentum, force, work, potential energy, and kinetic energy, but we barely mentioned heat energy. We introduced the principle of conservation of energy, stating that other forms of energy are often converted, ultimately, to heat. What is heat energy? If a piece of iron is warmed by repeated vigorous blows of a hammer, how does the iron when warm differ internally from when it was cold? How do repeated blows warm the iron?

We may best begin to understand how to answer these questions by examining kinetic and heat energy in gases and liquids. Many scientific ideas about the nature and behavior of atoms and molecules, some of which we have considered in the preceding chapter, were clarified or confirmed by studies of gases. Our examination in this chapter will help set the stage for a more detailed consideration of the concepts related to the structure and behavior of matter. Moreover, the picture of molecular behavior we shall develop is readily adaptable to explanations of osmosis, reaction rates, and other phenomena of interest to the biologist. We shall consider some of them in this chapter.

KINETIC THEORY OF GASES

What goes on inside a gas? Gases exert pressure in all directions, and expand into all parts of a closed container, no matter how large. What must be the nature of a gas to enable it to produce these and other phenomena? As we have seen, it is sometimes helpful to make a model—a simplified (and sometimes only partly accurate) picture of the workings of phenomena we wish to understand.

The kinetic model of a gas pictures it as consisting of a great number of molecules. Temporarily we can regard these molecules merely as very tiny, hard, elastic spheres. The molecules are in constant, violent motion (in the next section of this chapter we shall see some of the evidence which led scientists to conclude that the molecules move this way). They collide frequently with each other and with the walls of the vessel containing the gas. The pressure of the gas against the vessel's walls is, then, the net effect of a large number of rapid but small bombardments by the molecules. Figure 1 shows the

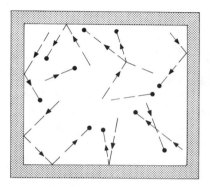

Figure 1

general sort of behavior of the gas molecules in a sketch which does not indicate their size or the great number of molecules present in even very low concentrations of gas. All collisions, we assume, are perfectly elastic; that is, no kinetic energy is lost during collisions. This means that a gas molecule rebounds from a collision with the walls of its container with just as much speed as it had on approach. Moreover, when two molecules collide in the interior of the container

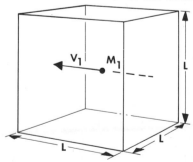

Figure 2

their relative speed—that is, the speed of one relative to the other will be the same after collision as before.

To understand how this model of a gas can account for pressure, let us assume first that we have a cubical box of dimensions L cm on an edge, as shown in Figure 2. The box is empty except for one molecule of mass m_1, which moves back and forth between the right and left faces of the box. In Figure 2, the molecule is seen moving with velocity v_1 toward the left face of the box. Pressure is defined as force per unit area, so we must see how this bombarding molecule can be regarded as exerting *force*.

It may be helpful to recall some of our discussion in Chapter 11. Remember that momentum is defined as the product of mass and velocity. Thus, a mass of 2 g moving with a velocity of 3 cm/sec to the east will have an eastward momentum of 6 g-cm/sec. Remember too that force is defined as the rate of change of momentum. Thus, by exerting a westward force on the 2-g mass we could diminish its velocity from 3 cm/sec to 0. If this took 4 sec, then the change of momentum would be 6 g-cm/sec *westward,* and the rate of change of momentum (the force) is

$$F = \frac{6 \text{ g-cm/sec}}{4 \text{ sec}} = 1.5 \text{ g-cm/sec}^2 \text{ westward.}$$

The unit of force, gram-centimeter per second squared, for convenience is given the name *dyne*. If the same 1.5-dyne force to the west continues to act for another 4 sec, the 2-g mass will acquire a velocity of 3 cm/sec westward, and its resultant momentum will be 6 g-cm/sec westward.

Consider the gas molecule in Figure 2. The molecule strikes the wall of the container with a velocity v_1 and rebounds with the same speed. (We ignore here any loss of kinetic energy, or transfer of kinetic energy, to the molecules constituting the wall; we also make use of a law similar to the law of conservation of energy, namely, the law of conservation of momentum.) The change in velocity is from v_1 leftward to v_1 rightward, so that the total change *toward the right* is 2 v_1. The change of momentum, then, since the mass of the molecule is m_1, is $2m_1v_1$. We are seeking, however, the *rate* of change of momentum, so we must ask how often molecule m_1 hits the left wall of the vessel with the change of momentum $2m_1v_1$. Let us suppose that the molecule leaves the left wall of the vessel with a velocity v_1 toward the right, travels a distance L which is the width of the container, strikes the right wall, from which it rebounds with velocity v_1 toward the left, and journeys back the distance L where it strikes the left wall again. The total distance in this round trip is $2L;$ hence the time required for a round trip is $2L/v_1 = t,$ and the number of round trips per second will be the reciprocal of this, or

$$\frac{1}{t} = \frac{v_1}{2L}.$$

Thus, the rate of change of momentum at the left wall will be given by the change of momentum at each impact times the number of impacts per second, or

$$\frac{m_1 v_1{}^2}{L} = \overline{F}.$$

This quantity is simply the average force against the left wall exerted by a single molecule as a series of pulses. The *average force* is, of course, only a concept; it is that imagined *steady* force which would produce the same net force effect as the series of pulses.

Now, how can we make the expression of the force against the left wall incorporate the effects of the other molecules in the box? This question could be answered rigorously if we took into account the fact that the molecules most generally do not strike the walls perpendicularly, and that the molecules frequently collide with one another in the interior of the vessel. We can greatly simplify the treatment, however, by making two fairly plausible assumptions.

The First Assumption

On the average, during the very short time interval, collisions that destroy velocities in a certain direction will be just about exactly balanced by other collisions that restore such velocities. This is illustrated in a crude way by Figure 3, where molecules A and B collide so that B no longer moves to the left, but at approximately the same time another collision occurs somewhere within the gas between molecules C and D so that C begins to move toward the left. It is not necessary that molecules B and C exactly balance each other, going in precisely the same direction, but only that the total momentum in any given direction before and after the collision be constant—that is, all momentum toward the left that ceases must just equal all that toward the left which commences. (Any physics textbook dealing with elementary classical mechanics will explain in more detail why this assumption is not arbitrary.) In essence, then, this first assumption suggests that we can treat the motion within a gas *as if intermolecular collisions never occur.*

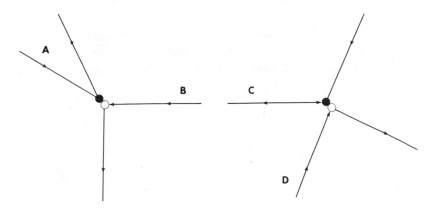

Figure 3

The Second Assumption.

In general, the velocity of a molecule might be in any direction. Since space is three-dimensional, however, we can assume that molecules would produce the same effects on the walls of the container if they were so restricted as to move only along three mutually perpendicular directions; we further assume that one-third of the molecules move along each direction—that is, one-third move up and down, one-third move right and left, and one-third move in and out. (A discussion on vector resolution and probabilities in any section on classical mechanics in a college physics text will show that this assumption is not arbitrary either.)

Although we shall not here attempt to justify these assumptions, we should note that, using them, we arrive at the same results as if we had used more rigorous methods.

From the previous discussion, we see that each molecule exerts a force on the wall of the vessel with which it collides equal to mv^2/L. The total force on any wall will be the sum of the forces exerted by each individual molecule as it hits, or

$$F_T = \frac{m_1 v_1^2}{L} + \frac{m_2 v_2^2}{L} + \frac{m_3 v_3^2}{L} + \cdots + \frac{m_n v_n^2}{L},$$

where $n = N/3$, and N is the total number of molecules in the container; the series of dots indicates that we have summed the quantities representing each molecular impact during a second of time, using as subscripts all integers from 1 to $N/3$.

N is a very large number. At ordinary pressures and temperatures, a liter of gas contains about 10^{22} molecules. It is only in thought, then, that we can add the series of contributions from individual molecules. We can, however, simplify the series in a number of plausible ways. First, in a pure gas we can assume that the masses of all the molecules are the same, and $m_1 = m_2 = m_3 = \ldots = m_n = m$. We ignore the existence of isotopes, which is unimportant for our purposes here. We can then collect m/L from the series, writing:

$$\frac{m}{L} (v_1{}^2 + v_2{}^2 + v_3{}^2 + \ldots + v_n{}^2).$$

Next, as is frequently the case when we deal wtih a large number of items, we can employ average values more effectively than individual values. Let $\overline{v^2}$ be the average value of all the $\overline{v^2}$'s within the parentheses. (Note that this is the *average of the squares* of the velocities, and not the square of the average velocity; $\overline{v^2}$ is not the same as $\overline{v}^2$.) Then we have

$$F_T = \frac{m}{L} (v_1{}^2 + v_2{}^2 + v_3{}^2 + \ldots + v_n{}^2) = \frac{m}{L} n\overline{v^2},$$

where F_T is the total force against the left wall. The *pressure* against the wall will be F_T divided by the area of the wall; in this case, since we have assumed that the box is a cube, the area of the wall is L^2. For the pressure, then, we obtain

$$p = \frac{F_T}{L^2} = \frac{N}{3} \frac{m}{L^3} \overline{v^2} = \frac{N}{3} \frac{m}{V} \overline{v^2},$$

since $L^3 = V$, the volume of the box.

The equation we have just written can be multiplied through by V; then

$$pV = \frac{N}{3} m\overline{v^2} = \tfrac{2}{3} N(\tfrac{1}{2} m\overline{v^2}).$$

The quantity in the parentheses is the *average kinetic energy* of the molecules of the gas. This equation is to be compared with the general gas law as derived from experiment. This law in equation form says that

$$pV = KT,$$

where p, V, and T are the pressure, volume, and absolute temperature of the gas, and K is a constant which depends on the quantity of gas used. By comparing these two equations, we see that

$$K \sim \tfrac{2}{3} N \quad \text{and} \quad T \sim \tfrac{1}{2} m\overline{v^2}.$$

N, the number of molecules in the closed box, and K are, indeed, both constant. The absolute temperature T, on the other hand, may vary, as may the average kinetic energy of the molecules. We conclude then that as the temperature of a gas increases the mean kinetic energy of its molecules also increases.

We know that a gas becomes hotter when we compress it quickly. Figure 4 shows a closed cylinder and a piston that compresses the

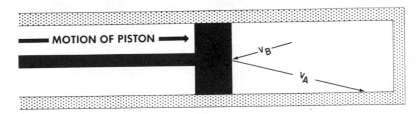

Figure 4

gas in it. We can explain the warming of a gas when compressed in terms of the kinetic model we have been discussing. As the piston advances, a molecule which approaches it with speed v_B will rebound with speed v_A, where $v_A > v_B$; this increase is something like the increase in speed of a batted ball over the incoming pitch. Since all molecules that strike the piston will have their speeds increased, the average kinetic energy of the molecules in the cylinder will be greater. Macroscopically, we can tell that the molecules are moving faster because the macroscopic manifestation of an increase in kinetic energy is a higher absolute temperature of the gas.

BROWNIAN MOTION

Many different kinds of evidence can be cited to support the statement that the molecules of a gas are in rapid and chaotic motion. We shall mention here only Brownian motion, which is perhaps the most striking evidence, and which is readily accessible to anyone who has a reasonably good microscope. Figure 5 illustrates the Brownian phenomenon, which was named after its discoverer.

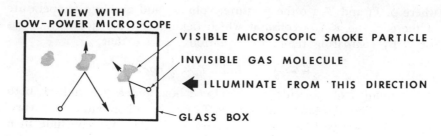

VIEW WITH
LOW-POWER MICROSCOPE

VISIBLE MICROSCOPIC SMOKE PARTICLE

INVISIBLE GAS MOLECULE

ILLUMINATE FROM THIS DIRECTION

GLASS BOX

Figure 5

To observe Brownian motion, we need very small, but visible, particles, such as smoke particles, which can be viewed through a microscope. These particles settle only slowly through the air, and can be viewed while suspended in the air for a considerable time. They are so small that not many gas molecules hit them in, say, a tenth of a second. When a gas molecule does strike a smoke particle, we observe

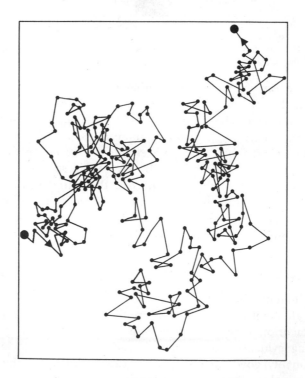

Figure 6

the rebound of the smoke particle from its collision with the fast-moving gas molecule. In Figure 5 the gas molecule's path before and after the collision is shown by the angled arrows. The short, straight arrows by the smoke particles represent their rebounds, which are much like the motion initiated in a stationary billiard ball by another colliding with it. We can, with the aid of several modern devices, watch or record the path of an individual particle over a considerable period of time and witness rebounds from several collisions. Figure 6 illustrates such a path.

One reason for using Brownian motion as our single citation of the evidence that molecules move is that Brownian motion is also visible in liquid samples. Biologists can frequently observe this motion in the fields of their microscopes; we discuss this in the next section.

THE LIQUID STATE

When a gas is cooled, the average kinetic energy of its molecules is reduced. In this condition the molecules are moving more slowly, and as two of them collide they spend more time near each other. In addition, if the gas is compressed, the molecules are closer together, so that for given speeds collisions are more frequent. Under these conditions slowly moving molecules may collide so lackadaisically that their *attractions* for one another make them stick together, rather than bounce apart elastically. This clumping of the slower molecules is the beginning of droplet formation, or condensation. If the process continues, the droplets may coalesce until a considerable amount of the liquefied gas may exist. The process is more rapid if small particles are present around which the droplets may clump.

The kinetic model of a liquid pictures the molecules as moving (still chaotically) within the mass, but only rarely do they move out of the attractive influences of neighboring molecules. That this motion exists is indicated by the Brownian phenomenon in liquids. When small solid impurities are observed in a drop of water, they may be seen to exhibit motion much like that illustrated in Figure 6. A suitable microscope slide for witnessing Brownian motion in a liquid can be prepared by diluting a drop of homogenized milk. If attention is focused on a small globule of butterfat, it may be seen to move or jiggle slightly.

EVAPORATION AND THE KINETIC MODEL

Cooling by evaporation is readily explained by the kinetic model of matter. It is a common experience that water on the hand makes

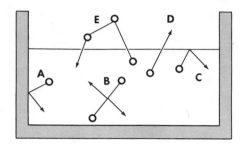

Figure 7

it feel cool. If a breeze is blowing, or the wet hand is waved rapidly in the air, the cool feeling is accentuated. What happens, according to the kinetic model, is illustrated in Figure 7. Several different kinds of interactions can happen as the molecules move about in the liquid. The molecule at A strikes the wall and rebounds. (We are here neglecting the fact that the wall, too, is composed of molecules. In Chapter 16 we discuss how we should view such collisions taking the structure of the material of the wall into account.) The molecules at B collide with each other. At C a molecule moves toward the surface of the liquid, tending to leave its fellows.

The attractive forces of the molecules in the mass of the liquid usually prevent such molecules from escaping from the liquid. The molecule's motion is something like that of a ball attached by a rubber band to a wooden paddle. As the ball, moving away after impact with the paddle, is slowed, and its direction reversed by the pull of the stretched rubber band, so the molecules, moving away from their fellows near the surface of the liquid, are pulled back toward them by attractive forces. We can think of these molecules as being "reflected" or "rebounding" from the surface of the liquid, but here again we must realize that at the molecular level we cannot consider the surface of the liquid to be a continuous skin, but must consider the molecular structure of the surface itself. We shall defer this point, however, until Chapter 16.

For cooling by evaporation, the molecules represented by D and E show the important cases. We have already noted that the absolute temperature is proportional to the mean kinetic energy of the molecules. Some of the molecules have much more than the average kinetic energy, whereas others have much less. Of those having more kinetic energy than the average, some will be near the surface of the liquid and moving toward it. From time to time—and very short times at that—some of these rapidly moving molecules, such as *D*, escape entirely from the body of the liquid. Since these molecules are *large contributors* to the average kinetic energy of the molecules in the liquid, *their removal causes the average kinetic energy to decrease.* That is, the temperature of the liquid falls.

The reason that blowing on hot soup or coffee hastens its cooling is illustrated by *E.* In this case, a highly energetic molecule has escaped from the surface of the liquid, only to suffer collisions that deflected it back into the liquid. A stream of air directed across the surface helps to remove these highly energetic molecules from the neighborhood of the surface so that they cannot be returned to the body of the liquid.

We can also see that the kinetic model will explain why heating a liquid makes it evaporate more quickly. Heating it raises the absolute temperature, and hence raises the kinetic energics of the molecules. If the molecules are, in general, moving faster, more molecules are able to escape from the surface of the liquid. Similarly, of course, cooling slows evaporation.

OSMOSIS

Our contemplation of the kinetic model of a liquid sets the stage nicely for explaining another phenomenon of great interest to the biologist, namely, *osmosis.* A traditional arrangement for demonstrating osmosis is shown in Figure 8. Here an inverted thistle tube *T* is closed at the large end by a semipermeable membrane *M.* A concentrated sugar solution *C* just fills the bell of the tube. The tube is suspended, as shown, in beaker *B,* which is filled with water *W* to the same level as the level of the sugar solution in the bell of the thistle tube. An observer sees then that the level of the solution in the tube rises above that of the water in the beaker, because water passes from the beaker through the semipermeable membrane into the tube.

According to the kinetic model, water molecules moving chaotically bombard the semipermeable membrane continually from both sides. Some of the water molecules, impinging on either side of the semipermeable membrane, will pass through it. The water molecules do not bombard the membrane as often, from the solution side, however, because the sugar molecules also participate in the bombardment, but cannot go through the membrane. (If several Cadillacs use a street, not so many Volkswagens can use it.) Therefore, more water molecules

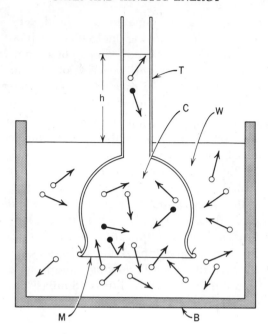

Figure 8

pass through from the water side than from the solution side. The net effect of this transfer is to raise the level of the solution in the narrow part of the thistle tube. In the diagram, this elevation is represented by h, and there are ways of calculating h in terms of the strength of the original solution and on the temperatures of the liquids.

For the osmotic process to operate, the solute molecules must not pass through the membrane; the solvent molecules are able to do so. Why do membranes allow some kinds of molecules to pass but not others? The answer to this question is not fully understood. Apparently, in some cases, semipermeable membranes have pores. For example, electron microscopy studies show pore-like surtctures in cell membranes. We must note, however, that the membrane itself is composed of molecules; hence part of the explanation of osmosis that depends on molecular structure and molecular forces must be postponed until we have talked about chemical bonding in Chapter 16.

SUMMARY

$$F = \frac{\Delta p}{\Delta t}$$

Force is equal to the change in momentum with respect to time.

$$F = \frac{m_1 v_1{}^2}{L}$$

The average force against one wall of a container exerted by one molecule of mass m_1 moving at velocity v_1 just prior to the impact.

$$F_T = \frac{m_1 v_1{}^2}{L} + \frac{m_2 v_2{}^2}{L} + \cdots + \frac{m_n v_n{}^2}{L} = \frac{m}{L}\, n\overline{v^2}$$

The total force on one wall of a container is equal to the sum of the force exerted by each molecule; treating masses of the molecules as equal, and considering the average of the squares of the velocities, the condensed version, where n is one-third the total number of molecules, is obtained.

$$p = \frac{F_T}{L^2} = \frac{N}{3}\frac{m}{V}\,\overline{v^2}$$

Pressure on a wall of a (cubical) container is equal to the total force divided by area of the wall, or one-third the number of molecules times molecular mass divided by volume times the average of the squares of the velocities of the molecules.

$$pV = \tfrac{2}{3}\,N(\tfrac{1}{2}\,\overline{mv^2}) = KT$$

Volume of a container times pressure of the gas in it is equal to the indicated quantity, with letters and symbols the same as for the preceding equation, which proves equal to the experimental value for pV, the product of the absolute temperature T of the gas and a concentration constant K.

CHAPTER

15

Atomic Structure

In Chapter 13 we examined some of the evidence that matter is composed of atoms and molecules, and we concluded that molecules of a particular element or compound are composed of groups of atoms combined in the same numerical ratio. Thus, water molecules always consist of two atoms of hydrogen combined with one atom of oxygen, chlorine molecules consist of two atoms of chlorine, and so forth. The kinetic model of a gas, which we then developed, depicts a gas as a chaos of many tiny, rapidly moving molecules. The myriad impacts of these randomly moving molecules with the walls of their container account for the pressure of the gas.

When using the kinetic model, we thought of its molecules as hard, elastic spheres, paying no attention to the fact that the molecules might be made up of two or more atoms. Although in our account of the work of Dalton, Gay-Lussac, and Avogadro, we saw that atoms *do* combine into molecules with certain fixed numerical ratios, we did not discuss *how* they are bound together or *why* the ratios have the value measurements have shown them to have. Why should two atoms of hydrogen combine with one of oxygen, rather than one of hydrogen with two of oxygen? What is it that makes one sort of atom differ from another? To answer these questions, we need to examine atomic structure. In this chapter we shall discuss basic features of some conceptual models used in modern atomic theory.

The most commonly known atomic models are structures composed of electrons circulating (rather like planets) around a nucleus of protons and neutrons. Electrons are negatively charged, protons are positively charged, and neutrons have no charge. To understand the arrangement of these particles into atoms, we first need to know something about electric charges and the forces of electric attraction and repulsion which hold particles together in an atom and provide for the combination of

512

atoms into molecules. Once we understand the forces between electric charges, we shall also see the evidence for the existence of atomic nuclei.

COULOMB'S LAW

In 1785 Charles Coulomb experimentally determined the law of force between electric point charges. By *point charge* we mean a charge occupying a region that is very small when compared with the distance between it and other charges that attract or repel it. It had been recognized before Coulomb's time that there are two kinds of charges, and these had been designated arbitrarily as *positive* and *negative*. Two positive electric charges repel each other, as do two negative charges, whereas a positive and a negative charge attract each other.

Essentially, what Coulomb did was measure carefully the force between two point charges as it changes with the distance separating them or with their magnitudes. The situation is illustrated schematically in Figure 1, where q_A and q_B are the charges and r is the separation between them.

Figure 1

Coulomb found that the force of either charge on the other is directly proportional to the product of their magnitudes, and inversely proportional to the square of the distance between them. Written as an equation, Coulomb's law says that

$$F = k\,\frac{q_A q_B}{r^2},$$

where F is the force (called "coulomb force") and k is a constant of proportionality. (The constant k depends on the units used and need not concern us further.) In the equation note that the product $q_A q_B$ will be positive if both charges have the same sign; thus a positive force is a *repulsive* force. If q_A and q_B are of different sign, their product is negative; thus a negative force means a force of *attraction*.

Figure 2 is a plot of $1/r^2$ against r. It shows how the force between two point charges varies with the separation between them. Note that the lower curve applies to negative (attractive) forces. From the equation we

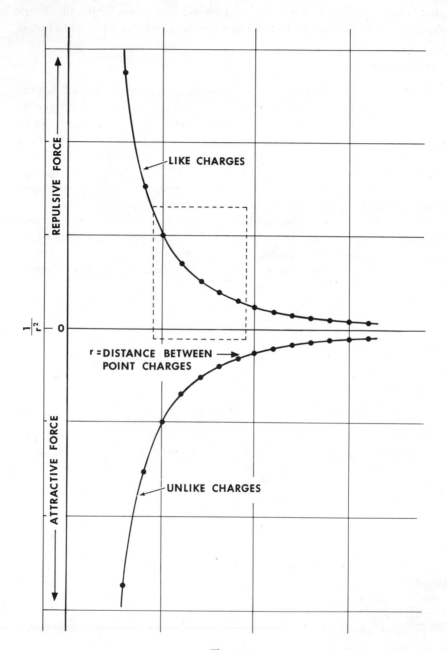

Figure 2

see that for any two particular charges q_A and q_B (as well as k) are constants. Thus we need only graph the variable $1/r^2$, and we will essentially be graphing the force, since the inclusion of constant terms will not change the graph's shape.

POTENTIAL ENERGY OF TWO POINT CHARGES

In our discussion of energy (Chapter 11) we defined energy by its relation to work. We saw that under some circumstances energy could be stored in a system for later retrieval. We called this stored energy *potential energy*. Our major example involved a brick lifted above the ground. The work we did against gravity in lifting the brick was stored as potential energy by maintaining the brick in an elevated position. This energy could be released wastefully, or employed usefully in a machine whenever the force of gravity did work in returning the brick to the ground. In other words, whenever the weight of the brick acted through the brick's displacement from a height to the floor, energy was released. A different example of potential energy showed us that the work we do in stretching a spring can be returned when we allow the spring to contract. Let us note that the mutual attractions and repulsions among them make electric charges systems which can store energy. Why is this so?

Consider Figure 3, which displays two positive point charges q_A and q_B. For convenience, let us regard q_A as fixed and q_B *as* movable. We will ignore the masses of the particles that hold both charges so that we need not concern ourselves with the forces required to accelerate the mass of the particle holding q_B. When q_B is at distance r_1 from q_A, the repulsive force of q_A on q_B is

$$F = k \frac{q_A q_B}{r_1{}^2} .$$

A force F_w that is equal and opposite to F is required either to hold q_B in place—to prevent its motion to the right—or to move q_B to the left. When q_B is moved ever so slightly to the left, r will be less, so the force of repulsion F will be greater. This means that F_w must increase as q_B is moved to the left. Calculating the work required to bring q_B from, say, $r = r_1$ to $r = r_5$ will show us how much potential energy could be retrieved by allowing q_A to repel q_B from $r = r_5$ to $r = r_1$. To calculate work, we multiply force by displacement. In this case the displacement is clear enough, but the force is variable.

Despite the variable force, this calculation turns out to be simple. Consider Figure 4, which is an enlargement of the dotted rectangle of

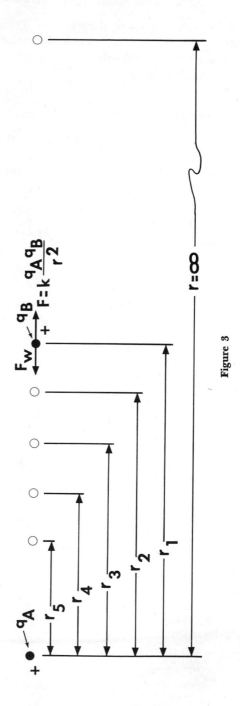

Figure 3

516

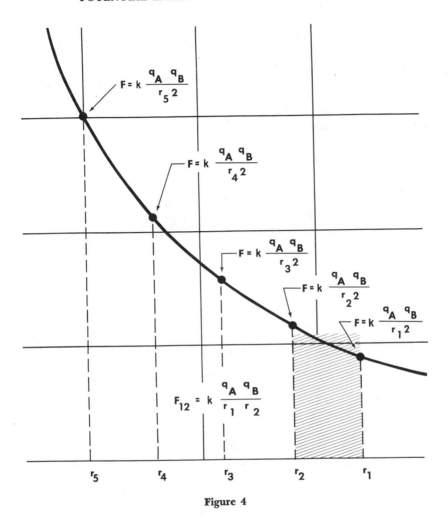

Figure 4

Figure 2. Keep in mind that $r_1 > r_2 > r_3 > r_4 > r_5$; the subscripts simply aid in distinguishing positions. Let us start with q_B at a distance r_1 from q_A and reduce the distance to r_2. In doing work against the repulsive force, we move q_B through the displacement $r_1 - r_2$. To compute the work, we must multiply this displacement by a suitable force. The force $F_1 = k(q_A q_B / r_1^2)$ is too small, and $F_2 = k(q_A q_B / r_2^2)$ is too large. Note, however, that we can specify a force value between these two extremes by writing $F_{1,2} = k(q_A q_B / r_1 r_2)$. Multiplying the displacement $(r_1 - r_2)$ by $F_{1,2}$, we obtain

$$F_{1,\,2}\,(r_1 - r_2) = \mathrm{k}\,\frac{q_A q_B}{r_1 r_2}\,(r_1 - r_2) = kq_A q_B\!\left(\frac{1}{r_2} - \frac{1}{r_1}\right) = W_{1,\,2}\,.$$

This is the amount of work represented by the cross-hatched rectangle in Figure 4. It is neither as small as if we had used F_1 nor as large as if we had used F_2.

We can calculate the work it would take to move q_B from r_1 to r_5 by dividing the total displacement into short steps, calculating the work for each step, and adding the individual calculations to get the total. Thus

$$W_{1,\,2} = kq_A q_B \left(\frac{1}{r_2} - \frac{1}{r_1}\right)$$

$$W_{2,\,3} = kq_A q_B \left(\frac{1}{r_3} - \frac{1}{r_2}\right)$$

$$W_{3,\,4} = kq_A q_B \left(\frac{1}{r_4} - \frac{1}{r_3}\right)$$

$$W_{4,\,5} = kq_A q_B \left(\frac{1}{r_5} - \frac{1}{r_4}\right).$$

The total work done in moving q_B *from* $r = r_1$ *to* $r = r_5$ *is simply*

$$W = W_{1,\,2} + W_{2,\,3} + W_{3,\,4} + W_{4,\,5}, \qquad \text{or}$$

$$W = kq_A q_B \left[\left(\frac{1}{r_2} - \frac{1}{r_1}\right) + \left(\frac{1}{r_3} - \frac{1}{r_2}\right) + \left(\frac{1}{r_4} - \frac{1}{r_3}\right) + \left(\frac{1}{r_5} - \frac{1}{r_4}\right)\right]$$

$$= kq_A q_B \left(\frac{1}{r_5} - \frac{1}{r_1}\right).$$

This is a satisfying result, because the work seems to depend only on the starting and finishing distances, and not on the path between r_1 and r_5. However, we might question the accuracy of using, say, $F_{2,\,3} = k\,(q_A q_B/ r_2 r_3)$ for an average force used in displacing q_B over $r_2 - r_3$. What assurance do we have that the true average force should not be slightly greater or slightly less than $F_{2,\,3} = k\,(q_A q_B/r_2 r_3)$? Still, whatever the error may be, we can make it smaller by making the steps smaller and taking more of them. Instead of dividing the distance $r_1 - r_5$ into four steps, we could divide it into four million, each only one-millionth as long as before. Then we would have four million terms of the form $1/r_2 - 1/r_1$, $(1/r_3 - 1/r_2)$, and so on, instead of four. *But the result of adding all these terms would be the same as before!* That is, only the first and the last would remain, and

$$W = kq_A q_B \left(\frac{1}{r_{4,000,000}} - \frac{1}{r_1}\right).$$

It is easy to see why we need only consider the first and the last terms if we examine how we summed the terms in our example of moving the charge from r_1 to r_5. We see that there is always a negative fraction equal to each positive fraction, except for the last fraction in the first term (which is negative) and the first fraction in the last term (which is positive).

Before proceeding to other matters, let us examine our original result, $W = kq_A q_B (1/r_5 - 1/r_1)$, in a special circumstance, namely, starting with q_B at $r = \infty$. Since $1/\infty = 0$, the work required to bring q_B from infinity to any distance r away from the charge q_A is $W = kq_A q_B (1/r - 1/\infty) = k(q_A q_B / r)$ (we can drop the subscripts for r as unnecessary). Figure 5 shows a graph of the potential energy of this system as a function of the final distance r between the two charges.

The concept of *negative potential energy* can be derived easily now. If either q_A or q_B is negative, then the system will do work for us as we allow q_B to move from $r = \infty$ (or from any distance $r_1 > r$) to $r = r$. Actually, the system will do work for us whenever a charge moves from $r = r_1$ to $r = r_2$, $r_1 > r_2$, whenever q_A and q_B are oppositely charged. It is reasonable to define the zero of potential energy at $r = \infty$, where the Coulomb force is zero no matter what the signs of the charges. It is customary to say that unlike charges at finite separations have negative potential energy, and like charges have positive potential energy.

ELECTRIC CHARGES AND MATTER

A solid, such as a pocket comb, is usually electrically neutral or uncharged. In dry weather, however, when the comb is used on clean, dry hair, comb and hair become oppositely charged   It is simple and plausible to designate the charge on the comb arbitrarily as negative, and to suppose that electrically neutral bodies contain equal numbers of positive and negative charges. We can then further suppose that the comb acquires electrically negative charges from the hair, or that the hair acquires positive charges from the comb, or both. But what kind of charge transfer really takes place?

Several experiments indicate that negative charges are usually more easily detached from matter than positive charges. Two such experiments are quite famous. The first involves the *photoelectric effect*. A plate of polished zinc is illuminated by ultraviolet light during the illumination, electrons (negative charges) leave the zinc, and the zinc acquires a positive charge. The other experiment involves *thermionic emission*. A wire, such as the filament of an incandescent light bulb, emits electrons when heated,

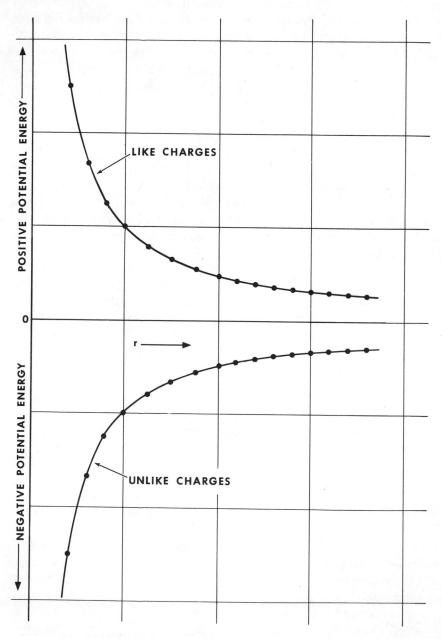

Figure 5

520

and under suitable experimental arrangements a consequent positive charge on the filament can be demonstrated.

There are situations where positive charges are emitted. For example, in some radioactive decay processes, alpha particles, which are positively charged, are emitted at high speeds. They were used by Lord Rutherford in 1911 in the famous experiment which established the notion of a nuclear atom.

Rutherford's Experiment and Atomic Mass Distribution

Although we shall not be especially concerned with radioactivity, Rutherford's experiment, which makes use of radioactive material, is important enough to be worth a brief description here. Figure 6 shows a simplified form of the arrangement Rutherford used.

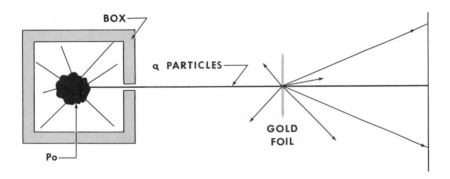

Figure 6

A sample of polonium (Po), a radioactive alpha-emitter, is placed in a box, so that only the alpha particles which hit a small hole in the side of the box can escape. The thin beam of escaping alpha particles is directed to a sheet of gold (Au) foil. Alpha particles emitted from polonium travel at about one-twentieth the speed of light, and in Rutherford's experiment the gold foil was about 10^{-5} cm thick. A vast majority of the particles went straight through the foil, but a few them were deflected at various angles, some nearly 180°. Rutherford interpreted this to mean that most of the foil was empty space, offering no impediment to the passage of the alpha particles. Occasionally an alpha particle would approach a strong positive charge within the foil, and the repulsive force between the charges in the foil and in the alpha particle would cause the

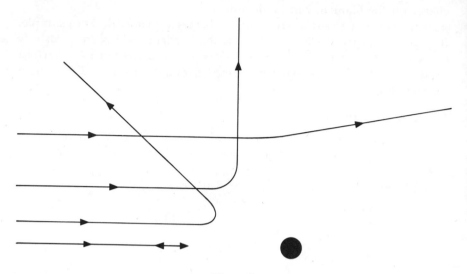

Figure 7

alpha particle to be deflected. The amount of deflection would depend on how close it came to the other positive charges, as we saw in the section on potential energy in this chapter. This is illustrated qualtitatively in Figure 7.

By careful analysis of the numbers of alpha particles scattered at different angles, Rutherford obtained measures of the size of the nuclei of atoms in various kinds of foil. He was able to show that even large nuclei are only about 10^{-12} cm in diameter, whereas atoms themselves are about 10^{-8} cm in diameter. Atomic diameters are about 10,000 times greater than atomic nuclei!

By far the greatest part of the mass of any matter composed of atoms is concentrated in the atomic nuclei. Consider, for example, a gold atom. Its nucleus contains 79 protons and 118 neutrons. Around this nucleus circulate 79 electrons. A neutron has nearly the same mass as a proton, however, and a proton has 1836 times the mass of an electron. Thus, 99.98% of the mass of the gold atom is concentrated in its nucleus.

Subatomic Particles—Charges and Masses

We shall not here trace the history of the discoveries of subatomic particles, but we shall indicate what their charges and masses are. We deal with electrons, protons, and neutrons in this section, though there are

many other subatomic particles, and new variants are still being discovered.

Electric charge is frequently measured in *Coulombs*. The Coulomb
is easily related to the *Ampere,* an electric unit of current. A current is
composed of a stream of electrons, and when 1 Amp flows in a wire, the
electrons carry 1 Coulomb of charge per sec through the wire. (This is
about the charge that passes each second through an ordinary 100-watt
light bulb.) One Coulomb of negative charge is the charge carried by
6.25×10^{18} electrons; one electron bears a negative charge of 1.6×10^{-19}
Coulombs. All electrons have been found to bear this small amount of
charge. The quantity 1.6×10^{-19} Coulombs is known as the *electronic
charge*. The mass of the electron has also been measured. It is $9.1 \times
10^{-28}$ g.

Any proton bears a positive charge of exactly the same magnitude as
the negative charge of an electron. Thus, if a proton and an electron are
brought quite close together, then at some distance from both, the
Coulomb forces from the *two* charges will be exactly equal and opposite;
the net force will be zero. The two charges *neutralize* each other. The
protonic mass is 1.67×10^{-24} g.

The neutron has no electric charge. Its mass is slightly greater than
the mass of a proton, but only by about 0.1%; therefore, to the accuracy
with which we are concerned, it mass too is 1.67×10^{-24} g.

THE HYDROGEN ATOM

We can consider a simple model of a hydrogen atom resembling a
single planet circling the sun. (Later we shall consider a somewhat better
and more complicated model that more closely represents the facts we
have.) The nucleus of the hydrogen atom consists of a single proton. At
a distance of about 0.53×10^{-8} cm, a single electron moves in an orbit
around the nucleus.

Since the electron and the proton exert a mutual attraction on each
other, why does the electron remain at this distance? Why doesn't it fall
all the way in to collide with the proton? The answer is illustrated by
Figure 8. Because of its inertia, the electron always tends to move off in a
straight line (tangent to its orbit), as indicated by the arrow v. The
attractive force f at each instant pulls at right angles to v, tending to
make the electron fall into the proton. During the "fall," inertia carries
the electron to *P,* where the arrows (indicated by dotted shafts) have
exactly the same relation as before. In other words, the electron does fall
toward the nucleus, but its fall only converts its inertial motion in a

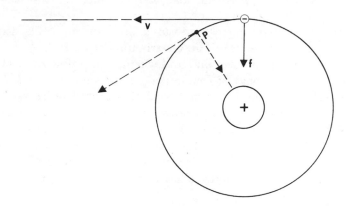

Figure 8

straight line into circular motion. (This is analogous to the way a planet moves around the sun, with gravitational attraction playing the role of charge attraction.)

We can calculate how much work would be required to remove the electron entirely away (that is, to infinity) from the proton. We use the expression for the potential energy of two charges which are close together. Recall that the equation for this is

$$W = k \frac{q_A q_B}{r}.$$

To use this equation, we need the constant $k = 9 \times 10^{11}$. (For completeness, we should say $k = 9 \times 10^{11}$ Joule — cm/Coulomb², but this detail need not bother us here.) The charges q_A and q_B are both 1.6×10^{-19} Coulombs, but one is negative. Substituting these values and performing the arithmetic, we obtain

$$W = - 9 \times 10^{11} \frac{(1.6 \times 10^{-19})^2}{0.53 \times 10^{-8}} = - 4.4 \times 10^{-18} \text{ Joules .}$$

This is a *negative* potential energy, and indicates how much work would need to be done to separate the charges to a distance where they would have no attraction for each other—a distance where the system would have zero potential energy.

The value we have just obtained is not quite correct, however, for it neglects the fact that the electron circling in orbit has a kinetic energy of $\frac{1}{2}$ mv². This energy would aid in removing the electron. When corrections are made, the proper value for energy which must be supplied to remove the electron from the hydrogen atom is just half what we calculated, or

$$W_T = 2.2 \times 10^{-18} \text{ Joules} .$$

Now consider an electron and a proton so far separated that there is no mutual attraction. These are the components of a hydrogen atom. If we bring them close together, so close that they *just begin* to attract each other, the force of attraction *can* move the electron all the way in to the 0.53×10^{-8} cm orbit. We shall see in a moment that there are intermediate way stations; hence we use the word "can." It should be noted, however, that under normal conditions the electron will move in to occupy this orbit, which is its lowest energy level, called *ground state*.

Some of the work produced by the attractive force acting through this displacement—namely 2.2×10^{-18} Joules—goes into the kinetic energy of the electron in its orbit. Since the electron, in moving in, pulled by the attraction of the nucleus, produces 4.4×10^{-18} Joules of energy, we must find out what happens to the other 2.2×10^{-18} Joules, for we know that this energy cannot be created nor destroyed. It has been found that this energy is radiated as light. Figure 9 shows an illustration of a very simplified situation during and after the fall of the electron.

When we examine the behavior of atoms whose energies are changing, we can actually find this radiated light. Moreover, we can measure its

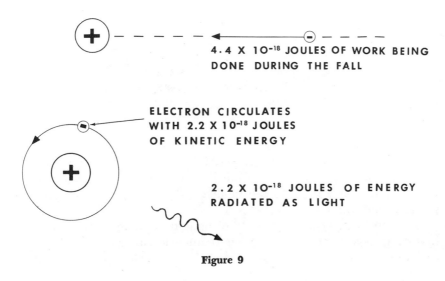

4.4 X 10⁻¹⁸ JOULES OF WORK BEING DONE DURING THE FALL

ELECTRON CIRCULATES WITH 2.2 X 10⁻¹⁸ JOULES OF KINETIC ENERGY

2.2 X 10⁻¹⁸ JOULES OF ENERGY RADIATED AS LIGHT

Figure 9

wavelength. This enables us to check some predictions made by the theory. The equation that relates the frequency of light to the energy is

$$W = h\nu,$$

where, if W is in Joules and ν the frequency is in oscillations per second, $h = 6.6 \times 10^{-34}$ Joule-sec. The frequency of light, then, corresponding to 2.2×10^{-18} Joules is

$$w = h\nu,$$

$$2.2 \times 10^{-18} = 6.6 \times 10^{-34}\,\nu,$$

$$\nu = 3.3 \times 10^{15} \text{ cycles/sec}.$$

The constant h is known as Planck's constant. This frequency corresponds to a wavelength

$$\lambda = \frac{c}{\nu} = \frac{3 \times 10^8}{3.3 \times 10_{15}} = 9.1 \times 10^{-8} \text{ m} = 9.1 \times 10^{-2}\,\mu\ .$$

TABLE 1

ν	λ
3.3×10^{15}	9.1×10^{-2}
3.2	9.4
3.16	9.5
3.09	9.7
2.93	10.2
2.47	12.1
0.82×10^{15}	36.6×10^{-2}
0.73	41.0
0.69	43.4
0.62	48.6
0.46	65.6

From our discussion of light energy in Chapter 12, we recall that the shortest waves we can see are about 0.4 μ; hence these much shorter (0.091-μ) waves are invisible. Though they are well in the ultraviolet region of the spectrum, they can be detected and measured with suitable spectrometers and other instruments.

Actually, a spectrometric study of light from atomic hydrogen shows that we receive light of many other frequencies, all less than 3.3×10^{15} cycles/sec. A few of these are given in Table 1, where ν is in cycles per

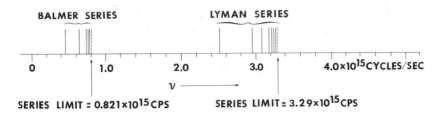

Figure 10

second and λ is in microns. A plot of the frequencies on a linear scale is shown in Figure 10.

Note that the lines in Figure 10 fall into two groups. The longer wavelength set is known as the Balmer series, after the physicist who first found a relation among these lines; the other set is known as the Lyman series, also after its discoverer. Niels Bohr interpreted the Balmer and Lyman data as indicating that there are other discrete orbits for the electron of a hydrogen atom than the one having a radius of 0.53×10^{-8} cm. We have seen that this orbit corresponds to a negative potential energy of 2.2×10^{-18} Joules (and an electron kinetic energy of the same amount). The other orbits, according to Bohr's analysis, will be found at distances which can be calculated from considering the wavelength of light the atom radiates (this, of course, can be observed and measured).

Figure 11 shows a vertical scale at the left, E, giving the potential energy of the atom in Joules. We see that at level 1 (which is called ground state, because it is the normal condition of the atom), the atom has a total energy of 2.2×10^{-18} Joules, which corresponds to what we saw before. At $E = 0$ on the potential scale, which corresponds to the state where the electron is just removed from the electrostatic field of the atom but has no kinetic energy, the system is at its highest energy level, because none of its energy is negative. Recall that in our discussion of energy and work in Chapter 11 we said that when a body could do work when released (because a force on it makes it move), it would be said to have positive potential energy. When work must be done on it to move it, it has *negative* potential. When the electron of the hydrogen atom occupies any of its possible energy levels (except $E = 0$, of course), we would have to do work against the attractive nuclear forces in order to remove it. As the electron moves in toward the nucleus, it does work, so its negative potential becomes greater and greater, since we would have to do more and more work to move it back out again.

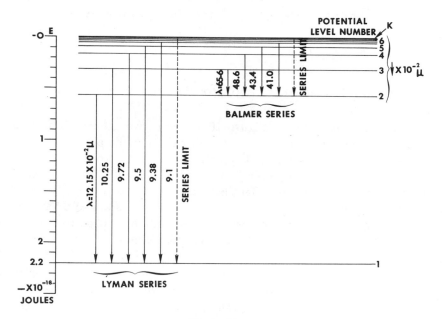

Figure 11

While the electron moves in toward the nucleus it may lodge at any of the negative potential energy levels indicated for hydrogen by the horizontal lines in Figure 11, but never between them. Jumps between levels we call *quantum jumps,* because the jump must be of a fixed size (quantity). Suppose an electron moves inward from, say, level 4 to level 2. We see from the vertical arrow between these two levels that to do so the atom would have to emit light having $\lambda = 48.6 \times 10^{-2} \ \mu$. The electron could jump directly from level 4 to level 1, emitting light of $\lambda = 9.7 \times 10^{-2} \ \mu$. It could also make the jump by stopping, say, at level 3, then going to 2, then to 1.

The numbers by the vertical arrows connecting the energy levels show us, then, the wavelengths of light that must be emitted if the atom is to make a transition from the state where it has an electron at one level to a state with the electron at another. Bracketed together are the vertical arrows that represent emissions taking the atom from $E = 0$ or other permissible potentials down to level 2; these are emissions of longer wavelength and fall in the Balmer series. At the left of the diagram emissions of light which take the atom from the zero potential level (or other levels) down to level 1 are bracketed together; they are the shorter wavelengths, and fall in the Lyman series.

As we would expect after our discussion in Chapter 12 on energy in light, the shorter wavelengths give the atom a higher *negative* potential energy (that is, they carry away from the atom more positive energy). Recall that the energy of light is given by hv, and that $\lambda = c/v$; we know then that light of the shorter wavelengths will have greater frequencies than light of longer wavelengths; hence its energies will be correspondingly greater.

Important things to remember in connection with quantum jumps are that they must be jumps to permissible levels, and that, if we observe the frequency of the emitted light, we can calculate energies gained or lost in jumps from one level to another, and hence know the state of the atom. We should note that nothing in *classical* (pre-quantum) physics will explain the existence of discrete negative potential levels; according to classical explanations, the electron would have to occupy (at least briefly) a continuum of positions on its way toward the nucleus. Thus the discovery that it did not and could not do so provided important evidence for the new theoretical explanations of the behavior of fundamental particles we are considering in this chapter. Note also that what we have said about "downward" transmissions and emissions of light may be reversed; if the atom is irradiated with light of certain frequencies, its electrons may move "outward" from the $n = 1$ level. The hydrogen atom whose electron occupies levels above $n = 1$ is said to be in an *excited* state.

The electron of a hydrogen atom in an orbit corresponding to the $n = 1$ state "traverses" this orbit about 10^{16} times per second. This is so rapid by ordinary standards that it is pointless to speak of where the electron is in relation to its nucleus—whether it is to this side, that side, or above or below it. In any case, it has been found that we *cannot* talk about an exact position and momentum for very small particles; hence, it is customary to speak of the electron as being "smeared out" into an electron cloud. This is represented in Figure 12. The cloud corresponding to the circulator orbit $n = 1$ for hydrogen is a fuzzy sphere. We shall also identify electron clouds of other shapes and associate their spatial interrelations with the geometric orientation of chemical bonds in Chapter 16.

Figure 12

THE HYDROGEN MOLECULE

In Chapter 14 we noted that hydrogen gas is composed of molecules having the formula H_2. The hydrogen atom just described is electrically neutral, so we can ask, What causes two hydrogen atoms to come together? What holds them together in a hydrogen molecule? In order to answer this, let us discuss what happens when a hydrogen atom and a proton are brought close together. Part of the time the electron will be "between" the proton which is its nucleus and the other proton—that is, a portion of the electron cloud will be between the two protons. In this state of affairs, the electron will be attracted by the independent proton, and this attraction will be stronger than the repulsion between the two protons. The electron cloud will tend to "become confused" and will shift so as to occupy a symmetrical position with respect to the two protons. The electron will be spending most of its time between the two protons, and, because of its attraction for them, it can provide a bond between them.

The two protons held together by a single electron can be symbolized by $H_2{}^+$. This assembly is not electrically neutral, and requires another electron for neutrality. Such non-neutral assemblies are called *ions*. From the general symmetries of the situation—two identical atoms, each composed of one electron and one proton combined into a single molecule—we might expect that the electron added to the $H_2{}^+$ would enter the combination in essentially the same way as the electron already there. This is true, with one qualification, which constitutes the first restriction on the number of electrons that can enter the same energy level.

What we must recognize is that electrons have *spin* or *intrinsic* angular momentum, like the rotating earth's momentum. Until now, we have considered electrons as virtually without size, as "point particles." If a stream of neutral hydrogen atoms is shot through an inhomogeneous magnetic field, however, the interaction of the stream with the field indicates that electrons cannot be considered simply as point charges, but must be regarded as extremely small current loops. Such current loops act as tiny electromagnets. From analysis of suitable experiments, we arrive at three significant points: (1) Electrons all spin at the same rate; (2) in a magnetic field the spin axes are always oriented in the same direction; (3) around these axes the spins may be either clockwise or counterclockwise.

Returning now to the problem of adding an electron to H_2^+, we can state the qualification that the new electron must enter the combination

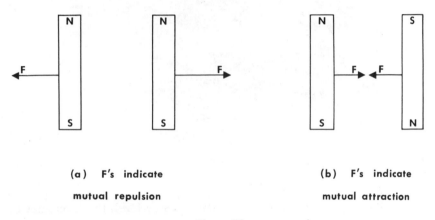

(a) F's indicate

mutual repulsion

(b) F's indicate

mutual attraction

Figure 13

with spin opposite to that of the electron already there. This restriction may seem more plausible if we look at an extremely oversimplified analogy from the macroscopic world. Figures 13a and 13b show pairs of simple bar magnets. When the magnets are arranged with their N ends pointing in the same direction (which corresponds to electrons having spin in the same direction), they repel each other, whereas when the N ends point in opposite directions (opposite spin) the magnets attract each other.

Our model of the hydrogen molecule, taking into account the spin restriction, becomes that illustrated by Figure 14. The two protons are located symmetrically, with two merged electron clouds. The electrons

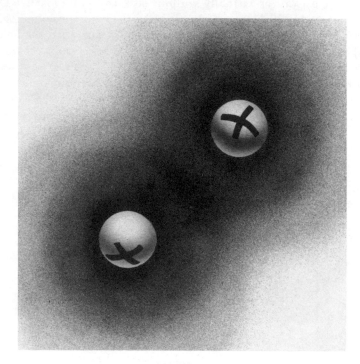

Figure 14

have opposite spins, and occupy the same energy level, the level that has quantum number $n = 1$

THE HELIUM ATOM

The helium atom, since it is constructed of two electrons and two protons (and two neutrons), can be understood in part as a modification of the hydrogen molecule. The neutrons are essential building blocks of atomic nuclei; however, we are concerned entirely with happenings outside the nucleus, so we shall note only that neutrons contribute to atomic mass and are electrically neutral. For the helium atom, we could, as for the hydrogen atom, draw potential energy curves, bringing an electron toward the nucleus and establishing its lowest energy level (its closest orbit) by considering analyses of spectra of emitted energy. Quantitative details such as the radius of the orbit and the frequencies of the spectral

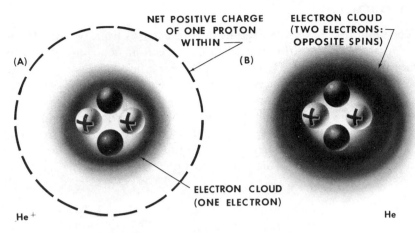

NET POSITIVE CHARGE
OF ONE PROTON
WITHIN →

ELECTRON CLOUD
(TWO ELECTRONS:
OPPOSITE SPINS)

(A)   (B)

ELECTRON CLOUD
(ONE ELECTRON)

He⁺   He

Figure 15

lines would differ, but in principle there would be nothing new except the fact that inclusion of the first electron would not produce an electrically neutral atom. There would still be a net positive charge equivalent to the charge on the proton. This system is illustrated in Figure 15*a*, and is called singly ionized helium: He^+. The helium ion can attract an additional electron to become a molecule, as shown in Figure 15*b*. Just as we found in building the hydrogen molecule, this electron must have spin opposite from that of the electron already orbiting the nucleus.

The helium atom and the hydrogen molecule seem quite similar at this point. Each has two electrons forming merged spherical clouds, which neutralize the charges on the two protons within. There is a significant difference, however; in helium, the two protons are tightly held in the same nucleus, while in the hydrogen molecule the two protons are separate nuclei. The repulsive force of the positive charges of the two nuclei of the hydrogen molecule tends to split the molecule into its two component atoms. By supplying comparatively little energy to hydrogen gas, we can break up the molecule. On the other hand, a large amount of energy would be required not only to ionize helium (to remove one of its electrons), but to disrupt the helium atom—to break apart its nucleus—because there are attractive nuclear forces much stronger than the electrical repulsions. We shall see in Chapter 16 why the easier separation of hydrogen molecules into atoms makes hydrogen chemically active, whereas helium atoms are chemically inert.

THE LITHIUM ATOM

The atom that has three protons in its nucleus is lithium. In its most abundant isotope, the lithium nucleus contains four neutrons. All atoms having the same number of protons in their nuclei and structures resembling one another in certain definite ways are atoms of the same element. Atoms having the same number of protons but different numbers of neutrons (and slightly different structures) are different isotopes of the same element. About 7.5% of the naturally occurring atoms of lithium have only three neutrons in their nuclei. A lithium ion, Li+, is illustrated in Figure 16a. This represents a nucleus of charge + 3 (that is,

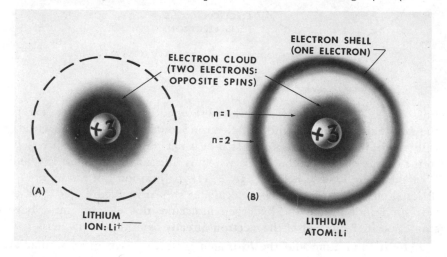

Figure 16

three protonic charges) surrounded by the merged spherical clouds of two electrons having opposite spins. The ion, however, has a net positive charge equivalent to that of one proton, and requires one more electron for electric neutrality.

The remaining electron cannot simply merge as a spherical cloud with the two already there. The reason for this is that the two electrons in the spherical cloud have opposite spins; a third would have the same spin as one of them. According to the Pauli exclusion principle, which will be touched on in more detail shortly, no two electrons can occupy the same *state* in the same atom; no two electrons can be in the same shell with the same spin. In this case, no two electrons can be in the innermost cloud with the same spin, and the two spin possibilities have already been used. (This is a quantum principle, and, while analogies from the macro-world may be intuitively helpful, they cannot explain the exclusion principle, which is established empirically.) Therefore the net positive charge of $+1$ must hold the third electron in a different orbit, somewhat farther out from the nucleus. This results in the establishment of an electron cloud *shell,* illustrated in Figure 16*b*.

Recall now our discussion of the spectra of atomic hydrogen. We designated the energy level corresponding to the smallest permissible orbit as $n = 1$, that corresponding to the next smallest as $n = 2$, out to $n = \infty$, corresponding to zero energy, the condition where the electron has been entirely removed. In the same manner as for hydrogen, we assign numbers to the levels at which electrons enter into the structure of more and more complex atoms. The two innermost electrons in lithium—those in the merged spherical clouds—are $n = 1$ electrons. The electron in the spherical shell cloud is an $n = 2$ electron. Traditionally, the $n = 1$ electrons are said to belong to the K shell and the $n = 2$ electrons to the L shell. More complicated atoms have electrons at the third level $(n = 3)$, and these are M shell electrons. The only further reference to this will be in Table 2.

THE BERYLLIUM ATOM: ATOMIC NUMBER

It is convenient for us to conform to the usual practice of designating the number of protons in an atom as its *atomic number.* Thus the atomic number of hydrogen is 1, that of helium 2, and that of lithium 3.

There are four protons in beryllium nuclei, and in the most common isotope there are five neutrons. It will take four orbital electrons to produce a neutral atom. The fourth electron can, of course, enter the $n = 2$ shell, since it can have opposite spin to the electron already there. The beryllium atom, then, would look very much the same as the lithium atom illustrated in Figure 16*b*, except that the nucleus is represented by $+4$, and the $n = 2$ shell is labeled "two electrons: opposite spins."

RELATIVE ATOMIC SIZES

Figures illustrating hydrogen, helium, lithium, and beryllium have been drawn large in order to illustrate clearly the details of atomic models, and for this reason the more complex atoms have been shown in larger figures. We should note, however, that, while the sizes of atoms do vary for different elements, the illustrations are misleading. More complex atoms are *smaller* than simpler ones, although in order to show details this has not been indicated. (In addition, the relative sizes of nuclei are greatly exaggerated in the figures.)

The reason why more complex atoms are more compact can be understood when we compare the $n = 1$ electron orbits in, say, hydrogen with those in beryllium. The nuclear charge in beryllium is four times greater than in hydrogen; hence, it is reasonable to suppose that its attractive forces will be greater and will hold the $n = 1$ electrons in much more closely. This turns out to be true. Consequently, the shell for the $n = 2$ electrons does not need to be nearly as large as might be suggested by a figure for beryllium similar to Figure 16b.

THE BORON ATOM

So far, we have seen that the two $n = 1$ electrons and the two $n = 2$ electrons go into spherically symmetrical clouds as the atomic models are built up from hydrogen through beryllium. For $n = 1$ electrons, we are limited to two because the introductions at the $n = 1$ level would put two electrons at the same energy level with the same spin, in violation of the exclusion principle. We avoided this by introducing the third electron at an energy level represented by $n = 2$. In beryllium, however, a second electron occurs at the $n = 2$ level, so again no more can be introduced without violating the spin requirement. As we construct a model for atomic number 5, boron, must we go, and should we, to $n = 3$? The answer to this question is *no*.

When the spectrum of boron is examined, it is found that there is another, slightly different energy level that might be interpreted as $n = 2$. In our model, this is handled by recognizing that the fifth electron—or the third electron in the $n = 2$ shell—can be in what is called a *penetrating orbital*. To see what this means, let us consult Figure 17. The nucleus has a charge of $+5$. Two electron clouds are in close, while two more constitute a spherical shell farther out. Electron 5 is in the dumbell-shaped cloud and partially shielded from the nucleus, since the electron clouds for electrons 1 and 2 partially intervene. It is also partially shielded from the attraction by the nucleus because of the electron clouds for electrons 3 and 4—the $n = 2$ shell; this shielding is less effective than that of the first shell. The result of this is that electron 5 moves in an elliptical orbit which lies sometimes inside and sometimes outside the spherical shell occupied by electrons 3 and 4. Its orbit is called a *penetrating* orbit, since it penetrates the $n = 2$ shell. Therefore, electron 5, too, is called an $n = 2$ electron, but we need to distinguish it from electrons 3 and 4. Elec-

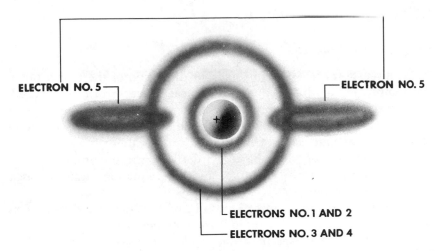

ELECTRON NO. 5

ELECTRON NO. 5

ELECTRONS NO. 1 AND 2

ELECTRONS NO. 3 AND 4

Figure 17

trons 3 and 4 are also partially shielded, but calculations based on the forms of the clouds show them less fully shielded than electron 5; hence they are more tightly bound. This is why the spherical ($n = 2$) clouds were filled before the dumbell.

We can find a method to distinguish the clouds in the fact that possible cloud positions around the nucleus are not spherically symmetric. Rather, as we imagine looking at electron 5 from outside the nucleus, there is a greater probability of seeing it in certain directions than in others. It has an angular dependence not found for electron 1, 2, 3, or 4. Just as the energy of the level is represented by the letter n ($n = 1, 2, \ldots$), so the angular dependence is indicated by a letter, the letter l. In general, $l = 0$ corresponds to a circular orbit, and the spherical clouds have $l = 0$. Moreover, $l = 1, 2$, or more corresponds to orbitals of increasing ellipticity. But there is a restriction on the values of l, as we shall see in the next section.

For atoms that have electrons of given n in elliptical orbits, that is, for atoms having electrons with $l \geqq 1$, there is an additional way in which the atom interacts with a magnetic field; it provides still another way to put additional electrons into the shells represented by the same n. This magnetic reaction can be represented by a set of integers; they are indicated by m_l, where the subscript indicates a dependence of this m_l on the ellipticity of the orbit. This quantum number specifies the spatial orientations of nonspherically symmetrical clouds.

THE QUANTUM NUMBERS

Our discussion of n, l, and m_l and our discussion of the fourth quantum number m_s in this section may appear rather arbitrary. Indeed, in the early days of modern atomic theory, the attempts to explain these phenomena in terms of certain relations between integers were pursued by trial and error. Only since 1925 has there been a general theoretical explanation; it is based on deriving the quantum numbers from Schroedinger's wave equation, and this equation itself is based on the concept of energy.

We have now discussed three kinds of numbers that are called *quantum numbers*. The term "quantum" specifies that these numbers must have certain discrete or specific values and cannot assume intermediate values. In the cases of n, l, and m_l the numbers are always integers.

The first quantum number, $n = 1, 2, 3, \ldots$, can never be zero, must always be positive, and is called the *principal quantum number*. The principal quantum number gives a rough idea of the average distance of the electron orbitals from the nucleus. The quantum number represented by l is the *angular momentum quantum number*, or the *azimuthal quantum number*. This quantum number can be zero, but never negative (because there can be no negative probabilities, and the physical basis of the number is a probability). It can never be greater than $n - 1$. Thus $l = 0, 1, 2, \ldots, n - 1$.

The letter m in m_l shows the magnetic dependence of this quantum number. It is called simply the *magnetic quantum number*. For a given electron in an atom, m_l can have only integral values from $-l$ to $+l$. Thus, if $n = 4$, we could be considering an electron with $l = 3, 2, 1,$ or 0. Suppose we are considering a particular electron with $n = 4$ and $l = 3$. That electron could be one of several with the same n and l, and its magnetic quantum number could be any one of the values $-3, -2, -1, 0, 1, 2,$ or 3.

The fourth kind of quantum number that we shall consider is the *spin* quantum number, referring to electron spin. Its symbol, m_s, again indicates its relation to magnetic interactions. This quantum number can have one of only two values: $+\frac{1}{2}$ or $-\frac{1}{2}$. The magnitude of $\frac{1}{2}$ for both cases corresponds to our earlier statement that electrons always spin at the same rate; the plus and minus signs refer to the clockwise or counterclockwise directions of spin.

PAULI EXCLUSION PRINCIPLE

The exclusion principle, discovered by Wolfgang Pauli, states that *no two electrons in any atom may have identical values of all four quantum numbers.* Table 2 shows how electrons are arranged into shells in the first twelve elements. The plus and minus signs stand for $m_s = +\frac{1}{2}$ and $m_s = -\frac{1}{2}$, respectively. In passing, note that for the biologically important elements carbon, nitrogen, and oxygen the total number of electrons with positive spins *exceeds by one the total* number with negative spins.

The three parts of Table 3 further illustrate the point. In carbon, electrons 5 and 6 both have the same spin; the exclusion principle is satisfied by the difference in m_l. The same comment can be made of electrons 5, 6, and 7 in nitrogen. Only with electron 8 in oxygen do

TABLE 2. *How Electrons Group into Atomic Shells*

		K Shell	L Shell				M Shell						
Principal quantum number n		1	2				3						
Angular quantum number l		0	0	1			0	1			2		
Magnetic quantum number m_l		0	0	−1	0	+1	0	−1	0	+1	−2	−1	0
At. No.	**Element**	m_s	m_s	m_s	m_s	m_s	m_s	m_s	m_s	m_s	m_s	m_s	m_s
1	Hydrogen	+											
2	Helium	+ −											
3	Lithium	+ −	+										
4	Beryllium	+ −	+ −										
5	Boron	+ −	+ −	+									
6	Carbon	+ −	+ −	+	+								
7	Nitrogen	+ −	+ −	+	+	+							
8	Oxygen	+ −	+ −	+ −	+	+							
9	Fluorine	+ −	+ −	+ −	+ −	+							
10	Neon	+ −	+ −	+ −	+ −	+ −							
11	Sodium	+ −	+ −	+ −	+ −	+ −	+						
12	Magnesium	+ −	+ −	+ −	+ −	+ −	+ −						

TABLE 3

Electron No.	Carbon				Electron No.	Nitrogen				Electron No.	Oxygen			
	n	l	m_l	m_s		n	l	m_l	m_s		n	l	m_l	m_s
1	1	0	0	+	1	1	0	0	+	1	1	0	0	+
2	1	0	0	−	2	1	0	0	−	2	1	0	0	−
3	2	0	0	+	3	2	0	0	+	3	2	0	0	+
4	2	0	0	−	4	2	0	0	−	4	2	0	0	−
5	2	1	−1	+	5	2	1	−1	+	5	2	1	−1	+
6	2	1	0	+	6	2	1	0	+	6	2	1	0	+
					7	2	1	1	+	7	2	1	1	+
										8	2	1	−1	−

we begin to finish off these partially filled orbitals by adding a second electron with opposite m_s. This detail turns out to be important for understanding bond directions, and for the geometrical aspects of atomic structure. We shall discuss this in Chapter 16.

ATOMIC NUMBERS 5 THROUGH 10

In our discussion of boron (atomic number 5), we introduced the notion of nonspherically symmetrical orbits. We can extend this notion for increasing atomic numbers, but since we are describing conceptual models of atoms we should look ahead to see how much use we can make of the idea of such orbitals.

The Pauli exclusion principle shows that we must go from $n = 2$ to $n = 3$ as we pass atomic number 10. At this point, there are no more allowable values for l, m_l, and m_s for combination with n = 1 or $n = 2$ without giving electrons the same four quantum numbers, which the exclusion principle will not permit. Therefore, we should like our conceptual model to exhaust the supply of possible nonsymmetrical orbits when they have been built through neon.

It turns out to be easy to establish this, when we observe that we are concerned with *six* elements—boron, carbon, nitrogen, oxygen, fluorine, and neon. According to the exclusion principle, these are the first six elements (consult Table 2) which have $l \neq 0$; and, as a group, they must have *three* possible values of m_l; −1, 0, +1. Moreover, each atomic model in the list of boron through neon has one additional electron, which must have one of two possible values of m_s;

$+\frac{1}{2}$, $-\frac{1}{2}$. So the three values of m_l and the two values of m_s suggest that the number of elements, six, should be related to three and two. It takes little imagination to fix on the arithmetical relation $6 = 3 \times 2$ for the models to reflect. (6 is the total number of orbitals with different values of m_l and m_s.)

This exercise may seem more like numerology than science. However, we find that this sort of thinking builds useful and consistent conceptual models, and many of the seemingly arbitrary or merely convenient assumptions can be derived from energy formulations by strict mathematical techniques provided in quantum mechanics.

The six diagrams in Figure 18 illustrate models of the atoms boron through neon. The x, y, and z axes are mutually perpendicular, as in our ordinary three-dimensional world. Only for boron are the spheres for $n = 1$, $l = 0$ and $n = 2$, $l = 0$ shown. These spheres are to be understood in the other five cases but they are omitted from the figures for simplification. The $l = 1$ orbitals are all dumbbell shaped. The $l = 1$ orbital for boron is oriented in a certain direction, which we designate as, say, the x axis. In building the next atom, carbon, the electron enters a new $n = 2$, $l = 1$ dumbbell-shaped orbital, whose axis is perpendicular to the one already present; for concreteness, we can call this the y axis. In like manner, the next electron, too, enters a new $n = 2$, $l = 1$ dumbbell-shaped orbital, whose axis is perpendicular to both of those already present. We have thus added three dumbbell-shaped orbitals oriented in three mutually perpendicular directions; this takes care of the "3" part of the arithmetic statement $6 = 3 \times 2$.

As we move on to oxygen, fluorine, and neon, we add successively an electron to the x dumbbell, to the y dumbbell, and to the z dumbbell. Since each of these orbitals is already occupied by an electron with, say, positive spin, the additions must all be made with electrons having negative spin. The double lines used in some of the orbitals for oxygen and fluorine and in all of the orbitals for neon indicate that both the $m_s = +\frac{1}{2}$ and $m_s = -\frac{1}{2}$ electrons are present; no further electrons can be added to orbitals where there is a double line.

Before leaving our discussion of these six atoms, we shall mention the more common isotopes of each. This will help set the stage for discussing radioactive dating in Chapter 17. Boron, in addition to its five protons, as a rule has either five or six neutrons in its nucleus. This gives boron an atomic mass of either ten or eleven units: B^{10} or B^{11}. The superscripts show the atomic masses of the atoms. Carbon occurs most commonly as C^{12}, but there is also the stable isotope C^{13};

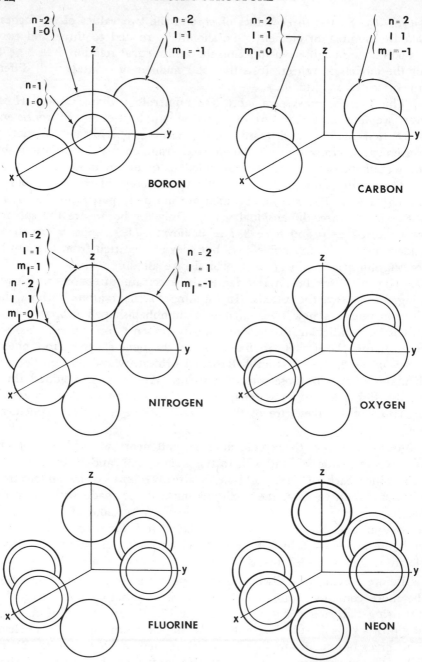

Figure 18

C^{14} is a radioactive isotope. Since C^{14} is produced when cosmic radiation hits the atmosphere, and behaves in plants and animals just as does stable carbon-12, except that it is radioactive, we shall find that this isotope is very important for dating organic materials.

The stable isotopes of nitrogen are N^{14} and N^{15}; those of oxygen are O^{16}, O^{17}, and O^{18}. Fluorine has one stable isotope, F^{19}, while neon has Ne^{20}, Ne^{21}, and Ne^{22}. (The massive protons and neutrons contribute such a large proportion of the weight of the atom that the electrons can be ignored. Each isotope of an element will have a different atomic weight (or mass) but the same atomic number.)

SODIUM ATOM

The next atom after neon, as we move to higher atomic numbers, is sodium. We want to use it as an example later, so it is included as the final atom whose structure we discuss in such detail. The first ten electrons around the sodium nucleus will be arranged as in neon. It takes one more electron for electric neutrality, and all the quantum number possibilities for n = 2 are filled; therefore, for the eleventh electron in sodium, n = 3. The other three quantum numbers must then be $l = 0$, $m_l = 0$, and $m_s = +\frac{1}{2}$. The number $l = 0$, as we have indicated, corresponds to a spherical shell. The eleventh electron, then, is found in a spherical shell which surrounds a kernel whose outer structure is essentially the same as that of the neon atom. As far as the outer electronic structure is considered, sodium, then, looks much like lithium. That is, in each there is one electron in the outermost spherical shell. This gives sodium and lithium a great similarity in their chemical properties, and for this reason they are classed as members of the same family or group (column) in the periodic table of the elements. In fact, the periodic table is so constructed (though at the time it was worked out, no one knew this) that elements with structures similar to one another (and hence with similar chemical properties) occupy the same columns.

We have discussed the structures of the atoms having atomic numbers 1 through 11. By applying the same methods, with a few complications, we could move on through the whole list of atoms. This would not serve our purpose, however, so we turn now to chemical bonds.

SUMMARY

$$F = k\ \frac{q_A q_B}{r^2}$$

> Coulomb's law for force between two point charges, q_A and q_B, separated by a distance r; k is a constant.

$$W_{1,2} = k q_A q_B \left(\frac{1}{r^2} - \frac{1}{r_1} \right)$$

> Work required to displace a point charge q_A by a distance $r_1 - r_2$ toward or away from another charge q_B. If the charges are of like sign, work must be done to move them together; work is gained when they are allowed to separate. If the charges are of opposite sign, the opposite relation holds.

$$W = k\ \frac{q_A q_B}{r}$$

> Work required to bring a change from infinity to a distance r from another charge of like sign.

$w = h\nu$ Energy (in Joules) is equal to frequency times Planck's constant, $h = 6.625 \times 10^{-34}$ Joule-sec.

$\lambda = \dfrac{c}{\nu}$ Wavelength of light equals speed of light divided by its frequency.

Pauli Exclusion Principle. No two electrons can occupy the state which would require that all four quantum numbers n, l, m_l, and m_s have identical values for both.

16

Chemical Bonds

What forces bind atoms together in a water molecule, a salt crystal, or a piece of wire? Each of these is bound by a different sort of bond, but, as we shall see, all the bonds depend on the electric forces represented in the atom by the charges on electrons and protons. The types of bonds are, respectively, the *covalent bond,* the *ionic bond,* and the *metallic bond.* Each type has different properties, which arise from the outer electronic structure of the atoms themselves. We shall discuss each type of bond, slighting the metallic bond, since metals are relatively unimportant in biology.

Before considering the bonds themselves, it is necessary to review atomic models and consider some properties of *closed shells.* The reason for this is that the atoms in combination seem to prefer combinations that give each a closed shell.

CLOSED SHELLS

In building the atoms from neutrons, protons, and electrons, going from atomic number 1 to higher atomic numbers, we found that the Pauli exclusion principle was very useful. The principle clearly indicates when we must change from $n = 1$ to $n = 2$, from $n = 2$ to $n = 3$, and so on. This is shown by Table 2 in Chapter 15. Reference to this table shows that hydrogen and helium *exhaust all the possibilities* for varying the quantum numbers of electrons for $n = 1$, and that lithium through neon employ all the allowed quantum number combinations for $n = 2$. Helium and neon, then, are said to have *closed shells,* because they contain a number of electrons that is the maximum allowable for some shell of a particular n. These atomic structures are extraordinarily stable and do not usually form molecules.

We know that both hydrogen and helium atoms are electrically neutral. It is possible, however, for one hydrogen atom to join with another, *if* they share the two electrons, and *if* the electrons have opposite spins. In a sense, we may then regard each hydrogen nucleus as having two electrons—its own, which it shares with the other nucleus, and another, which the other nucleus shares with it. Thus we may regard each of the hydrogen atoms as having reached closed-shell status.

Helium, however, already has two electrons with opposite spins in the shell around its nucleus. The Pauli exclusion principle forbids the occupation of the $n = 1$ shell by any other electron. Moreover, the spherically symmetric, negatively charged shell effectively shields the positive charges of the nucleus from the negative parts of other atoms (or from free electrons) so that there is no tendency for electrons to attach in an $n = 2$ shell. Therefore, helium is an inert gas. (We ignore here as inappropriate the attraction that helium atoms have for one another at very low temperatures. These attractions are chemically unimportant, and play a part only in the liquefaction of helium.)

We account for the inertness of neon in a similar fashion. The $n = 2$ shell is completely filled, both the symmetrical cloud and the three mutually perpendicular dumbbells, each with two electrons having opposite spins. This is enough to keep neon atoms from attaching to any other atoms. No further electrons can be added to the $n = 1$ or $n = 2$ shells, and the electrons present shield the positive charges in the nucleus and reject the approach of other electrons by both electric and spin repulsion.

Fluorine, on the other hand, is missing one electron from what would be a closed shell configuration like neon's. Two of the dumbbells have two electrons of opposite spin, but the third dumbbell has only one electron, so it can add an electron before it has achieved a closed-shell status.

We shall see in the rest of this chapter that the open and closed shells play an important part in determining the theory of how chemical bonds are formed.

THE COVALENT BOND

The simplest example of a covalent bond is, perhaps, the bond of the atoms of the hydrogen molecule, which has already been discussed. Here the two atoms are bound together by their continual sharing of two electrons. This electron sharing is the distinctive feature of the covalent bond.

The water molecule is another example of atoms bound together by a covalent bond. Our discussion of the oxygen molecule was summarized in part of Figure 18 of Chapter 15. There we saw that two of oxygen's dumbbells have only one electron each, while the third has two. The oxygen atom, then, needs two electrons to reach the closed-shell status. Figure 1 shows how hydrogen atoms, by sharing, can supply these electrons. The dotted circles in the y and z dumbbells indicate that these are now filled by sharing the electrons from the two hydrogen atoms, H. The dotted circles in the representations of the hydrogen atoms show that the hydrogen atoms now share the electrons which, before bonding, were the sole property of the y and z axis dumbbells of the oxygen atoms. Hence, the oxygen atom has reached its closed-shell status by having all six $n = 2$, $l = 1$ electrons; and both hydrogen atoms have reached their closed-shell state by having both $n = 1$, $l = 0$ electrons.

The portions of Figure 1 marked O, which represent only the oxygen atom, with dotted circles showing shared electrons in the two dumbbells, should show these circles greatly reduced in size. This end of the dumbbell (recall that these representations are simply different directions from which we can look at the atom) should become

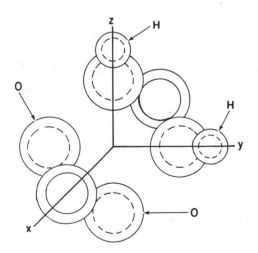

Figure 1

degenerate when an atom attaches by covalent bond to the other end. This is a complication, however, we need not consider further.

The model of the water molecule provides the first example of *bond directions*. Figure 1 shows that lines from the oxygen nucleus to each of the hydrogen nuclei should make an angle of 90°. It is possible to measure such angles, and in water molecules the angle between the bonds is 105°. This discrepancy is accounted for by noting that the mutual repulsion between the two hydrogen molecules might distort the angle. That such distortion occurs is evident from consideration of larger molecules, say H_2S, where the hydrogen atoms are held farther apart and the distortion is much less.

The carbon atom is extremely important for biology, since it is one of the basic building blocks of all organic matter. The carbon atom, according to our model, has two dumbbell-shaped orbitals, each with only one electron. Four more electrons are required to bring the carbon atom to closed-shell status. Thus, in a molecule, the carbon atom displays four covalent bonds. But in what directions are these bonds oriented? By a process known as hybridization, which we do not discuss here, the carbon atom converts the two $n = 1$, $l = 0$ electrons and the two $n = 1$, $l = 1$ electrons to four identical, half-filled orbitals, which make mutual angles at 109° 28′. Each of these half-filled orbitals can participate in a covalent bonding of the carbon atom to some other suitable atom. Again, for each covalent bond there will be two shared electrons, as was the case with water molecules. The way that these bonds occur in methane (CH_4), and in carbon tetrafluoride (CF_4) is shown in Figures 2*a* and 2*b*. Figure 2*c* shows that the four atoms joined to the carbon atom lie at the apices of a regular tetrahedron.

A given carbon atom can be bonded covalently to one, two, three, or four other carbon atoms. It is possible for carbon chains to be formed

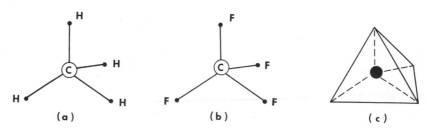

Figure 2

of thousands of carbon atoms joined together. When we deal with chemical combinations of particular interest to biochemistry, we see the formations of some such chains, and we see also that the carbon atoms so joined are also able to form bonds with other atoms, since carbon-carbon bonds need not give the atoms closed shells. Sometimes one carbon atom may be joined to another by a double or triple bond, sharing four of six electrons, respectively. Of course, when a double or triple bond exists, the bond angles will no longer be as they are shown in Figure 2.

THE IONIC BOND

The ionic bond is important for holding together molecules in some crystalline structures; it also holds "together" (loosely, as we shall see) ionic compounds dissolved in solution. As an example of an ionic bond, let us consider sodium fluoride, NaF. Recall that the sodium atom has a single electron in the $n = 3$ shell. This shell lies outside a neon-like $n = 2$ closed shell, which constitutes a sort of kernel around which the lone electron circulates. The ten electrons making up the $n = 1$ and $n = 2$ shells almost shield the eleventh electron from the attraction exerted by the eleven protons in the nucleus. As a result, the eleventh electron is rather easily lost by the atom. We then have a fragment of the atom, composed in this case of a nucleus with a charge of $+11$ and ten $n = 1$ and $n = 2$ electrons in shells around the nucleus. Such a fragment is known, as has been remarked earlier, as an *ion,* in this case a sodium ion, Na^+. Flourine, on the other hand, lacks only one electron of having a neon-like closed shell. It can easily attach an electron with proper spin to achieve the closed-shell status. It then is a negative ion and can be written F^-. (We shall see later in this chapter why ions are important for understanding fundamental material inter-actions.)

Since the sodium ion is positively charged and the fluorine ion nega-tively charged, they exert a mutual attraction on one another. Since each of these ions has a high degree of symmetry, each tends to surround itself with as many ions of opposite charge as possible. In the case of Na^+ and F^- this turns out to be six. The crystalline form is cubic and is illustrated in Figure 3. This crystal lattice has a high degree of organization, so high that (as with most crystals) when it is stressed it tends to fracture, rather than to bend.

In solution the sodium and fluoride ions become somewhat more mobile. They are free to move with respect to one another, if one ion

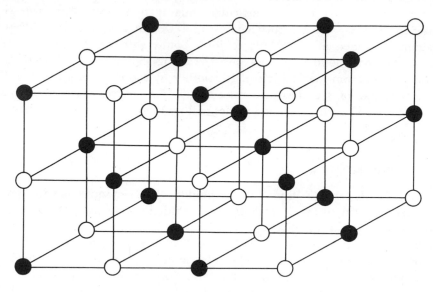

Figure 3

never moves far from those of opposite sign. Thus, if two electrodes connected to a battery (or other external circuit) are inserted into a solution of sodium fluoride (or other ionic solutions) a current will flow through the solution. The Na$^+$ ions will carry positive charge in the direction of the current, and F$^-$ ions will carry negative charge in the opposite direction. (The directions of current flow are based on a convention which is itself based on an early mistake; namely, that only positive charges "flowed." Still, it is useful for everyone to agree on a direction; hence current flow is defined—by convention only— as the direction of the movement of the *positive* charges; in metallic conductors which are conducting nonionic currents of electrons, however, it is actually the negatively charged electrons that move). In currents set up by the motion of ions in solution, the ions of opposite sign move past one another somewhat like the men and women in "right- and-left-grand" at a square dance.

Not all crystals are held by ionic bonds, but many are. The essen- tial feature of an ionic bond is that electrons are not shared; atoms that become positive ions really *give up* electrons to other atoms that then become negative ions.

THE METALLIC BOND

Since sodium has one $n = 3$, $l = 0$, $m_l = 0$ electron, we could give it a closed subshell by adding an electron of opposite spin. This can be done by letting it share electrons with another sodium atom. What happens then is similar to the events which occur when two hydrogen atoms join into a hydrogen molecule. The difference is that the two-electron cloud does not so nicely enclose the nuclei. The joining of two sodium atoms into a sodium molecule is illustrated at the top of Figure 4. The whole assembly is electrically neutral, but the ends are positive and the middle negative. The positive ends of one pair then tend to "stick to" the middles of other pairs, as is illustrated by the

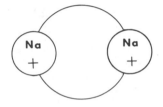

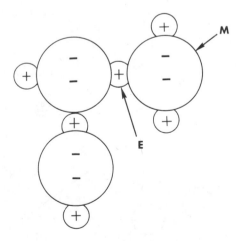

Figure 4

bottom of Figure 4. Thus sodium metal will be an aggregate of such pairs.

In general, there is little to keep the end E in Figure 4 from moving to a different position around the middle $M;$ this is to say that the directional properties of the metallic bond are not nearly as well defined as those of the other bonds. This property of the model of the structure of a metal thus can be made to account for the ductility and malleability of metals.

We have here examined only one simple case to illustrate the nature of metallic bonds. They are of little interest for elementary biology, so we shall not consider them further.

WALLS AND SURFACES

In Chapter 14 we noted that kinetic behavior of molecules in gases could not be regarded as entirely analogous to interactions among perfectly elastic billiard balls, nor could interactions between molecules and container or liquid surfaces be regarded as being like those between a billiard ball and a macroscopic wall. Now we shall begin to see why.

In microscopic detail, the surface of a solid will look something like Figure 5; the white circles represent atoms (or molecules) of the material constituting the walls of a vessel. These atoms (or molecules) are held together by electric attraction in the case of atoms or by a bond in the case of molecules. The black dot represents a gas molecule moving toward the wall. As it moves into the neighborhood of the atoms constituting the wall, it feels strong electric repulsions, which deflect it back into the interior of the vessel. The collision cannot

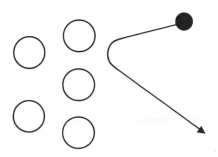

Figure 5

be thought of as a miniature replica of the collision between billiard balls, but only as an interaction of electric fields as the approach progresses.

Similarly, the body and surface of a liquid must be thought of as clumps of sluggishly colliding molecules. An energetic molecule moving upward very near the surface of the liquid may begin to move away from the molecules constituting the liquid's body. However, the same attractive electric fields which hold most of the molecules in the liquid state will slow down, and possibly pull back, the escaping molecule. Again, at a molecular level, we cannot think of the surface as a geometric plane; the retention of the energetic molecules must be accounted for by interactions of fields.

THE BOHR MODEL: A SIMPLER REPRESENTATION

Our discussion of the structure of atoms and molecules has utilized pictures showing them as composed of fuzzy spheres, fuzzy shells, dumbbells, and tetrahedra. These are hard to draw, especially if we attempt

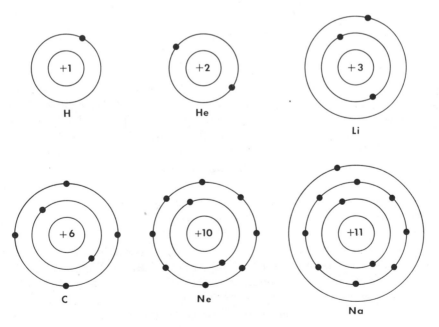

Figure 6

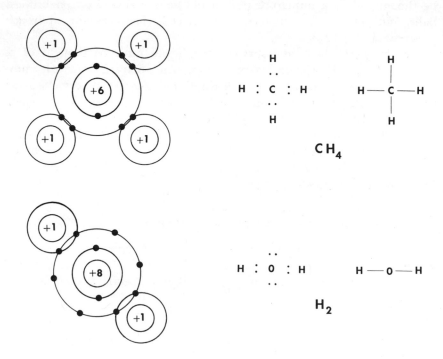

Figure 7

to make two-dimensional figures try to show three dimensions. For many purposes, especially in biology, earlier models of atoms and molecules (especially those conceived by Niels Bohr in 1913) will be adequate. Figure 6 shows six atoms arranged according to Bohr's scheme. Electrons move in planetary orbits around their nuclei. The first orbit can take only one or two electrons; it is filled for helium, which, as we have noted, has two electrons in its $n = 1$ shell. As the atoms are built up, electrons 3 through 10 go into the second orbit; this orbit when it is filled contains eight electrons. The eleventh electron, as in sodium, goes into the third orbit.

The top half of Figure 7 represents methane (CH_4) in three different ways, based on Bohr's model. The bottom half of Figure 7 depicts water by the same three methods. Note that in each half the second and third methods subsume the first or inner orbit in the letters themselves. Note also that there is no pretense in the figures of representing a three-dimensional arrangement.

STRUCTURES AND PROCESSES

Several common chemical processes are of interest to the biologist. Two important classes of these are *oxidation-reduction* reactions and *acid-base* reactions. Having discussed some models of atoms and the ways in which they form bonds, we can now look at some samples of these processes to see how they can be interpreted in terms of our models.

Compounds of oxygen with metals are generally called *oxides*. Examples are sodium oxide (Na_2O), magnesium oxide (MgO), and cupric oxide (CuO). The processes by which oxides are formed are classed as *oxidations*. Burning hydrogen with oxygen is also an oxidation. The idea may be extended, as we shall see, to reactions not involving oxygen at all. These reactions are written in equation form showing atoms on the left, rather than molecules, for we wish to consider what happens, according to our models, when one atom combines with another of a different sort. It is most usual to show atoms on the right, molecules on the left in depicting such reactions as we now turn to.

$$2Na + O \rightarrow Na_2O \qquad (1)$$

$$Mg + O \rightarrow MgO \qquad (2)$$

$$2H + O \rightarrow H_2O \qquad (3)$$

$$Na + Cl \rightarrow NaCl \qquad (4)$$

The outer shell ($n = 3$) of the magnesium (Mg) atom looks in all essential respects like that of beryllium, and that of chlorine (Cl) like that of fluorine. The equations could equally well be written with Be replacing Mg and F replacing Cl. The bond formed in reaction 1 is ionic. Each sodium atom gives up one electron to the oxygen atom. Each sodium ion then has a neon-like closed shell for its outermost electronic structure, and the oxygen atom, having gained two electrons, becomes an oxygen ion, also with a neon-like closed shell. Reaction 2 also forms an ionic bond, but in this case both electrons come from a single magnesium atom. These two reactions may then be written as

$$2Na + O \rightarrow 2Na^+ + O^= \qquad (1')$$

$$Mg + O \rightarrow Mg^{++} + O^= \qquad (2')$$

in order to show the ionic constitution of the compounds.

When water is formed by reaction 3 a covalent bond is established. In such a bond, electrons are *shared*. However, the sharing is not quite symmetrical; the oxygen atom has a strong affinity for the shared electrons and pulls them close to it, so that in effect the hydrogen atoms *donate* electrons, which were solely their property, to the bond, and the oxygen *accepts* these electrons. We can therefore write reaction 3 after the fashion of 1' and 2', but we must realize that *ions are not produced*. Thus we obtain

$$2H + O \rightarrow 2H^+ + O^=. \tag{3'}$$

In these reactions oxygen is the *oxidizing agent;* its characteristic is that it *gains* electrons during the process of oxidation. Sodium, magnesium, and hydrogen are called *reducing agents*. The numbers of electrons attached to the sodium, magnesium, or hydrogen nuclei are diminished. In oxidation reactions involving these elements as reducing agents, their characteristic is that they *lose* electrons. Whenever there is oxidation, there is always an accompanying reduction. In the reactions cited sodium, magnesium, and hydrogen are oxidized, while oxygen is reduced.

What about a reaction like 4? Here, sodium gives up an electron and becomes an ion with a neon-like closed shell, while chlorine gains an electron to become an ion with an argon-like shell. Reaction 4 may then be written

$$Na + Cl \rightarrow Na^+ + Cl^-. \tag{4'}$$

Chlorine here plays the same role as oxygen in 1', 2', and 3', so chlorine, too, is said to be an oxidizing agent.

Reaction 4' forms an ionic bond, but this is irrelevant for classing it as an oxidation-reduction reaction. The important characteristic of oxidation-reduction reactions is the source and destination of the electrons, not the particular type of transaction.

The same elements may under some circumstances be oxidizing agents and under other circumstances reducing agents. For example, consider

$$2H + S \rightarrow 2H^+ + S^= \tag{5'}$$

and

$$S + 2O \rightarrow S^{4+} + 2O^=. \tag{6'}$$

Both hydrogen sulfide and sulfur dioxide are covalent compounds. In reaction 5' sulfur oxidizes hydrogen, gaining two electrons, one from each hydrogen atom, and is itself reduced. In reaction 6' sulfur reduces oxygen, giving up four electrons, two from each oxygen atom, and is itself oxidized.

VALENCE

If we consider the assignment of the plus and minus superscripts on the symbols to the right in the chemical equations 1' through 6', we can construct the following table:

$$H = +1 \qquad O = -2$$

$$Na = +1 \qquad Cl = -1$$

$$Mg = +2 \qquad S = +4 \text{ or } -2$$

These numbers, and their signs, are called the *valences* of the atoms.

Magnesium, with a valence of $+2$, is an atom which can achieve a closed-shell status by *losing* 2 electrons; it then becomes an ion with a charge of $+2$. Chlorine, by *gaining* one electron, can reach a closed-shell status and will then be an ion with a charge of -1. Sulfur can obtain a closed-shell status either by gaining two electrons or by losing four.

Although we could explain the assignments of valences by referring back to our discussion of atomic states and quantum numbers, it is convenient to refer to the Bohr atom. We may note that for the simpler atoms closed-shell status is obtained with two electrons in the $n = 1$ shell and eight electrons in the $n = 2$ shell. However, closed-shell status for more complicated atoms becomes less easy to determine by the Bohr model (the modifications required led to its eventual replacement), since an electron goes into the available space of lowest energy content. Thus the two electrons that sulfur gains go into the $n = 3$, $l = 0$ shell, to give it closed-shell status; but it can also lose four electrons from the $n = 3$, $l = 1$ dumbbell orbitals for closed-shell status.

The concept of valence is not simple; the discussion has emphasized the easiest cases, those exhibiting the most regularity. Nitrogen and oxygen form five different compounds, however, displaying in these combinations several different valences. The interested reader who wishes to go beyond the necessarily abbreviated discussion presented here will find a more complete explanation in any moden college text on general inorganic chemistry.

ACIDS AND BASES

Acids are ionic compounds that contain a hydrogen ion. Some simple acids are H^+Cl^-, H^+F^-, and H^+I^-, which are known as hydrochloric acid, hydrofluoric acid, and hydroiodic acid. An example of more complexity is nitric acid: $H^+(NO_3)^-$. In this formula the symbol $(NO_3)^-$ is called a *radical* and is an ion. The three oxygen atoms are held to the nitrogen atom by covalent bonds. This satisfies all the unfilled dumbbell-shaped orbitals except for one in one of the oxygen atoms. This one unfilled orbital can then capture an electron to make the whole a negatively charged ion. Sometimes this ion is represented by $(ONO_2)^-$ to indicate the differences in the roles the oxygen atoms can play in reactions of the nitrate radical with other substances.

Bases are also ionic compounds; they always involve a metal and a radical ion compounded of one atom of hydrogen and one atom of oxygen. These definitions require some modification, especially for organic chemistry. A more rigorous and modern treatment defines acids as substances which in some reactions can donate hydrogen ions, and bases as substances which can accept them. Examples of bases are $Li^+(OH)^-$ and $Mg^{++}(OH)_2^-$. The oxygen atom going into the $(OH)^-$ radical has two half-filled dumbbell-shaped orbitals. By covalent bond with the hydrogen atom one of these orbitals is filled. The other half-filled orbital captures an electron, so the complex becomes a negative ion.

Acids always react with bases to form a salt and water. The term "salt" here refers to a whole class of ionic compounds, one of which is sodium chloride or ordinary table salt. Typical reactions of acids and bases are shown in equations 7–10; by convention, the acid is always given first, on the left side of the equation, the base next, the salt first on the right side, and the water last (any other byproducts of the reaction usually precede the water).

$$H^+F^- + Li^+(OH)^- \rightarrow Li^+F^- + H_2O \qquad (7)$$

$$H^+Cl^- + Na^+(OH)^- \rightarrow Na^+Cl^- + H_2O \qquad (8)$$

$$H^+(NO_3)^- + Na^+(OH)^- \rightarrow Na^+(NO_3)^- + H_2O \qquad (9)$$

$$(H^+)_2(SO_4)^= + Mg^{++}[(OH)^-]_2 \rightarrow Mg^{++}(SO_4)^= + 2H_2O \qquad (10)$$

In each of these reactions we see that there is a reaction of some quantity of hydrogen and hydroxyl radical involved which gives $H^+ +$ $(OH)^- \rightarrow H_2O$. This is characteristic of *neutralization*, a special name given to the reaction between acids and bases. When the acid and the base entering into such a reaction are supplied in the proper proportions, all the H^+ ions can combine with the $(OH)^-$ ions and become covalent water molecules. Since there are no longer any H^+ ions, the mixture will not display acidic properties; conversely, since there are no longer any $(OH)^-$ ions, the mixture is no longer basic. We say when this has taken place that the acid and the base have neutralized each other.

Usually if the salt is soluble and enough water is present (or is formed in the reaction), the salt occurs only as the ions that compose it. As the water is evaporated, however, the salt will precipitate in crystalline form. Thus, form the reactions depicted in equations 7–10, we would obtain crystals of LiF, $NaNO_3$ and $MgSO_4$, respectively.

We know that energy is often liberated or absorbed by the chemical structures of the atoms (or molecules) that combine to form new molecules or ions, or in dissociations of molecules. When energy is *liberated,* it appears as heat, and we call the reaction *exothermic.* When heat is *absorbed,* we call the reaction endothermic.

CHAPTER
17

Dating by Radioactive Isotopes

OLDER METHODS

The history of the earth and the time scale of the many events associated with it are fascinating subjects. Geochronology, the dating of these events, has developed rapidly over the last few decades, largely because of the introduction of radioactive methods of dating. Some of the basic principles on which these methods are based have been touched on in Chapters 15 and 16. We shall deal with them more specifically here. Other methods have also been used, and we shall mention them, although only briefly. In general, these latter methods were not especially useful, or had only a very rough accuracy, when used to date such objects as the biologist might be most interested in—organic substances and organisms of relatively recent origin. Hence the new methods have been a great boon to biologists interested in evolutionary development.

Estimates of the passage of time taken from effects of natural processes on geological formations can be made by methods based on erosion and sedimentation rates. For example, the first estimates of the age of the earth were based on the salt content of the ocean and the amount of sediments on its floor. The present annual input of salt and sediments from rivers is known approximately, so that the time necessary to bring the ocean to its present salt content and to accumulate the amount of sediments observed can be obtained by simple division. This method gives an age for the earth (really, for the oceans) of several hundred million years.

This figure—and the method by which it was reached—is based on the assumption that the present rate of deposit of salts and sediments in the oceans was also the rate in the past. We know that this is not true. Since the last glaciation land has eroded considerably faster than it did in earlier times. Hence the age of several million years is far too low a figure, but we cannot determine by this method just how much too low it is. Radio-dating methods give us a figure of several *thousand* million years for the earth's age, which indicates how far in error the geological figure may be.

Generally, geological methods involve doubtful assumptions about the resemblance of the past and the present. Hence the figures obtained may have a *relative* value—they may be able to tell us, for example, which of two comparable areas or objects is older or younger—but they do not yield dependable absolute values.

Two methods of dating do give absolute ages without using radio-activity. One is dendrochronology, the analysis of tree rings. This method can give absolute dates in areas of the world where there are large differ-ences between summer and winter rainfall, so that the trees produce one definite ring of growth per year. However, this method is extremely limited. It is applicable only to the last few thousand years, and has been widely used only in the southwestern United States.

The second nonradioactive method for determining absolute dates involves counting "varves." These are annual layers of sediment in lakes deposited there by water from summer glacier melt. This method can yield absolute figures as the use of total sediment in oceans cannot, for the varves may vary widely in thickness from year to year. Hence a series of varves will display a characteristic pattern. This pattern can be identi-fied wherever it appears, and the cross-identification it makes possible en-ables us to combine data from different areas. This method, however, like dendrochronology, is limited in its usefulness. It can be used only for dating over the ten thousand years since the last glaciation.

By contrast to these methods, radiodating has two outstanding ad-vantages. In the first place, the method in general can be applied to a vastly greater span of time. No one radioactive substance gives us data for the whole period in which we are interested. One substance may be useful for a time period of only a few hundred years, another for tens of thousands of years, and another for millions of years. Taken all together, however, the radioactive elements we have learned to work with yield data applicable over many years of geological time.

The second great advantage of radiodating is that the rate of the process on which it depends—radioactive *decay* of certain elements—is unaffected by any external factor that could have come about on earth since the element was formed. The rate of decay is a characteristic of the structure of matter itself, and of the particular structure of the radioactive atoms employed. That is, uranium 238, radium 226, and carbon 14, are each substances with a characteristic rate of decay. The relatively rapid rate of decay of C^{14} is a consequence of its atomic structure and according to some of the most fundamental physical theories was the same when life first began as it is today. The slow rate of decay of U^{238} is likewise a function of the structure of its substance and independent of external conditions. All radioactive substances have similarly constant rates of decay.

The rate of decay of a radioactive isotope can be expressed in several ways. It is helpful to think of it first as the tendency of each atom of a radioisotope to change by ejecting some of the particles of which it is composed. (This way of thinking is not quite accurate—as we shall see, the atom actually ejects from its nucleus particles of which it is not "composed" in any ordinary sense.) The idea of a "tendency" can be made more definite by considering the tendency as a probability. That is, given a large number of atoms of some radioisotope, the tendency toward decay is of such magnitude that a certain proportion of the atoms will decay in a given period of time. We could express rate of decay as the proportion of atoms originally present which will decay in some chosen unit of time.

More common and more useful is an inversion of this expression. Instead of indicating what proportion of the atoms of a radioisotope will decay in a given time, we choose some convenient fraction of them and state the probable time required for that fraction of them to decay. The fraction which has proven most convenient is one-half, and this choice of fraction has given us the familiar expression "half-life." "Half-life" is simply the period of time during which it is probable that half of the atoms present in a mass of radioactive material of a certain element will have ejected a particle, thus decaying. Note that this has nothing to do with the number of atoms present to begin with, nor with the amount of the substance. Thus, if the half-life of some element is thirty years, then, at the end of thirty years, one-half of the original number of atoms will remain. Of these remaining atoms, one-half will decay in the *next* thirty years, and so on.

The constant rate of radioactive decay stands in marked contrast to the variable or uncertain rates characteristic of other processes on which dating methods have been based. It is this constancy which gives radiodating its dependability. Nevertheless, as we shall see, our methods of measurement and our ways of interpreting data depend for their reliability on more than the assumption of constant rates of radioactive decay. Radioactive methods, like the geological methods, rest on assumptions about geological conditions and geological history. Hence, the results of particular methods of radiodating have factors leading to inaccuracy and doubt, as do other methods. These will be mentioned in descriptions of particular methods.

ISOTOPES AND RADIOACTIVE DECAY

Let us begin with a brief examination of the behavior of radioactive isotopes, for it is this behavior on which radiodating depends. You may want to review Chapters 15 and 16 first.

An isotope is a particular form of a chemical element having the same chemical properties, but a different atomic weight, because it has more or fewer neutrons in the nucleus. A *radioactive* isotope is one whose atoms tend to eject one or another of three particles (there is also a form of "decay" by capture, which we shall not go into here). Some atoms eject *alpha* (α) particles, which are identical with helium nuclei, having two protons bound together. Others eject *beta* (β) particles, which are identical with electrons. Some eject *gamma* (γ) particles, x-rays of varying energies. Combinations are also possible, and some decays involve ejections of other particles, which we shall not discuss here.

Alpha and beta decay are the kinds that will concern us, for with the loss of these particles the atom is changed in an easily identifiable way. It becomes an isotope of another element. The reason for this is that as we have seen the chemical properties of atoms depend on their electronic structure—that is, their charged components are the important determinants. When an atom ejects an alpha particle, it is effectively getting rid of two protons, and diminishing the number of its nuclear charges by two. It thus becomes an isotope of whatever element has an atomic number corresponding to the new atomic number of the decayed atom. Similarly, when it ejects an electron from the nucleus, this may be conveniently thought of as the decay of a neutron into a proton plus the ejected electron (though this is not quite what happens). Hence the ejection leaves the atom with one more positive charge in its nucleus than before, and it becomes an isotope of the element whose atomic number is one greater than that of the original atom.

The isotope resulting from alpha or beta decay may also be radioactive. In that case its atoms will also tend to eject an alpha or beta particle, resulting in a change to yet another isotope. Eventually, this series of changes ends in a stable isotope, one which is not radioactive.

The process of transformation from one isotope to another provides the basis for radiodating. We measure the amount of the original radioactive (parent) isotope. We also measure the amount of stable, end product (daughter) isotope which is present. From these two figures we can calculate the time elapsed since the parent isotope became lodged in the sample studied. This is possible because the *rate* at which the daughter isotope has been formed is a constant rate.

This is the basic principle of radiodating; note that it involves certain assumptions. We have to assume that: None of the daughter isotope from the parent has been lost from the sample; none of the daughter isotope was incorporated into the sample from other sources; there has been no loss (except by decay) or gain of the parent isotope since the sample was formed. These assumptions render radioactive dating less than infallible. However, in applications of the basic principle, there may be ways of checking the reliability of any necessary assumptions. In such cases, there can be an estimate of the accuracy of the dating and of the confidence we can place in it.

A number of variants of this basic method are used in radiodating. The best-known method, that using C^{14}, is a case in point. This method can determine the time which has passed since the *death* of a living organism. In this method we assume that a *living* thing—whenever it lived—always contained the same proportion of radioactive C^{14} to non-radioactive carbon. We also assume that after death radioactive C^{14} continues to decay to nonradioactive forms at its regular rate, but that no further radioactive C^{14} enters the dead body from the atmosphere. Thus, the less radioactive C^{14} we find, the longer the time since the sample was part of a living organism.

These examples show how applications of the general method of radiodating involve further assumptions. In the C^{14} method we must assume that the relative amount of C^{14} to other forms of carbon available to living organisms has been the same throughout the period in which life has existed on the earth. This requires not only a constant decay rate for C^{14} in the atmosphere and waters of the earth, but also a constant replenishment of it by bombardment of stable carbon particles in the upper atmosphere by cosmic rays. The former assumption, we have noted, is a consequence of fundamental and well-established physical theory. The latter has more complex origins and is less firmly established.

Using similar modifications of the general method, three major classes of radioisotopes are used for dating purposes:

1. Isotopes that have existed since the formation of the earth, for example, uranium, thorium, potassium, and rubidium. Obviously, their half-lives must be long—thousands of millions of years—if some of the isotope is still present.

2. Isotopes that are produced by the decay of long-lived parent isotopes, for example, ionium, radium.

3. Isotopes that are being produced continuously by cosmic rays entering the atmosphere, for example, C^{14}, tritium (hydrogen with two neutrons, as well as a proton, in its nuclei).

Each of the dating methods we shall discuss is applicable to only certain types of materials and over certain time periods. There is no universal dating method for any material over all time ranges.

Let us review briefly the basic concept and notations of atomic structure which will be involved in an explanation of radiodating. An atom consists of a dense central nucleus and surrounding electrons. Only the nucleus will concern us here. Recall from our discussion in Chapter 15 that the nucleus is formed of positively charged and neutral particles, which comprise most of the mass of the atom. The number of these nuclear particles (nucleons) characteristic of a kind of atom is called its atomic weight; variations in the weight because of the presence of more or fewer neutrons result in isotopes of the same element. To express the name of a kind of element two symbols are used. The abbreviation of the atom's name is taken from classical chemistry and is given a superscript to indicate its atomic weight. Thus the radioactive isotope of carbon that has fourteen nucleons is written C^{14}, and the nonradioactive isotopes with twelve and thirteen nucleons are written C^{12} and C^{13}.

In addition to the total number of nucleons characteristic of a kind of atom, its characteristic number of protons is also important, since this number determines the chemical characteristics of the atom in question. It is also a measure of the units of positive charge in the nucleus. To provide this information, the appropriate number is given as a subscript *before* the traditional chemical symbol of the element. Thus carbon is written $_6C$ to indicate that it has six protons. Similarly, we can write $_{26}Fe$ (iron), $_{90}Th$ (thorium), and $_{92}U$ (uranium).

Thus, the complete symbol for an isotope would contain the symbol for the element of which it is an isotope, the subscript for the atomic number of that element, and the superscript for its atomic weight. The four isotopes of carbon would be written

$_6C^{11}$—radioactive, half-life 20 minutes

$_6C^{12}$—stable

$_6C^{13}$—stable

$_6C^{14}$—radioactive, half-life 5600 years .

We can use this notation to be more explicit about the atomic events that hide behind the word "decay." We have noted that some radioisotopes emit alpha particles and others emit beta particles. An alpha particle is a helium atom's nucleus. Using this notation it may be written $_2He^4$. With this notation we can write a model of a "decay" event in which an *alpha* particle is emitted. We use the familiar radioactive isotope U^{238} as our example:

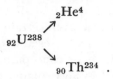

This says that by the emission of an alpha particle an atom of U^{238} loses a total of four nucleons, thus reducing its atomic weight by 4 to 234. Two of these nucleons are protons; hence the atomic number which indicates the chemical properties of uranium (92) is diminished by 2 to the number (90) characteristic of the properties of thorium.

In *beta decay,* an electron is emitted from one of the neutrons in the atom, and a neutron in the nucleus becomes a proton, with a positive charge, instead of no charge. Such an event, using as an example the Th^{234} which arose from the alpha decay described, would be written:

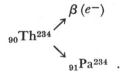

Notice that there is no change of atomic *weight* in this second decay, **as** there was in the alpha decay, since there has been *no loss of nucleons.* But there has been an *increase of one* in the number of protons, *the atomic number,* since one of the neutrons was transformed into a proton. Since the number of protons determines the chemical properties, and the chemical properties determine the classification of a substance as one or another element, the product of this decay is an isotope of another element, protactinium. An incidental point to remember is that a neutron should not be thought of as a proton and and electron stuck together. The transformation of a neutron into a proton (or the conversion of a proton into a neutron) is a genuine case of decay, although the language we have available to describe the process (ejection, emission) is somewhat misleading. The electron in beta decay "comes" from the necessity of charge conservation, which is derived from the fundamental energy conservation principle. As you can see in examining the schematized decay of thorium, if no electron were "emitted," a positive charge would exist where **none** existed before. The electron's existence is necessitated by charge conservation.

The first two steps of the uranium decay chain taken together underline a point made earlier: that the offspring of one decay may also be radioactive and decay again into still a third isotope. There may be many such intermediate radioactive steps between the first radioactive isotope and a stable daughter. In such cases we speak of a radioactive series. We shall examine three such series briefly.

Table 1 indicates that the decay process may produce the same element (in different isotopes) more than once. In Table 1 we see this feature of some decay chains in the repetition of uranium ($_{92}U^{238}$, $_{92}U^{234}$), polonium ($_{84}Po^{218}$, $_{84}Po^{214}$, $_{84}Po^{210}$), and lead ($_{82}Pb^{214}$, $_{82}Pb^{206}$). Atomic *number,* the preceding subscripts, rather than atomic *weight,* identifies the element.

Note that Table 2 shows a decay to stability of a different isotope of uranium than is shown in Table 1. In the first three steps, the elements produced by the decays are the same (though different isotopes) as in the U^{238} series; then the decay chain goes to different elements beginning with actinium, $_{89}Ac^{227}$. In Table 3 we may note that different isotopes of the same element may have widely varying half-lives. Compare the half-life of Th^{234} (24.1 days) from Table 1 with that of Th^{232} (about 10.5 billion years) from Table 3. This may suggest some difficulties for radiodating which will be discussed in more detail shortly.

LEAD ISOTOPE METHODS OF DATING

Methods of radiodating that use lead isotopes depend on the decay of uranium and thorium, radioactive elements present in the earth since

TABLE 1. *The Uranium (U^{238}) Series*

Element	Isotope	Particle Emitted	Half-Life	
Uranium	$_{92}U^{238}$	α	4.50×10^9	years
Thorium	$_{90}Th^{234}$	β, γ	24.1	days
Protactinium	$_{91}Pa^{234}$	β, γ	1.17	minutes
Uranium	$_{92}U^{234}$	α	2.47×10^5	years
Ionium	$_{90}Io^{230}$	α	8.0×10^4	years
Radium	$_{88}Ra^{226}$	α, γ	1620	years
Radon	$_{86}Rn^{222}$	α	3.82	days
Polonium	$_{84}Po^{218}$	α	3.05	minutes
Lead	$_{82}Pb^{214}$	β, γ	26.8	minutes
Bismuth	$_{83}Bi^{214}$	β, α, γ	19.7	minutes
Polonium	$_{84}Po^{214}$	α	1.6×10^{-4}	seconds
Tantalum	$_{81}Tl^{210}$	β	1.32	minutes
Lead	$_{82}Pb^{210}$	β, γ	22.5	years
Bismuth	$_{83}Bi^{210}$	β	5.0	days
Polonium	$_{84}Po^{210}$	α	138	days
Lead	$_{82}Pb^{206}$		Stable	

TABLE 2. *The Actinium (U^{235}) Series*

Element	Isotope	Particle Emitted	Half-Life	
Uranium	$_{92}U^{235}$	α	7.13×10^{8}	years
Thorium	$_{90}Th^{231}$	β	25.6	hours
Protactinium	$_{91}Pa^{231}$	α	3.43×10^{4}	years
Actinium	$_{89}Ac^{227}$	β	22.0	years
Thorium	$_{90}Th^{227}$	α	18.6	hours
Radium	$_{88}Ra^{223}$	α	11.2	days
Radon	$_{86}Rn^{219}$	α	3.92	seconds
Polonium	$_{84}Po^{215}$	α	1.83×10^{-3}	seconds
Lead	$_{82}Pb^{211}$	β	36.1	minutes
Bismuth	$_{83}Bi^{211}$	β, α	2.16	minutes
Polonium	$_{84}Po^{211}$	α	0.52	seconds
Tantalum	$_{81}Tl^{207}$	β	4.79	minutes
Lead	$_{82}Pb^{207}$		Stable	

TABLE 3. *The Thorium (Th^{232}) Series*

Element	Isotope	Particle Emitted	Half-Life	
Thorium	$_{90}Th^{232}$	α	1.39×10^{10}	years
Radium	$_{88}Ra^{228}$	β	6.7	years
Actinium	$_{89}Ac^{228}$	β	6.13	hours
Thorium	$_{90}Th^{228}$	α	1.90	years
Radium	$_{88}Ra^{224}$	α	3.64	days
Radon	$_{86}Rn^{220}$	α	54.3	seconds
Polonium	$_{84}Po^{216}$	α	0.158	seconds
Lead	$_{82}Pb^{212}$	β	10.6	hours
Bismuth	$_{83}Bi^{212}$	β, α	60.5	minutes
Polonium	$_{84}Po^{212}$	α	2.9×10^{-7}	seconds
Tantalum	$_{81}Tl^{208}$	β	3.1	minutes
Lead	$_{82}Pb^{208}$		Stable	

its formation, since they have very long half-lives. Three radioactive isotopes of these two elements are of major interest—the two uranium isotopes U^{235} and U^{238} and the thorium isotope Th^{232}. The complete series of each as it goes to the stable element lead is shown in Tables 1-3. In each series the stable end product is lead—but a different isotope of lead in each case.

Most of the data so far published on the ages of rocks and minerals have been obtained by radiodating using lead and its isotopes. It may well be asked what is meant by the age of a given rock, found and expressed in these terms, for all the atoms present in this rock have always been on earth. For igneous rocks, age relates to the time when the rock solidified; for sedimentary rocks, it relates to the time when the rock became sufficiently compacted to prevent any more movement of the incorporated elements. From these dates the accumulation of daughter elements in the rock begins.

If the rock contains a reasonable amount of uranium and thorium, then the time lapse since formation can be determined by analysis of the uranium and thorium content, together with determination of the various amounts of lead isotopes that are present. Most useful in determining ages is the mathematical notation involved in the statement of decay rate. The basic expression uses N to refer to the number of atoms of the element involved. The fundamental notation of the calculus is also used: $-du/dt$; this represents the change (downward, to a smaller number, as indicated by the minus sign) of some quantity u with the passage of time t. Using this notation, we obtain

$$-\frac{dN}{dt} = \lambda N \ . \tag{1}$$

This simply says that the decrease in the number of radioactive atoms (the left side of the equation) is equal to the number of atoms present times a constant λ, which is called the decay constant for the isotope in question.

Equation 1 can, by integration, yield an equation for the number of atoms to be expected after any lapsed time t. The equation is

$$N_t = N_o e^{-\lambda t} \ . \tag{2}$$

Here N_0 is the number of atoms present initially, N_t is the number of atoms present after time t, and e is the base of the natural logarithms. Note that we cannot tell *which* atoms will decay, but only that over time t some number $N_0 - N_t$ of them will decay.

The equations used for calculating age can be derived fairly simply from these basic assumptions, as they have been expressed in equation form. Consider a sample of a mineral of age t that has been analyzed for U^{238} and Pb^{206}.

Let $N_t{}^{238}$ = number of atoms of U^{238} now

$N_0{}^{238}$ = number of atoms of U^{238} at time zero

$N_t{}^{206}$ = number of atoms of Pb^{206} now

λ = decay constant for U^{238} (1.54×10^{-10} years).

Table 1 shows that for each atom of U^{238} that distintegrates one atom of Pb^{206} is eventually formed. Thus the atoms of Pb^{206} that have accumulated in the material over time t will be equal in number to those of U^{238} that have decayed during this period. Therefore

$$N_t{}^{206} = N_0{}^{238} - N_t{}^{238} .$$

From the basic law of radioactive decay (equation 2), we have

$$N_0{}^{238} = N_t{}^{238} e^{\lambda t}$$

so that

$$N_t{}^{206} = N_t{}^{238}[e^{\lambda t} - 1] . \qquad (3)$$

Solving for t, we have

$$t = \frac{1}{\lambda} \log_e \left(1 + \frac{N_t{}^{206}}{N_t{}^{238}} \right) \text{ years} \qquad (4)$$

$$t = 1.5 \times 10^{10} \log_{10} \left(1 + \frac{N_t{}^{206}}{N_t{}^{238}} \right) \text{ years} . \qquad (5)$$

Tables 2 and 3 now show that the decay of U^{235} and Th^{232} forms Pb^{207} and Pb^{208}, respectively. Thus the age of a mineral can be calculated from the ratios Pb^{207}/U^{235} and Pb^{208}/Th^{232}, using equation 4 with appropriate changes, as well as from the $Pb^{206}/{}^{238}$ ratio. Age may also be caculated from the Pb^{207}/Pb^{206} ratio, since the U^{235}/U^{238} ratio has a constant value of 0.0073 over the earth, but the half-lives of U^{235} and U^{238} differ by a factor of 6.3. If the ages as computed from these ratios are in close agreement, it is an excellent indication that the measured age is a reliable value.

In using the uranium-lead series for dating, we must remember that a certain amount of time is necessary to establish equilibrium between the decay of uranium and thorium and the consequent production of lead, because of the half-lives of the intermediary isotopes in the radioactive series. This time is approximately 10^6 years for the U^{238} series, 10^5 years for the U^{235} series, and 20 years for the Th^{232} series. However, since the applicable age range for the lead isotope method is greater than 50×10^6 years, no significant error results from this time lag.

The amount of the lead isotope substituted for N_t in equation 4 should, of course, be only that associated with the decay of its parent.

Obviously, if the sample contained ordinary lead at the time of its formation, this lead (common lead) would be measured together with that formed by the decay of uranium and thorium (radiogenic lead). Fortunately, the two can be distinguished. Common lead contains Pb^{204}, which is known to be nonradiogenic. We measure the ratios of the four lead isotopes Pb^{204}, Pb^{206}, Pb^{207}, and Pb^{208} in the sample and in common lead from the same geographical region as the sample. The amount of Pb^{206}, Pb^{207}, and Pb^{208} of common lead origin in the sample can then be determined so that the amount of radiogenic Pb^{206}, Pb^{207}, and Pb^{208} in the sample is then known. Still, it is desirable to have this correction be small, and samples for lead isotope age measurement are therefore deliberately selected to have a low content of common lead.

Sometimes ages calculated from the various isotope ratios do not agree. Reference to the radioactive series in Tables 1, 2, and 3, will show why this is so in some cases. Suppose that the rock has been subject to chemical leaching during part of its history so that some of the intermediate elements between uranium and thorium and the end product, lead, have been lost. Since the half-lives of the intermediate products are different for the three series, such leaching will have a different effect on the apparent age as computed from each series. Moreover, in each series

TABLE 4. *Published Lead Isotope Ages on Radioactive Minerals*

Locality	Isotope Ratio			
	206/238	207/235	207/206	208/232
	Age: Millions of Years			
Joachimstal (pitchblende)	244 ± 5	249 ± 22	242 ± 20	—
Great Bear Lake, Canada (pitchblende)	1220 ± 10	1282 ± 15	1400 ± 20	—
Parry Sound, Canada (uraninite)	970 ± 5	1015 ± 10	1050 ± 30	955 ± 30
Ebonite Tantalum Claims, Bikita, So. Rhodesia (monazite from pegmatite)	2675 ± 30	2680 ± 70	2680 ± 30	2645 ± 30
Soafia, Madagascar (thorianite)	524 ± 5	493 ± 3	450 ± 10	465 ± 5

an inert gas Rn^{222}, Rn^{219}, or Rn^{220}, different isotopic forms of radon, is one of the intermediary elements. The half-lives of the radon isotopes are very different, so that the possibility of loss by diffusion of the gas out of the rock before it decays is different for each series. Rn^{222} has a half-life of 3.8 days, while the half-lives of Rn^{219} and Rn^{220} are of the order of seconds. Only Rn^{222} will have an appreciable chance of diffusing out of the mineral—when it does, the age computed by finding the ratio of Pb^{206}/U^{238} will be too low, for the loss of Rn^{222} will mean a small production of Pb^{206}. Fortunately, a study of the history of the rock or mineral since its formation will usually reveal the cause of the discrepancy and so indicate which ages are the most nearly correct.

Table 4 shows some ages for different ratios of isotopes. Note the agreement obtained for ages of rocks in different locations, and also the indications of the accuracy of each method.

THE HELIUM METHOD

Table 1 shows that when U^{238} decays into Pb^{206} eight alpha particles are emitted by the various radioisotopes in the series. Alpha particles are the nuclei of helium atoms, so when the alpha particles come to rest they acquire two electrons each and become atoms of helium. Thus the disintegration of one atom of U^{238} eventually produces eight atoms of helium. Similarly, the distegration of one atom of U^{235} produces seven atoms of helium, and of Th^{232}, six atoms. If the helium formed in these disintegrations is retained in the rock or mineral, then the age of the rock can be determined from a measurement of the amounts of helium, uranium, and thorium.

Ages determined by the helium method are generally too low, because helium is lost from the rock. Hence ages so obtained are always regarded as minimums. Helium is a small atom, and thus diffuses rapidly, so that it is difficult to find minerals which have retained all the helium formed in them over a period of millions of years.

Study of the helium method is still being pursued, because it is applicable to rocks and minerals with very low contents of uranium and thorium. Indeed, in contrast to the lead method, which requires strongly radioactive minerals for its application, the radioactivity of a sample must be low if the helium method is to be used. Otherwise the amount of helium produced over millions of years causes a build-up of pressure inside the rock which facilitates the escape of the gas. In highly radioactive samples the escape of helium is also facilitated by damage to the original crystal structure from the continual bombardment of its components by alpha particles, which are relatively massive.

POTASSIUM-ARGON METHOD

Considerable investigation of the potassium-argon method of dating has been undertaken during the last few years because potassium is an element present in most rocks, and the method is therefore applicable to a greater number of rock types than the lead method.

The radioactivity of potassium is due to the potassium 40 (K^{40}) isotope, which occurs as 1 part in 10^4 potassium. It has a half-life of 1.35×10^9 years. Potassium 40 has two modes of decay, one by beta emission to Ca^{40}, and one method we shall not deal with here, called K electron capture. K electron capture results in the formation of argon 40 (A^{40}). The number of disintegrations that produce A^{40} as compared to those that produce K^{40} is called the branching ratio and has a value of about 0.12. Argon is an inert gas and thus will retain its gaseous form in rocks, since it cannot form part of any chemical compound. If all the argon formed in the rock has been retained, then its age can be calculated after its potassium and argon content has been measured.

The recent interest in the potassium-argon method is mainly a result of advances in mass spectrometric techniques for measuring isotopic composition of extremely small amounts of gas. The small amounts of argon that must be measured can be shown in a typical example.

Consider a 100-g rock sample that is 10 million years old and is composed of 3.5% potassium.

$$\text{The } K^{40} \text{ content} = 100 \times 3.5 \times 10^{-2} \times 10^{-4}$$
$$= 3.5 \times 10^{-4} \text{g } K^{40} \ .$$

$$\text{The } A^{40} \text{ content} = 3.5 \times 10^{-4} \times \frac{10^7}{1.35 \times 10^9} \times 0.693 \times 0.12$$
$$= 2 \times 10^{-7} \text{ g } A^{40}$$
$$= 1 \times 10^{-4} \text{ cc } A^{40} \text{ at normal temperature and}$$
pressure .

The technique currently used for this type of measurement is called isotope dilution. It involves adding a known quantity of an isotope that is not present in the sample before measurements are taken. The isotopic composition of the argon is always determined, so that a correction can be made for any argon of atmospheric origin, rather than radiogenic origin, that was trapped in the rock at the time of its formation.

One of the main problems of the potassium-argon method that remains to be clarified is the question of just which minerals do retain all the argon that is formed in them by potassium decay. It should be noted, however, that this is a much less serious problem than the retention of helium in a mineral, because argon diffuses more slowly than helium.

STRONTIUM METHOD

The decay of radioactive rubidium 87 (which has a half-life of 6×10^{10} years) to strontium 87 has been used for dating some minerals. Results obtained are of small importance compared to those obtained by the lead and potassium methods, mainly because of analytical difficulties. It is not easy to find suitable samples, either, for they must be high in strontium and have a low natural rubidium content. The long half-life of Ru^{87} means that this method is applicable only to very old rocks, of the order of 10^9 years.

IONIUM METHOD

Though the ionium method of dating has not been especially useful, it will be mentioned briefly here for two reasons: It is a method based on a daughter product of uranium, and it is the only radioactive dating method known now that can be used for the time period from 50,000 to 1 million years ago.

Ionium is one of the intermediate radioactive elements of the U^{238} series and has a half-life of 80,000 years. Ionium dating is useful only for deep-sea sediments laid down within the last 400,000 years. The basic fact that makes the method possible is that sedimentary material sinking to the bottom of the sea absorbs ionium, but not uranium, on its way down. Thus the freshly deposited sediments contain more ionium than the amount which would be produced by uranium decay in the sediments. After deposition, this excess ionium decays with time, so a measurement of the ionium content of the sediment layers will tell the time that has elapsed since their deposition.

THE CARBON 14 METHOD

Carbon 14 dating is probably now the most widely known of all the dating methods, even though only fifteen years have passed since it was first shown practical by Willard Libby, who won a Nobel Prize for his efforts. The method is applicable to carbon-containing material, and covers the time range of 100 to 50,000 years. This time period is one of great interest to many disciplines—history, archeology, geology—since during this time major events in the history of mankind have occurred. The C^{14} method is ideally suited to unraveling the absolute time scale associated with these events; it is also of interest to biologists, since carbon is a basic building block in all terrestrial living organisms.

The C^{14} method of dating depends on the fact that atoms of C^{14} are continually produced in the upper atmosphere by cosmic rays. These atoms are soon oxidized to $C^{14}O_2$ and then mixed thoroughly throughout the atmosphere by natural circulation. They then enter the biosphere and the life cycles of all living organisms and are also dissolved in the oceans. If the production rate of C^{14} is constant over time, and $C^{14}O_2$ is mixed thoroughly and rapidly in the ocean, when compared with its half-life time, then the specific activity of C^{14} in the atmosphere and biosphere will be constant over time, because the production of new C^{14} atoms will be balanced by the decay of others. So far as can be determined now, these assumptions are correct; therefore we can assume that C^{14} activity is the same now as it was 50,000 years ago.

After the death of an organism the exchange of CO_2 between the organism and its environment ceases. The C^{14} present in the organism then decreases with time according to the decay rate of C^{14} (which has a half-life of 5568 $\pm$ 30 years). Thus measurement of the C^{14} specific activity in a sample of unknown age will tell when that organism died—by showing how much time has passed since carbon exchange was taking place in the living matter.

From these measurements the age t of a sample can be computed.

Let I_0 = activity of C^{14} in the living biosphere

I_t = activity of C^{14} in a sample of unknown age

λ = disintegration constant of C^{14} (1.245×10^{-4}/years) .

Then from equation 2, given earlier, we have

$$I_t = I_0 e^{-\lambda t}$$

$$t = \frac{1}{\lambda} \log_e \frac{I_0}{I_t} \text{ years} \tag{6}$$

$$= 8030 \log_e \frac{I_0}{I_t} \text{ years} . \tag{7}$$

High-energy cosmic rays collide with the nitrogen and oxygen atoms of the atmosphere when they enter it, and one of the products of this reaction is neutrons. These neutrons are then absorbed by nitrogen atoms to produce C^{14}:

$$_7N^{14} + _0n^1 \rightarrow {_6}C^{14} + {_1}H^1 .$$

Figure 1

Counter for low-level concentrations of radioisotopes. (Courtesy of G. J. Fergusson, Institute of Geophysics and Planetary Physics, University of California, Los Angeles.)

Eight atoms of C^{14} are formed for each cosmic ray entering the earth's atmosphere, giving a production rate of about two atoms of C^{14} per square centimeter of the earth's surface. The total amount of C^{14} present on earth is very small (80 tons) when compared with the total amount of all the carbon in the atmosphere, biosphere, and ocean, so that the ratio of C^{14} to normal carbon atoms (C^{12}) in living organisms is small, only 1 part in 10^{12}. This corresponds to a disintegration rate of 15/min/g of carbon (dpm/g). For a sample 18,000 years old, the C^{14} activity is 1.5 dpm/g, and only 0.15 dpm/g for a sample 37,000 years old.

Special techniques have been developed for measuring natural C^{14}, because its low levels of activity are undetectable on conventional equipment for measuring radioactivity. Large, sensitive counters operating in a large iron shield to reduce the background radioactivity from nearby material and the counts from cosmic radiation are used. Figure 1 shows a typical piece of equipment for low-level measurements of radioactivity. This equipment consists of a 10-ton iron shield

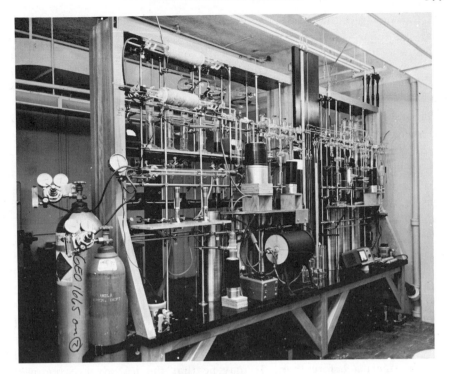

Figure 2

Equipment for preparing a sample for radiodating. (Courtesy of G. J. Fergusson, Institute of Geophysics and Planetary Physics, University of California, Los Angeles.)

and, placed inside it, a sample counter, surrounded by a ring of counters that detect the penetrating component of the cosmic radiation that reaches sea level. The pulses from the sample counter are stored and added by the associated electronic equipment.

Unfortunately, a sample submitted for C^{14} dating is destroyed in the dating procedure, for the carbon atoms in the sample have to be incorporated into a chemical compound suitable for counting. Usually, this will be a gas such as carbon dioxide, methane, or acetylene. The carbon atoms in the sample are first converted to carbon dioxide—by controlled combustion, if the sample is organic—or by evolution of acid, if the sample is a carbonate. Further treatment of the carbon dioxide depends on the type of counting desired. A typical labarotory apparatus for performing these operations is shown in Figure 2. This apparatus, together with the counting equipment, constitutes the basic equipment necessary for C^{14} dating.

Carbon 14 dating has been checked experimentally by dating material of known age, such as samples from the ancient Near East civilization, where dates are known historically back to about 3000 B.C.. The C^{14} content has also been checked for wood containing tree rings, because these can establish tree ages of up to 4000 years. The results of these checks agree within the accuracy limits obtainable for the laboratory assay, so the validity of the method is established. The dates of many important events have been fixed by the C^{14} method, but space permits mention of only a few.

Perhaps the most valuable is the dating of the last ice age. It has been shown that the time of the maximum of the last great ice advance was reached about 11,000 years ago, and that this happened simultaneously in North America and Europe and in the southern hemisphere.

In Europe and the ancient Near East evidence of many past cultures has been discovered by archeologists. Among the remaining evidence of such people and their activities is usually carbon-containing material (such as charcoal from campfires, wood from buildings). Carbon 14 dates on these materials give ages going back to the limit of the C^{14} method (50,000 years). In North America, however, no evidence of mankind has been found which precedes the last evidence of the ice sheet, 11,000 years ago, so it is still uncertain whether man inhabited North America before then. It may be that the ice advance removed all traces of mankind as men moved south.

Carbon 14 dates have also been used to determine erosion and sedimentation rates, and to show that erosion of the land surface has been proceeding more rapidly since the last ice advance—indeed, three times as rapidly as previously—so that now sediments are being deposited on the floor of the deep ocean at the rate of a few inches per 10,000 years.

Radiocarbon dating can also determine eruption dates of volcanoes. Some of the vegetation that surrounded the volcano is reduced to charcoal by the hot lava and ashes and trapped under this ejected material. Radiocarbon dating of the organic debris has shown that the explosion of Mount Mazuma, which created Crater Lake in Oregon, happened 6500 years ago.

TRITIUM

Tritium (H^3) is a radioactive isotope of hydrogen, with a half-life of 12.5 years. It is produced in the atmosphere by cosmic rays in a fashion similar to the production of C^{14}. The tritium atoms soon join into water molecules, so that all rain water contains some tritium. By knowing the decay rate of tritium water can be dated over a time range of 10–15 years since it has fallen as rain. The amount of tritium in rain water is extremely small—1 to 10 tritium atoms per 10^{18} hydrogen atoms.

Tritium activity in recent rain water varies slightly in different regions of the world, depending on meteorological factors and the type of climate. For a given geographical area, however, a measurement of the tritium in a sample, whether from the local wine or from an underground water supply, provides data for calculating the time since the water fell as rain.

The small concentration of tritium in water means that the ratio of tritium to hydrogen atoms must be increased by electrolytic methods before the activity can be measured even by the low-level counting equipment shown in Figure 1.

SUMMARY

$$_{92}U^{238} \nearrow {}_{2}He^{4} \atop \searrow {}_{90}Th^{234}$$

An example of alpha decay. The U^{238} nucleus ejects an alpha particle, thus losing four units of mass and two of charge and becoming an isotope of thorium.

$$_{90}Th^{234} \nearrow \beta(e^{-}) \atop \searrow {}_{91}Pa^{234}$$

An example of beta decay. The thorium nucleus ejects an electron; one of its neutrons thereby becomes a proton. Its atomic mass is unchanged, but its atomic number (of protons) is increased by one, and it becomes an isotope of protactinium.

$$N_t = N_0 e^{-\lambda t}$$

The basic equation used in radiodating. The number of atoms undecayed after time t is equal to the original number times the base e of natural logarithms to the negative power of the decay constant times t.

$$I_t = I_0 e^{-\lambda t}$$

The basic equation used in C^{14} radiodating, derived from the one just given. Here I_0 and I_t are specific activities of C^{14} originally, and after time t; when they are known, t can be determined.

SUGGESTIONS FOR FURTHER READING

Bowen, R. N. C., 1958. *The exploration of time.* Clowes, London.

Faul, H. (editor), 1954. *Nuclear geology.* Wiley, New York.

Kulp, J. L. (editor), 1961. Geochronology of rock systems. *Annals of New York Academy of Sciences,* Vol. 91, Art. 2, pp. 159–594.

Landsberg, H. (editor), 1955. *Advances in geophysics.* Academic Press, New York, Vol. II, pp. 179–219.

Libby, W. F., 1955. *Radiocarbon dating.* Second ed. Univ. of Chicago Press, Chicago.

Rankama, K., 1954. *Isotope geology.* McGraw-Hill, New York.

Zeuner, F. E., 1952. *Dating the past.* Third ed. Methuen and Co., London.

CHAPTER

18

Biochemistry

Biochemistry, as perhaps no other biology-related field, exemplifies the rapid growth and revision of scientific knowledge. In the five years since the first edition of the *BSCS biology teachers' handbook,* many changes have occurred in the field of biochemistry and hence much of the material in that edition is now out of date. Also, because of all the work now being done to elucidate synthetic processes, cell control, energy conversion, and the like, it is highly probable that, five years from now, today's glittering new concepts will seem just as outmoded as do some of the ideas of the early 1960's now.

Another important consideration in providing a background of biochemistry for biology teachers is to present it in such a way that the relationships between current generalizations and concepts and the data, hypotheses, experimental designs, and assumptions on which they are based are shown. For example, what sort of experiments have been done to elucidate protein synthesis? Are the ideas presented facts or tentative hypotheses? Which are growing points, which are "dying" areas in the field of biochemistry?

For these two reasons, this edition presents recent references that may aid the biology teacher in keeping abreast of the changing times. These are minimal source materials that provide much of the necessary information, although the individual teacher must, of course, remain alert to new information and concepts. The references have been selected on the grounds of easy availability, relatively recent publication, readability, and a clear intent to provide observational background as well as inferences drawn from experiments. The references are grouped into biochemical topics of concern to the biology teacher.

THE NATURE AND CLASSIFICATION OF BIOCHEMICAL COMPOUNDS

This is a field that is changing relatively slowly, although it is indeed changing. For instance, there is nearly explosive change in the understanding of the structure of nucleotides, and of the roles of phospholipids and complex polysaccharides. Any up-to-date introductory biochemistry text will give much background information. Additional references include:

Paperbacks

Baker, J. J. W., and G. E. Allen. *Matter, energy, and life,* 1965. Addison-Wesley, Palo Alto, Calif. Chs. 7–11.

Watson, J. D. *Molecular biology of the gene,* 1965. W. A. Benjamin, New York, Chs. 2–4.

Guthe, K. F. *The physiology of cells,* 1968. Macmillan, New York, Ch. 1.

Article

Holley, R. W. The nucleotide sequence of a nucleic acid. *Scientific American,* **214** (2), 30, 1966.

ENZYMES AND CATALYSTS

Again, much of the general background information as to enzyme-substrate interactions is stabilized. The treatment of this background is adequate in the appropriate sections of the revised BSCS Versions. The exciting growing areas in enzyme functions involve control and integration of enzymatically catalyzed reactions, as they occur in cells. Control is exerted at the level of reactivity of the existing enzyme molecule (for example, by allosteric interaction with hormones and other controlling substances), at the level of gene function (rate and specificity of enzyme synthesis), and probably often at the level of enzyme inactivation. Some appropriate references include:

Articles

Phillips, D. C. The three-dimensional structure of an enzyme molecule. *Scientific American,* **215** (5), 78, 1966.

Changeux, J. P. The control of biochemical reactions. *Scientific American,* **212** (4), 36, 1965.

Davidson, E.H. Hormones and genes. *Scientific American,* **212** (6), 36, 1965.

van Overbook, J. The control of plant growth. *Scientific American,* **219** (1), 75, 1968.

ENERGY TRANSFORMATION-TRANSDUCTION

This is a complex area of explosive growth and also an area in which teachers have often been given inadequate background for developing basic concepts. Whereas the first law of thermodynamics appears to be taught nearly universally in an effective way, there is much less consistent appreciation of the nature and implications of the second law of thermodynamics (and its relevance to fields ranging from biochemistry and genetics to ecology and evolution). Moreover, detailed understanding of photosynthesis, oxidative phosphorylation, mechanical movement, etc., is undergoing continuing change. Thus, the teacher should not only work toward understanding of the basic concepts but should also watch for important new ideas as these emerge. Some appropriate references include:

Paperbacks

Goldsby, R. A. *Cells and energy,* 1967. Macmillan, New York.
Guthe, K. F. *The physiology of cells,* 1968. Macmillan, New York.
Watson, J. D. *Molecular biology of the gene,* 1965, W. A. Benjamin, New York, Ch. 5.
Lehninger, A. L. *Bioenergetics,* 1965. W. A. Benjamin, New York.
watson, J. D. *Molecular biology of the gene,* 1965, Ch. 5.

Somewhat more detailed treatment will be found in:

Klotz, I. *Energy changes in biochemical reactions,* 1967. Academic Press, New York.
Racker, E. *Mechanisms of bioenergetics,* 1965. Academic Press, New York.

Articles

Rabinowitch, E. I., and Govindjee. The role of light in photosynthesis. *Scientific American,* **213** (1), 74, 1965.
Racker, E. *Mechanisms of bioenergetics,* 1965. Academic Press, New York. 32, 1968.

SYNTHETIC PROCESSES

Again, this is a rapidly changing area, with continuing expansion in detail of knowledge about nuclear and extranuclear sites and roles of DNA, the nature of the messenger and the code by which it operates, the structure and manner of functioning of ribosomes, the structure of transfer RNA and the roles of the many enzymes, and the energy sources that participate in protein synthesis. There are exciting findings in immunology with important theoretical and practical implications. Teachers should also be aware of the regulation of DNA synthesis through the action of interferons (of importance as an approach to understanding some aspects of cellular control, and with potential for combatting viruses and cancer). Some relevant references are:

Paperbacks

Watson, J. D. *Molecular biology of the gene,* 1965, W. A. Benjamin, New York, Ch. 6–16.
Ingram, V. M. *The biosynthesis of macromolecules,* 1965. W. A. Benjamin, New York.

Articles

Clark, B. C. F., and K. A. Marcker. How proteins start. *Scientific American,* **218** (1), 36, 1968.
Merrifield, R. B. The automatic synthesis of proteins. *Scientific American,* **218** (3), 56, 1968.
Crick, F. H. C. The genetic code. *Scientific American,* **215** (4), 55, 1966.
Yanofsky, C. Gene structure and protein structure. *Scientific American,* **216** (5), 80, 1967.
Kornberg, A. The synthesis of DNA. *Scientific American,* **219** (4), 64, 1968.
Hanawalt, P. C., and R. H. Haynes. The repair of DNA. *Scientific American,* **216** (2), 36, 1967.
Mirsky, A. E. The discovery of DNA. *Scientific American,* **219** (6), 78, 1968.
Nossal, G. J. W. How cells make antibodies. *Scientific American,* **211** (6), 106, 1964.
Porter, R. R. The structure of antibodies. *Scientific American,* **217** (4), 81, 1967.
Smithies, O. Antibody variability. *Science,* **157,** 267, 1967.
Isaacs, A. Foreign nucleic acids. *Scientific American,* **209** (4), 46, 1963.

QUESTIONS AS TO THE ORIGIN AND EVOLUTION OF LIFE

This area is an exciting aspect of modern biochemistry. It is represented in the BSCS Versions, and, perhaps more importantly, it is fruitful in stimulating enquiry and speculation.

Books

Fox, S. W. *The origin of prebiological systems,* 1965. Academic Press, New York.
Bryson, V., and H. J. Vogel (Editors). *Evolving genes and proteins,* 1965. Academic Press, New York.

Articles

Eglinton, G., and M. Calvin. Chemical fossils. *Scientific American,* **216** (1),
    32, 1967.
Echlin, P. Blue-green algae. *Scientific American,* **214** (6), 74, 1966.
Zuckerkandl, E. The evolution of hemoglobin. *Scientific American,* **212** (5),
    110, 1965.

Statistics

Statistics are widely misunderstood. Among many people, the misunderstanding takes the form of sweeping cynicism, as in the remark: "Figures never lie, but statistics always do." Among others, the misunderstanding takes the form of supposing that statistics can provide *definitive* answers to very important questions, such as whether a difference between two populations is due to chance or not.

To help remove such misunderstandings, we shall begin with a definition of statistics for our purposes: Statistics are a series of decisions and inventions for dealing with three kinds of problems often encountered in scientific research. The three classes of problems are: (1) the problem of providing *economical descriptions* of populations; (2) the problem of obtaining useful *estimates* about a population when all we have is measurements of a sample of it; (3) the problem of deciding how much confidence to put on *decisions* we make about a population or about differences between two or more populations.

It is important to note that these three problems do *not* lend themselves to definitive solutions. There may be many economical descriptions of populations; the one we choose depends on how economical we want to be, among other things. Estimations are always estimates, inferences about the character of the population; there is no way of knowing with certainty and precision what a population is like, even from many samples. Hence what we mean by enough confidence will always be in some doubt.

The tentative and necessarily inexact solutions that statistics offers to these three problems are briefly described in three sections in this chapter.

ECONOMICAL DESCRIPTIONS: PARAMETERS

First, let us agree on what we mean by "a population." We do not mean a group of *things,* such as horses, *Drosophila,* men, or machines. We do mean a group of numbers, values, or measurements, which our theory tells us should be treated as a single significant whole. Thus, the measurements representing the heights of all adult human males would, from the point of view of genetics, constitute a biologically meaningful population. In the same fashion, test scores representing the arithmetic achievement of all human beings constitute a meaningful population. The sciences of demography and economics might lead us to treat the yearly incomes of all wage earners in the United States, or the ages of all persons in continental America, as meaningful populations.

Each of these is not only an example of a population, but also of a particular sort of population very common among biological phenomena. It is the kind of population which arises when the population variable (such as age, height, or yearly income) is one affected by a great many causal determiners, each of which acts to increase or decrease the magnitude of the factor in each instance by a small amount, and independently of the other determiners. Thus we treat the height of any one individual as the summed effect of many different genes and of many different environmental determiners.

The population of such a variable factor as height has the characteristic shape we have come to know as the *normal distribution*. Let us assume we have measured the heights of all human males. Suppose we plot these measurements, placing the heights along the horizontal axis (abscissa) and erecting on the vertical axis distances representing the frequency of occurrence of each height. The resulting normal curve would look like Figure 1.

Obviously, this plot of a normal distribution is a most complete and, in one sense, highly economical description of the population of heights. However, it has one grave shortcoming as a description. It is a drawing, and not one or a few simple numbers which can be used in calculations. It is the search for one or a few simple, summary numbers which constitutes the first problem of statistics.

One of the simplest and best known items for economical description of such populations is the familiar *arithmetic mean*. For our present purpose, let us think of the arithmetic mean as nothing more than the position on the horizontal axis of a line dropped from the highest point on the normal curve to the horizontal axis.

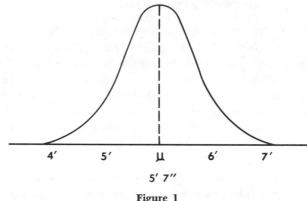

4'   5'   μ   6'   7'

5' 7"

Figure 1

The mean is useful, but it is not a sufficient description of a normal population. Figure 2 illustrates its deficiency. In Figure 2, we see two normal distributions, each of which has the same arithmetic mean, but each of which is clearly different from the other.

The differences between the distributions shown in Figure 2, a difference the mean does not reveal, is revealed if we add only one further descriptive measure. This measure, called the *standard deviation*, is a measure of the "spread" (dispersion) of a normal curve. Figure 3 shows the line which represents this measure of spread, or *dispersion*. For our present purpose, let us consider a standard deviation to be the length of the line in abscissa units which connects the position of the mean to that point on the curve (in either direction) where the curve ceases to be concave and becomes convex. (This point is called the *inflection point*.)

Such descriptive measures as the mean and standard deviation are called *parameters* of a distribution, when the distribution is an ideal

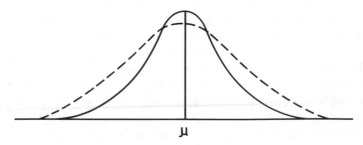

μ

Figure 2

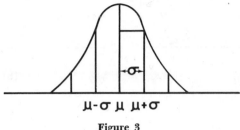

$$\mu-\sigma \quad \mu \quad \mu+\sigma$$

Figure 3

one, that is, for example, when it is such an impossible distribution as that of the height of *all* adult human males who ever lived and ever will live, and when the height is the resultant of a multitude of equally and independently acting small causes. Furthermore, for the ideal *normal* distribution, these two parameters provide a complete description. That is, one and only one normal distribution will fit a given mean and standard deviation.

DESCRIPTIVE STATISTICS: ESTIMATES OF PARAMETERS

Of course, distributions of real samples of populations will rarely, if ever, be perfectly normal. They may vary from the ideal in many ways. They will be irregular in shape, rather than smooth curves, for such populations as we can deal with in samples are finite, and the smooth curve is generated by an infinite number of members of a population. The curves may be asymmetrical (certain kinds of asymmetry are known as "skews"), they may have more than one hump or *mode,* they may be more peaked or flattened than a normal distribution. Figure 4 shows some departures from the ideal normal distributions. In such cases, obviously, we need additional descriptive measures besides the mean and the standard deviation, measures which indicate the degree of flattening or peaking, the skewness of asymmetry, or the number and locations of the modes. Many of the populations dealt with in biology, however, may be described adequately by assuming the normality of the distribution. Hence, the mean and standard deviation constitute, for our purposes, an adequate, economical description. In many cases, moreover, although we can plainly see that the plots of our data do not give us an ideal normal curve, it is possible to establish mathematically that it is legitimate to treat the distribution as a normal curve when we are making certain kinds of useful calculations.

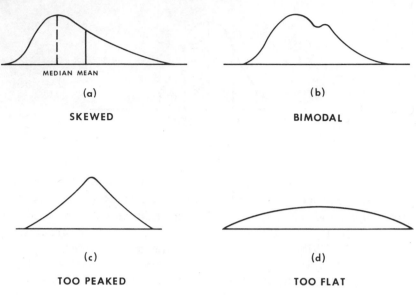

Figure 4

The Sample Mean and Median

One of the most useful estimates of the true mean of a population (or its central tendency) is the *sample mean*. This estimate is familiar to us under the name of *arithmetic average*. It is the sum of the measured values of our sample distribution divided by the number of individuals whose values make up the sample. The sample mean is expressed by $\overline{x}$.

$$\overline{x} = \frac{\Sigma x}{n}$$

where Σ is a summation sign.

There are occasions when the sample mean is not the best estimator of the true mean of a population. If the population itself is not normally distributed, or if we believe the sample may contain some unusual values, the mean is not a good indicator. Two examples will show why this is so.

First, consider an attempt to estimate the average (mean) height of American males. Suppose you take a small sample, the heights of five men. Included in this group there happen to be two men with heights of over 8 feet. When you average the heights (supposing the other heights to be more "normal"), you will find the sample mean much too large, perhaps nearly 7 feet. (In a larger sample, these unusually large values would be balanced out by others unusually small.)

In the case of an asymmetric or skewed distribution, as in Figure 4a, the arithmetic mean of the distribution will similarly estimate the true mean badly. In this curve it is clear that if we took a number, n, of points at equal distances from one another along the curve, added their values (taken from ordinates at each point), and then divided by n, the result would lie considerably to the right of the most common value of points on the curve.

We can improve our estimates in these cases by using another measure, the median, as an estimator. As you probably know, the *median* is obtained by simply listing the values of all the members of the sample in order of increasing size; we then take as median the one in the middle of the array, the $n/(2)$th value. For our height sample, we would list the heights of the men as, say, 5', 5'2", 5'7", 8', 8'. The median would be 5'7". *For a normal distribution the population's mean and median coincide; but for asymmetric and skewed distributions they do not.* We see that for a small sample, and for skewed samples, the *sample median* may be a better estimator of the *population mean* than is the *sample mean*.

Mean Deviation and Standard Deviation

Let us turn now to sample estimates of the standard deviation, remembering that this parameter of a normal distribution is a measure of dispersion. (Figure 2 shows curves for two distributions of which the dashed curve has a greater dispersion than the solid-line curve.)

One estimate of dispersion is the *sample mean deviation*. This estimate is, in effect, no more than the average of all deviations from the mean (the sum of all deviations, devided by the number of deviations). However, there is a catch to this estimate, for if calculated by the ordinary arithmetic rules, its value would be 0 for a symmetrical distribution. This oddity comes about from the fact that deviations from the mean leftward would be negative, whereas deviations to the right of the mean would be positive. Hence their sum would be 0.

This difficulty is avoided by ignoring the sign of the deviations. Hence, the formula of a mean deviation is written as follows (the vertical bars meaning, "without regard to sign"):

$$\text{Mean deviation} = \frac{\Sigma |x - \overline{x}|}{n}$$

Because the mean deviation arbitrarily ignores sign, it makes for considerable awkwardness when used in other calculations. A much more useful estimate would be one in which differences of sign were dispensed with by legitimate methods. One such method is merely to *square* each deviation, since, by arithmetic rules, the square of a negative number is positive (as is the square of a positive number).

From this simple device comes the estimate called the *sample standard deviation*. Its formula is written as follows:

$$s = \sqrt{\Sigma(x - x)^2 / (n - 1)}$$

Note that this estimate is not radically different in meaning from that afforded by the mean deviation. First, we compute each deviation $x - x$ by subtracting the mean from each value. Then we square each deviation (to rid ourselves of differences of sign). We then sum these squares and divide the sum by $n - 1$. (We will indicate later why the denominator is $n - 1$ rather than n.) Finally, we take the square root of our mean of squares in order to return the calculated value to the original order of magnitude.

The standard deviation has a special usefulness. Because of the geometry of a normal curve, and the way the standard deviation is constructed, the area under that portion of the curve lying one standard deviation above or below the mean is 34.13% of the total area under the curve. One standard deviation in both directions from the mean will therefore include 68.26% of the population. Figure 5 shows this graphically. Note that in curves *(a)* and *(b)* of the figure the ranges of values covered differ drastically, but in each case one standard deviation in each direction tells us the ranges within which 68.26% of the population lies. Similarly, 2σ from the mean in both directions will mark off a value range that will include 95% of the population.

Table 1 gives a numerical example. The value $n - 1$ is referred to as the *"degrees of freedom"* for the equation for determining the standard deviation shown in Table 1. We wish in our denominator to have a number representing the number of values *which contribute independently to* the deviations which constitute the numerator. At first, this number appears to be n, the number of values. In fact, however, it is one less. An example will·show this. Suppose we have a distribution of one each of six values: 3, 5, 6, 7, 9, 12 (that is, $n = 6$). Calculating the mean to use in the numerator of s we obtain

$$\frac{3 + 5 + 6 + 7 + 9 + 12}{6} = 7$$

If we now subtract each of these values from the mean calculated from them, as in the numerator of s, not all the values contribute to the total deviation so computed. For the calculated mean incorporates these values

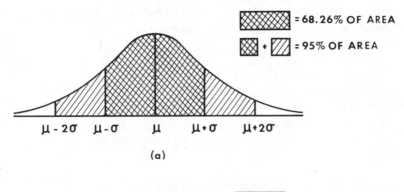

(a)

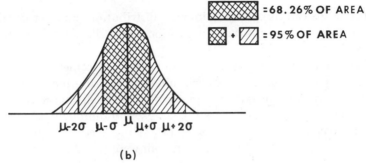

(b)

Figure 5

TABLE 1. *Some Statistics from a Small Sample of Heights*

Heights Sampled; x	$\lvert x - \overline{x} \rvert$	$(x - \overline{x})^2$
5′	−6″	36
5′2″	−4″	16
5′5″	−1″	1
5′7″	1″	1
6′2″	8″	64
$\dfrac{\Sigma x}{n} = $ mean: $\overline{x} = 5'6''$	$\lvert x - \overline{x} \rvert = 20''$	$\Sigma (x - \overline{x})^2 = 118$
	Mean deviation: $\dfrac{\Sigma \lvert x - \overline{x} \rvert}{n} = 4''$	Standard deviation: $\sqrt{\Sigma (x - \overline{x})^2 / (n-1)}$ $s = 5.8''$

in such a way that *one of them* is no longer free to deviate from the mean, but rather is "fixed" by it. That is, given

$$\frac{3 + 5 + 6 + 7 + 9 + x}{6} = 7$$

x can haves one and only one value—in this case, 12.

Thus in this equation, the number of degrees of freedom is one less than n. In other cases, the number of degrees of freedom may differ from n by more than 1.

Probability and the Binomial Distribution

Before we can discuss the binomial distribution (another distribution very useful in biology) a brief discussion about the statistical concept of probability is in order.

Let us take an event with only two possible outcomes, such as flipping a coin, as a model for generating a binomial distribution. The coin can come up only heads or tails (we ignore the very small chance that it might stand on edge). If the coin is well balanced, and thoroughly shaken before a toss, it has a *probability (p)* of ½ of landing heads up, and $1 - $ ½, or ½, of landing tails up. That is, if the coin were tossed an infinite number of times, we would expect it to come up heads on exactly half the tosses. Note that when probabilities are assigned to the possible outcomes, the sum of the probabilities of all these possible outcomes must be 1; that is, by definition, it is a *certainty* that one or another of these possibilities must happen.

We can find the probability of a combination of events by the simple rule: *The probability of two independent events occurring is the product of the probabilities of each occurring separately.* Thus the probability of getting heads for two coin tosses in a row is:

$$p \; heads = \tfrac{1}{2} \times \tfrac{1}{2} = \tfrac{1}{4}$$

Applying the joint probability rule, suppose we toss three coins, or one coin three times. There are four possible outcomes: *(a)* 3 heads; *(b)* 1 head, 2 tails; *(c)* 2 heads, 1 tail; *(d)* 3 tails. How do we arrive at the probability of the occurrence of each outcome? We do so by extending the joint probability rule into a binomial expansion, using the probabilities of "heads" and of "tails" as factors, and the number of coins or tosses as the exponent to which the expression is to be raised: $(\tfrac{1}{2}H + \tfrac{1}{2}T)^3$ is the expression for three tosses. Expanding this, we obtain

$$\frac{1}{2}H + \frac{1}{2}T$$
$$\times \frac{1}{2}H + \frac{1}{2}T$$

$$\frac{1}{4}HH + \frac{1}{4}HT$$
$$\frac{1}{4}HT + \frac{1}{4}TT$$

$$\frac{1}{4}HH + \frac{1}{2}HT + \frac{1}{4}TT$$
$$\times \qquad \frac{1}{2}H + \frac{1}{2}T$$

$$\frac{1}{8}HHH + \frac{1}{4}HHT + \frac{1}{8}HTT$$
$$\frac{1}{8}HHT + \frac{1}{4}HTT + \frac{1}{8}TTT$$

$$\frac{1}{8}HHH + \frac{3}{8}HHT + \frac{3}{8}HTT + \frac{1}{8}TTT$$

This situation is graphed in Figure 6b. As we increase the number of coins tossed, the step curve of parts (a), (b), and (c) of Figure 6 grows more and more like the bell-shaped curve of the normal distribution. When the number of tosses is infinite, the curve is a normal curve.

We should be clear about what this curve means and can tell us. The area under the curve, as we can see from the way it was constructed, must be 1, and must represent a certainty. If we draw a vertical line from the abscissa intersecting any part of the curve, the area to the left of the line represents the probability that any one randomly chosen member of the population will have a value to the left of the intersecting line. This situation is illustrated in Figure 3a. There the line dividing the curve is the mean, μ. There is a probability of ½ that a randomly chosen individual will have a value to the left of the mean, and ½ that the individual will have a value to the right of it.

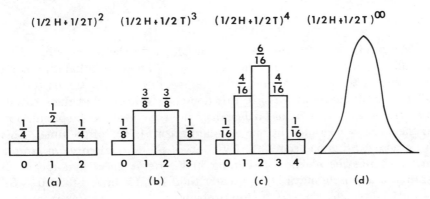

Figure 6

Many biological problems are or can be treated as binomials, that is, as cases with only two alternatives. For example, if we were testing the effectiveness of certain chemicals as plant fertilizers, the chemicals could have more or less effect on different individuals, and different species might respond differently. But we might choose to consider only whether plants treated with a chemical do or do not grow larger than untreated plants. Similarly in genetic studies of the colors of mice; for example, mouse offspring might be white, black, or many shades of gray. For some investigations, however, we could ask merely: Are they white or not-white?

ANALYTICAL STATISTICS

Reliability of a Sample Mean

We know that the true mean is an essential parameter of a population, but we have found that usually we can only estimate it with samples. Since different samples will be likely to give somewhat different values for sample means, we need a measure of the reliability of the sample mean as an estimator of the true mean.

If we take many samples of the same size from a known population, find the mean of each, and then plot these sample means, they will form a normal curve with its mean at or very near the mean of the original population, but with much less dispersion than the original population. The standard deviation of this distribution of sample means is known as the *standard error,* or standard error of the mean. It is sometimes written SE, and it is found by dividing the sample standard deviation by the square root of the number in one sample. For the sample given in Table 1, the standard error would be:

$$\text{SE} = \frac{s}{\sqrt{n}} = \frac{5.8}{\sqrt{5}} = 2.6$$

We rarely can take sample after sample, as in the hypothetical case. Usually we take one sample, and we want to know its reliability as an estimator of the population mean. When we calculate the standard error of the sample mean, as we have just done, we estimate that the mean of our sample will be as far or farther from the true mean of the population as the distance measured by one standard error 31.74% of the time. This is because 1 SE is one standard deviation of the normal curve of sample means—hence the patches of shading in Figure 7a show that the probability of a sample mean being greater than $\mu + \text{SE}$ or less than $\mu - \text{SE}$ is 100.00 — 68.26. Conversely, the probability that the sample mean will

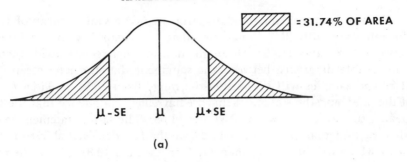

(a)

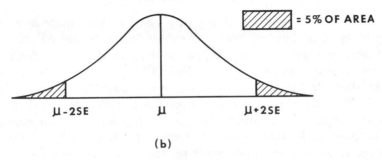

(b)

Figure 7

be as close to the true mean as any point within the unshaded area is 68.26. Similarly, we know that the sample mean will lie as far away, or farther, than 2 SE from the true mean approximately 5% of the time, as shown in Figure 7b.

The sample mean is commonly given with its SE after it, as: 11.23 ± 0.21. This indicates that the sample mean is 11.23 and the SE is 0.21. It means that if the sampling were done over and over, only 31.74% of all the sample means would be likely to be less than 11.02 (11.23 − 0.21), or greater than 11.44 (11.23 + 0.21). Only 5% are likely to be less than 10.81 or greater than 11.65.

There is one difficulty with the use of standard errors described. The sample mean is taken to be the true mean, and the sampling distribution is calculated about the sample mean as if it were a parameter of the distribution, instead of an estimator. To avoid this difficulty, we use confidence intervals, instead of standard errors. A confidence interval is a range of values over the distribution in which we assert the true value to lie. It is called a confidence interval because we assert our confidence that the value does lie in that interval by stating the probability that it lies in that interval.

From the example for standard error, we had a sample mean of 11.23. We can state with 95% confidence that the mean lies in the interval between 10.81 and 11.65 Although the same figures are used, there is considerable difference between the statement that the true mean is at 11.23 and samples will lie outside the range from 10.81 to 11.65 in 5% of the cases, and the statement that we have 95% confidence that the true mean is somewhere between 10.81 and 11.65. The latter statement means when we find a sample mean of 11.23, we shall be deceived about the true mean, which will be greater than 11.65 or less than 10.81, 5% of the time.

"Significant" Difference: The Null Hypothesis

When we toss a coin 100 times, from the fact that there is a probability of ½ of its landing heads we expect 50 heads and 50 tails. We would not, however, be surprised to find some deviations from this expectation. For example, we might find 48 heads and 52 tails. We pass over this slight deviation as of no importance and attribute it to *sampling error*. In this case the error comes from the fact that the prediction is based on a population of an infinite number of tosses, and our sample is comparatively small. By applying the term "sampling error" to deviations from expectation in one sample, we imply that we believe that the average of many samples will be close to the expected 50% heads and 50% tails, that is, that in the long run, low values for heads in some samples will be balanced by high values in others.

If in one sample of 100 we obtain 90 heads and 10 tails, we might begin to wonder if the coin is a fair one, weighted equally on both sides. If we obtained 54 heads and 46 tails 100 times in a row, or 5400 heads and 4600 tails in 10,000 tosses, we might doubt the coin's fairness. Hence we see that we need some method of deciding just when we should start questioning the accuracy of a hypothesis, such as "The coin will land heads 50% of the time," rather than when we should ignore discrepancies as mere sampling errors.

When 100 tosses are made, we can calculate the probability of obtaining 100 heads and 0 tails, 99 tails and 1 head, and so on, by the methods discussed earlier in the section on binomial distributions. We find that the probability of an extreme departure from 50-50 (such as 99-1) is extremely small, but finite nevertheless. We find that the probability of a combination such as 51-49 is relatively large, though smaller than the probability of 50-50. Hence, the possible cause of a departure from the expected value of 50% heads could be sampling error, or it could be an unbalanced coin. We can never know with certainty by statistical methods alone. If we know, however, that a large departure has only a very low probability of occurring by sampling error, we may decide that departures of this magnitude are more likely to be due to an unbalanced coin.

Let us look at the problem in more detail. What we are asking is whether the difference between an observed value and a theoretical value is due to sampling error or to error in the theory. To answer this, we start with some hypothesis which allows us to calculate the probability of sampling errors of various sizes. We call this the *null hypothesis* because it states that the expected difference between the theoretical value and that obtained by observation is zero. This apparently backhanded way of starting is actually necessary, since it gives us a basis for calculating how often sampling error will give departures from the expected theoretical values as great as, or greater than, the values obtained in the sample. If we used the approach of assuming that our samples should differ somewhat from the theoretical value, and then asked how much, we would have to make a calculation for each of the infinite number of possible differences.

To calculate the probability of obtaining a value with a great or greater departure from the expected value due merely to sampling requires that we know the distribution of statistics from samples about the true value, or parameter. For the moment we shall consider only normal distributions, but we should bear in mind that other shapes are possible.

If our calculation shows only a small probability that the departure of our sample from the theoretical value is due to sampling error, we may decide that the cause of the departure is an error in the theory; that is, we would reject the null hypothesis. The commonly used level for rejecting a hypothesis is 0.05, the 5% *level of significance*. It is so called because we state that the difference between observed and theoretical values is *statistically* significant. This has nothing to do with other common meanings of the word "significance," so the word "statistically" should always be used in this context. The 5% level of significance tells us that only 5% of the time will the difference we have observed be due to sampling error.

But this, in turn, tells us that we can still make an error in judgment. There are four possible outcomes, after a statistical significance test:

1. Hypothesis is true, we accept it.
2. Hypothesis is true, we reject it. (Error of the first kind)
3. Hypothesis is false, we accept it. (Error of the second kind)
4. Hypothesis is false, we reject it.

We would like to avoid situations 2 and 3, the two kinds of errors. Setting the level of significance automatically determines how often we shall make each kind of error. At the 5% level, we shall make an error of the first kind (reject a true hypothesis) 5% of the time, because we shall automatically reject any hypothesis if departures from the theory have only a 5% chance of occurring through sampling error.

We could reduce the number of errors of the first kind by lowering the level of significance, say, to 1%. By so doing, instead of rejecting a true hypothesis 5% of the time, we should do so 1% of the time. As we lessen errors of the first kind, however, we automatically *increase* errors of the second kind—accepting the null hypothesis when it is really false.

Let us now make a test of how the theory applies. Suppose we mate a *Drosophila* heterozygous *Bb* (grey) with the homozygous recessive *bb* (black) and obtain 400 offspring. From theory, we expect 200 grey *(Bb)* and 200 black *(bb)*. We obtain 218 grey and 182 black. The difference between observed and theory is: grey (200 − 218), black (200 − 182) = 18. Note that we do not worry about the negative value, since it equals the positive value, and for a value to be too large is exactly as likely as for it to be too small. We proceed as previously outlined to measure the departure from the expected (theoretical) in standard deviation units; that is, we divide the departure from theory, 18, by the standard deviation of the theoretical distribution. In this case, σ for the binomial (1/2 homozygotes + 1/2 heterozygotes) $^{400} = \sqrt{1/2 \times 1/2 \times 400} = 10$. The observed value is then $18/10 = 1.8$ standard deviations $= \tilde{u}$ from the expected value. We can now turn to Table 2 for the probability that our observed value

TABLE 2. *Probability* of Obtaining a Discrepancy from Predictions of Null Hypothesis: P_N of $\tilde{u} > u$ from Sampling Error*

u	$P_N (u > u)$	u	$P_N (u > u)$
0.1	0.920	1.2	0.230
0.2	0.841	1.3	0.194
0.3	0.764	1.4	0.162
0.4	0.689	1.5	0.134
0.5	0.617	1.6	0.110
0.6	0.549	1.7	0.089
0.7	0.484	1.8	0.072
0.8	0.424	1.9	0.058
0.9	0.368	2.0	0.046
1.0	0.317	2.1	0.036
1.1	0.271	2.2	0.028

* This table is for "two-tailed" tests. That is, it applies when we are asking whether our observed and expected values are the same or not *irrespective* of whether the deviation is more or less than the expected.

$\tilde{u}$ is due to sampling error. This table gives us the areas of the "tail ends" of normal distributions, that is, the areas under the curve which are some distance u, or greater, from the mean in both directions. The distance u is expressed in multiples of σ. The area under these tail ends represents the probability that we will get an observed deviation $\tilde{u}$ in either direction from the mean as great or greater than one of value u through sampling error. Our deviation is 1.8σ, so we look for $u = 1.8$, and read $P_N = 0.072$. Since 0.072 is greater than 0.05, we accept the null hypothesis; that is, the parents of the offspring *Drosophila* are ½ homozygotes, ½ heterozygotes.

Had we obtained 221 grey and 179 black, we would have had a deviation of 21. Dividing by the standard deviation gives $u = 2.1\sigma$, and the corresponding probability is 0.036. This is less than the arbitrary 0.05 level of significance, and if that was our decision point, we would reject the null hypothesis.

"Significant" Difference: The t Test

When the population is normally distributed but the sample contains 30 or fewer members, we use a different method for accepting or rejecting the null hypothesis, called t test, or sometimes "Student's t," after its discoverer W. S. Gosset, who wrote under the pseudonym of Student. If our hypothesis says the population mean should be μ, and our sample mean is $\bar{x}$, then t is measured by:

$$t = \frac{\bar{x} - \mu}{s/\sqrt{n}}$$

where s is the standard deviation of the sample and n the number of individuals in the sample. Having gotten a value for t, we look at a t table (Table 3 is a short version), in a row with degrees of freedom equal to $n - 1$. In that row we will find a t value close to the one we have calculated. At the head of the column is the probability that the value (or a greater value) is due to sampling error.

Suppose we have 10 *Drosophila* offspring, with 6 grey and 4 black. Then $s = \sqrt{½ \times ½ \times 10} = 1.6$, $n = 10$, and

$$t = \frac{6 - 5}{1.6/\sqrt{10}} = \frac{1}{1.6/3.3} = 2$$

We have $n - 1 = 9$ degrees of freedom, and, from Table 3, we see that our value for t lies between $t = 2.26$ and 1.83; since $t = 2.26$ has a probability of 0.05, our value of t must have a larger probability. So we accept the null hypothesis. But notice how close we come to rejecting it, using

TABLE 3. *Table* of t*

Degrees of Free-dom	P		Degrees of Free-dom	P	
	0.05	0.01		0.05	0.01
1	12.71	63.66	16	2.12	2.92
2	4.30	9.93	17	2.11	2.90
3	3.18	5.84	18	2.10	2.88
4	2.78	4.60	19	2.09	2.86
5	2.57	4.03	20	2.09	2.85
6	2.45	3.71	21	2.08	2.83
7	2.37	3.50	22	2.07	2.82
8	2.31	3.36	23	2.07	2.81
9	2.26	3.25	24	2.06	2.80
10	2.23	3.17	25	2.06	2.79
11	2.20	3.11	26	2.06	2.78
12	2.18	3.06	27	2.05	2.77
13	2.16	3.01	28	2.05	2.76
14	2.15	2.98	29	2.05	2.76
15	2.13	2.95	30	2.04	2.75
			∞	1.96	2.58

* This table is for "two-tailed" tests. It applies when we are asking whether our observed and expected values are the same or not, *irrespective* of whether the deviation is more or les than the expected.

this test, and the small sample. By contrast, if we were to use the other method, we should have $u = 1/1.6 = 0.6$ and P_N from Table 2 would be 0.548. This would tell us that there was greater than 50% chance that our hypothesis is *right*, whereas the *t* test tells us only (for this small sample) that the observed deviation will be due to sample error slightly more than 5% of the time. The *t* test, then, is much more sensitive for small samples.

"Significant" Difference: χ^2 Test

When our data can have three or more values, rather than two, we cannot test all the observed values against theoretical values by the *t* test or the *u* test. For this situation, we use the χ^2 (chi-square) test, which is based on different distributions for each number of degrees of freedom. χ^2 is the sum of the squares of the differences between observed and calculated values divided by calculated values:

$$\chi^2 = \Sigma \frac{(O - C)^2}{C}$$

Note that the division is performed *before* summation, unlike the rest of our estimators. The number of degrees of freedom is usually $n - 1$ (although in some cases we subtract some number greater than 1 from n, we shall not consider such cases). We use 1 here because the calculated value of *aabb* is determined by the fact that the total of the column of calculated values must equal 102; hence, though the calculated values in the other three rows may be whatever our theory calls for, one value in the column must make the total equal to that in the observed column. For an example of the use of χ^2 we can calculate the F_2 of a dihybrid cross; see Table 4. Knowing that we have four classes, we have $4 - 1 = 3$ degrees of freedom, so we turn to Table 5, and look for a value for χ^2 in the row for 3 degrees of freedom. We find our calculated χ^2 to have a probability between 0.5 and 0.3, far larger than 0.05, so we accept the null hypothesis, that the observed offspring come from a dihybrid cross, with dominants and recessives distributed according to the first column in Table 4. Had we found χ^2 greater than 7.81, that is, with a probability of 0.05 or less, we should have rejected the null hypothesis.

"Significant" Difference: Populations

A common question asked in both descriptive and experimental biology is whether two samples are from the same or different populations. In an experiment where half the individuals are treated in some way, and half are "controls," we may think we have samples from two different populations. If the treatment is without effect, however, then we really have two samples from the same population. Of course, the very thing we are trying to find out is whether the treatment has an effect.

Our null hypothesis will be that there is no difference (no effect), and that the two samples come from the same population. Since we have no

TABLE 4

Class	Observed, O	Calculated, C		$O - C$	$(O - C)^2$	$(O - C)^2/C$
AB	59	$\frac{9}{16}$ of $102 =$	57.375	1.625	2.64	0.046
Abb	16	$\frac{3}{16}$ of $102 =$	19.125	−3.125	9.77	0.511
aaB	23	$\frac{3}{16}$ of $102 =$	19.125	3.875	15.02	0.785
aabb	4	$\frac{1}{16}$ of $102 =$	6.375	−2.375	5.64	0.885
Total	102		102.000	0.000		$\chi^2 = 2.227$

TABLE 5. *Table of* χ^2

Degrees of Freedom	P		Degrees of Freedom	P	
	0.05	0.01		0.05	0.01
1	3.84	6.64	16	26.30	32.00
2	5.99	9.21	17	27.59	33.41
3	7.82	11.34	18	28.87	34.81
4	9.49	13.28	19	30.14	36.19
5	11.07	15.09	20	31.41	37.57
6	12.59	16.81	21	32.67	38.93
7	14.07	18.48	22	33.92	40.29
8	15.51	20.09	23	35.17	41.64
9	16.92	21.67	24	36.42	42.98
10	18.31	23.21	25	37.65	44.31
11	19.68	24.73	26	38.89	45.64
12	21.03	26.22	27	40.11	46.96
13	22.36	27.69	28	41.34	48.28
14	23.69	29.14	29	42.56	49.59
15	25.00	30.58	30	43.77	50.89

theoretical values for the mean or standard deviation of what we are measuring, we must estimate these parameters. At first it might appear logical to take sample values from the untreated control group as the best estimate of the two parameters. We hypothesize, however, that there is no difference between the populations from which we have taken our samples; hence, *any* differences between means of the two samples and sample standard deviations must be attributed to sampling error, unless the probabilities are too low, in which case we must reject our null hypothesis. Before we have rejected it, however, we cannot select one sample (the control) as the better for estimating parameters. To do so is to state, in essence, that the control has no sampling error and all error lies in the experimental group.

To avoid this indefensible position, we use *all* the data from both samples to estimate the parameters. This has the advantage (if the null hypothesis is true) of giving better estimates of the parameter, since the larger the sample, the smaller its sampling error. (When the sample is as large as the population, there is no error, and the statistic becomes the parameter.)

We are usually more interested in finding out whether the means from two samples can be considered to have come from the same population than whether the sample standard deviations are from the same population. To test this we use a t test, but before we can apply it, we must first show that the standard deviations of the samples do not differ by a statistically significant amount. To do this, we use an F test. If we have two samples s_1 and s_2, with degrees of freedom f_1 and f_2, then the value of F will be

$$\frac{f_1}{f_2}F = \frac{s_1^2}{s_2^2}$$

where each s is a standard deviation of the samples. We always arrange the samples so that s_1 is larger than s_2, since the numbering is arbitrary. This makes F always larger than 1.0. Each row in the F table (Table 6) is for a different number of degrees of freedom in the denominator; each column lists degrees of freedom in the numerator. There is a separate table for each important probability level. Ordinarily we must look from table to table until we find in the row and column for our two degree of freedom indicators the nearest value to the calculated F. The probability for this table is the one which applies to our test. If this is above, say, the 0.05 level, we accept the hypothesis that the two samples come from the same population. We then turn to the t test to see if the sample means come from the same population. If we reject the null hypothesis on the basis of the F test, then we need not test the means, for we already know that the samples do not come from the same population.

To test for means, we use a different formula for t:

$$t = \frac{\overline{x_1} - \overline{x_2}}{s\sqrt{1/n_1 + 1/n_2}}$$

where x_1 is the mean of sample 1, x_2 the mean of sample 2, and n_1 and n_2 the numbers of individuals in each sample. The sample standard deviation, found by *combining* the data for two samples, is designated s, and calculated:

$$s = \sqrt{f_1s_1^2 + f_2s_2^2/f_1 + f_2}$$

where $f_1 = n_1 - 1$ and $f_2 = n_2 - 1$, and s_1 and s_2 are standard deviations of each sample. In the t test the number of degrees of freedom is $f_1 + f_2$. Again, we either accept or reject our hypothesis that the two sample means were from the same population depending on the probability we find above the appropriate column in the t table.

There are many other statistics, and kinds of statistical tests. All such tests are based on attempts to find out, by using samples of finite size, parameters that will describe complete populations. The sizes of these populations can range from small to infinitely large. Other statistical tests

TABLE 6. *F* (*Points of Variance*) *Distribution*

5% Level

f_1

f_2	1	2	3	4	5	10	20	∞
1	161.0	200.0	216.0	225.0	230.0	242.0	248.0	254.0
2	18.5	19.0	19.2	19.3	19.3	19.4	19.5	19.5
3	10.1	9.6	9.3	9.1	9.0	8.8	8.7	8.5
4	7.7	6.9	6.6	6.4	6.3	6.0	5.8	5.6
5	6.6	5.8	5.4	5.2	5.1	4.7	4.6	4.4
10	5.0	4.1	3.7	3.5	3.3	3.0	2.8	2.5
20	4.4	3.5	3.1	2.9	2.7	2.4	2.1	1.8
∞	3.8	3.0	2.6	2.4	2.2	1.8	1.5	1.0

1% Level

f_1

f_2	1	2	3	4	5	10	20	∞
1	4052.0	4999.0	5403.0	5625.0	5764.0	6056.0	6208.0	6366.0
2	98.5	99.0	99.2	99.3	99.3	99.4	99.5	99.5
3	34.1	30.8	29.5	28.7	28.2	27.2	26.7	26.1
4	21.2	18.0	16.7	16.0	15.5	14.5	14.0	13.5
5	16.3	13.3	12.1	11.4	11.0	10.1	9.6	9.0
10	10.0	7.6	6.6	6.0	5.6	4.9	4.4	3.9
20	8.1	5.9	4.9	4.4	4.1	3.4	2.9	2.4
∞	6.6	4.6	3.8	3.3	3.0	2.3	1.9	1.0

are designed, as are the χ^2, *F*, and *t* tests, to aid us in accepting or rejecting certain hypotheses, and to help us estimate errors that may arise because of incorrect acceptance or rejection.

SUMMARY

$$\bar{x} = \frac{\Sigma x}{n}$$

The arithmetic mean of a sample is the sum of the measured values of its members divided by the number of members. The population mean μ is arrived at in the same fashion, summing every member of the population. Since the population is often infinite, $\bar{x}$ is used to estimate μ.

$$s = \sqrt{ \Sigma\,(x - \overline{x})^2/(n - 1)}$$

The standard deviation measures the tendency of members of a population to cluster more or less closely about the mean. It is used with normal distributions.

$$\sigma = \sqrt{\overline{pqN}}$$

The standard deviation of a binomial population (such as coin tosses or either/or treatment of hereditary factors). p and q are the probabilities of the alternatives (for example, $\frac{1}{2}$ and $\frac{1}{2}$ in the case of coins) and N is the number of cases.

$$SE = \frac{s}{\sqrt{n}}$$

The standard error of the mean, shows us how often the mean of the sample we take will be as far or farther from the true mean of the population. A 95% confidence interval is determined over the range covered by the distance 2 SE from the mean on either isde. The true mean will lie outside this interval 5% of the time.

$$\frac{\mu - \overline{x}}{\sigma} = \tilde{u}$$

The formula for finding by the use of Table 2 whether departures from theoretical values are statistically significant or due to sampling error.

$$t = \frac{\overline{x} - \mu}{s/\sqrt{n}}$$

t for a t test for normally distributed populations containing fewer than 30 individuals.

$$\chi^2 = \Sigma\,\frac{(O - C)^2}{C}$$

The χ^2 test for testing hypotheses about theoretical values of characteristics which can have more than two values. O stands for "observed values" (in the sample), and C for calculated (theoretical) values.

$$\frac{f_1}{f_2}\,F = \frac{s_1{}^2}{s_2{}^2}$$

The F test for showing that standard deviations of two samples do not differ by a statistically significant amount and hence may come from the same population. f_1 is equal to $n_1 - 1$, f_2 to $n_2 - 1$, and $s_1{}^2$ and $s_2{}^2$ are the mean squared deviations from each of the two samples; by convention, the subscript 1 is always assigned the sample with the larger standard deviation.

$$t = -\frac{\overline{x_1} - \overline{x_2}}{s\sqrt{1/n_1 + 1/n_2}}$$

The t test for determining whether two sample means are from the same population. $\overline{x_1}$ and $\overline{x_2}$ are means of the two samples; s is the standard deviation of the sample formed by combining the two samples.

$$s = \sqrt{(f_1 s_1{}^2 + f_2 s_2{}^2)/(f_1 + f_2)}$$

The formula for finding the standard deviation of two samples combined into one, used in the t test.

SECTION 5

Appendices

Republished Research Papers in Biology

Adelberg, E. A., ed. *Papers on bacterial genetics*. Little, Brown, Boston, 1960.

Beaumont, W. *Experiments and observations on the gastric juice and the physiology of digestion* (1833). Dover Publications, New York (paperback), 1959.

Boyer, S. H., ed. *Papers in human genetics*. Prentice-Hall, Englewood Cliffs, N. J., 1963.

Camac, C. N. B. *Classics of medicine and surgery*. Dover Publications, New York, 1960.

Clendening, L. *Sourcebook of medical history*. Dover Publications, New York, 1960.

Darwin, C. R. *On the origin of the species by means of natural selection*. Numerous editions, including Modern Library, New York.

Darwin, C. R. *The structure and distribution of coral reefs*. University of California Press, 1962.

Darwin, C. R. *The voyage of the beagle*. Reprint by Bantam Books, New York, 1960.

Dobell, C., ed. *Antony van Leeuwenhoek and his "little animals."* Dover Publications, New York, 1960.

Doetsch, R. N. *Microbiology: historical contributions from 1776 to 1908*. Rutgers University Press, New Brunswick, N. J., 1961.

Fulton, J. F. *Selected readings in the history of physiology*. Charles C. Thomas, Springfield, Ill., 1966.

Gabriel, M. L. and S. Fogel. *Great experiments in biology*. Prentice-Hall, Englewood Cliffs, N. J., 1955.

Haldane, J. B. S. *The inequality of man*. Penguin, Baltimore, 1938.

Hall, T. S. *A sourcebook in animal biology*. McGraw-Hill, New York, 1951.

Harvey, W. *Anatomical studies on the motion of the heart and blood*. Charles C. Thomas, Springfield, Ill., 1958.

Hazen, W. E. *Readings in population and community ecology*. W. B. Saunders, Philadelphia, 1964.

Huxley, J. S. *The living thoughts of Darwin*. Fawcett Books, New York, 1958.

611

Knobloch, I. *Readings in plant science.* Prentice-Hall, Englewood Cliffs, N. J., 1963.

Lack, D. *Darwin's finches.* Harper & Brothers, New York, 1961.

Long, E. R. *Selected readings in pathology.* Charles C. Thomas, Springfield, Ill., 1961.

McGill, T. E., ed. *Readings in animal behavior.* Holt, Rinehart and Winston, New York, 1965.

Major, R. *Classic descriptions of disease.* Charles C. Thomas, Springfield, Ill., 3rd ed., 1959.

Malthus, T., J. S. Huxley, and F. Osborn. *On population: three essays.* New American Library of World Literature, New York (paperback), 1960.

Mendel, G. *Experiments in plant hybridization.* Harvard University Press, Cambridge, Mass., 1961.

Peters, J. A. *Classic papers in genetics.* Prentice-Hall, Englewood Cliffs, N. J., 1959.

Schwartz, G., and P. W. Bishop. *Moments of discovery.* Vols. I and II. Basic Books, New York, 1958.

Stent, G. S. *Papers on bacterial viruses.* Little, Brown, Boston, 1960.

Suner, A. P. *Classics in biology.* Philosophical Library, New York, 1955.

APPENDIX

2

Selected Bibliography for Teachers

LEVELS OF ORGANIZATION—MOLECULES: BIOCHEMISTRY AND BIOPHYSICS

Baldwin, E. *Dynamic aspects of biochemistry.* Cambridge University Press, London, 1957.

Barry, J. M. *Molecular biology: genes and the chemical control of living cells.* Prentice-Hall, Englewood Cliffs, N. J., 1964.

Campbell, P. N. and G. D. Grenville. *Essays in biochemistry.* 2 vols. Academic Press, New York, 1966.

Cheldelin, V. H. and R. W. Newburgh. *The chemistry of some life processes.* Reinhold, New York, 1964.

Crick, F. H. C. *Of molecules and men.* University of Washington Press, Seattle, 1966.

Edsall, J. T. and J. Wyman. *Biophysical chemistry,* Vol. I (Vol. II in preparation). Academic Press, New York, 1958.

Fox, S. W., ed. *The origins of prebiological systems and of their molecular matrices.* Academic Press, New York, 1965.

Giese, A. C. *Cell physiology.* W. B. Saunders, Philadelphia, 1962.

Grunwald, E., and R. H. Johnson. *Atoms, molecules and chemical change.* Prentice-Hall, Englewood Cliffs, N. J., 1960.

Lagowski, J. J. *The chemical bond.* Houghton Mifflin, Boston, 1966.

Lessing, L. *Understanding chemistry.* New American Library, New York, 1959.

Setlow, R. B. and E. C. Pollard. *Molecular biophysics.* Addison-Wesley, Reading, Mass., 1962.

Whittingham, C. P. *The chemistry of plant processes.* Philosophical Library, New York, 1965.

LEVELS OF ORGANIZATION—CELLS AND TISSUES

Butler, J. A. V. *The life of the cell: its nature, origin, and development.* Basic Books, New York, 1964.

Clark, W. E. L. *Tissues of the body.* Clarendon Press, Oxford, 1958.

Gerard, R. W. *Unresting cells.* Harper and Row, New York, 1961.

Giese, A. C. *Cell physiology.* W. B. Saunders, Philadelphia, 2d ed., 1962.

Loewy, A. and P. Siekevitz. *Cell structure and function.* Holt, Rinehart & Winston, New York, 1963.

McElroy, W. D. *Cell physiology and biochemistry.* Foundations of Modern Biology Series, Prentice-Hall, Englewood Cliffs, N. J., 1964.

Mercer, E. H. *Cells, their structure and function.* Doubleday, New York, 1962.

Monroy, A. *Chemistry and physiology of fertilization.* Holt, Rinehart and Winston, New York, 1965.

Porter, K. R., and M. A. Bonneville. *An introduction to the fine structure of cells and tissues.* Lea & Febiger, Philadelphia, 2d ed., 1964.

Robertis, E., de, and others. *General cytology.* W. B. Saunders, Philadelphia, 1960.

Spratt, N. T. *Introduction to cell differentiation.* Reinhold, New York, 1964.

Swanson, Carl P. *The cell.* Prentice-Hall, Englewood Cliffs, N. J., 1964.

LEVELS OF ORGANIZATION—ORGANISMS: PLANTS

Bonner, J. and A. Galston. *Principles of plant physiology.* W. H. Freeman, San Francisco, 1952.

Chapman, V. J. *The algae.* St. Martin's Press, New York, 1962.

Christensen, C. M. *Molds and man: an introduction to the fungi.* University of Minnesota Press, Minneapolis, 1961.

Doyle, W. T. *Nonvascular plants: form and function.* Wadsworth, Belmont, Calif., 1964.

Esau, K. *Plant anatomy.* 2d ed., Wiley, New York, 1965.

Galston, A. *The life of the green plant.* Prentice-Hall, Englewood Cliffs, N. J., 1964.

Greulach, V. A., and J. E. Adams. *Plants: an introduction to modern botany.* 2d ed., Wiley, New York, 1967.

James, W. O. *An introduction to plant physiology.* 6th ed., Oxford University Press, New York, 1963.

Jensen, W. A., and L. G. Kavaljian, eds. *Plant biology today.* Wadsworth, Belmont, Calif., 1963.

Lee, A. E., and C. Heimsch. *Development and structure of plants.* Holt, Rinehart and Winston, New York, 1962.

Steward, F. C. *Plant physiology: a treatise.* Academic Press, New York, 1965.

Went, F. W. and editors of Life. *The plants.* Time, New York, 1963.

LEVELS OF ORGANIZATION—ORGANISMS: ANIMALS

Asimov, I. *The human brain.* Houghton Mifflin, Boston, 1964.

Barnes, R. D. *Invertebrate zoology.* Saunders, Philadelphia, 1964.

Best, C. H., and N. B. Taylor. *The human body: its anatomy and physiology.* 4th ed., Holt Rinehart and Winston, New York, 1963.

Carlson, E., and V. Johnson. *The machinery of the body.* 5th ed. University of Chicago Press, Chicago, 1961.

Darling, F. F. *A herd of red deer.* The Natural History Library, Anchor Books, Garden City, N. Y., 1964.

Dethier V. *To know a fly.* Holden-Day, San Francisco, 1962.

Eimerl, S., and I. DeVore, and editors of Life. *The Primates.* Time, New York, 1965.

Farb, P. and editors of Life. *The insects.* Time, New York, 1962.

Graubard, M. *Circulation and respiration: the evolution of an idea.* Harcourt, Brace & World, New York, 1964.

Gray, J. *How animals move.* Rev. Ed., Cambridge University Press, London, 1959.

Griffin, D. R. *Animal structure and function.* Holt, Rinehart and Winston, New York, 1962.

Houssay, B. A., and others. *Human physiology.* McGraw-Hill, New York, 1955.

Jacob, S. W. and C. A. Francone. *Structure and function in man.* W. B. Saunders, Philadelphia, 1965.

Michelmore, S. *Sexual reproduction.* The American Museum of Natural History, New York, 1966.

Moment, G. B. *General zoology.* 2d ed., Houghton Mifflin, Boston, 1967.

Moore, J. A. *Principles of zoology.* Oxford University Press, New York, 1957.

Morley, D. W. *The ant world.* Penguin Books, Baltimore, 1953.

Nalbandov, A. V. *Reproductive physiology.* 2d ed., Freeman, San Francisco, 1964.

Nottingham University School of Agriculture. *Soil zoology.* D. K. McE. Kevan, ed. Butterworths, London, 1955.

Ommanney, F. D., and editors of Life. *The fishes.* Time, New York, 1963.

Peterson, R. T., and editors of Life. *The birds.* Time, New York, 1963.

Prosser, C. L. and F. A. Brown, Jr. *Comparative animal physiology.* W. B. Saunders, Philadelphia, 1961.

Riedman, S. R. *Our hormones and how they work.* Collier Books, New York, 1962.

Romer, A. S. *The vertebrate body.* 3d ed. W. B. Saunders, Philadelphia, 1962.

Romer, A. S. *The vertebrate story.* 4th ed. University of Chicago Press, Chicago, 1969.

Ruch, T. C., and J. F. Fulton. *Medical physiology and biophysics.* W. B. Saunders, Philadelphia, 1960.

Schmidt, K. P., and R. F. Inger. *Living reptiles of the world.* Doubleday, New York, 1957.

Schmidt-Nielson, K. *Animal physiology.* Prentice-Hall, Englewood Cliffs, N. J., 1964.

Storer, T. I., and R. L. Usinger. *General zoology.* 4th ed., McGraw-Hill, New York, 1965.

Tinbergen, N. *The herring gull's world.* Basic Books, New York, 1961.

Turner, C. D. *General endocrinology.* 4th ed., W. B. Saunders, Philadelphia, 1966.

Van Lyne, J., and A. J. Berger. *Fundamentals of ornithology.* Wiley, Science Editions, New York, 1959.

Yapp, W. B. *Vertebrates: their structure and life.* Oxford University Press, New York, 1965.

Young, J. Z. *The life of vertebrates.* 2d ed. Oxford University Press, 1962.

LEVELS OF ORGANIZATION—BIOSPHERE: PAST AND PRESENT

Amos, W. H. *The life of the seashore.* McGraw-Hill, New York, 1966.

Bourliere, F. *The natural history of mammals.* Alfred A. Knopf, New York, 1964.

Brouwer, A. *General palaeontology.* University of Chicago Press, Chicago, 1967.

Carson, R. *The sea around us.* Simon and Schuster, New York, 1958.

Coker, R. E. *Streams, lakes, ponds.* University of North Carolina Press, Chapel Hill, N. C., 1954.

Colbert, E. H. *Men and dinosaurs.* E. P. Dutton, New York, 1968.

Darlington, P. J., Jr. *Zoogeography: the geographical distribution of animals.* Wiley, New York, 1957.

Darrah, W. C. *Principles of paleobotany.* The Ronald Press, New York, 1960.

Fenton, C. L. and M. A. Fenton. *The fossil book: a record of prehistoric life.* Doubleday, Garden City, N. Y., 1958.

Gleason, H. A. and A. Cronquist. *The natural geography of plants.* Columbia University Press, New York, 1964.

Goldring, W. *Handbook of paleontology for beginners and amateurs.* Part I, "The Fossils." Paleontological Research Institution, Ithaca, N. Y., 1960.

Good, R. O. *Geography of the flowering plants.* Wiley, New York, 1964.

Hanson, H. C. and E. D. Churchill. *The plant community.* Reinhold, New York, 1961.

Hardy, A. C. *The open sea: its natural history.* Part I., "The World of Plankton." Houghton Mifflin, Boston, 1957.

Hesse, R., W. C. Allee, and K. P. Schmidt. *Ecological animal geography.* Wiley, New York, 1951.

Matthews, W. H. *Fossils: an introduction to prehistoric life.* Barnes and Noble, New York, 1962.

McCormick, J. *The life of the forest.* McGraw-Hill, New York, 1966.

Niering, W. A. *The life of the marsh.* McGraw-Hill, New York, 1967.

Ramson, J. E. *Fossils in america.* Harper and Row, New York, 1965.

Ricketts, E. F., and J. Calvin. *Between pacific tides.* 3d ed. rev. by J. W. Hedgpeth. Stanford University Press, 1962.

Romer, A. S. *Vertebrate paleontology.* 3d ed., University of Chicago Press, Chicago, 1966.

Russell, E. J. *The world of the soil.* Longmans, Green, London, 1950.

Simpson, G. G. *Life of the past: an introduction to paleontology.* Yale University Press, New Haven, Conn., 1953.

Sverdrup, H. U., and others. *The oceans: their physics, chemistry, and general biology.* Prentice-Hall, Englewood Cliffs, N. J., 1942.

DIMENSIONS OF BIOLOGICAL INVESTIGATION—BEHAVIOR

Ardrey, R. *The territorial imperative.* Atheneum, New York, 1966.

Barnett, S. A. *Instinct and intelligence: behavior of animals and man.* Prentice-Hall, Englewood Cliffs, N. J., 1967.

Carr, A. *Guideposts of animal navigation.* D. C. Heath, Boston, 1962.

Case, J. *Sensory mechanisms.* Macmillan, New York, 1966.

Davis, D. E. *Integral animal behavior.* Macmillan, New York, 1966.

Dethier, V. G., and E. Stellar. *Animal behavior: its evolutionary and neurological basis.* 2d ed., Prentice- Hall, Englewood Cliffs, N. J., 1964.

Etkin, W. *Social behavior and organization among vertebrates.* University of Chicago Press, Chicago, 1964.

Frings, H., and M. Frings. *Animal communication.* Blaisdell, New York, 1964.

Huxley, J. and L. Koch. *Animal language.* Grosset & Dunlap, New York, 1964.

Klopfer, P. H. *Behavioral aspects of ecology.* Prentice-Hall, Englewood Cliffs, N. J., 1962.

Klopfer, P. H., and J. P. Hailman. *An introduction to animal behavior.* Prentice-Hall, Englewood Cliffs, N. J., 1967.

Lorenz, K. Z. *King solomon's ring.* Thomas Y. Crowell, New York, 1952.

Lorenz, K. Z. *On aggression.* Harcourt, Brace & World, New York, 1966.

Marler, P. R., and W. J. Hamilton III. *Mechanisms of animal behavior.* Wiley, New York, 1966.

McGaugh, J. L., and others. *Psychobiology.* W. H. Freeman, San Francisco, 1967.

Portmann, A. *Animals as social beings.* The Viking Press, New York, 1961.

Rheingold, H. *Maternal behavior in mammals.* Wiley, New York, 1963.

Roe, A., and G. G. Simpson, *Behavior and evolution.* Yale University Press, New Haven, 1958.

Scott, J. P. *Animal behavior.* Doubleday, New York, 1963. Originally published in 1958 by University of Chicago Press.

Southwick, C. H. *Primate social behavior: an enduring problem.* D. Van Nostrand, Princeton, N. J., 1963.

Thorpe, W. H. *Learning and instinct in animals.* Harvard University Press, Cambridge, 1958.

Tinbergen, N., and editors of Life. *Animal behavior.* Time, New York, 1965.

Tinbergen, N. *Social behavior in animals.* Butler & Tanner, London, 1965.

Tinbergen, N. *A study of instinct.* Clarendon Press, Fair Lawn, N. J., 1951.

Van Der Kloot, W. G. *Behavior.* Holt, Rinehart and Winston, New York, 1968.

Wynne-Edwards, V. C. *Animal dispersion in relation to social behavior.* Oliver & Body, London, 1962.

DIMENSIONS OF BIOLOGICAL INVESTIGATION—HEREDITY AND DEVELOPMENT

Beadle, G., and M. Beadle. *The language of life.* Doubleday, New York, 1966.

Bonner, D. *Heredity.* Prentice-Hall, Englewood Cliffs, N. J., 1961.

Borek, E. *The code of life.* Columbia University Press, New York, 1965.

Boyd, W. *Genetics and the races of man.* Little, Brown, Boston, 1958.

Brewbaker, J. L. *Agricultural genetics.* Prentice-Hall, Englewood Cliffs, N. J., 1964.

Comfort, A. *The process of ageing.* New American Library, New York, 1964.

Golding, W. *The inheritors.* Harcourt, Brace & World, New York, 1963.

Hartman, P. E., and S. R. Suskind. *Gene action.* Prentice-Hall, Englewood Cliffs, N. J., 1965.

Kendrew, J. C. *The thread of life.* Harvard University Press, Cambridge, Mass., 1966.

King, R. C. *Genetics.* 2d ed., Oxford University Press, New York, 1965.

Levine, R. P. *Genetics,* Holt, Rinehart and Winston, New York, 1963. Contains sections on gene action, gene transmission and gene chemistry.

McKusick, V. A. *Human genetics.* Prentice-Hall, Englewood Cliffs, N. J., 1964.

Moore, J. A. *Heredity and development.* Oxford University Press, New York, 1963.

Moore, R. *The coil of life.* Alfred A. Knopf, New York, 1961.

Roslansky, J. O., ed. *Genetics and the future of man.* Appleton-Century-Crofts, New York, 1966.

Rugh, R. *Vertebrate embryology: the dynamics of development.* Harcourt, Brace & World, New York, 1965.

Stahl, F. W. *The mechanics of inheritance.* Prentice-Hall, Englewood Cliffs, N. J., 1964.

Stern, C., and E. R. Sherwood. *The origin of genetics.* W. H. Freeman, San Francisco, 1966.

Strehler, B. L. *Time, cells and aging.* Academic Press, New York, 1962.

Strickberger, M. W. *Experiments in genetics with drosophila.* Wiley, New York, 1962.

Sturtevant, A. H. *A history of genetics.* Harper & Row, New York, 1965.

Sussman, M. *Animal growth and development.* 2d ed., Prentice-Hall, Englewood Cliffs, N. J., 1964.

Waddington, C. H. *Principles of development and differentiation.* Macmillan, New York, 1966.

Watson, J. D. *The double helix.* Atheneum, New York, 1968.

DIMENSIONS OF BIOLOGICAL INVESTIGATION—ECOLOGY

Ager, D. V. *Principles of paleoecology.* McGraw-Hill, New York, 1963.

Andrewartha, H. G. *Introduction to the study of animal populations.* University of Chicago Press, Chicago, 1961.

Bates, M. *The forest and the sea.* Random House, Toronto, 1960.

Bates, M. *Man in nature.* Prentice-Hall, Englewood Cliffs, N. J., 1961.

Becker, H. F. *Resources for tomorrow.* Holt, Rinehart & Winston, New York, 1964.

Billings, W. D. *Plants and the ecosystem.* Macmillan, New York, 1965.

Brown, Harrison. *The challenge of man's future.* Viking Press, New York, 1956. Available in paperback.

Browning, T. O. *Animal populations.* Harper & Row, New York, 1963.

Cloudsley-Thompson, J. L. *Microecology.* St. Martin's Press, New York, 1967.

Carson, R. *Silent Spring.* Houghton Mifflin, Boston, 1962.

Dowdeswell, W. H. *Animal ecology.* Harper & Row, New York, 1961.

Elton, C. S. *The ecology of invasions* by *animal and plants.* Wiley, New York, 1958.

Fiennes, R. *Man, nature and disease.* New American Library, New York, 1965.

Frey, D. G. *Limnology in North America.* University of Wisconsin Press, Madison, 1963.

Hanson, H. C. and E. D. Churchill. *The plant community.* Reinhold, New York, 1961.

Hardin, G. *Nature and man's fate.* New American Library, New York, 1961.

Lack, D. *Population studies of birds.* Oxford University Press, Oxford, 1966.

National Academy of Sciences. *The growth of world population.* National Research Council Publication, 1963.

Odum, E. P. *Ecology.* Holt, Rinehart and Winston, New York, 1963.

Oosting, H. J. *The study of plant communities.* W. H. Freeman, San Francisco, 1956.

Phillips, E. A. *Field ecology.* (A BSCS Laboratory Block.) D. C. Health, Boston, 1964.

Phillipson, J. *Ecological energetics.* St. Martin's Press, New York, 1966.

Ruttner, F. *Fundamentals of limnology.* 3d ed., University of Toronto Press, Toronto, 1963.

Sears, M. *Oceanography.* American Association for the Advancement of Science, Publication No. 67, Washington, D. C., 1961.

Sears, P. B. *Where there is life.* Dell, New York, 1962.

Shapiro, H. L. *Man, culture and society.* Oxford University Press (Galaxy Books), New York, 1960.

Storer, J. H. *The web of life, a first book of ecology.* The Devin-Adair, New York, 1967.

Tinbergen, N. *Curious naturalists.* Basic Books, New York, 1958.

Weaver, J. E., and F. E. Clements. *Plant ecology.* McGraw-Hill, New York, 1938.

DIMENSIONS OF BIOLOGICAL INVESTIGATION—EVOLUTION

Barrington, E. J. W. *Hormones and evolution.* Van Nostrand, Princeton, N. J., 1964.

Brodrick, A. H. *Man and his ancestry.* Fawcett Publications, New York, 1964.

Colbert, E. H. *Evolution of the vertebrates.* Wiley, New York, 1961.

Coon, C. S. *The story of man.* 2d ed. rev., Knopf, New York, 1962.

Dart, R., and D. Craig. *Adventures with the missing link.* Harper and Row, New York, 1959.

Delevoryas, T. *Plant diversification.* Modern Biology Series, Holt, Rinehart and Winston, New York, 1966.

Dobzhansky, T. *Mankind evolving.* Yale University Press, New Haven, 1962.

Eiseley, L. *Darwin's century.* Doubleday, New York, 1961.

Hanson, E. D. *Animal diversity.* 2d ed., Prentice-Hall, Englewood Cliffs, N. J., 1964.

Huxley, J. *Evolution in action.* New American Library, New York, 1957.

Lasker, G. W. *Human evolution.* Holt, Rinehart and Winston, New York, 1963.

Montagu, A. *Man: his first million years.* New American Library of World Literature, New York, 1962.

Moore, R., and editors of life. *Evolution.* Time, New York, 1962.

Simpson, G. C. *The meaning of evolution.* New American Library, New York, 1960.

Srb, A. M., and B. Wallace. *Adaptation.* 2d ed. Prentice-Hall, Englewood Cliffs, N. J., 1964.

Stebbins, G. L. *Processes of organic evolution.* Prentice-Hall, Englewood Cliffs. N. J., 1966.

Stirton, R. A. *Time, life, and man.* Wiley, New York, 1959.

Wendt, H. *Before the deluge.* Doubleday, Garden City, N. Y., 1968.

Volpe, E. P. *Understanding evolution.* William C. Brown, Dubuque, Iowa, 1967.

DIMENSIONS OF BIOLOGICAL INVESTIGATION—TAXONOMY

Benson, L. *Plant classification.* D. C. Health, Boston, 1957.

Blair, W. F. and others. *Vertebrates of the United States.* McGraw-Hill, New York, 1957.

Edmonson, W. T., ed. *Fresh-water biology.* Wiley, New York, 1959. Second Edition, the revised edition of the classic book by H. B. Ward and G. C. Whipple.

Lawrence, G. H. *Taxonomy of vascular plants.* Macmillan, London, 1952.

Lawrence, G. H. *Introduction to plant taxonomy.* Macmillan, New York, 1955.

Porter, C. L. *Taxonomy of flowering plants.* W. H. Freeman, San Francisco, 1961.

Rothschild, N. M. V. *A classification of living animals.* Wiley, New York, 1961.

Savory, T. *Naming the living world.* Wiley, New York, 1963.

Scagel, R. F., and others. *An evolutionary survey of the plant kingdom.* Wadsworth, Belmont, Calif., 1965.

Simpson, G. G. *Principles of animal taxonomy.* Columbia University Press, New York, 1961.

Sokal, R. R., and P. H. A. Sneath. *Principles of numerical taxonomy.* W. H. Freeman, San Francisco, 1963.

OTHER PERSPECTIVES ON BIOLOGY—PHILOSOPHY OF SCIENCE

Bates, M. *The nature of natural history.* rev. ed., Charles Scribner's, New York, 1962.

Beveridge, W. I. B. *The art of scientific investigation.* Norton, New York, 1957.

Bonner, J. T. *Cells and societies.* Princeton University Press, Princeton, 1955.

Bonner, J. T. *The ideas of biology.* Harper and Row, New York, 1962.

Cohen, M. R., and E. Nagel. *An introduction to logic and scientific method.* Harcourt, Brace & World, New York, 1934.

Conant, J. B. *Science and common sense.* Yale University Press, New Haven, Conn., 1961.

Conant, J. B. *On understanding science.* Yale University Press, New Haven, Conn., 1947.

King, A. C., and C. B. Read. *Pathways to probability.* Holt, Rinehart and Winston, New York, 1963.

Lwoff, A. *Biological order.* The MIT Press, Cambridge, Mass., 1962.

Newman, J. R. *What is science?* Washington Square Press, New York, 1961.

Platt, J. R. *The excitement of science.* Houghton Mifflin, Boston, 1962.

Pyke, M. *The boundaries of science.* Penguin Books, Baltimore, 1963.

Simpson, G. G. *This view of life: the world of an evolutionist.* Harcourt, Brace & World, New York, 1964.

Singer, C. J. *A history of biology to about the year 1900,* 3d rev. ed., Abelard-Schuman, New York, 1959.

Sinnott, E. W. *Cell and psyche.* Harper and Row, New York, 1961.

OTHER PERSPECTIVES ON BIOLOGY—MATTER, ENERGY AND THE UNIVERSE

Baker, J. J. W., and G. E. Allen. *Matter, energy, and life.* Addison-Wesley, Palo Alto, Calif., 1965.

Gamow, G. *The creation of the universe,* rev. ed. New American Library, New York, 1965.

Gates, D. M. *Energy exchange in the biosphere.* Harper and Row, New York, 1962.

Hoyle, F. *The nature of the universe,* rev. ed. Harper, New York, 1960.

Lehninger, A. L. *Bioenergetics.* W. A. Benjamin, New York, 1965.

Shapley, H. *Of stars and men,* rev. ed. Beacon Press, Boston, 1964.

Watson, F. G. *Between the planets.* Doubleday, New York, 1962.

OTHER PERSPECTIVES ON BIOLOGY—GENERAL BIOLOGY

Hardin, G. *Biology: its principles and implications,* 2d ed. W. H. Freeman, San Francisco, 1966.

Johnson, W. H., and W. C. Steere. *This is life: essays in modern biology.* Holt, Rinehart, and Winston, New York, 1962.

Simpson, G. G. and others. *Life: an introduction to biology,* 2d ed. Harcourt, Brace & World, New York, 1965.

Weisz, P. B. *The science of biology,* 2d ed. McGraw-Hill, New York, 1963.

SCIENCE EDUCATION—RESEARCH REPORTS

ERIC Information Analysis Center for Science Education. 1460 West Lane Avenue, Columbus, Ohio 43221.

Lee, Addison E., editor. *Research and curriculum development in science education,* Vol. 1. "The New Programs in High School Biology." The University of Texas at Austin, 1968.

3

Laboratory Facilities for BSCS Biology

THE LABORATORY CLASSROOM

Four basic floor plans, each designed to accommodate thirty students, have been found suitable for the teaching of BSCS High School Biology.

Plan *A*

This plan begins with dual-purpose laboratory tables arranged in the center of the room and equipped with electrical outlets. These serve either as laboratory work benches or as writing desks. Sinks are placed around the periphery of the room as part of a complete counter system, with both upper and lower wall cabinets. This provides space for additional work and storage. This plan can be realized in a room 30 × 35 ft large (see Figure 1).

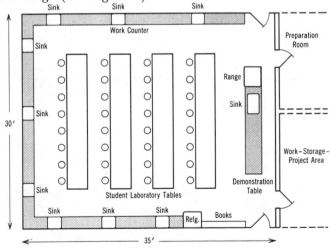

Figure 1

623

Plan *B*

This plan is based on U-shaped work areas arranged about the periphery of the room. Writing desks are then arranged in the center and are movable to facilitate lectures and discussions. Plan *B* requires a larger room than the one in plan *A*, that is, a space of at least 30 × 45 ft (see Figure 2).

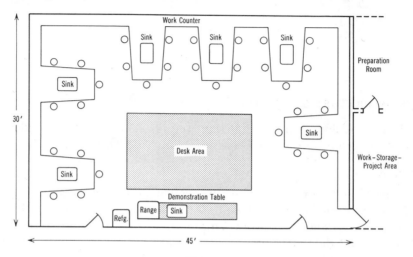

Figure 2

Plan *C*

In plan *C* island laboratory benches are arranged throughout a room of 25 × 30 ft. Lectures are given in separate rooms (see Figure 3).

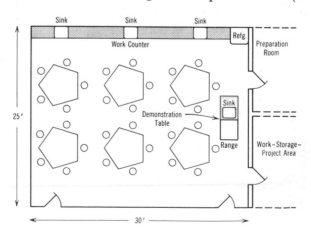

Figure 3

Plan *D*

Here the island laboratory benches are arranged at the rear of a room 30 × 50 ft. Lecture and discussion seating is at the front (see Figure 4).

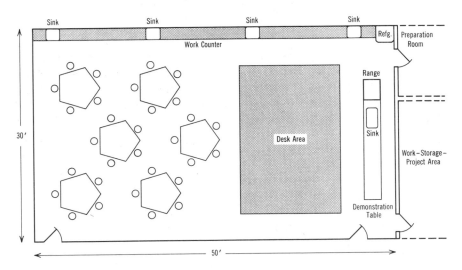

Figure 4

Other laboratory arrangements have also proved highly satisfactory, such as fixed utility areas with movable tables to permit flexibility. Some of the innovations in team teaching may have significant bearing on laboratory design. Schools planning specifically for multipurpose flexibility should provide for it in their building plans.

In considering a basic floor plan for a laboratory, provision should be made for:

1. A laboratory containing *at least* four work areas, with each work area equipped with sink, water, gas, and electricity.

2. Sufficient storage space so that students using the laboratory can maintain an experiment over a long period of time. (This is economically feasible with extensive installation of wall cabinets.)

3. An adequate preparation space (at least 6 × 10 ft) for the teacher, with storage areas for supplies and equipment. (An additional dispensing area located in the center of the laboratory has proved highly advantageous for issuing supplies and equipment.)

4. An overflow space or project area for students to carry out experiments beyond the regularly assigned work. (This may be part of the regular laboratory or in a separate project room.)

5. A life alcove or area set aside for the raising and care of plants and animals, both aquatic and terrestrial. This area should be large enough to provide elbow room for at least ten students. Temperature and humidity control in this facility (separate from the main heating-cooling plant) are desirable.

6. A demonstration desk equipped with a sink and utilities.

7. *At least* 12 linear feet each of chalkboard and tackboard space. (Tackboard panel on the cabinet doors of wall storage area gains bulletin-board space without sacrificing wall space.)

8. A student science library easily accessible within the confines of the science department. Such small libraries have proved very beneficial in many schools; they should include a minimum of fifty titles covering a wide range of biological and related science fields, so that students can be encouraged to go beyond the limits of the textbook. (See Appendices 1 and 2.)

MAJOR EQUIPMENT

1. Refrigerator or preferably refrigerator-freezer combination
2. Autoclave or large pressure cooker
3. One or more incubators
4. Aquaria
5. Binocular dissecting microscopes (one for every two students)
6. Monocular microscopes (one for every two students)
7. Several balances sensitive to at least $\frac{1}{100}$ g
8. A kitchen range for cooking of bacteriological media and use as a drying oven
9. Portable ac-dc power supply units
10. A wide selection of chemical glassware *in abundance*. (With increasing emphasis on biochemistry, relatively complete chemical laboratory and glassware stores are needed.)
11. A good stock of chemical reagents
12. A centrifuge
13. One rolling table on large wheels, for transportation of materials from storage area
14. A fume hood for handling radioactive or toxic and volatile materials

BUDGETARY REQUIREMENTS

First, at least $500 per year per biology teacher for perishable items such as chemicals, glassware, and living organisms is necessary. *The budget should be flexible enough so that teachers may purchase supplies as they are needed throughout the year.*

Second, additional funds must be provided for the purchase of capital outlay equipment and reference books. The amount will depend on the extent of existing equipment and reference materials.

BUILDING OR REMODELING FACILITIES

Costly mistakes, which are easily made, detract from the operation of good laboratories and remain as monuments to inadequate thought. If consultant service is not available, visits to schools that have most or all of the recommendations in use are recommended. A few such schools are:

The Melbourne High School, Melbourne, Florida
The Henry Ford High School, Detroit, Michigan
The Northport High School, Northport, Long Island
Wichita High School Southeast, Wichita, Kansas
The Palo Alto High School, Palo Alto, California
Mills Union High School, Millbrane, California

Schools unable to build new facilities may be able to remodel existing rooms. The following are actual modifications that have been undertaken by some schools to meet laboratory needs of the BSCS courses more fully.

A school having three small classrooms in a row converted the center room into a combination storeroom-project area. Work counters were constructed along three walls in the central room with upper and lower case cabinets. Several inexpensive laundry sinks were installed and a second-hand kitchen range and refrigerator added. Additional counter space and wall storage cabinets were constructed in the two classrooms.

Several old bathtubs provided one school with large aquaria that were used for algae and protozoan cultures as well as the keeping of turtles, alligators, and frogs.

Another school obtained old chemistry laboratory tables no longer in use, refinished the tops, and installed them in an ordinary classroom for use in biology.

A janitor's storeroom was converted into a science library.

Locally installed softwood cabinets, ordinary kitchen sinks, and linoleum counter tops are adequate for most biological work and result in large savings. (Formica tops are often more expensive and are damaged by alkali and easily burned by hot glass.) Some home economics laboratories would serve admirably.

Plastic substitutes for glassware reduce breakage.

LABORATORY FACILITIES CHECK LIST

To facilitate evaluation of biology laboratory facilities in a school, a check list is proposed for a laboratory used daily by four classes of thirty students each. This check list is not to be construed as a definitive standard.

In the check list facilities have been grouped in categories, to permit a separate evaluation of each category, since the pedagogic and monetary values of the various categories cannot be equated. For example, the fixed laboratory installations category is the most costly and is of paramount importance. A "B" rating in this category would be more significant than "A" in the demonstration aids category.

The check list in Table 1 (p. 00) may be used as follows:

TABLE 1. *Laboratory Facilities Check List**

Facility	Point Value				
Category A	16	12	8	4	*Your School*
Fixed Laboratory Installations (maximum possible score 144)					
Demonstration table	1	—	—	—	___
Work counter (peripheral) —feet	120	60	30	15	___
Sinks/utilities	4	3	2	1	___
Shelf storage—sq ft	450	300	200	100	___
Preparation room	large	med.	small	—	___
Life alcove	large	med.	small	—	___
Work project area	large	med.	small	—	___
Science library—min 50 vols.	large	med.	small	—	___
Display cases (in halls)	2	1	—	—	___
Subtotal points					___
Budget Considerations (maximum possible score 48)					
Funds for perishables, glassware, chemicals, specimens, etc.	$500/yr	$250/yr	$125/yr	$50/yr	___

Funds available *during year* as needed	yes	—	—	—	—
Capital outlay funds	$500/yr	$250/yr	$125/yr	$50/yr	—
Subtotal points					
Microscopes (maximum possible score 32)	30	15	8	4	—
Compound microscopes					
Binocular stereoscopic microscopes	30	15	8	4	—
Subtotal points					
Laboratory Assistants (maximum possible score 16)					
Paid laboratory assistants—5 hr per week per section	1	—	—	—	—
Subtotal points					—

Category B	12	9	6	3	*Your School*
Major Equipment (maximum possible score 84)					
Refrigerator	1	—	—	—	—
Gas range/oven	1	—	—	—	—
Incubator	2	1	—	—	—
Balances (0.01-g sensitivity)	4	3	2	1	—
Pressure cooker	2	1	—	—	—
Centrifuge	2	1	—	—	—
Power supply units	2	1	—	—	—
Subtotal points					—

Category C	4	3	2	1	*Your School*
Small Equipment and Supplies (maximum possible score 52)					
Thermometers (centigrade)	15	10	5	2	—
Filter flasks	2	1	—	—	—
Bunsen burners	8	4	2	1	—
Aquaria	4	3	2	1	—
Terraria	4	3	2	1	—
Dissecting sets	30	20	10	2	—
Laboratory specimens for dissection (pref. fresh)	many	adeq.	few	sparse	—
Laboratory organisms (going cultures)	5	4	3	2	—
Chemical reagents	many	adeq.	few	sparse	—

Pipettes (calibrated in 3 sizes)	36	24	15	6	—
Glassware, misc.	many	adeq.	few	sparse	—
Collecting equipment (nets, vials, buckets, etc.)	many	adeq.	few	—	—
Animal cages	8	6	4	2	—
Subtotal points					—

Demonstration Aids
(maximum possible score 12)

Preserved specimen sets (not for dissection)	many	adeq.	few	—	—
Models and charts	many	adeq.	few	—	—
Prepared slides, microscope	many	adeq.	few	—	—
Subtotal points					—

All Facilities

(maximum posible score 388)
Your school—total score ___

* Based on thirty students for a four-class day.

For each facility listed, circle the category that best describes the laboratory. In the last column on that line write in the point value for the circled item; that is, the point value at the head of the column in which the circled item appears.

Where the laboratory has none of the facilities mentioned (where a dash is circled on the table) the response value is zero.

Find the area subtotal for the school in each of the seven areas and compare it with the maximum possible score for that area.

Obtain a grand total for all areas and compare it with the rating scale in Table 2.

TABLE 2. *Rating Scale*

Rating	Points	Percent of Optimal
A	330-388	85-100
B	273-329	70-84
C	215-272	55-69
D	155-214	40-54
E	97-154	25-39
F	0-96	0-24

SOME REFERENCES ON SCHOOL FACILITIES

Barthelemy, Richard E., James R. Dawson, Jr., and Addison E. Lee, 1964. *Innovations in equipment and techniques for the biology teaching laboratory.* D. C. Health and Co., Boston.

CCSSO, 1959. *Purchase guide for programs in science, mathematics, modern foreign languages.* Ginn and Company, Boston, Mass.

Grobman, Arnold, Paul DeH. Hurd, Paul Klinge, Margaret McKibben Lawler, Elra Palmer, 1964. *BSCS biology. Implementation in the schools.* BSCS Bulletin No. 3.

Hurd, P. DeH., 1954. *Science facilities for the modern high school.* Education Administration Monograph No. 2, Stanford University Press, Stanford, Calif.

Johnson, P., 1952. *Science facilities for secondary schools.* U.S. Office of Education Misc. Bull. No. 17, U.S. Government Printing Office, Washington, D.C.

Lerner, Morris R. "Facilities, Materials, Resources and Their Effects," *The Science Teacher* **35** (7), October, 1968, p. 56.

Richardson, J. (editor), 1954. *School facilities for science instruction.* NSTA, Washington, D.C.

The Science Teacher 27, No. 1 (February, 1960). Science Facilities Issue.

APPENDIX

4

Techniques and Materials for the Biology Laboratory

This appendix of the Revised Edition of the BSCS *Biology teachers' handbook* includes material from three appendices in the original edition—"Sources of Laboratory Supplies," "Preparation of Solutions, Stains and Media for Laboratory Biology," and "Culture of Microorganisms and Small Invertebrates." Also some new material on organisms, frequently used in BSCS courses, has been added and the material has been reorganized into sections according to the major organisms used in high school biology. The Table of Contents below indicates the sequence of the material in this appendix. Following the Table of Contents, there is an Index, which will facilitate quick reference to particular preparations.

CONTENTS

INDEX

GENERAL INFORMATION AND BASIC STAINS AND REAGENTS

I. Measurement and Conversion Tables [1]

The basic unit in the metric system is the *meter*. From this unit all others are derived. The meter is the distance between two scratches on a platinum-iridium bar that is kept in the vaults of the International Bureau of Measures near Paris. The United States Bureau of Standards has copies of this bar. Fractions and multiples of the meter are designated by prefixes: *milli* ($\times$ 0.001) *centi-* ($\times$ 0.01), *deci-* ($\times$ 0.1), *deka-* ($\times$ 10), *hecto-* ($\times$ 100), *kilo-* ($\times$ 1000). See Table 1.

The metric unit of volume is the *liter,* which is defined as the volume of a cube having an edge 10 centimeters long. The metric unit of weight is the *gram,* which is defined as the weight of a milliliter of pure water at 4 degrees celsius (4°C). The metric unit of surface is the

---

[1] *High school biology, BSCS green version,* Student's Manual, Chicago: Rand McNally, 1963, pp. 372-374.

TABLE 1. *Metric Units*

Scale	Length	Volume	Weight	Surface
0.001	millimeter (mm)	milliliter (ml)	milligram (mg)	
0.01	centimeter (cm)			
0.1	decimeter (dm)			
1	meter (m)	liter (l)	gram (g)	are
10	dekameter (dkm)			
100	hectometer (hm)			hectare (ha)
1000	kilometer (km)		kilogram (kg)	

are, which is defined as an area of 100 square meters. Areas and volumes may also be indicated by the squares and cubes of linear units.

The following units are not properly a part of the metric system but are derived from it:

calorie   the quantity of heat needed to raise the temperature of 1 g of water 1°C

Calorie: equal to a kilocalorie

curie: the amount of any radioactive substance that emits the same number of alpha rays per unit of time as does 1 g of radium

micron (μ):   a unit of length equal to one millionth of a meter; $1000\mu = 1$ mm

Some equivalents between the British and Metric Systems are listed below:

1 centimeter =	.3937 inches	1 inch =	2.54 cm
1 meter =	39.37 inches	1 yard =	.914 m
1 kilometer =	.62 miles	1 mile =	1.6 km
1 liter =	1.057 liquid quarts	1 quart =	.95 l
1 gram =	.035 ounces	1 ounce =	28.34 g
1 kilogram =	2.2 pounds	1 pound =	.453 kg
1 hectare =	2.47 acres	1 acre =	.405 ha

In countries using British units of measure, temperature is expressed on a scale devised by Gabriel Daniel Fahrenheit (1686–1736). (The British Weather Service has recently abandoned this scale.) In countries using the metric system, temperature is expressed on a scale devised by Anders Celsius (1701–1744). The Celsius scale has been called *centigrade*. The figure below compares the two scales.

There is only one system of units for measuring time, though in many parts of the world (and in the armed forces of the United States) hours are designated from midnight to midnight with one set of twenty-four numbers—thus making "A.M." and "P.M." unnecessary.

II. Preparation of Solutions

Four different ways of specifying solutions are commonly used—percentage by weight, percentage by volume, normal *(N)*, and molar *(M)*. Formulas for solutions should be made up with as much precision as can be attained, except when otherwise directed. The balances available should permit reasonably accurate weighing of quantities as small as 0.1 g.

Caution: Solutions that contain chemicals with toxic vapors [marked with an asterisk (*)] should be prepared under a fume hood.

(A) Percentage by Weight. Most biological solutions use water as the solvent. For aqueous solutions of low-percentage concentration of solutes (less than 10%) it is sufficiently accurate to use water measured in a graduated cylinder and to assume that 100 ml of water weigh 100 g. With higher-percentage solutions it is desirable to weigh both solvent and solute. Table 2 gives the quantities of solute and solvent (in ml) required to make up solutions of some commonly used percentage strengths.

(B) Percentage by Volume. This technique is used to prepare dilute solutions of alcohols, formaldehyde, and similar liquids from more concentrated solutions. This may be done volumetrically by using graduated cylinders since most solutions are water solutions which will be diluted with water. For example, a 70% solution of ethyl alcohol is wanted and must be prepared from a stock bottle containing 95% alcohol. Measure 70 ml of 95% alcohol in a graduated cylinder and add 25 ml of water to bring the total volume up to 95 ml (the percentage of the original solution). Use Table 3 to determine quantities quickly.

The general rule is: Find the concentration of the dilute solution required; then add an amount of diluent to bring the total volume up to the numerical value of the concentrated solution.

Absolute (100%) alcohol is much too expensive to use for making dilute solutions of alcohol, so 95% alcohol is used as a stock solution.

In preparing formalin solutions* it is necessary to note that 40% formaldehyde gas dissolved in water constitutes a 100% formalin solution.

TABLE 2. *Percentage Solution Table, by Weight*

Strength of Solution: % by Weight	Grams of Solute Required to Make:			
	100 ml	250 ml	500 ml	1000 ml (1 l)
0.1	0.1	0.25	0.5	1.0
0.5	0.5	1.25	2.5	5.0
1.0	1.0	2.5	5.0	10.0
2.0	2.0	5.0	10.0	20.0
3.0	3.0	7.5	15.0	30.0
4.0	4.0	10.0	20.0	40.0
5.0	5.0	12.5	25.0	50.0
10.0	10.0	25.0	50.0	100.0
15.0	15.0	37.5	75.0	150.0
20.0	20.0	50.0	100.0	200.0
25.0	25.0	62.5	125.0	250.0

The generalized formula for the figures given in the table is:

$$\text{Grams of solute} = \% \text{ by weight of solution} \times \frac{\text{desired quantity of solvent (ml)}}{100}$$

TABLE 3. *Table for Dilution of Liquids*

		Percentage Strength of Concentrated Solutions												
		100	95	90	85	80	75	70	65	60	55	50	40	30
Percentage Strength of Desired Dilute Solution and Volume of Concentrated Solution To Be Used	95	5												
	90	10	5											
	85	15	10											
	80	20	15	10	5									
	75	25	20	15	10	5								
	70	30	25	20	15	10	5							
	65	35	30	25	20	15	10	5						
	60	40	35	30	25	20	15	10	5					
	55	45	40	35	30	25	20	15	10	5				
	50	50	45	40	35	30	25	20	15	10	5			
	40	60	55	50	45	40	35	30	25	20	15	10		
	30	70	65	60	55	50	45	40	35	30	25	20	10	
	20	80	75	70	65	60	55	50	45	40	35	30	20	10

Volume of Diluent To Be Added

To make a 10% formalin solution add 90 ml of distilled water to 10 ml of the 100% formalin.

(C) Molar Solutions (M). A molar solution contains one gram-molecular weight of the dissolved substance in one liter of solution. For example, $1M$ HCl contains 36.5 g of HCl in one liter of solution (atomic weight of $H = 1$, of $Cl = 35.5$, therefore molecular weight of HCl = 36.5). Similarly, $1M$ H_2SO_4 contains 98 g H_2SO_4 in one liter of solution; $1M$ H_3PO_4 contains 98 g of H_3PO_4 in one liter of solution; and 1M NaOH contains 40 g of NaOH in one liter of solution, and so on. To make up molar solutions of greater dilutions use Table 4.

TABLE 4. *Dilution of Normal and Molar Solutions*

To make up 1 l of HCl, H_2SO_4, NaOH NaCl, Na_2HPO_4, Ca $(OH)_2$, or KH_2PO_4 of specified molarity (M) and normality (N) *from 1M solution* of each of these substances, use the specified volumes from the table and add to 1 l of water

	Volume of 1M Solution Required (Milliliters)		
Specified M or N	HCl, NaOH, NaCl	H_2SO_4, Ca $(OH)_2$	Na_2HPO_4, KH_2PO_4
0.5M	500	500	500
0.2M	200	200	200
0.1M	100	100	100
0.01M	10	10	10
0.5N	500	250	167
0.2N	200	100	67
0.1N	100	50	33
0.01N	10	5	3

The general formula for 1 l of dilute *molar* solutions is:

ml of 1 M solution required for 1 l $= 1000 \times$ desired molarity

The general formula for 1 l of dilute normal solutions is:

$$\text{ml of 1M solution required for 1 l} = 1000 \times \frac{\text{desired normality}}{\text{total positive } or \text{ negative ions}}$$

For example, for 1 l of 0.2N H_2SO_4:

$$\text{ml of 1M solution required} = 1000 \times \frac{0.2}{2} = 1000 \times 0.1 = 100$$

(D) Normal Solutions (N). A normal solution contains 1 gram-equivalent weight of the compound in a liter of solution. A gram-equivalent weight of a compound is equal to the gram-molecular weight divided by the total valence of its positive or negative ions. Thus a $1N$ solution of HCl contains

$$\frac{1 \text{ g-mol wt of HCl } (36.5)}{\text{total valence of positive ions } (1)}$$

or 36.5 g of HCl in one liter of acid; a $1N$ solution of H_2SO_4 contains 49 g of H_2SO_4 in one liter of acid,

$$\frac{\text{g-mol wt } (98)}{\text{total positive ions } (2)}$$

and a $1N$ solution of phosphoric acid contains just a fraction less than 33 g of H_3PO_4 in one liter of the acid:

$$\frac{98}{3}$$

Concentrated HCl is $11.6N$. To make $1N$ HCl from concentrated reagent grade HCl add 93 ml of concentrated HCl to water to make one liter.

> *Do not add water to concentrated acid. Acid must be added carefully and slowly to water to avoid violent bubbling and splashing.* To make $0.1N$ HCl add 9.3 ml of concentrated HCl to enough water to make one liter of the acid.

See Table 5 for molarity and normality of commercial acids and bases.

III. Buffer Solutions (from ph 5.3 to 8.0)

Prepare the following stock solutions to produce buffer solutions in the range indicated.

(A) Disodium Hydrogen Phosphate ($Na_2HPO_4 \cdot 12H_2O$). To produce a $M/15$ solution dissolve 23.88 g of $Na_2HPO_4 \cdot 12H_2O$ in distilled water to make one liter of solution.

(B) Potassium Dihydrogen Phosphate (KH_2PO_4). To produce a $M/15$ solution dissolve 9.08 g of KH_2PO_4 in distilled water to make one liter of solution.

Mix the two stock solutions in the amounts indicated in Table 6 (or in multiples of these quantities) to produce the required amount of buffer solution of the specified pH.

TABLE 5. *Molarity and Normality of Commercial Acids and Bases*

Reagent	Formula	Molarity (Formality)	Normality	Percent Solute	Quantity of Solution in Usual Container
Acetic acid, glacial	$HC_2H_3O_2$	17 M	17 N	99.5%	5 lb
Acetic acid, dilute		6 M	6 N	34	
Hydrochloric acid, conc.	HCl	12 M	12 N	36	6 lb
Hydrochloric acid, dil.		6 M	6 N	20	
Nitric acid, conc.	HNO_3	16 M	16 N	72	7 lb
Nitric acid, dil.		6 M	6 N	32	
Sulfuric acid, conc.	H_2SO_4	18 M	36 N	96	9 lb
Sulfuric acid, dil.		3 M	6 N	25	
Ammonium hydroxide, conc.	NH_4OH	15 M	15 N	58 (28% NH_3)	4 lb
Ammonium hydroxide, conc.		6 M	6 N	23	
Soduim hydroxide, dil.	$NaOH$	6 M	6 N	20	solid

Adapted from Joseph, Alexander, et al. *A sourcebook for the physical sciences.* Harcourt, Brace and World, Inc., 1961, p. 170.

IV. Basic Stains and Reagents

(Additional information may be found in Morholt et al , *A sourcebook for the biological sciences,* second edition, New York: Harcourt, Brace, and World, 1966.)

(A) Benedict's Reagent (Test for Glucose). Dissolve 173 g of crystalline sodium citrate ($C_6H_5Na_3O_7$) and 100 g of anhydrous sodium carbonate (Na_2CO_3) in about 800 ml of water. Stir thoroughly and filter. Add to the filtered solution 17.3 g of copper sulfate ($CuSO_4$) dissolved in 100 ml of water. Make up to one liter with distilled water.

(B) Biuret Reagent (Test for Protein). Make up the following two solutions.

TABLE 6. *Buffer Solutions*

pH	M/15 Na$_2$HPO$_4$ (ml)	M/15 KH$_2$PO$_4$ (ml)
5.4	3.0	97.0
5.6	5.0	95.0
5.8	7.8	92.2
6.0	12.0	88.0
6.2	18.5	81.5
6.4	26.5	73.5
6.6	37.5	62.5
6.8	50.0	50.0
7.0	61.1	38.9
7.2	71.5	28.5
7.4	80.4	19.6
7.6	86.8	13.2
7.8	91.4	8.6
8.0	94.5	5.5

From Harrow, and others, 1952. *Laboratory manual of biochemistry.* Third edition. W. B. Saunders Co., Philadelphia.

1. 0.01*M* copper sulfate (CuSO$_4$). Dissolve 2.5 g of copper sulfate in one liter of water.

2. 10M sodium hydroxide (NaOH). Dissolve 440 g of sodium hydroxide in sufficient water and make up to one liter.

(C) **Bromthymol Blue.** An indicator with a narrow range, from pH 6.0 (yellow) to pH 7.6 (blue). Prepare a 0.1% stock solution by dissolving 0.5 g of bromthymol blue in 500 ml of distilled water. Add a trace of ammonium hydroxide (NH$_4$OH) to turn the solution a deep blue. Test the solution by breathing through a soda straw into a test tube containing a small quantity of solution, until a yellow color is produced. If it is too alkaline, add a drop of extremely dilute HCl.

Bromthymol blue is sometimes used as a 0.04% solution. Prepare this by grinding 0.1 g of dry bromthymol blue with 16 ml of 0.1N sodium hydroxide in a mortar *(handle NaOH with care)*. Dilute to 250 ml with distilled water.

(D) **Congo Red Vital Stain (pH Indicator).** Add 0.1 g of Congo red to 100 ml of distilled water and stir thoroughly so that all the dye is dissolved. Filter, if necessary, to remove undissolved dye.

(E) Lactophenol* (for Fixing and Mounting Small Specimens)

Phenol (pure crystal) (C_6H_5OH)	10 g
(*Caution*: Do not touch crystals; phenol burns.)	
Lactic acid ($C_3H_6O_3$)	10 g
Glycerol ($C_3H_8O_3$)	20 g
Distilled water	10 g

Warm the water and phenol slightly until the phenol is dissolved. Then add the glycerol and lactic acid.

(F) Lugol's Iodine Solution (Test for Starch). First, dissolve 6.0 g of potassium iodide (KI) in 100 of distilled water. Then dissolve 4.0 g of iodine crystals in the KI solution.

(G) Ringer's Solution (Isotonic Salt Solution)

1. *For general use*

Potassium chloride (KCl)	0.42 g
Calcium chloride ($CaCl_2$)	0.24 g
Sodium bicarbonate ($NaHCO_3$)	0.20 g
Distilled water	1000.00 ml

2. *For mammalian tissue (except blood).* Add to the first solution:

Sodium chloride (NaCl)	9.0 g

3. *For frog tissue (except blood).* Add to the first solution:

Sodium chloride (NaCl)	7.5 g

4. *For invertebrate tissue and vertebrate blood.* Add to the first solution:

Sodium chloride (NaCl)	6.0 g

(H) Isotonic Sodium Chloride Solutions (May Be Used in Place of Ringer's Solution)

1. *For mammalian tissue (except blood)*

Sodium chloride (NaCl)	9.0 g
Distilled water	1000.0 ml

2. *For frog tissue (expect blood)*

Sodium chloride (NaCl)	7.5 g
Distilled water	1000.0 ml

3. *For invertebrate tissue and vertebrate blood*

Sodium chloride (NaCl)	6.0 g
Distilled water	1000.0 ml

V. Radiation Techniques

Certain radiation techniques provide excellent means for investigating some biological principles. One such technique involves the use of radioisotopes as "tracers" in studying the use of various chemicals in living systems. These chemicals are available from several suppliers, some of whom are listed in the last section of this appendix. The prospective user of radioactive materials must be familiar with both local and federal regulations and with proper handling procedures before attempting to use them. Such information may be obtained from the Division of Materials licensing, U. S. Atomic Energy Commission, Washington, D. C. 20545. Also, local agencies, such as state radiation offices or medical associations, should be contacted.

Radioautography. The technique of radioautography is a relatively easy method of detecting radioactive substances in organisms that have previously taken in these chemicals by either injection or natural means. The specimen being analyzed is secured next to a piece of no-screen x-ray photographic film in the dark. The film is then stored in a film holder while the nuclear radiation exposes it. (The list of suppliers at the end of this appendix includes one company that supplies film kits that can be used in normal or subdued light, making a darkroom unnecessary.) At the end of the period of exposure, the film is developed and rinsed. The film may then be analyzed for the presence and amount of radiation being emitted from the specimen. More detailed information on this technique and on other uses of radiation in the high school laboratory may be found in the BSCS Laboratory Block on Radiation (William V. Mayer, D. C. Heath and Co., in press).

MAINTENANCE AND HANDLING OF LABORATORY ORGANISMS

Teaching biology effectively requires a dependable supply of living animals. The cost of purchasing such supplies is sometimes prohibitive, so that the teacher has two alternatives in providing living organisms for instruction—he makes field collections of the forms as needed or he maintains cultures of the required types. Successful culture of organisms depends on the choice of a suitable growth medium and on the proper care and handling of the cultures.

I. Protists.

A few principles apply generally to the culturing of microorganisms.

The quality of the water used is of critical importance. Tap water generally contains toxic materials which make culturing difficult, if not impossible. Spring water or boiled pond water, which can be prepared in quantity in advance, is preferable. Distilled water is satisfactory when artificial pond media are used as culture solutions. If tap water is used, allow it to season by standing in a clean container for a few days and then boil and cool. Heat all solutions used in culturing to about 85°F, cool, and aerate by shaking before using.

Culture solutions must be prepared carefully and .with the specified ingredients in proper amounts. Dilute accurately according to instructions. Label all solutions.

The pH of the medium should be neutral (pH = 7) or very slightly alkaline.

The optimal temperature for all microorganisms except bacteria is about 70°F.

Culture dishes—finger bowls ($4\frac{1}{2} \times 2$ in.) are ideal—should be covered to exclude contaminants and dust but should admit air. This can be achieved with the finger bowls which are designed to be stacked. The top dish should be empty. If finger bowls are used, the ideal amount of culture fluid is about 200 ml. Other glass containers of all types can easily be utilized if the budget does not permit the purchase of finger bowls.

Algae, of course, require light; all other microorganisms should be kept in dim light.

Subculture each culture about the time it reaches its peak (usually every 4 to 6 weeks) to maintain optimal cultures throughout the school year.

Make certain that cultures are kept away from noxious fumes such as the fumes of acids, illuminating gas, and other volatile chemicals.

Be sure that all glass vessels and pipettes used are thoroughly washed to remove all traces of chemicals, soaps, or detergents. Wash with plenty of water and use some of the culture fluid as a final rinse.

(A) Algae. A number of algae and flagellates can be utilized in the laboratory and in student projects. These include *Spirogyra, Cladophora,* and *Oedogonium* among the filamentous forms; the blue-green alga *Oscillatoria;* desmids such as *Closterium; Volvox* and other Volvocales; *Euglena; Chara;* and *Nitella;* among others. These may often be collected from ponds, lakes, swamps, and streams but they do not generally keep for more than a few days—hence the need to culture them.

Pure culture techniques have been developed for the cultivation of algae, but these are not necessary for maintaining the organisms needed for class use. Excellent student projects, however, can be developed using these pure culture techniques which were developed by Pringsheim and described by him in *Pure cultures of algaes: their preparation and maintenance* (Cambridge University Press, 1946).

Three main factors must be considered in the culture of algae:

temperature, light, and a satisfactory culture medium. The optimum temperature for the growth and maintenance of most algae is about 68°F. Many algae tolerate a range of temperatures above and below this optimum, but it is desirable that some way be found to keep the temperature of the laboratory in which the algae are being cultured as close to 68°F as possible. In general, the higher temperatures are more likely to damage or destroy cultures than temperatures below the optimum.

If north light is available, it is generally best, since the plants will not be in direct sunlight. They should not be kept in direct sunlight under any circumstances since this can elevate temperatures to 90°F and over. If it is necessary to provide more light than is available from windows, a 40-watt fluorescent lamp, a few feet from the culture dishes, will supplement the natural light enough to encourage growth. In the winter months it is desirable to use the light beyond the end of the school day to proivde about 16 hours of light.

Many culture media have been developed for growing algae in general, and for specific algae. For ordinary purposes it is not necessary to know and use more than a few of these. The literature can be consulted for additional techniques if these are needed for projects.

A soil extract was found by Pringsheim in 1913 to provide elements necessary for algal growth which were not provided by inorganic media alone. Two variations of this technique are described since most algae can be grown by these methods. A few other techniques of less general utility are also described.

1. *Basal medium for algae*

Potassium nitrate (KNO_3)	1.0 g
Magnesium sulfate ($MgSO_4$)	0.25 g
Dipotassium phosphate (K_2HPO_4)	1.25 g
Distilled water	1000.0 ml

2. *Bold's basal medium (sometimes called "modified Bristol's solution,"* Bischoff and Bold, 1963). Prepare 6 stock solutions 400 ml in volume, each containing one of the following salts in the concentration listed:

$NaNO_3$	10.0 g	K_2HPO_4	3.0 g
$CaCl_2 . 2H_2O$	1.0 g	KH_2PO_4	7.0 g
$MgSO_4 . 7H_2O$	3.0 g	$NaCl$	1.0 g

To 940 ml of distilled water add 10 ml of each stock solution and 1.0 ml of each of the 4 stock, trace-element solutions prepared as follows:

(a) 50 g EDTA and 31 g KOH dissolved in one liter H_2O.
(b) 4.98 g $FeSO_4 . 7H_2O$ dissolved in one liter H_2O. (Acidified H_2O: 1.0 ml H_2SO_4 added to 999 ml distilled water.)
(c) 11.42 g H_3BO_3 dissolved in one liter H_2O.
(d) The following, in amounts indicated, all dissolved in one liter H_2O:

$ZnSO_4 . 7H_2O$, 8.82 g; $MnCl_2 . 4H_2O$, 1.44 g; MoO_3, 0.71 g; $CuSO_4 . 5H_2O$, 1.57 g; $Co(NO_3) . 6H_2O$, 0.49 g.

Sterilize in the autoclave or a pressure cooker. (From Bold, Harold C. "'The Neglected Cryptograms," *The american biology teacher*, **27** (2:), 101-103 (February, 1965).

3. *Bristol's solution (for culture of algae).* Make up stock solutions of each of the following salts in 400 ml of distilled water:

$NaNO_3$	10.0 g
$CaCl_2$	1.0 g
$MgSO_4 . 7H_2O$	3.0 g
K_2HPO_4	3.0 g
KH_2PO_4	7.0 g
NaCl	1.0 g

Use 10 ml of each of these stock solutions to 940 ml of distilled water. Then add 1 drop of a 1% $FeCl_3$ solution and 2 ml of a trace element solution made up as follows:

Distilled water	1000.0	ml
$ZnSO_4 . 7H_2O$	0.1	g
H_3BO_3	0.1	g
$MnSO_4 . 4H_2O$	0.15	g
$CuSO_4 . 5H_2O$	0.03	g

4. *Erd-Schreiber solution (Plymouth formula): 1000*

1000 ml filtered sea water
50 ml soil extract (One kg finely sieved garden soil to one liter tap water; autoclave for one hour at 10 lb. pressure. If there is time, allow soil to settle and use clear liquid, but, if needed, quickly centrifuge to clear.)

0.03 g $Na_2HPO_4 . 12 H_2O$
0.2 g $NaNO_3$

1st day: Filter sea water through No. 1 Whatman filter paper and then heat to 73° C.

2nd day: (a) Again heat sea water to 73° C.

(b) Autoclave soil extract at 15 lb pressure for 30 min.

(c) Autoclave salt solutions (made up in distilled water so that 1 ml of each solution gives required amount for one liter of culture solution).

3rd day: Add cold salt solutions to cold soil extract, and then add soil extract to cold sea water. Allow culture solution to reach temperature of cultures to be subcultured before using.

This medium should be stored in the refrigerator until used. (From Bold, *American Biology Teacher,* **27,** 103.)

5. *Fish meal medium.* This is a successful medium and procedure for most algae. Add 0.2 g of commercial fish meal to one l of spring water and heat to 80–90°C. Filter through filter paper while hot and add 0.5 ml. of 1% $FeCl_3$ solution, freshly made. Shake and pour into finger bowls while it is still hot. When it is cool add algae (Volvocales, *Spirogyra,* and so on).

Prepare 4–10–4 or 5–10–5 commercial fertilizer in the same manner except use 1 g instead of 0.2 g. Use to culture the same organisms.

6. *Inorganic salt media.* A number of inorganic salt media such as Bristol's solution and Knop's solution have been used, with some success, with algae. Use in a 1:1 dilution in finger bowls or similar containers and use from 100 to 200 ml of culture solution. Provide light and temperature conditions as indicated.

7. *Kleb's solution, modified (for culture of Euglena)*

KNO_3	0.25 g
$MgSO_4$	0.25 g
KH_2PO_4	0.25 g
$Ca(NO_3)_2 \cdot 4H_2O$	1.00 g
Bacto-tryptophane broth	0.01 g
Water to	1.00 l

To use, dilute 1:10 with remineralized water or distilled water.

8. *Knop's solution.* Divide a liter of distilled water into four 250-ml portions. Dissolve each of the following salts in one of the 250-ml portions.

$MgSO_4 \cdot 7H_2O$	0.10 g
K_2HPO_4	0.20 g
KNO_3	1.00 g
$Ca(NO_3)_2 \cdot 4H_2O$	0.10 g

Combine the four solutions, adding $Ca(NO_3)_2$ solution last. Add one drop 4% FeCl₃ solution (freshly made) to each liter. Use diluted 1:1 with distilled water. This has a pH of about 7.6. Culture algae as directed.

There have been many modifications of the original Knop's solution formula for the cultivation of algae and other plants. They reveal the empirical nature of many such solutions—that which works best is the one used thereafter. Two variants are here given (Bold and Garnett) in the hope that some project work for students might reveal the formula which works best for you.

9. *Modified Knop's solution 1% (after Bold; for the cultivation of algae)*

Calcium nitrate [Ca(NO₃)₂]	4.0 g
Monobasic potassium phosphate (K₂HPO₄)	1.0 g
Potassium nitrate (KNO₃)	1.0 g
Ferric chloride (FeCl₃)—1% sol	2.0 drops
Glass distilled water	700.0 ml

Dissolve each of the ingredients separately in portions of water in clean glass containers, then mix together adding the calcium nitrate last. This is a 1% Knop's solution. For Volvocales use 1 ml of the 1% solution to each 20 ml of distilled or spring water (yields a 0.05% solution).

10. *Modified Knop's solution (Garnett).* This modification uses the following ingredients:

Potassium nitrate (KNO₃)	2.0 g
Magnesium sulfate (MgSO₄)	2.0 g
Calcium nitrate [Ca(NO₃)₂]—crystalline	6.0 g
Potassium phosphate (K₂HPO₄)	2.0 g

Dissolve in water to 2000 ml. This yields a 0.6% stock solution. Use measured small quantities—1 drop, 10 drops, 20 drops per liter—in making up culture solutions for algae.

11. *Pringsheim's soil extract technique.* This technique requires rich garden soil with a fair amount of humus and little clay. It is also desirable that no commercial fertilizers have been used recently. Large test tubes, small flasks, or even jars serve well as containers. From ½ to 1 in. of soil is put into each, and distilled water is added until the container is about ¾ full. The containers are plugged with cotton or capped loosely and heated (steamed) in a large vessel with a cover. (Place 2 or 3 in. of water in the large vessel and heat, below boiling, for 1 or 2 hr on 2 successive days.) This heating (below boiling) accom-

plishes two results; it dissolves humus and other materials and it destroys the majority of soil and water organisms already present. Allow the containers to cool and the sediments to settle (a day should be sufficient time), and inoculate with the algae to be cultured. Many algae require slightly alkaline conditions which can be provided by placing a small amount of powdered calcium carbonate (enough to cover the tip of a scalpel blade) on the bottom of the container below the soil and the water. Most common fresh-water algae such as *Spirogyra* (some species), *Closterium* (desmid), *Hydrodictyon, Volvox* and other Volvocales, and *Cladophora* will do well on this medium.

12. *Soil Extract Medium (Bold's modification).* This solution is preferred by many phycologists over solutions of mineral salts. The stock solution is prepared by autoclaving 500 g of good field or garden soil in one liter of glass-distilled water or demineralized water at 15 lb pressure for 2 hr. Cool, decant, and filter a number of times until the filtrate is clear. This is the "stock soil solution." It may be diluted according to the following table (dilutions A, B, C, D). It will be necessary to have ready a stock solution of 5% aqueous KNO_3 (potassium nitrate) made by dissolving 5g of KNO_3 in 100 ml of distilled water.

Ingredients	A	B	C	D
Distilled water	94	84	74	64
Stock soil solution	5	15	25	35
5% KNO_3 solution	1	1	1	1

Dilutions (ml)

Pour into shallow containers (100 ml) which are kept in a cool, well-lighted area. Inoculate with algae and cover the containers. Experiment with the different dilutions recommended and with different algae as available. Keep data on pH (measured with hydrion paper) and perhaps use 5 ml of phosphate buffer to different pH values for each container.

(B) Bacteria.[2] After having settled on a good medium for the growth of an organism, it is relatively straightforward to maintain it in culture for extended periods of time. Before attempting any of the procedures outlined below, the organisms must be permitted to develop on the medium until there is evidence of vigorous growth. In most cases two

[2] Sussman, Alfred. *Microbes: their growth, nutrition and interaction,* D. C. Heath. 1964.

or three days' growth after inoculation should suffice, but certain of the slow-growing organisms should be kept longer. When the bacterium has grown sufficiently, one of the following techniques can be used to preserve the culture.

1. *Storage at cold temperatures.* A temperature of 4°C will maintain many bacteria in a viable form for months. However, some bacteria will not survive under these conditions, so their viability at cold temperatures should be checked.

2. *Storage under reduced oxygen tension.* Dip a cork which fits into the culture tube into alcohol and flame it so that it is sterilized. Push it down the test tube containing the culture until it rests about an inch above the culture. Replace the cotton plug. Cultures will last for six months if the cork fits tightly enough and thereby restricts evaporation and access of oxygen.

3. *"Pliofilm-coating."* Push the cotton stopper into the neck of the test tube so that a small weft remains for you to hold onto in order to withdraw it. Wrap "pliofilm" or the equivalent around the mouth of the test tube in order to prevent drying out of the agar.

4. *Soil-preservation.* Autoclave about 1/4 inch of soil in a test tube for at least half an hour. Transfer a rich suspension of the bacterium you want to store and mix it with the soil. This technique is one of the most effective and easiest to use and works with a large number of organisms, including actinomycetes, fungi, and some algae.

5. *Storage under mineral oil.* Autoclave a bottle of "Nujol," or some other fine-grade mineral oil, for about an hour. Carefully pipette enough of this sterile oil over the surface of your culture so that about 1/2 inch of oil rests above the highest part of the agar. This technique, as well as the one using soil, will preserve the viability of microorganisms for years in some cases.

The techniques above are recommended as being the most universally applicable and convenient.

Any electric or gas oven will serve for the sterilizing of glassware or metal instruments. Petri dishes can be placed in empty coffee cans or large fruit juice cans, the tops of which are covered with paper towelling. Such dishes can also be wrapped separately in paper towelling and sterilized. Pipettes can also be wrapped separately in towelling and placed in the oven. Dry heat at 165 to 175°C for one hour is usually enough to sterilize these materials because the paper toweling will char and disintegrate.

In microbiological work it is useful to have a range of temperatures available for variations in the rate of growth. The influence of this variable is so marked and general that many meaningful experiments can be designed around it. Most schools have a refrigerator which can serve as an incubator set at low temperatures (4 to 8°C). A cool basement is sometimes available for temperatures around 15°C. As for temperatures above 22 to 24°C (room temperature), an inexpensive incubator can be constructed from household materials, as illustrated in *Innovations in Equipment and Techniques for the Biology Teaching Laboratory* by Barthelemy et al (D. C. Heath and Co., 1964). Any box which is reasonably well insulated may serve as the body of the incubator. An inexpensive container may be purchased at hardware stores in the form of rectangular bake ovens which fit on the burners of stoves. The temperature control may be made from an expansion type thermoregulator used in chick brooders and available at farm stores. The unit may be bolted to the top or side of the interior of the oven and connected to an ordinary light bulb which serves as the heater. Control elements are available for several different temperatures and may be replaced very easily with a screwdriver by loosening a nut.

Often it is desired to isolate a pure culture of a bacterial species from the mixed population of a natural sample. The material to be studied may be a liquid culture or a suspension in sterile tap water of colony growth from an agar culture. A *pure culture* refers to a culture containing the cells of only one species. Two methods for obtaining a pure culture are recommended, and the first is preferred since less glassware is involved, and it has other advantages.

1. *Pour nutrient agar plates and allow medium to harden.* It will be better if the plates are allowed to dry for a few hours after pouring. They should be inverted during this time. Charge the inoculation loop by dipping in liquid culture or suspension and drain the liquid from the loop by touching the inside of the culture tube above the level of the liquid. Then make parallel streaks, without cutting into the agar, across about one third of the agar surface. Flame the loop, cool it by touching the agar, and repeat streaks over half of the remaining area, taking care to streak across a portion of the area first inoculated. Flame needle again and repeat the process with the remaining area on the plate. Isolated colonies should appear on one of the sectors following incubation. Such colonies should be emulsified in sterile tap water and a new plate streaked and this, upon incubation, should produce colonies of uniform shape.

2. *Using tubes or bottles of a size to contain about 20 ml of medium, dispense the medium, and sterilize.* When needed, liquefy the agar by placing it in a pan of water which has been heated to boiling. Cool,

in a pan of water to 40 to 42°C, three tubes per culture to be studied. Use a sterilized inoculation loop to transfer one loopful from the culture suspension to Tube 1 while the agar is still liquid. Flame the needle and lay it on a desk. Rotate inoculated Tube 1 between the palms of the hands and transfer two loopfuls from it to Tube 2. Repeat the process given above, transferring one loopful from Tube 2 to Tube 3. Before the agar column cools and solidifies, pour the contents of each tube, after flaming tube lip, into a separate sterile petri dish. Upon incubation, both surface and subsurface colonies should appear. These may appear to be different, even with a pure culture, for the subsurface colonies are usually either lens-shaped or irregular wooly balls, whereas the morphology of surface colonies can be quite varied with different species.

From the plates prepared by these methods, pure cultures may be established by transferring bacteria from an isolated colony to a tube of broth or an agar slant appropriate to the type being studied. By these methods it is not difficult to isolate bacteria for class exercises or special projects.

The following preparations are commonly used in working with bacteria.

1. *Gram's iodine*

Iodine (I_2)	1.0 g
Potassium iodide (KI)	3.5 g
Distilled water	300.0 ml

To the iodine and potassium iodide add a few milliliters of water. Be sure the iodine is in solution before adding the remaining water.

2. *Gram's staining technique—reagents*

Ammonium Oxalate-Crystal Violet Stain
  Solution (a)
    2 g crystal violet (gentian violet) dissolved in 20 ml ethyl alcohol (C_2H_5OH) (95%)
  Solution (b)
    1.0 g ammonium oxalate [$(NH_4)_2C_2O_4 \cdot H_2O$] dissolved in 100 ml distilled water

Mix solutions (a) and (b). It is sometimes found that this gives so concentrated a stain that some Gram-negative bacteria do not decolorize properly. To avoid this, solution (a) may be made to contain half the amount of dye.

3. *Nitrogen-free agar: stock salt mixture*

KCl	10.0 g
K_2HPO_4	2.5 g
$CaSO_4 \cdot 2H_2O$	2.5 g
$MgSO_4 \cdot 7H_2O$	2.5 g
$Ca_3(PO_4)_2$	2.5 g
$FeSO_4$	0.1 g

Mix salts, grind to powder, place 1.5 g in one liter of water. Use 8 g of agar per liter for growth of plants.

4. *Nitrogen-free nutrient medium*

K_2HPO_4	0.8 g
KH_2PO_4	0.2 g
$MgSO_4$	0.2 g
$CaSO_4$	0.1 g
$Fe_2(SO_4)_3$ Trace—1 drop of 1% solution	
Distilled wate	1000.0 ml

5. *Nutrient agar*

Peptone	5.0 g
Beef extract	3.0 g
Agar	15.0 g
Distilled water	1000.0 ml

Heat water (below boiling) and dissolve agar first, then add peptone and beef extract. Sterilize in autoclave (pressure cooker) for 15 min at 15 lb pressure (121°C). When it is still hot pour it into sterile Petri dishes.

If a sugar nutrient agar is required, add 10 g of sucrose (1%) when making up original mixture.

6. *Nutrient broth (beef broth)*

Peptone (Difco B120)	5.0 g
Beef extract (Difco B126)	3.0 g
Distilled water	1000.0 ml

Dissolve peptone and beef extract in the water in a large flask. Pour into smaller flasks or containers in which the broth will be used and sterilize in the autoclave (pressure cooker) for 15 min at 15 lb pressure (121°C).

If the nutrient broth medium (Difco B3) is used, rehydrate by dissolving 8 g in 1000 ml of distilled water and sterilize as above.

It is possible to make up a satisfactory beef broth by dissolving a beef bouillon cube in a liter of distilled water and sterilizing as before.

7. *Potato dextrose agar*

Potato infusion	200.0 ml
Dextrose ($C_6H_{12}O_6$)	20.0 g
Agar	15.0 g
Distilled water	1000.0 ml

Finely cut 40 g of peeled potatoes and boil for 15 to 20 min in 200 ml of water. Strain the liquid and add other ingredients. Sterilize in autoclave (pressure cooker) for 15 min at 15 lb pressure (121°C).

8. *Safranin counterstain (basic dye for staining acid structures).* Dissolve 0.5 g safranin in 20 ml ethyl alcohol (95%); add 200 ml distilled water.

9. *Starch nutrient agar*

Soluble starch	2.0 g
Nutrient agar (liquefied)	1000.0 ml

10. *Sugar nutrient agar*

Specific sugar, sucrose (cane sugar), glucose (dextrose), or lactose (milk sugar)	10.0 g
Nutrient agar (liquefied)	1000.0 ml

(C) **Fungi.** Fungi or their spores are found in most environmental situations so that their collection, isolation, and culture present no difficulties. The problem, rather, is to keep fungal contaminants out of cultures of other organisms especially bacteria.

Fungi can be maintained and grown in the laboratory in three basic ways: on the original substrate, for example, bread, fruit, decaying leaves, insect bodies; in or on fluid media of known composition; and on solid nutrient media with agar as the solidifying agent.

1. *Solid media.* Plating techniques using culture media containing agar are utilized for the production of pure cultures of fungi. Formulas, other than those in this appendix, may be found in the Difco and Baltimore Biological Laboratories manuals.

(A) ISOLATION AND CULTURE OF FUNGI. If a fungus is developing rapidly, and fruiting bodies are visible, a few spores can be picked off the colony with the aid of a fine sterile needle and a hand lens or a dissecting microscope. Under these circumstances a pure culture can result from the first transfer. It is infrequent, however, that molds will be found growing free of other organisms, so contaminants are often transferred with the fungus. Under these conditions several expedients are available, including the following one.

(B) PLATING OUT. Melt a few tubes of agar medium containing about 10 to 12 ml, and keep these until needed in a water bath maintained at 45°C. At the same time, the mixture of organisms from which the mold is to be isolated is subdivided as finely as possible and a few fragments dropped in one of the tubes of melted agar by means of sterile forceps. After reinsertion of the plug, the tube is gently rotated between the palms of the hand in order to mix the contents. After mixing, the contents of the tube are poured under sterile conditions into a sterile Petri plate. The medium from another tube is poured into the first, mixed as before, and poured into a second Petri dish. The contents of a third tube are poured into the same tube and the mixing and pouring repeated. This procedure is carried out for as long as necessary with the particular material, and only your experience will determine how many transfers to make. Usually 3 to 5 should suffice. As soon as the agar hardens in the Petri dishes, the dishes are inverted and incubated. Aften incubation, discrete colonies should arise from which transfers can be made. The rationale of this technique is based on the fact that the small amount of liquefied agar left in the tube contains very few of the cells of the organisms being suspended, and each successive transfer dilutes them further.

2. *Bread mold* (Rhizopus). Place a slice of bread on the bottom of a container which has been lined with moist blotting paper. Scatter some dust over the surface of the bread and place the container, loosely covered, in a dark place. Within a week the surface of the bread will be covered with the mycelia and many sporangia. These spores may be used to seed other pieces of bread to continue the culture. Maintain the wetness of the blotting paper and the *Rhizopus* will be succeeded on the bread by *Aspergillis, Penicillium,* and other fungal types which can then be cultured by some of the techniques described next.

3. *Corn meal agar (used to maintain stock cultures of a wide variety of fungi)*

Corn meal	60.0 g
Agar	15.0 g
Distilled water	1000.0 ml

Mix the corn meal to a smooth cream with the water, simmer for 1 hr, filter through cheesecloth, and add the agar. Heat until the agar is dissolved, make up to one liter with water, and sterilize in the autoclave (pressure cooker) for 15 min. at 15 lb pressure (121°C). Pour into sterile Petri dishes.

This medium is available from Difco as Corn Meal Agar (B386). This may be rehydrated by suspending 17 g of Bacto-Corn Meal Agar in one liter of cold distilled water; then heating to boiling to dissolve the medium completely. Pour into tubes or flasks and sterilize in autoclave for 15 min at 15 lb pressure (121°C).

4. Neurospora *complete medium*. Difco bacto *Neurospora* culture agar (B321) is best to use here. Prepare by dissolving 65 g of the dehydrate in one liter of distilled water and sterilize in the autoclave (pressure cooker) for 15 min at 15 lb pressure (121°C).

5. Neurospora *minimal medium*. It is best to use a prepared minimal medium for growing *Neurospora*. Difco choline assay medium (B460) is available in dehydrated form. To rehydrate, dissolve 57 g of bacto-choline assay medium in one liter of distilled water. Sterilize in the autoclave (pressure cooker) for *10 min* at 15 lb pressure. *Do not over-sterilize.*

6. *Oatmeal agar medium (for growth of* Physarum polycephalum *plasmodium).* Soak 6 g of oatmeal in 200 ml of water for 1 hr. Strain through cheesecloth. To 100 ml of the clear filtrate add 1.5 g of non-nutrient agar. Heat until agar is completely dissolved. Pour a thin layer into Petri dishes. Do not sterilize, since the slime mold feeds on bacteria. This quantity will provide 5 plates.

7. *Pasteur's solution (for the cultivation of fungi)*

Cane sugar ($C_{12}H_{22}O_{11}$)	150.0 g
Ammonium tartrate ($C_4H_{12}N_2O_6$)	10.0 g
K_2HPO_4	2.0 g
$Ca_3(PO_4)_2$	0.2 g
$Mg_3(PO_4)_2$	0.2 g
Distilled water	1000.0 ml

8. *Sabourand's dextrose agar (an excellent medium for the cultivation of fungi and yeast)*

Peptone	10.0 g
Glucose	40.0 g
Agar	15.0 g
Distilled water	1000.0 ml

Mix peptone, agar, and water and bring to a boil for complete solution.

Then add the glucose and sterilize in autoclave (pressure cooker) at 15 lb pressure for 15 min (121°C). Pour into sterile Petri dishes and culture tubes for use.

A slight modification is to add 3 g of yeast extract to these ingredients. The dehydrated agar mixture may be purchased (Difco B109) and rehydrated by suspending 65 g in one liter of cold distilled water and heated to boiling to dissolve. Then autoclave at 15 lb for 15 min and use as before.

Omitting the agar produces a *Sabouraud's glucose broth* which is prepared in the same way and used also for the cultivation of molds, yeasts, and fungi.

9. *Slime mold culture medium. Physarum* is generally cultivated from sclerotia rather than from spores. In the form of sclerotia on paper towels, the slime mold can be kept for a year or more and restored to activity by putting a small bit of the sclerotium on the surface of a Petri dish with oatmeal agar medium. When the plasmodium streams out over the surface of the dish, it may be transferred to fresh Petri dishes with oatmeal agar.

If microscopic examination of the plasmodium is intended, it is easier to add a small bit of sclerotium to the surface of a Petri dish in which there is a thin layer of 1½% *nonnutrient agar* (made by dissolving 1½ g of agar in 100 ml of distilled water). No sterilization is needed.

Another method for growing the plasmodium and for obtaining sclerotia for future use utilizes dried, powdered oatmeal as a food for the plasmodium. Cover an inverted finger bowl with a single piece of paper towel and place it in a 12- to 15-in. battery jar. Put 1 to 1½ in. of water in the battery jar so that the water reaches to about ⅔ of the height of the finger bowl. The towel will become wet and will continue to receive water by capillary action. Place a small bit of *Physarum* sclerotium in the middle of the toweling along with a pinch of powdered oatmeal. As the plasmodium forms and spreads, add oatmeal daily. Keep the battery jar covered to prevent too rapid evaporation of the water. When the plasmodium covers the surface of the finger bowl, line the inside of the battery jar with paper towel and remove the cover. As the water evaporates and the jar dries, the plasmodium will crawl on the lining paper and form sclerotia. When the paper is completely dry, cut the towel into small pieces and store in envelopes.

10. *Water molds.* Aquatic molds such as *Achlya* and *Dictyuchus* are frequently found in protozoa cultures maintained by the artificial pond medium method with wheat or rice grains. *Saprolegnia* can be grown

in finger bowls in which artificial pond medium is used and in which the bodies of dead insects—particularly aquatic insects—are placed.

11. *Yeast culture medium* (Saccharomyces cerevisiae). Prepare a 10% solution of molasses in distilled water (20 ml of molasses and 180 ml of water). Put into a flask and add ¼ of a package of dried commercial yeast pellets. Plug the flask with nonabsorbent cotton. Put in a warm place (about 20° to 30°C) for 12 to 24 hr. Adding 0.5 g of commercial peptone to the fluid will accelerate the fermentation process. Large numbers of budding yeast cells will be available for microscopic examination.

Another effective procedure is to add the same amount of dried yeast pellets to 200 ml of freshly boiled, cooled, Pasteur's solution.

(D) Protozoa. A compendium of culture techniques, J. G. Needham's *Culture Methods for Invertebrate Animals,* is available as a paperback published by Dover Publications and may be used as a source of many procedures beyond those recommended here.

Protozoa do best in moderate light, at temperatures of about 70°F and at pH 7 (neutral) or slightly alkaline. Check the pH with hydrion paper and if alkalinity is required add a drop of 1N NaOH solution. Ordinarily, finger bowls which stack easily, make good containers although small jars and similar glass containers will serve just as well. The culture dishes should be kept covered and the volume of solution maintained by adding pasteurized spring water (heated to about 85°F). pond water, or additional culture solution to replace that lost by evaporation. About 200 ml of solution is a practical quantity which should yield large numbers of organisms. When the culture is ready, it should be inoculated with one or two pipettes full of water containing the organisms to be cultured—obtained by collection in the field or by purchase. A stereoscopic binocular microscope is invaluable for locating organisms and should be used to check the condition of the cultures and for removing organisms for use. To avoid contamination use a different pipette for each type of culture.

1. *Artificial pond water medium.* A number of methods that utilize a synthetic pond water of specific chemical composition have been described in the literature. Formulas for two of these solutions are described below. These should be prepared in a concentrated form and diluted as directed for use. Organisms such as *Amoeba, Paramecium, Stentor, Blepharisma, Spirostomum, Colpoda, Colpidium, Euplotes,* and *Actino-sphaerium* may be maintained with these fluids.

Pour about 200 ml of the diluted solution into each finger bowl and add 5 to 8 preheated, dry rice grains to the culture. Space the grains equally to provide centers of concentration. Allow several days for the rice grains to begin to decay before adding the inoculant. Subculture every month or 6 weeks.

The pH of the culture may be maintained at a value of 7 by adding 10 to 20 ml of a phosphate buffer solution made up to pH 7.

(A) BRANDWEIN'S SOLUTION (FOR CULTURING PROTOZOA AND SMALL INVERTEBRATES). Make up the following concentrated stock solution:

Sodium chloride (NaCl)	12.0 g
Potassium chloride (KCl)	0.3 g
Calcium chloride (CaCl$_2$)	0.4 g
Sodium bicarbonate (NaHCO$_3$)	0.2 g
Phosphate buffer (pH 7)	100.0 ml
Distilled water to	1000.0 ml

To use this stock solution use 1 part of the concentrate to each 1000 parts of distilled water.

(B) CHALKEY'S SOLUTION (FOR CULTURING PROTOZOA). Mix the following materials to produce a concentrated stock solution.

NaCl	11.0 g
KCl	0.4 g
CaCl$_2$	0.6 g
Distilled water to	1000.0 ml

Dilute 1 part of stock solution with 100 parts of distilled water.

2. *Egg yolk medium.* Make a thin, smooth paste by grinding a pea-sized piece of boiled egg yolk with pond water or distilled water. Add water up to 500 ml and mix thoroughly. Pour into a half-gallon battery jar and allow to stand for two days before inoculating with *Chilomonas.* If the jar is left uncovered, it will be inoculated with cysts of *Chilomonas,* which seem to be ubiquitous. After the *Chilomonas* culture has been established it can be inoculated with *Paramecium, Colpidium,* or *Colpoda* since these larger ciliates feed on *Chilomonas.* Subculture about once a month.

3. *Fertilizer medium.* Mix 1 g of a commercial fertilizer (formula 4–10–4 or 5–10–5) in one liter of pond or spring water and heat to 80–90°C. Filter while hot and pour into finger bowls or a battery jar. When cool, inoculate with chlorophyll-bearing flagellates such as the Volvocales (in finger bowls) or *Euglena* (in battery jar). These cultures require light and should be kept in a well-lighted location (although not in direct

sunlight for long). Some biologists prefer to use a medium of known chemical composition such as a modified Kleb's solution for this type of protozoan.

4. *Hay infusion.* Boil 10 g of chopped timothy hay (other types of hay will work) in a liter of spring water for about 30 min. Filter and add 1 or 2 drops of *1N* NaOH. Cool and inoculate with a pinch of soil. Then add 2 crushed, cooked wheat grains (wheat seeds boiled for 10 min). After a day, inoculate with *Amoeba, Paramecium, Stentor, Blepharisma, Spirostomun, Colpoda, Colpidium, Euplotes,* or *Actinosphaerium,* all of which can be cultured with this medium. Subculture once every month or 6 weeks. Each week add 1 or 2 boiled wheat grains. A variation which has worked successfully is to preheat uncooked rice grains in place of wheat grains. (Place dry rice grains in a test tube of boiling water for 5 min.) If rice grains are used, allow a few days before inoculating with the protozoan to be cultured.

5. *Methyl cellulose solution (for slowing down protozoan for microscopic examination).* Dissolve 10 g of dry methyl cellulose in 90 ml of distilled water. The resulting mixture has the consistency of a syrup. Use 1 drop of methyl cellulose solution to 1 drop of protozoan culture.

6. *Nonfat dried milk medium.* Dissolve 1 g of nonfat dried milk in one liter of pasteurized pond water or spring water (or well-seasoned tap water). This solution may be used unbuffered or it may be buffered to pH 7 with 10 to 20 ml of a phosphate buffer solution. Inoculate with organisms to be cultured. This method has been used with considerable success for *Paramecium, Chilomonas, Colpoda,* and *Colpidium.*

II. Invertebrates

The small invertebrates most valuable in teaching consist mainly of forms found in fresh water. With appropriate collecting techniques they may be found throughout most of the year in ponds, swamps, and lakes, and brought into the laboratory. If collected in sufficient numbers, they may be maintained for a month or more by observing two precautions: (1) they should be kept in a sizable volume of clear pond (or spring) water, and (2) they should be kept at a fairly low temperature —the bottom shelf of a refrigerator is excellent for this purpose. It must be emphasized that this is a maintenance technique and does not generally result in a noticeable increase in numbers. Some of these organisms can be cultured for growth by procedures recommended below.

(A) Flatworms. Planarians are found in a wide variety of fresh-water habitats. Some species occur in running waters while others commonly occur in ponds and lakes. They may be collected by using small cubes of raw beef or raw liver to which a piece of string has been attached. In 15 min to a half hour after the cubes have been lowered into the water, they are raised and examined for planarians. Shake the collected animals into a white enamel pan of clear water. The planarians will be seen settled on the bottom and moving slowly. They may be picked up with a medicine dropper. Ponds or lakes with *Elodea* or similar plants are good sources of planarians. Use a white enameled pan with an inch or two of clear water and shake plants from the pond into the water. The worms may be seen against the pan bottom and collected as before.

Culture planarians in enamel pans with pond water or L solution (see *Hydra*). Feed them with small pieces of raw beef or liver. After an hour or two, remove the food and replace the water with fresh water. It is essential that the food not be left in the culture pan for any length of time. Since planarians are photophobic, it will help to cover the pans with an opaque cover—a piece of cardboard is fine for this purpose.

(B) Fruit Flies *(Drosophila)*. Many media have been used for culturing Drosophila. Two commonly used ones are described below.

1. *Media*

(A) Dry-mix medium (developed by W. F. Hollander, Genetic Department, Iowa State College.) It may be kept in dry storage until needed, at which time water is added to make the desired amount. No mold inhibitor is used by Hollander, although 5 to 6 ml of 0.5% propionic acid, or some other inhibitor in the right proportion, may be added.

Sugar	3 parts by volume
Corn meal	2 parts by volume
Brewer's yeast	1 part by volume
Granulated agar	1 part by volume

All these ingredients are mixed thoroughly. Add about 1¾ cups of the dry mix to one liter of cold water, stirring to prevent lumpiness. Boil gently for about 5 min or until the foaming stops. Transfer the mixture, after boiling, from the flask to a glass funnel mounted on a ring stand. A rubber tube should be attached to the funnel stem. Extend the rubber tube into the culture bottle (½ pint milk bottle or the equivalent) or vial and regulate the flow of medium by using a clamp on the rubber tubing. Fill each bottle to a depth of 2½ cm (1 in.) and each vial to a depth of about 2 cm. With a glass rod or a pencil, push one end of a doubled strip of paper towel to the bottom of the medium while it is still soft. Plug the bottles with a cotton stopper wrapped with cheesecloth. (These stoppers may be used over and over if sterilized each time.) A fermentation vent extending to the bottom of the medium is sometimes needed when live yeast is introduced; otherwise CO_2 may build up rapidly under the medium and push it towards the top of the bottle or vial, trapping the flies. It may not be necessary to gouge a fermentation vent to the bottom of the medium if the paper strip is inserted deeply. The paper strips also provide additional surface for egg laying and pupation.

After sterilization and cooling, and before use, inoculate a half dozen drops of a milky suspension of living baking yeast into each culture container. The dry, packaged yeast obtainable in grocery stores is quite satisfactory. The yeast should be allowed to grow for about 24 hr before the flies are introduced.

After the culture containers have been filled and plugged, it is important to sterilize them in a pressure cooker or autoclave. The standard time of sterilization is 15 min at about 15 lb pressure. This procedure will usually kill bacteria and mold spores that are present in the medium. After sterilization, the culture containers may be stored in a refrigerator for several weeks. They may also be stored in a deep freeze. If a freezer is available, it may be possible to prepare, at one time, all the culture containers needed for a semester. Simply allow the containers to thaw before yeasting.

(B) CORN MEAL-AGAR MEDIUM FOR CULTURE OF *Drosophila*

Corn meal	95.0 g
Agar	15.0 g
Brewer's yeast	5.0 g
Water	850.0 ml
Corn syrup	75.0 ml
Sulfur free molasses	75.0 ml

Mix dry ingredients with small quantity of water to make a paste. Add remaining fluid ingredients slowly, mixing to avoid lumps. Simmer until the mixture thickens. Autoclave (pressure cook) for 15 min at 15 lb. Add small quantity of mold inhibitor if desired.

2. *Pure strains of* Drosophila *may be obtained free from certain geneticists or may be purchased from biological supply houses.* The flies received from stock centers will be in small vials. Use these to start your own stock cultures. Before transferring the flies from the vials to your stock bottles, be sure there is no excess moisture on the

food or on the sides of the bottles; flies become stuck rather easily. Place a funnel in the new stock bottles. Before removing the cotton stopper from the vial, shake the flies to the bottom by tapping the vial on the table. When the flies have collected on the bottom of the vial, quickly remove the cotton stopper, invert the vial into the funnel, and transfer the flies to the fresh culture bottle by gently tapping the culture bottle on a rubber pad.

Do not discard the stock center vial, but add a little yeast suspension and attach the vial to the side of the culture bottle with a rubber band. Eggs and larvae still in the vial will thus be saved to continue developing and to provide more stock material.

Stock and experimental cultures may be kept on shelves or on the desks in the laboratory. They should not be kept in the refrigerator, near a radiator, or on a window ledge in the sun. If the room is quite warm (25°C), they will develop rapidly; but if it is cool (15°C), a longer period of time is required to complete the life cycle. The temperature of *Drosophila* cultures should not exceed 28° (81°F) for any period of time or the flies will then become sterile.

In the southern states it has been found that some species of ants will crawl through the cotton plugs and chew up the developing flies. If this is likely to happen, all culture bottle should be placed on blocks in a pan of shallow water.

3. *How to maintain stock cultures.* At least 5 to 10 pairs of flies should be used as parents in making stock cultures. Stocks should be changed once every 3 or 4 weeks to avoid contamination by mites (minute parasites that live on flies) and molds. Stocks should never be allowed to dry out, and water or yeast suspension should be added when moisture is required. At least two sets of cultures of each stock should be maintained in case one culture fails. Cultures may be changed on a rotating basis. It is important to remember that cultures should not be kept too long, and that old culture bottles should be sterilized and cleaned soon after they are no longer needed.

4. *How to collect virgin females.* *Drosophila* females store the sperms they receive during a mating, and use them to fertilize a large number of eggs over a period of time. It is therefore necessary that the females used in any experimental cross be virgin. Flies used in maintaining stock cultures need not be virgin, since the females could only have been impregnated by males from the same genetic stock.

Females usually do not mate within 12 hr after hatching. If *all* the adult flies are shaken from a culture bottle, the females which emerge during the next 12 hr are likely to be virgin when collected. Some geneticists recommend collecting the females after a 10-hr period. Virgin females may be collected over several periods of time and kept in a feed vial. Two to three collections may be made per day from a culture bottle. Separate the males from the females and keep those wanted. It is sometimes difficult to distinguish the sex of very young flies; use the sex comb characteristic in males if you are in doubt. One stray male in a vial of (originally) virgin females will exclude the entire vial of flies from being used in an experiment. As a rule, put no more than 25 virgin females in a single vial.

5. *How to make an experimental mating.* When an experimental mating between virgin females and males of the desired genetic strains is made, the flies are put directly into a culture vial containing medium. If the flies are still etherized, place them on the side of the vial, which in turn should be placed on its side until the flies recover. A safe method for introducing etherized flies into a new vial is to invert the food vial over the etherized flies; they will crawl up to the food when they recover, and the cotton plug can then be inserted. In either case there should be no excess moisture in the new culture vial.

The parent flies in the experimental mating should breed, the females lay eggs, and the larvae develop to a good size within a week at room temperature. After 7 or 8 days the parent flies should be shaken out of the culture vessel. This procedure is necessary in order that the parent flies not be confused with the offspring when the counts are made. Once flies begin to emerge, they may be classified and counted for 10 days. Counting longer than this once again runs into the danger of overlapping generations. The first F_1 flies may have mated before being counted and hence have left some eggs to develop in the same culture vessel.

(C) **Hydra.** W. F. Loomis [*Science*, **117**, 565–566 (1953)] has developed a remarkably effective method for cultivating *Hydra* under controlled conditions. His method yields hundreds of actively reproducing individuals with certainty. It avoids the depression normally encountered in methods using *Daphnia* as a source of food. He uses Littoralis (L) solution instead of pond water and uses the dried eggs of the brine shrimp, *Artemia,* as a source of living crustacea.

Loomis cultures *Hydra* in finger bowls (similar low containers may certainly be used). *Hydra* is grown at room temperature in L solution. Use about 200 ml of culture solution per container. This should provide water to a depth of about 2 in. If larger containers are used, fill them to a depth of 2 or $2\frac{1}{2}$ in. with L solution.

Brine shrimp larvae are the right size about 48 hr after the eggs are put into saline solution. To feed *Hydra* collect the larvae with a fine mesh net and wash with L solution before adding them to the *Hydra* cultures. The larvae can be separated from unhatched eggs by shining a light at one end of the container. The larvae will move towards the light and be concentrated there.

Within 24 hr of the time the hydras are fed, decant the old and fouled culture solution and replace with fresh L solution. The hydras are generally attached to the glass sides and bottom of the culture dish and will not be poured away. This culturing procedure encourages reproductive activity, and the colonies will increase rapidly.

If active increase is not desired, leave the hydras in clean L solution without feeding them. They will survive for several weeks at room temperature or for several months in a refrigerator. These individuals will bud again within 48 hr after being returned to room temperature and fed daily with an excess of brine shrimp larvae.

1. Artemia—*brine shrimp.* Brine shrimp larvae can be produced from the dried eggs (available in supply houses or tropical fish shops) in 48 hr. About 0.5 g of the dried eggs is dusted on the surface of 500 ml of 3.5-g/l NaCl solution in a large shallow pyrex dish (Corning 3-qt utility dish). The salt solution may be prepared readily by diluting a concentrated saturated solution of NaCl 1:100 with spring or distilled water. When needed, the larvae may be attracted to one end of the dish with a lamp and removed with a fine mesh net. Wash with pond water or L solution before use.

2. *Littoralis (L) solution (for the culture of* Hydra). Use distilled or deionized water in making up the following stock solutions:

(a) NaCl		133.0 g
CaCl$_2$		26.6 g
Demineralized water to		1000.0 ml
(b) NaHCO$_3$		38.0 g
Demineralized water to		1000.0 ml

Use 10 ml of each of these stock solutions and dilute up to a gallon with demineralized water.

(D) Water Fleas *(Daphnia).* There are no completely reliable methods for culturing *Daphnia* although a large number of procedures have been described in the literature. Tropical fish shops now offer live *Daphnia* through much of the year. When purchased, these may be placed in a larger container of pond water and kept for a month in a refrigerator. Some success in culturing *Daphnia* has been reported using the egg yolk medium recommended for protozoa.

A fresh moist suspension of yeast may be added to the *Daphnia* culture until it is slightly milky. This procedure has been successful for some workers. The *Daphnia* feed on the yeast and clear the culture solution. Additional yeast may then be added as needed.

III. Vertebrates [3]

Although the modern trend in biology teaching emphasizes the use of living materials, there are many points of view concerning the use of live animals for experimental purposes in the high school laboratory.

The Institute of Animal Resources of the National Research Council, National Academy of Sciences, as well as other groups, has considered this problem. The National Research Council Institute, recognizing the importance of the use of laboratory animals for research and teaching, does not favor eliminating the use of laboratory animals and has designed a set of general standards for their use as follows:

NATIONAL ACADEMY OF SCIENCES
NATIONAL RESEARCH COUNCIL
INSTITUTE OF LABORATORY
ANIMAL RESOURCES
Guiding Principles in the Use of Animals by
Secondary School Students and Science
Club Members

1. The basic aim of scientific studies that involve animals is to achieve an understanding of life and to advance our knowledge of the processes of life. Such studies lead to a respect for life.

2. A qualified adult supervisor must assume primary responsibility for the purposes and conditions of any experiment that involves living animals.

3. No experiment should be undertaken that involves anesthetic drugs, surgical procedures, pathogenic organisms, toxicological products, carcinogens, or radiation, unless a biologist, physician, dentist, or veterinarian trained in the experimental procedure assumes direct responsibility for the proper conduct of the experiment.

4. Any experiment must be performed with the animal under appropriate anesthesia if the pain involved is greater than that attending anesthetization.

5. The comfort of the animal used in any study shall be a prime concern of the student investigator. Gentle handling, proper feeding, and provision of appropriate sanitary quarters shall be strictly observed at all times. Any experiment in nutritional deficiency may proceed only to the point where symptoms of the deficiency appear. Appropriate measures shall then be taken to correct the deficiency, or the animal killed by humane methods.

---

[3] Follansbee, Harper. *Animal behavior,* D. C. Heath, 1965.

6. All animals used must be lawfully acquired in accordance with state and local laws.

7. Experimental animals should not be carried over school vacation periods unless adequate housing is provided and a qualified caretaker is assigned specific duties of care and feeding.

(A) Incubating Chicken Eggs.[4] Use only fertile hen eggs. These may be obtained from commercial hatcheries or from poultry farmers who keep roosters with their flocks. After eggs are obtained, they may be stored at a cool temperature, but *not below 10°C,* until time to incubate them. However, viability of embryos is much less in eggs which are kept more than one week after they are laid.

Two important environmental factors which must be controlled when incubating chicken eggs are temperature and humidity. The optimum temperature for development of chicken embryos is 38°C. However, other temperatures may be desired such as when the effect of temperature on embryo development is being investigated.

The equipment used to incubate the eggs should be tested to determine if it will maintain a constant temperature over an extended period since chicken eggs require approximately three weeks to hatch. Racks or trays placed in the incubator will hold the eggs in place and will make it possible to put more eggs in the incubator. However, ample air space must surround the eggs so that the developing embryos can obtain sufficient oxygen.

The humidity in the incubator should be within a 50% to 70% range, with 58% as an optimum. Specific suggestions for building incubators and for controlling temperature and humidity may be found in Section VII of *Innovations in equipment and techniques for the biology teaching laboratory* (Barthelemy, Dawson and Lee, D. C. Heath, 1964).

(B) Maintaining Baby Chicks.[5] A custom brooder is best for this, however, baby chicks may be kept in almost any enclosure where they may be kept warm. An electric light bulb will provide adequate warmth for most cages or brooders. However, light bulbs should not be used in experiments which include the study of hormones since continuous light depresses the activity of some hormones. Chicks must be fed and watered daily. Feed stores can supply chick starter which is the best food; however,

[4] Moog, Florence. *Animal growth and development,* D. C. Heath, 1963.
[5] *Ibid.*

finely ground cereals such as corn meal, rolled oats or dried baby food are

(C) Solutions Used in Investigations with Chicks or Chicken Embryos [6]

1. *Physiological saline solution, 0.9%.* Dissolve 9 g of NaCl in one liter of distilled water.

2. *Sex hormones. (a)* Chorionic gonadotrophin—with a sterile hypodermic syringe and needles, inject 10 ml of sterilized 0.9% saline solution into a sealed vial of chorionic gonadotrophin (500 Cortland-Nelson units or 10,000 international units).

(b) Testosterone—Dissolve 250 mg of Testosterone in 10 ml of sesame oil. Place in a serum vial and pasteurize.

(D) Frogs. Frog cages should include a large open water supply which is readily accessible to the frogs. The water should be circulated or changed daily in order to minimize odors and bacterial growth.

Frogs may be fed earthworms, mealworms, pill bugs or most kinds of crawling insects if live frogs are to be maintained for more than a few days. Suggestions on the housing of frogs may be found in Section IV of *Innovations in equipment and techniques for the biology teaching laboratory* (Barthelemy, Dawson and Lee, D. C. Heath, 1964).

(E) Tadpoles. [6] Tadpoles should be kept in an aquarium or similar container with plenty of water. The water used may be either pond water or tap water which has been allowed to set several days for the chlorine to escape.

A good source of food for tadpoles is algae. Other foods such as boiled lettuce, spinach leaves, or any canned, strained, green baby food vegetables are also satisfactory. However, with these latter foods, the water should be changed about thirty minutes after each feeding. The tadpoles should be fed three times a week and may be fed up to one teaspoon of vegetable matter per feeding depending on the size of the tadpoles.

(F) Frog Embryos. [6] Frog embryos should be examined daily and any dead ones removed. If the water becomes cloudy or develops a foul odor, it should be changed. The water may be changed by pouring it through a strainer or fine mesh aquarium net to prevent loss of embryos. Put in fresh water (near the temperature of original water) and replace the embryos which have been caught in the strainer or net. Frog embryos need no food until they have grown to well-developed tadpoles.

[6] Moog, Florence. *Animal growth and development*, D. C. Heath, 1963.

(G) Holtfreter's Solution (May be Used to Grow Frog Eggs or Tadpoles or as a Substitute for Pond Water). Dissolve 3.5 g of NaCl, 0.05 g KCl, 0.1 g $CaCl_2$, and 0.02 g $NaHCO_3$ in one liter of distilled or deionized water.

IV. Plants [7]

This section is intended to provide some generally used techniques for working with various plants in the high school biology laboratory.

(A) Techniques for Germinating Seeds and Growing Seedlings. Seeds generally require only water and oxygen for germination. Seeds may be soaked in a container for a limited time, a few hours often being sufficient, although a day is usually not too long. Aeration is desirable for seeds that are soaked for long periods of time. A continuous source of air may be supplied by either an aquarium pump or a continuous stream of water running over the seeds with or without the use of an aspirator. Here it is best to package the seeds loosely in cheesecloth or a light wire screen held securely in place with string or rubber bands.

Seeds may also be germinated by placing them between several layers of moist paper towels. This may be done in a closed container called a germination tray or box which will reduce the loss of water by evaporation. If flower pots are used for germinating seeds, then the disposable paper pots are most convenient. When using the common clay pot, place a curved piece of broken pot over the drain hole convex side up, and fill the pot with soil, sand, or vermiculite (or a 1:1:1 mix of the three) to about one inch from the top of the pot. After moistening the soil, place the seeds individually in holes to a depth of about twice the size of the seed, carefully covering the seeds afterwards. When transplanting or removing seedlings, be sure to take enough soil so as not to damage the roots, or keep the roots wrapped in a damp paper towel while working with them.

The maintenance of plants requires regular brief visits rather than irregular long sessions. Watering is the most important daily activity, temperature and lighting being the other main factors. For most plants, the sunlight of a usual day period and a comfortable room temperature are adequate, although certain plants may require special handling. More complete suggestions for maintaining plants are offered in Section V of *Innovations in equipment and techniques for the biology teaching laboratory* (Barthelemy, Dawson and Lee, D. C. Heath, 1964).

[7] Lee, Addison E. *Plant growth and development*, D. C. Heath, 1963.

(B) Solutions for Plant Nutrition. In carrying out various nutrition experiments, it is usually desirable to make up stock solutions. These are more concentrated than the final solutions and can be added in small quantities to a large quantity of distilled or deionized water to make up the final test solution. This is a useful procedure as it is easier and more accurate to weigh a large quantity of salt for the stock solution than the very minute quantity required for the final solution.

For mineral nutrition experiments, it is absolutely necessary to use chemically clean glassware in the preparation of all solutions and distilled or deionized water for rinsing all glassware and preparing all solutions. Below are the quantities required for the stock solutions. The quantity of each salt should be added to 100 ml of distilled water for each stock solution.

$Ca(NO_3)_2 \cdot 4H_2O$	11.8 g
KNO_3	5.0 g
$MgSO_4 \cdot 7H_2O$	1.4 g
KH_2PO_4	1.4 g
$CaCl_2$	5.6 g
KCl	0.8 g

Iron chelate
   (iron ethylenediamine-
   tetra acetic acid)
	1.0 g

For the trace elements, the dilution series below should be followed in preparing one liter of the stock solution.

	Dilution No. 1	Final Amount	
$MnCl_2 \cdot 4H_2O$			1.8 g
H_3BO_3			2.8 g
$ZnSO_4 \cdot 7H_2O$	2.2 g/100 ml	10 ml dil. no. 1 =	0.22 g
$CuSO_4 \cdot 5H_2O$	0.8 g/100 ml	10 ml dil. no. 1 =	0.08 g
$NaMoO_4 \cdot 2H_2O$	2.5 g/100 ml	1 ml dil. no. 1 =	0.025 g
Water			979.0 ml
		Final Stock Volume	1000.0 ml

To prepare a deficient medium, such as nitrogen-free, the following combination of the stock solutions should be made:

$MgSO_4 \cdot 7H_2O$	10 ml
KH_2PO_4	10 ml
$CaCl_2$	10 ml
KCl	10 ml
Iron chelate	10 ml
Trace elements	10 ml
Water	940 ml
Total	1000 ml

(C) Paper Chromatography.[8] By the technique of paper chromatography one can separate quite similar molecules from each other. Only small amounts of material are required and for this reason it is widely used in biochemical research. Separation of two compounds using paper chromatography depends on their different rates of migration on filter paper. This in turn depends on the type of solvent used and the strength of the forces absorbing the molecules to the paper. A wide range of special chromatogram papers are available, although Whatman's No. 1 filter paper is generally quite adequate for simple seprations.

The molecules to be studied are spotted at the base of a strip or square of chromatogram paper. Then the paper is placed in a closed chamber and allowed to touch and stand in a solvent which moves up the paper for a time carrying the substances with it. It is important to avoid handling the chromatogram paper since the fingers usually have chemicals on them, particularly amino acids. Do not allow the solvent to run completely off the upper edge of the paper, but take the chromatogram out of the solvent just before it reaches the top.

The identity of particular molecules may be determined either by their color (in the case of some pigments) or by their position on the chromatogram relative to the distance travelled by the solvent as compared with a known sample. It will be necessary to add a reagent such as ninhydrin to identify amino acids and to use an ultraviolet light source in identifying nucleic acids. In using ninhydrin, the mixture should

[8] Albersheim, Peter, John Dowling and Johns Hopkins, III. *The molecular basis of metabolism*, D. C. Heath, 1968.

either be sprayed on with an atomizer or the whole chromatogram dipped into the solution. After the paper has been thoroughly dried, mark the various spots with a pencil as they tend to fade with time.

A more detailed discussion of the techniques involved is offered in Smith and Feinberg's *Paper and thin layer chromatography and electrophoresis* (second edition, Shandon Scientific Co., Ltd., London, 1965). Also, the BSCS Laboratory Block on *The molecular basis of metabolism* has a well outlined series of flow charts on paper chromatography.

1. *Acetone-ether solvent** (for plant pigment separation). Mix 8 parts acetone with 92 parts petroleum ether. For various plant pigment extracts, the acetone concentration may be varied from 5 to 30%.

2. *Amino-acid solvents.** *(a)* Mix 10 parts formic acid, 70 parts isopropanol and 20 parts distilled water. *(b)* Mix 38 ml n-butanol, 26 ml propionic acid (99%) and 36 ml distilled water. *(c)* Mix 80 ml 95% ethanol, 8 ml ammonium hydroxide (ACS 28-30%) and 12 ml distilled water. For two-way separation use solvent (c) first, then use either solvent (a) or (b) second.

3. *Nucleic acid solvent.** Mix 15 parts acetic acid, 60 parts butanol and 25 parts distilled water.

4. *Ninhydrin reagent.** Dissolve 5 g ninhydrin in one liter acetone.

(D) Solutions for Investigating Auxins. One basic substance used in work with auxins is indoleacetic acid (IAA) which usually is in a crystalline form. To prepare a basic stock solution of 100 mg/liter, dissolve 100 mg of IAA in 1 to 2 ml of absolute ethyl alcohol and add approximately 900 ml of distilled water. Warm the mixture gently on a hot plate or steam bath to evaporate the alcohol and then dilute it with distilled water to one liter. This solution should be made up fresh and it should not be used after two weeks. It should be stored in a refrigerator, preferably in a dark bottle or flask covered with aluminum foil to exclude light.

Another common preparation of IAA is in a lanolin base so it may be applied directly to a plant. An approximately 1% IAA mixture can be made by adding 0.1 g IAA to 10 g lanolin. Weigh the needed amount of lanolin directly in a glass vial with a screw cap which can be used for storing the mixture (weigh the empty vial first, of course). Heat this in a hot water bath until the lanolin melts, then add the 0.1 g IAA to the melted lanolin and stir well for several minutes.

(E) Slide Preparations of Plant Tissues. In order to observe details of cellular organization it is often desirable to clear the tissue being studied. This means that the cellular contents are removed or made transparent. To do this, transfer small pieces of plant material to be cleared into vials containing 25% aqueous solution of NaOH and leave them overnight. If the tissue is not clear enough, repeat this procedure using an oven to increase the temperature. Cleared material may be stored in a 50% aqueous solution of alcohol.

1. *Acetocarmine stain* (for fixing and staining nuclear materials, especially chromosomes).* Dissolve 1 g orcein in 45 ml acetic acid which has been warmed to near boiling (CAUTION). Allow to cool, then add 55 ml distilled water; shake well and filter.

2. *Formaldyhyde acetic acid (FAA)* (for killing and fixing plant tissue).* Mix 90 ml of 50% ethyl or grain alcohol, 5 ml of commercial formalin and 5 ml of glacial acetic acid.

3. *Gibberellic acid solution (plant growth stimulator).* Prepare a one part per million solution as follows: dissolve 100 mg of gibberellic acid in 1 to 2 ml absolute ethyl alcohol and add approximately 900 ml of water. Warm the mixture gently on a hot plate or steam bath to evaporate the alcohol and then dilute it with distilled water to one liter. This solution should be made up fresh, and it should not be used after two weeks. Store in a refrigerator in a dark bottle or a flask covered with aluminum foil to exclude light.

4. *Methylene blue stain (for staining nucleus and cytoplasmic granules).* Dissolve 0.3 g methylene blue in 30 ml of 95% ethyl alcohol. Then add 100 ml distilled water.

5. *Tetrazolium test solution (for testing seed viability).* Dissolve 0.5 g of 2, 3, 5-triplenyltetrazolium chloride in 10 ml of distilled water. This solution should be used within 3 to 6 hours after preparation. When viable seeds are treated with tetrazolium, they exhibit a pink to reddish color. Seeds with dead embryos remain the original yellow color.

V. Source Books—Field and Laboratory Techniques

Barthelemy et al. *Innovations in equipment and techniques for the biology teaching laboratory.* D. C. Health, Boston, Mass. 1964.

Benton, A. H. and W. E. Werner. *Workbook for field biology and ecology.* Burgess Publishing Co., Minneapolis, Minn. 1957.

Denerec and Kaufman. *Drosophila guide.* Carnegie Institute, Washington, D.C. 1961.

Gray, Peter. *The dictionary of the biological sciences.* Reinhold Publishing Corp., New York. 1967.

Hall, Thomas S. *A sourcebook in animal biology.* McGraw-Hill, New York. 1951.

Jaeger, Edmund, C. D. Sc. *A source book of biological names and terms.* Charles C. Thomas Publisher, Springfield, Ill. 1959.

Morgan, Ann H. *Field book of ponds and streams.* Putnam, New York. 1930.

Morholt, Evelyn, Paul F. Brandwin, Alexander Joseph. *Sourcebook for biological sciences.* Harcourt, Brace & World, New York. 2nd Ed. 1966.

Needham, J. G. et al. *Culture methods for invertebrate animals.* Dover, New York. 1937.

Needham, J. G., P. R. Needham. *A guide to the study of freshwater biology.* Holden Day, Inc., San Francisco, Calif., 5th Ed. 1962.

Pimentel, R. A. *Invertebrate identification manual.* Reinhold Publishing Corp., New York. 1967.

Stehli, G. J. *The microscope and how to use it.* Sterling Publishing Co., New York. 1961.

White, P. R. *The cultivation of animal and plant cells.* The Ronald Press, New York, 2nd Ed. 1963.

SOURCES OF LABORATORY SUPPLIES [a]

The following list of sources of materials is not complete, but does include sources for all materials required by any BSCS laboratory investigation.

Aloe Scientific, 1831 Olive, St. Louis, Mo. 63103, Laboratory supplies, equipment, chemicals, etc.

American Type Culture Collection, 2029 M. Street, N.W., Washington, D.C. 20036. Universally authoritative source of all bacteria and fungi.

Ames Company, Inc., 819 McNaughton Ave., Elkhart, Ind. 46514. Manufactures of Clinistix and Clintest tablets for glucose test and Combistix for testing for urinary proteins, glucose, and pH. Write for sources of supply.

Baltimore Biological Laboratory, Inc., 1640 Gorsuch Ave., Baltimore, Md., 21218. Culture media, materials, and apparatus for microbiology. In addition, write for their book of products; and excellent handbook of microbiological techniques and procedures.

Bausch & Lomb, Inc., 77666 Bausch St., Rochester, N.Y. 14602.

Biological Research Products Co., 243 West Root St., Stockyards Station, Chicago, Ill., 60698. Injected fetal pigs for mammalian anatomy; sheep brains, eyes.

Calbiochem, 3625 Medford St., Los Angeles, Calif. 90063.
    161 W. 231 St., New York, New York. 10463
    4930 Cordell Ave., Bethesda, Md., 20014

Cambosco Scientific Co., Inc., 37 Antwerp St., Brighton Station, Boston, Mass., 02135. Chemicals, apparatus, glassware, preserved specimens, prepared microscope slides, dissecting instruments, collecting equipment.

[a] There are no "official" or "recommended" BSCS suppliers.

Carolina Biological Supply Co., Elon College, Burlington, N.C., 27215, and Powell Laboratories Division, Gladstone, Ore., 97027. A source of almost everything needed in biology teaching. Their catalog is over 600 pages long.

Central Scientific, 1700 Irving Park Rd., Chicago, Ill., 60613.

    3232 Eleventh Ave., N., Birmingham, Ala., 35234

    160 Washington St., Somerville, Mass., 02143

    6446 Telegraph Rd., Los Angeles, Calif., 90022

    1040 Martin Ave., Santa Clara, Calif., 95050

    237 Sheffield St., Mountainside, N.J., 07092

    6610 Stillwell, Houston, Tex., 77017

    6901 E. 12th St., Tulsa, Okla., 74112

Apparatus and equipment for all sciences. Good for biological supplies.

Clay Adams, Inc., 141 East 25th St., New York, N.Y., 10010. Supplies for microscopy, dissecting instruments, skeletons, anatomical models, charts.

Clinton Misco Corp., P.O. Box 1005, Ann Arbor, Mich. 48106.

Colab Laboratories, Inc., Chicago Heights, Ill., 60411. Antibiotic sensitivity disks and plastic disposable sterile Petri dishes.

Colorado Serum Co., 4950 York St., Denver, Colo., 80216. Blood sera, antisera, and tissue culture supplies.

Conant, George H., Ripon, Wisc., 54971. Unusually fine prepared microscope slides—large botanical material with some animal material available.

Connecticut Valley Biological Supply Co., Valley Road, Southhampton, Mass., 01073. Living and preserved plant and animal materials, particularly protozoans, algae, invertebrates, insectivorous plants.

Culture Collection of Algae, Dr. Richard C. Starr, Dept. of Botany, Indiana University, Bloomington, Inc., 47401. Excellent source of a wide range of algae.

Dale Scientific Co., Box 1721, Ann Arbor, Mich. 48106.

Distillation Products Industries (Division of Eastman Kodak Co.,) Rochester, N.Y., 14603. 3800 organic chemicals.

Dow Chemical Co., Midland, Mich., 48640. Dowex resins for water demineralization—Dowex 21K (20-50) anion exchange resin and Dowex 50 wx cation exchange desin.

Elgett Optical Co., Inc., 303 Child St., Rochester, N.Y. 14611.

Falcon Plastics, 5500 West 83rd St., Los Angeles, Calif., 90045. Sterile and inexpensive plastic items.

Faust Scientic Supply Ltd., 5108 Gordon Ave., Madison, Wis., 53716.

Fisher Scientific Co., 633 Greenwich St., New York, N.Y., 10014, and 711 Forbes Ave., Pittsburgh, Pa. 15219. Instruments, apparatus, furniture, and chemicals.

General Biochemicals, Inc., 677 Laboratory Park, Chagrin Falls, O., 44022. Amino acids, vitamins, enzymes, fatty acids, and carbohydrates.

General Biological Supply House, Inc. (Turtox), 8200 South Hoyne Ave., Chicago, Ill., 60620. One of the best known biological supply houses, The Catalog is over 900 pages long. Turtox service leaflets (sixty titles) provide instruction on a wide variety of activities.

General Scientific Equipment Co., 3055 Dixwell Ave., Hamden, Conn. 06514. Laboratory equipment, glassware, supplies, and chemicals.

Goldschmidt Chemical Corp., 153 Waverly Place, New York, N.Y., 10014. Tegosept M, a mold inhibiter used in culturing *Drosophila*.

Gradwohl Laboratory Supply Co., 3512 Lucas Ave., St. Louis, Mo. 63103. Laboratory equipment, glassware, and reagents.

Hawshaw Scientific, 1945 E. 97th St., Cleveland, O., 44106. Laboratory supplies and equipment.

Harvard Apparatus Co., Inc., Dover, Mass. 02054.

Hyland Laboratories, 4501 Colorado Blvd., Los Angeles, Calif. 90039. Suppliers of blood diagnostic reagents and biological specialties, including materials for tissue culture.

Lemberger Co., 1222 W. South Park Ave., P.O. Box 482, Oshkosh, Wis., 54901. Living amphibians, reptiles, and mammals, as well as some invertebrates and plants.

Macalaster Scientific Corp., Waltham Research & Development Park, 186 3rd Ave., Waltham, Mass., 02154. Particularly good source for Lab Blocks.

Michigan Scientific Co., P.O. Box 1005 or 6780 Jackson Rd., Ann Arbor, Mich., 48103. Living materials, preserved specimens, chemicals, apparatus, etc.

Nalco Chemical Co., 6216 W. 66th Place, Chicago, Ill. 60638. Dowex ion exchange resins for water conditioning. Write for Bulletin 2-5 on water conditioning with Nalcite.

New York Scientific Supply Co., 28 West 30th St., New York, N.Y., 10001. Biological supplies including charts and models.

Northern Biological Supply, P.O. Box 222, New Richmond, Wis. 54017.

Nutritional Biochemicals Corp., 26201 Niles Rd., Cleveland, O, 44128. Biochemicals of all types.

A. J. Nystrom and Co., 3333 Elson Ave., Chicago, Ill. 60618. Charts and models.

Pacific Bio-Marine Supply Co., P.O. Box 285, Venice, Calif. 60618. Live and preserved marine organisms.

Quivira Specialties, 4202 West 21st Street, Topeka, Kan., 66604. A wide variety of living vertebrates.

Research Specialties Co., 200 South Garrard Boulevard, Richmond, Calif. 94801. Supplies for paper chromatography. Small chromatobox and all reagents.

E. H. Sargent and Co., 4647 West Foster Avenue, Chicago, Ill., 60630. Instruments, apparatus, glassware, chemicals, and culture media.

Schlueter Scientific Supplies, 8609 Lincoln Avenue, Morton Grove, Ill., 60053. Instruments, apparatus, chemicals, and glassware.

Schwarz BioResearch, Inc., 230 Washington Street, Mt. Vernon, N.Y., 10553. Biochemicals.

Science Education Products Co., Inc., Redwood City, Calif., 94063. Bacteriology kits for performing twenty-seven different student experiments.

Science Supplies Co., 600 Spokane Street, Seattle, Wash. 89134. 2 x 2 color slides including photomicrographs.

Scientific Products, Division of American Hospital Supply Corp., 1210 Leon Pl., Evanston, Ill., 62202.

    5056 Peachtree Rd., Chamblee, Ga. 30005

    101 Third Ave., Waltham, Mass. 02154

    3713 N. Davidson St., Charlotte, N.C. 28205

    1586 Frebis Lane, Columbus, Ohio. 43206

    2505 Butler St., Dallas, Tex. 75235

    17150 Southfield Rd., Allen Park, Mich. 48101

    12th Ave. and Gentry St., N. Kansas City, Mo. 64116

    3815 Valhalia Drive, Burbank, Calif. 91504

    1951 Delaware Pkwy., Miami, Fla. 33125

    3846 Washington Ave., N. Minneapolis, Minn. 55412

    4408 Catherine Ave., Metairie, La. 70001

    4005 168 St., Flushing, Long Island, N.Y. 11358

    150 Jefferson Dr., Menlo Park, Calif. 94025

    14850 N.E. 36 St., Bellevue, Wash. 98004

    3175 V. St., N.E., Washington, D.C. 20018

Sherwin Scientific Co., N. 1112 Ruby St., Spokane, Wash. 99202

Sigma Chemical Co., 3500 DeKalb St., St. Louis, Mo., 63118. Biochemicals and organic compounds.

Singleton, Ray, Interbay Station, Tampa, Fla., 33611. Live mammals, reptiles, and amphibians.

Southwestern Biological Supply Co., P. O. Box 4084, Dallas, Tex., 75208. Living and preserved animals and some preserved plants.

Standard Scientific Supply Corporation, 808 Broadway, New York, N.Y., 10003. Laboratory supplies, equipment, and chemicals.

E. G. Steinhilber and Co., Oshkosh, Wis., 54901. Live frogs, turtles, some invertebrates, white mice, and rats.

Arthur H. Thomas Co., Vine Street at Third, P. O. Box 779, Philadelphia, Pa., 19105. Chemicals, biochemicals, apparatus, and other supplies.

Travel-Lab Science Co., P. O. Drawer M, Manor, Tex., 78653. Portable science laboratory and special laboratory apparatus.

Turtox (See General Biological Supply House, Inc.).

Van Waters & Rogers, Inc., P.O. Box 5287, Denver, Colo. 80217.

    BKH Division:

        P.O. Box 3200, San Francisco, Calif. 94119

        850 S. River Rd., W., Sacramento, Calif. 95691

        650 W. 8 St., S., Salt Lake City, Utah 84110

        313 Kamakee St., Honolulu, Hawaii 96814

    Scientific Supplies Co. Division:

        600 S. Spokane St., Seattle, Wash. 98134

        3950 N.W. Yeon Ave., Portland, Oreg. 97210

Braun Division:
  1363 S. Bonnie Beach Pl., Los Angeles, Calif. 90023
  P.O. Box 1391, San Diego, Calif. 92112
  P.O. Box 143, Phoenix, Ariz. 85001
  2030 E. Broadway, Tucson, Ariz. 85719
  324 Industrial Ave., N.E., Albuquerque, N.W. 87107
  6980 Market Ave., El Paso, Tex. 79915
Visual Sciences, Box 599B, Suffern, N.Y., 10901. Film strips in biology.
Ward's Natural Science Establishment, Inc., P. O. Box 1712, Rochester, N.Y., 14603 and P. O. Box 1749, Monterey, Calif., 93940. Living and preserved materials, bioplastic specimens, skeletons, prepared microscope slides, models, color slides, collecting equipment, aquaria, and fossil specimens.
Welch Scientific Co., 1515 Sedgwick, Chicago, Ill. 60610.
  331 E. 38th St., New York, N.Y. 10016
  7300 N. Linder Ave., Skokie, Ill. 60076
  General laboratory supplies, including irradiated seeds.
Wilkins-Anderson Co., 4525 W. Division St., Chicago, Ill., 60651. Wide range of laboratory equipment and glassware.
Will Corporation, Rochester, N.Y., 14603. Complete supplies for chromatography.
Worthington Biochemicals, Freehold, N.J., 07728. Source of enzymes and other biochemicals.

Microscopes

American Optical Co., Instrument Division, Buffalo, N.Y., 14215.
Bausch and Lomb, Inc., 82917 Bausch Street, Rochester, N.Y., 14602.
Cooke, Troughton and Simms, 91 Waite Street, Malden, Mass., 02148.
Edmund Scientific Co., Barrington, N.J., 08007.
Elgeet Optical Co., Inc., 838 Smith Street, Rochester, N.Y., 14606.
Graf-Apsco Co., 5868 Broadway, Chicago 40, Ill., 60626.
Pacifica Biological Laboratories, Box 63, Albany Station, Berkeley, Calif. 97416.
Swift Instruments, Inc., 1572 North Fourth Street, San Jose, Calif., 95112.
Technical Instrument Co., 98 Golden Gate Avenue, San Francisco 2, Calif., 94102. Kyawa compound microscopes.
Testa Manufacturing Co., 10126 East Rush Street, El Monte, Calif., 91733.
Unitron Instrument Co., Microscope Sales Division, 66 Needham Street, Newton Highlands 61, Mass., 02161.
Carl Zeiss, Inc., 444 Fifth Ave., New York, N.Y., 10018.

Radiation Biology

Atomic Corporation of America, 7901 San Fernando Rd., Sun Valley, Calif., 91352. Supplier of radio-chemicals and film for radioautography which does not require darkroom.
Baird-Atomic Inc., Technical Director, Radioactive Source Division, 33 University Rd., Cambridge, Mass., 02138. Excellent source of radioisotopes and various radiochemicals and radiation equipment.

Nuclear-Chicago, 333 Howard Avenue, Des Plaines, Ill., 60018. Another fine source of radiochemicals and radiation equipment.

Oak Ridge Atom Industries, Inc., Educational Products Division, 500 Elza Dr., P.O. Box 429, Oak Ridge, Tenn., 37831. Provide irradiation services and stock irradiated seeds.

Sources of Films

MOTION PICTURE FILM LIBRARIES

The following is a partial listing of film rental libraries. Teachers who cannot call on the services of an audio-visual specialist may wish to contact these sources directly.

Alabama

Auburn University, Main Library, Auburn, 36830.
A-V Film Service, 2114 Eighth Avenue, North, Birmingham, 35203.

Alaska

Department of Health and Welfare, Box 3–2000, Juneau, 99801.
University of Alaska, AV Communications, College, 99735

Arizona

University of Arizona, AV Services, Tucson, 85721.

Arkansas

Arkansas State College, AV Center, State College, 72467.

California

Moody Institute of Science, 12000 E. Washington Blvd., Whittier, 90606.
Pacific Film Library, 104 Fountain Ave., Pacific Grove, 93950.
University of California, Ext. Media Center, Berkeley, 94720.

Colorado

Colorado State College, Inst. Materials Center, Greeley, 80631.
University of Colorado, Stadium 348, Boulder, 80302.

Connecticut

University of Connecticut, AV Center, Storrs, 06268.

District of Columbia

Film Center, 915 12th St., N.W., Washington, 20005.
Smithsonian Institution, AV Library, Washington, 20560.

Florida

Florida State University, Educational Media Center, Tallahassee, 32306.

Georgia

University of Georgia, Film Library, Athens, 30601.

Hawaii

University of Hawaii, AV Activities, Honolulu, 96822.

Idaho

Idaho State University, Educ. Film Library, Pocatello, 83201.
University of Idaho, AV Center, Moscow, 83843.

Illinois

Southern Illinois Univ., AV Service, Carbondale, 62901.
University of Illinois, 704 S. 6th., Champaign, 61820.

Indiana

Indiana State University, Stalker Hall, Terre Haute, 47809.
Indiana University, AV Center, Bloomington, 47401.
Purdue University, AV Center, Lafayette, 47907.

Iowa

Iowa State University, Pearson Hall, 121, Ames, 50010.
State University of Iowa, Bureau of AV Instruction, Iowa City, 52240.

Kansas

University of Kansas, Bur. Visual Instr., Lawrence, 66044.

Kentucky

University of Kentucky, Taylor Educ. Bldg., Lexington, 40506.

Louisiana

Louisiana State Univ. Agr. Ext. Serv., Baton Rouge, 70803.

Maine

University of Maine, Stevens Hall So., Orono, 04473.

Maryland

University of Maryland, Film Library, College Park, 20742.

Massachusetts

State College at Bridgewater, AV Dept., Bridgewater, 02324.

Michigan

University of Michigan, Frieze Bldg., Ann Arbor, 48103.
Wayne State Univ., 5448 Cass Ave., Detroit, 48202.
Western Michigan Univ., AV Center, Kalamazoo, 49001.

Minnesota

AV Center and South's Fm. Libr., 6422 W. Lake St. Minneapolis, 55426.
University of Minnesota, 2037 Univ. Ave., S.E., Minneapolis, 55455.

Mississippi

University of Mississippi, Educ. Film Libr., University, 38677.

Missouri

University of Missouri, AV Library, Columbia, 65201.

Nebraska

University of Nebraska, Bur. of AV Instr., Lincoln, 68508.

Nevada

University of Nevada, AV Communctns. Center, Reno, 89507.

New Hampshire

University of New Hampshire, AV Dept., Durham, 03824.

New Jersey

Department of Education, State House Annex, Trenton, 08652.

New Mexico

University of New Mexico, AV Aids, Albuquerque, 87106.

New York

Australian News & Info. Bur., 636 Fifth Ave., NYC, 10020.
N.Y. University Film Library, 26 Washington Pl., NYC, 10003.
State University of N.Y. at Buffalo, Communctns, Ctr., Buffalo, 14214.

North Carolina

N.C. State Board of Health, Film Libr., Raleigh, 27602.

North Dakota

State University, Fargo, 58102.

Ohio

Kent State University, AV Center, Kent, 44240.

Oklahoma

University of Oklahoma, Educ. Mat. Serv., Norman, 73069.

Oregon

Oregon State System of Higher Education, AV Instr., Corvallis, 97331.

Pennsylvania

Pennsylvania State Univ., AV Serv., University Park, 16802.
Psychological Cinema Register, AV Aids Libr., Univ. Park, 16802.

Puerto Rico

University of Puerto Rico, AV Educ. Ctr., Rio Piedras, 00931.

Rhode Island

State Department of Education, Park and Hayes St., Providence, 02908.

South Carolina

University of South Carolina, AV Div., Columbia, 29208.

South Dakota

South Dakota State University, Film Library, Brookings, 57006.

Tennessee

Tennessee Visual Education Service, 416A Broad St., Nashville, 37203.
University of Tennessee, Film Serv., Knoxville, 37916.

Texas

Av Services, 2310 Austin St., Houston, 77004
T.E.A. Film Library, 4006 Live Oak St., Dallas, 75204
University of Texas, Visual Instr. Bur., Main Univ., Austin, 78712.

Utah

University of Utah, AV Bur., Salt Lake City, 84112.
Utah State Univ., AV Aids, Logan, 84321.

Vermont

University of Vermont, Film Libr. & AV Serv., Burlington, 05401.

Virginia

A-V Center of Tidewater, 135–137 E. Little Creek Rd., Norfolk, 23505.

Washington

University of Washington, AV Services, Lewis Hall, Seattle, 98105.
Washington State Univ., AV Center, Pullman, 99163.

West Virginia

State Department of Health, Bureau of Public Health, Charleston, 25305.

Wisconsin

Milwaukee Public Museum, AV Center, Milwaukee, 53203.
University of Wisconsin, Bur. of AV Instr., Madison, 53706.

Wyoming

University of Wyoming, AV Services, Laramie, 82070.

MOTION PICTURE DISTRIBUTORS—SALES

There are many distributors of motion pictures from whom films may be purchased. The following is a partial list of distributors who sell films on biological subjects.

California

Film Associates of California, 11559 Sta. Monica Blvd., L. A., 90025.
Moody Institute of Science, 12000 E. Washington Blvd., Whittier, 90606.

Colorado

Thorne Films, Inc., 1229 University Ave., Boulder, 80302.

Illinois

Coronet Instructional Films, 65 E. S. Water St., Chicago, 60601.
Encycl. Brit. Educ. Corp., 425 N. Michigan, Chicago, 60611.
Int. Film Bureau, Inc., 332 S. Michigan Ave., Chicago, 60604.
Soc. for Visual Educ., Inc., 1345 Diversey Pkwy., Chicago, 60614.
Rand McNally & Co., P. O. Box 7600, Chicago, 60680.

Indiana

Indiana University, AV Center, Bloomington, 47401.

Iowa

Iowa State University, Pearson Hall 121, Ames, 50010.

Massachusetts

Ealing Corp., 2225 Massachusetts Ave., Cambridge, 02140.
Houghton Mifflin Co., 110 Tremont St., Boston, 02107.

Minnesota

University of Minnesota, 2037 Univ. Avc., S. E., Minneapolis, 55455.

New York

Life Filmstrips, Time and Life Bldg., New York, N.Y., 10020.
Modern Talking Picture Serv., 1212 Ave. of the Americas, New York, N.Y., 10036.
Popular Sci. Pub. Co., AV Div., 355 Lexington Ave., New York, N.Y., 10027.
Visual Sciences, Suffern, 10901.
McGraw Hill Films, 330 W. 42nd St., New York, N.Y., 10036.
Harcourt, Brace & World, Inc., 757 Third Ave., New York, N.Y., 10017.

Pennsylvania

Curtis Audio-Visual Materials, Independence Sq., Philadelphia, 19105.

APPENDIX

6

Career Opportunities in the Biological Sciences

Agriculture

A science career for you in agriculture. 1960. Southern Regional Education Board, 130 6th St., N.W., Atlanta, Ga.

Career service opportunities in the U.S. department of agriculture. Agriculture Handbook No. 45, U.S. Department of Agriculture. U.S. Government Printing Office, Washington, D.C.

Research careers in agriculture. Careers Research Monograph No. 210. The Institute for Research, 537 S. Dearborn St., Chicago, Ill.

Agronomy

Soil science career. 1956. Miscellaneous Publication No. 716. Soil Conservation Service, U.S. Department of Agriculture, U.S. Government Printing Office, Washington, D.C.

Animal Biology

Careers in animal biology. 1959. American Society of Zoologists, Goucher College, Baltimore, Md. 25 cents.

Animal Husbandry

Animal husbandry as a career. 1948. Careers Research Monograph No. 22. The Institute for Research, 537 S. Dearborn St., Chicago, Ill.

Anthropologist

Anthropology as a career. 1960. By William C. Sturtevant. Publications, Smithsonian Institution, Washington, D.C. 20 cents.

Archaeology

Archaeology as a career. 1955. Smithsonian Institution, Washington, D.C.

686

Artist

Medical and scientific illustrators. Science Research Associates, 57 W. Grand Ave., Chicago, Ill. 45 cents.

Bacteriology

A career in bacteriology. 1958. Society of American Bacteriologists, Business Office, Mount Royal Ave. and Guilford Ave., Baltimore, Md.

Biochemistry

A career for biochemist, bacteriologist, serologist. Veterans Administration, Washington, D.C.

Opportunities in biochemistry. American Society of Biological Chemists, 9650 Wisconsin Ave., Washington, D.C.

Biology—General

Biological scientists. Occupational Briefs, No. 131. Science Research Associates, 57 W. Grand Ave., Chicago, Ill.

Biologist. 1958. Careers, Largo, Fla. 25 cents.

"Careers in Biology Education," *The American Biology Teacher,* Vol. 30, No. 4 (April, 1968).

Careers in the biological sciences. 1963. By William Wellington Fox. H. Q. Walck, New York, N.Y.

Career opportunities in biology: the challenge of the life sciences. By Russel B. Stevens. National Academy of Science-National Research Council, 2101 Constitution Ave., N.W., Washington, D.C. $1.00.

Careers in biological sciences. Series No. 505. B'nai B'rith Vocational Service, 1640 Rhode Island Ave., N.W., Washington, D.C. 35 cents.

Should your child be a biologist? By Bentley Glass. New York Life Insurance Company, Career Information Service, Box 51, Madison Square Station, New York, N.Y.

The outlook for women in the biological sciences. Bulletin 222–3. U.S. Department of Labor, Washington, D.C.

Botany

A career in botany. 1959. Chicago Natural History Museum, Roosevelt Rd. and Lakeshore Dr., Chicago, Ill.

Botany as a career. Careers Research Monograph No. 204. The Institute for Research, 537 S. Dearborn St., Chicago, Ill.

Conservation

Find a career in conservation. 1959. By Jean Smith. G. P. Putnam's Sons, 200 Madison Ave., New York, N.Y. $2.75.

Dairy Husbandry

This is the dairy industry. American Dairy Association, 20 N. Wacker Dr., Chicago, Ill.

Dietetics

Dietetics as a profession. American Dietetic Association, 620 N. Michigan Ave., Chicago, Ill. 35 cents.

Entomology

Entomologist. 1960. Careers, Largo, Fla. 15 cents.

Exploring

Exploring as a career. Careers Research Monograph No. 82. The Institute for Research, 537 S. Dearborn St., Chicago, Ill.

Farming

Cattle farmers. Science Research Associates, 57 W. Grand Ave., Chicago, Ill. 45 cents.

Dairy farmers. Science Research Associates, 57 W. Grand Ave., Chicago, Ill. 45 cents.

Should you be a farmer? By R. I. Throckmorton. New York Life Insurance Company, Career Information Service, Box 51, Madison Square Station, New York, N.Y.

Vegetable farming as a career. Careers Research Monograph No. 175. The Institute for Research, 537 S. Dearborn St., Chicago, Ill.

Fishery Biology

Biological positions in the division of fishery biology. Fishery Booklet No. 96. U.S. Department of the Interior, Washington, D.C.

Opportunities for the ichthyologist and fishery biologist. American Society of Ichthyologists and Herpetologists, 34th St. and Girard Ave., Philadelphia, Pa.

Food Chemistry

Career as a food chemist. Careers Research Monograph No. 203. The Institute for Research, 537 S. Dearborn St., Chicago, Ill.

Forestry

So you want to work for the forest service. 1961. Forest Service, U.S. Department of Agriculture, Washington, D.C.

Health

Health careers guidebook. National Health Council, 1790 Broadway, New York, N.Y.

Herpetology

Opportunities for the herpetologist. American Society of Ichthyologists and Herpetologists, Florida State Museum, Gainesville, Fla.

Histology

Histology technician. 1959. Career Summary. Careers, Largo, Fla.

Home Economist

A career dedicated to better living—home economics. American Home Economics Association, 1600 20th St., N.W., Washington, D.C. 20 cents.

Horticulturist

Horticulturist. 1960. By Robert L. Love. Chronicle Occupational Briefs, 203. Chronicle Guidance Publications, Moravia, N.Y. 35 cents.

Librarian

Career as a Librarian in special fields. Careers Research Monograph No. 209. Institute for Research, 537 S. Dearborn St., Chicago, Ill.

Medical Fields

Decision for research. American Heart Association, 44 E. 23rd St., New York, N.Y.

Medical research career. Careers Research Monograph No. 148. Institute for Research, 537 S. Dearborn St., Chicago, Ill.

Medicine as a career. American Medical Association, 535 N. Dearborn St., Chicago, Ill.

The medical secretary. 1957. Chronicle Occupational Briefs. Chronicle Guidance Service, Moravia, N.Y. 35 cents.

The profession of medical technology, a career of service in science. Registry of Medical Technologists, Muncie, Ind.

Museum Work

Careers in museum work. Institute for Research, 537 S. Dearborn St., Chicago, Ill. 75 cents.

Mycologist

A career in mycology. By R. U. Benjamin. Mycological Society of America, Rancho Santa Ana Botanic Garden, 1500 N. College Ave., Claremont, Calif.

Naturalist

Park naturalist. 1960. Career Summary. Careers, Largo, Fla.

Nursing Careers

Let's be practical about a nursing career. Committee on Careers, National League for Nursing, 10 Columbus Circle, New York, N.Y.

Nurse anesthetists. American Association of Nurse Anesthetists, 116 S. Michigan Ave., Chicago, Ill.

Nurseryman

Careers as a landscape nurseryman. Institute for Research, 537 S. Dearborn St., Chicago, Ill. $1.00.

Careers in the nursery industry. California Association of Nurserymen, University of California, Davis, Calif.

Oceanography

A reader's guide to oceanography. 1960. Woods Hole Oceanographic Institution, Woods Hole, Mass.

Oceanographer. 1960. Chronicle Occupational Briefs. Chronicle Guidance Service, Moravia, N.Y. 35 cents.

Park Service

Your opportunity in the national park system. National Park Service, U.S. Department of the Interior, Washington, D.C.

Pathology

Clinical pathologist. 1960. Career Summary. Careers, Largo, Fla.

Plant Pathologist

Plant pathologist. 1959. Career Summary. Careers, Largo, Fla.

Public Health

Public health—a career with a future. American Public Health Association, 1790 Broadway, New York, N.Y.

Physiology

Careers in physiology. Council of the American Physiological Society, 9650 Wisconsin Ave., Washington, D.C.

Plant physiology as a career. American Society of Plant Physiologists, Department of Botany, University of Florida, Gainesville, Fla.

Science—General

Careers and opportunities in science. 1968. Rev. Ed. By Philip Pollack. E. P. Dutton and Co., 201 Park Ave. S., New York, N.Y.

Careers for women in scientific fields. 1957. Careers Research Monograph No. 228. The Institute for Research, 537 S. Dearborn St., Chicago, Ill.

Encouraging future scientists: keys to careers. National Science Teachers Association, 1201 16th St., N.W., Washington, D.C.

Great American scientists. 1961. By the editors of *Fortune.* Prentice-Hall, Englewood Cliffs, N.J. $1.95.

Jobs in science. 1958. Science Research Associates, 57 W. Grand Ave., Chicago, Ill.

Lives in science. 1957. Scientific American, Simon and Schuster, 630 Fifth Ave., New York, N.Y.

Science and your career. 1966. Pub. No. 7677. U.S. Department of Labor. Government Printing Office, Washington, D.C.

Should you be a scientist? By Edward Teller. New York Life Insurance Company, Career Information Service, Box 51, Madison Square Station, New York, N.Y.

So you're thinking about a career in science. Frontiers of Science Foundation of Oklahoma, 1701 Republic Building, Oklahoma City, Okla.

Your future in science. 1958. By Morris Meister and Paul F. Brandwein. Science Research Associates, 57 W. Grand Ave., Chicago, Ill.

Soil Conservation

Careers in soil conservation service. 1960. Miscellaneous Publication No. 717, Soil Conservation Service, U.S. Department of Agriculture. U.S. Government Printing Office, Washington, D.C. 10 cents.

Surgery

Surgery as a career. Careers Research Monograph No. 104. The Institute for Research, 537 S. Dearborn St., Chicago, Ill.

Teaching Science

Careers in science teaching. 1959. Future Scientists of America Foundation, National Science Teachers Association, 1201 16th St., N.W., Washington, D.C.

Tree Surgeon

Tree Surgeon. 1960. Career Summary. Careers, Largo, Fla.

Technical Writing

A new profession—technical writing. Simmons College Bulletin. Simmons College, 300 The Fenway, Boston, Mass.

Technician

Career as a laboratory technician. Careers Research Monograph No. 68. The Institute for Research, 537 S. Dearborn St., Chicago, Ill.

Veterinary Medicine

Veterinary medicine as a career. 1957. American Veterinary Medical Association, 600 S. Michigan Ave., Chicago, Ill. 25 cents.

Wildlife Careers

Employment opportunities in bureau of sports, fisheries and wildlife. National Wildlife Federation, 1418 16th St., N.W., Washington, D.C.

Zoologist

Zoologist. 1958. Career Summary. Careers, Largo, Fla.

Sources of Occupational Information—General

Books about occupations, a reading list for high school students. By M. H. Anderson, L. R. Gerakis, and O. M. Haugh. *Kansas Studies in Education,* Vol. 7, No. 2 (1957). School of Education, Lawrence, Kan.

Books of the traveling high school science library. By Hilary J. Deason. American Association for the Advancement of Science, 1515 Massachusetts Ave., N.W., Washington, D.C. 25 cents.

Careers in engineering, mathematics, science and related fields. 1961. A. Neal Shedd, Anita K. Scott, and James M. McCullough. A selected bibliography. U.S. Department of Health, Education and Welfare. U.S. Government Printing Office, Washington, D.C. 25 cents.

Guide to career information. 1957. By Devereux C. Josephs. New York Life Insurance Company, Career Information Service. Harper and Row, 49 E. 3rd St., New York, N.Y. $3.00.

NGVA bibliography of current occupational literature. 1959. Guidance Information Review Service Committee. National Vocational Guidance Association, Division of the American Personnel and Guidance Association, 1605 New Hampshire Ave., N.W., Washington, D.C.

Occupational information. 1957. By Robert Hoppock. McGraw-Hill Book Co., 330 W. 42nd St., New York, N.Y.

Occupational information. 1958. Second edition. By Max F. Baer and Edward C. Roeber. Science Research Associates, 57 W. Grand Ave., Chicago, Ill.

Occupational information for counselors: an annotated bibliography. 1958. U.S. Department of Labor, U.S. Government Printing Office, Washington, D.C. 15 cents.

Occupational literature: an annotated bibliography. 1958. By Gertrude Forrester. H. W. Wilson Co., 950 University Ave., New York, N.Y.

Occupational outlook handbook. 1959. Bureau of Labor Statistics. U.S. Government Printing Office, Washington, D.C. $4.25.

Sources of information on careers in the science fields. 1961. Manufacturing Chemists' Association, 1825 Connecticut Ave., N.W., Washington, D.C.